AF360729

Advances in Visible Light Communication

Advances in Visible Light Communication

Editors

Cuiwei He
Wajahat Ali

Basel • Beijing • Wuhan • Barcelona • Belgrade • Novi Sad • Cluj • Manchester

Editors
Cuiwei He
School of Information Science
Japan Advanced Institute of
Science and Technology
Nomi
Japan

Wajahat Ali
Department of Engineering
University of Cambridge
Cambridge
United Kingdom

Editorial Office
MDPI
St. Alban-Anlage 66
4052 Basel, Switzerland

This is a reprint of articles from the Special Issue published online in the open access journal *Photonics* (ISSN 2304-6732) (available at: www.mdpi.com/journal/photonics/special_issues/V_L_C).

For citation purposes, cite each article independently as indicated on the article page online and as indicated below:

Lastname, A.A.; Lastname, B.B. Article Title. *Journal Name* **Year**, *Volume Number*, Page Range.

ISBN 978-3-0365-9836-9 (Hbk)
ISBN 978-3-0365-9835-2 (PDF)
doi.org/10.3390/books978-3-0365-9835-2

Contents

MDPI

Editorial

Advances in Visible Light Communication

Cuiwei He [1,*] and **Wajahat Ali** [2]

1   School of Information Science, Japan Advanced Institute of Science and Technology, Nomi 923-1211, Japan
2   Department of Engineering, University of Cambridge, Cambridge CB3 0FA, UK; wa279@cam.ac.uk
*   Correspondence: cuiweihe@jaist.ac.jp

Citation: He, C.; Ali, W. Advances in Visible Light Communication. *Photonics* **2023**, *10*, 1277. https://doi.org/10.3390/photonics10111277

Received: 26 October 2023
Accepted: 13 November 2023
Published: 17 November 2023

1. Introduction

Visible light communications (VLC) have been a highly popular area of research in recent years [1]. With the ongoing advancement of various crucial technologies, the research community anticipates that VLC will play a significant role in future wireless communication systems [2]. The rapid recognition of VLC can be attributed to the development of various important technologies. First, thanks to the invention of semiconductor-based light sources, either light-emitting diodes (LEDs) or lasers, high-transmission bandwidths can be achieved to support Gbps data rates [3,4]. The use of state-of-the-art photodetectors and advanced optics at the receiver also contribute significantly to the development of VLC systems [5]. Furthermore, smart signal processing algorithms and multiplexing techniques have been widely investigated to boost the data rate and/or improve the reliability of transmission [6,7]. In recent research, we also see a trend of using data-driven machine learning techniques in different VLC scenarios, showing very promising performances [8].

Regarding their applications, VLC and optical wireless communications (OWC) show great potential in many distinct application scenarios, such as in the following: indoor optical wireless, which has the potential to greatly enhance existing WiFi infrastructures [1]; underwater OWC, which facilitates otherwise impossible high-speed data transmission in aquatic environments [9]; vehicle-to-vehicle (V2V) communication, which enables wireless communication between vehicles with minimal interference [10]; and indoor visible light positioning (VLP), which can be utilized to build highly accurate localization or positioning systems in indoor environments where global positioning system (GPS) signals may be unreliable [11].

Despite these many application scenarios for VLC techniques, there remains a noticeable gap between the highly encouraging results achieved in lab-based environments and their practical application in real-world settings. In this Special Issue, we aimed to gather papers that exhibit high academic quality and/or have a strong connection to real-world applications.

2. An Overview of Published Articles

In this Special Issue, we have published 21 papers covering the aforementioned important topics. These papers are listed in the 'List of Contributions' attached at the end of this editorial. Among these 21 papers, 14 are research articles and 7 are review papers. Notably, authors from various countries and well-known universities have contributed. Among these papers, some provide very detailed theoretical analysis and simulation results, while many others present high-quality experimental findings. In comparison to many other Special Issues in *Photonics*, a noteworthy aspect of this Special Issue is that we have invited and published many high-quality reviews, offering valuable insights into recent advancements in VLC or OWC. In the following, we briefly introduce these papers based on their specific focuses and topics, as well as their article types. The classified categories are signal modulation, photon counting detection, visible light positioning, transmission systems, and review articles.

Signal modulation: In recent years, non-orthogonal multiple access (NOMA) has become a popular modulation technique in VLC for supporting multiple users [12]. In Alqahtani et al.'s article (contribution 1), NOMA is studied for the VLC downlink. In particular, the authors calculate the symbol error rate (SER) considering the influence of multi-user interference, which is the primary interference issue in NOMA. Subsequently, a decoding-order-based power allocation (DOPA) method is proposed to determine the optimal power allocation that minimizes the SER. In Li et al.'s article (contribution 2), a hybrid form of optical orthogonal frequency-division multiplexing (OFDM) is introduced and integrated with NOMA. This work makes a contribution by introducing a signal modulation scheme which is shown to effectively mitigate the error propagation effect commonly encountered in both hybrid OFDM and NOMA systems. Multiple-input and multiple-output (MIMO) is another important signal transmission technique for boosting the transmission data rate [13]. In Zhong et al.'s article (contribution 3), a deep-learning-based MIMO system named DeepGOMIMO is proposed. In this technique, the transmitted information bits are mapped onto both the index of the active LEDs and the signals directly transmitted by these active LEDs. Moreover, a deep neural network (DNN) is incorporated into the system for signal detection so that the channel state information (CSI) is not required at the receiver. In Wu et al.'s article (contribution 4), a liquid crystal (LC)-based optical receiver is analyzed for a multiple-input single-output (MISO) transmission system via simulations. The main contribution of this work is the formation of an optimization task to obtain the refractive index of LC which can maximize the MISO capacity. In Saied et al.'s article (contribution 5), a modified form of OFDM modulation, named discrete Fourier transform spread-optical pulse amplitude modulation (DFTS-OPAM), is proposed and its application is discussed for a visible light network for connecting vehicles.

Photon counting detection: The performance of a VLC transmission link highly depends on the sensitivity of the photodetector. The most sensitive possible photodetector is one that can accurately count the number of photons arriving at the sensor within a short period of time. These types of sensors are commonly known as photon-counting detectors, which are now widely considered for detecting light signals in environments with low light intensity levels, such as underwater OWC and eye-safe laser-based systems. Recently, one specific type of photon counting sensor, known as silicon photomultipliers (SiPM), consisting of a large array of single-photon avalanche diodes (SPADs), has shown a very promising performance [14]. In Matthews and Collins's article (contribution 6), both experiments and Monte Carlo simulations are used to study the non-linearity of a SiPM sensor. This allows for the determination of the maximum photon counting rate of a SiPM and also an evaluation of the impact of its non-linearity on transmission performance. In Zhang et al.'s article (contribution 7), the performance of a SiPM is analyzed by considering OFDM modulation. Moreover, the SiPM's bandwidth is studied via an equivalent circuit model of the SiPM. It provides very useful information for guiding the use of SiPMs with OFDM modulation. In Yang et al.'s article (contribution 8), a neural-network-based synchronous clock recovery method is proposed for an underwater OWC system when SPADs are used. The effectiveness of this method is validated through experimental measurements.

Visible light positioning: VLP is another important research topic that employs LEDs as positioning beacons for object positioning or localization, particularly in indoor environments where signals may be unreliable [11,15]. In Yang et al.'s article (contribution 9), a camera-based VLP system was examined, utilizing a single square LED luminaire. The use of the square LED enables the extraction of the receiver's rotation angle, consequently correcting the geomagnetic angle acquired from a geomagnetic sensor, and thereby enhancing the positioning accuracy. In recent studies, there has been a notable trend towards utilizing machine learning (ML) techniques in VLP for improving positioning accuracy. For instance, both Yang et al.'s article (contribution 10) and Deng et al.'s article (contribution 11) investigate ML-based three-dimensional VLP systems, incorporating multiple LED lights and multiple photodetectors. Moreover, both studies account for the influence of both line-of-sight (LOS) an·d non-line-of-sight (NLOS) links on the received signals. Yang et al.'s article (contribution 10) employs gated recurrent unit (GRU) neural networks, whereas Deng et al.'s article (contribution 11) uses convolutional neural networks (CNN). Notably, both approaches demonstrate positioning accuracy at the centimeter level. While many studies on VLP primarily concentrated on analyzing the positioning accuracy performance, they often neglected to consider the illumination aspects of these integrated systems. In Menéndez and Steendam's article (contribution 12), a noteworthy contribution was made by also examining the illumination aspects, based on the main illumination characteristics defined in the European Standard EN 12464-1. This research provides invaluable insights for the development of a combined illumination and positioning system that complies with established illumination standards.

Transmission systems: In Yang et al.'s article (contribution 13), a very interesting laser-based OWC system is constructed. On the transmitter side, an all-fiber configuration is considered by connecting the single-mode fiber (SMF) directly from the fiber access network to a multi-mode fiber (MMF). The MMF then emits the light into the free space. At the receiver, light first passes through a collimator, and is subsequently coupled into another MMF, which is linked to a photodiode. To mitigate fiber coupling losses at the receiver, the collimator's focus is adjusted. Additionally, to maximize the received optical power, controlled perturbations are applied to the MMF at the transmitter side for beam shaping. In Li et al.'s article (contribution 14), the performance of a real-time display-camera communication (DCC) system is studied. Compared to the previous approaches, the transmission distance of this work is increased by clustering multiple adjacent LED display points for information transmission.

Review articles: This Special Issue also includes seven high-quality review articles. In Shi et al.'s article (contribution 15), a comprehensive review is presented on the application of various artificial intelligence (AI) techniques in VLC systems, addressing diverse transmission challenges, particularly those resulting from nonlinearity issues in electronic devices. Both Geng et al.'s article (contribution 16) and Loureiro et al.'s article (contribution 17) focus on reviewing popular VLC techniques, specifically different modulation methods and application scenarios. In Liu et al.'s article (contribution 18), recent significant advancements in optical injection locking for visible light communication applications are summarized. In He and Chen's article (contribution 19), a thorough review is provided on summarizing the different transmitter and receiver technologies used in LED-based VLC systems. Fang et al.'s article (contribution 20) surveys various high-speed underwater OWC systems, highlighting how ML techniques can enhance signal processing for improved transmission efficiency. Lastly, in Huang and Yamazato's article (contribution 21), a comprehensive survey is presented on imaging sensor-based VLC systems. It particularly emphasizes techniques based on rolling shutter cameras and global shutter high-speed cameras.

3. Conclusions

VLC or OWC will certainly play a pivotal role in future networks. Nevertheless, there remain many technical and regulatory challenges to overcome. We hope this Special Issue will be a valuable contribution to the growing VLC research community, driving the use of VLC techniques in practical real-life applications.

Author Contributions: Writing—original draft preparation, C.H.; writing—review and editing, C.H. and W.A. All authors have read and agreed to the published version of the manuscript.

Funding: The work of Cuiwei He was supported by the Japan Society for the Promotion of Science (JSPS) KAKENHI, Grant Number JP23K13332.

Acknowledgments: We extend our sincere gratitude to all the authors who submitted their work to this Special Issue and to the dedicated *Photonics* editorial staff for their invaluable assistance.

Conflicts of Interest: The authors declare no conflicts of interest.

List of Contributions

1. Alqahtani, A.H.; Almohimmah, E.M.; Alresheedi, M.T.; Abas, A.F.; Qidan, A.A.; Elmirghani, J. Decoding-Order-Based Power Allocation (DOPA) Scheme for Non-Orthogonal Multiple Access (NOMA) Visible Light Communication Systems. *Photonics* **2022**, *9*, 718. https://doi.org/10.3390/photonics9100718.
2. Li, B.; Shi, J.; Feng, S. Reconstructed Hybrid Optical OFDM-NOMA for Multiuser VLC Systems. *Photonics* **2022**, *9*, 857. https://doi.org/10.3390/photonics9110857
3. Zhong, X.; Chen, C.; Fu, S.; Zeng, Z.; Liu, M. DeepGOMIMO: Deep Learning-Aided Generalized Optical MIMO with CSI-Free Detection. *Photonics* **2022**, *9*, 940. https://doi.org/10.3390/photonics9120940.
4. Wu, Q.; Zhang, J.; Zhang, Y.; Xin, G.; Guo, D. Asymptotic Capacity Maximization for MISO Visible Light Communication Systems with a Liquid Crystal RIS-Based Receiver. *Photonics* **2023**, *10*, 128. https://doi.org/10.3390/photonics10020128.
5. Saied, O.; Kaiwartya, O.; Aljaidi, M.; Kumar, S.; Mahmud, M.; Kharel, R.; Al-Sallami, F.; Tsimenidis, C.C. LiNEV: Visible Light Networking for Connected Vehicles. *Photonics* **2023**, *10*, 925. https://doi.org/10.3390/photonics10080925.
6. Matthews, W.; Collins, S. An Experimental and Numerical Study of the Impact of Ambient Light of SiPMs in VLC Receivers. *Photonics* **2022**, *9*, 888. https://doi.org/10.3390/photonics9120888.
7. Zhang, L.; Jiang, R.; Tang, X.; Chen, Z.; Li, Z.; Chen, J. Performance Estimation and Selection Guideline of SiPM Chip within SiPM-Based OFDM-OWC System. *Photonics* **2022**, *9*, 637. https://doi.org/10.3390/photonics9090637.
8. Yang, H.; Yan, Q.; Wang, M.; Wang, Y.; Li, P.; Wang, W. Synchronous clock recovery of photon-counting underwater optical wireless communication based on deep learning. *Photonics* **2022**, *9*, 884. https://doi.org/10.3390/photonics9110884.
9. Yang, C.; Wen, S.; Yuan, D.; Chen, J.; Huang, J.; Guan, W. CGA-VLP: High accuracy visible light positioning algorithm using single square LED with geomagnetic angle correction. *Photonics* **2022**, *9*, 653. https://doi.org/10.3390/photonics9090653.
10. Yang, W.; Qin, L.; Hu, X.; Zhao, D. Indoor Visible-Light 3D Positioning System Based on GRU Neural Network. *Photonics* **2023**, *10*, 633. https://doi.org/10.3390/photonics10060633.
11. Deng, B.; Wang, F.; Qin, L.; Hu, X. A Visible Light 3D Positioning System for Underground Mines Based on Convolutional Neural Network Combining Inception Module and Attention Mechanism. *Photonics* **2023**, *10*, 918. https://doi.org/10.3390/photonics10080918.
12. Menéndez, J.M.; Steendam, H. On the optimisation of illumination LEDs for VLP systems. *Photonics* **2022**, *9*, 750. https://doi.org/10.3390/photonics9100750.
13. Yan, X.; Wang, Y.; Li, C.; Li, F.; Cao, Z.; Tangdiongga, E. Two-Stage Link Loss Optimization of Divergent Gaussian Beams for Narrow Field-of-View Receivers in Line-of-Sight Indoor Downlink Optical Wireless Communication. *Photonics* **2023**, *10*, 815. https://doi.org/10.3390/photonics10070815.
14. Li, J.; Pan, C.; Pan, J.; Lin, J.; Lu, M.; Jiang, Z.L.; Fang, J. Long-Distance, Real-Time LED Display-Camera Communication System Based on LED Point Clustering and Lightweight Image Processing. *Photonics* **2022**, *9*, 721. https://doi.org/10.3390/photonics9100721.
15. Shi, J.; Niu, W.; Ha, Y.; Xu, Z.; Li, Z.; Yu, S.; Chi, N. AI-enabled intelligent visible light communications: Challenges, progress, and future. *Photonics* **2022**, *9*, 529. https://doi.org/10.3390/photonics9080529.
16. Geng, Z.; Khan, F.N.; Guan, X.; Dong, Y. Advances in visible light communication technologies and applications. *Photonics* **2022**, *9*, 893. https://doi.org/10.3390/photonics9120893.
17. Loureiro, P.A.; Guiomar, F.P.; Monteiro, P.P. Visible Light Communications: A Survey on Recent High-Capacity Demonstrations and Digital Modulation Techniques. *Photonics* **2023**, *10*, 993. https://doi.org/10.3390/photonics10090993

18. Liu, X.; Hu, J.; Bian, Q.; Yi, S.; Ma, Y.; Shi, J.; Li, Z.; Zhang, J.; Chi, N.; Shen, C. Recent Advances in Optical Injection Locking for Visible Light Communication Applications. *Photonics* **2023**, *10*, 291. https://doi.org/10.3390/photonics10030291.
19. He, C.; Chen, C. A Review of Advanced Transceiver Technologies in Visible Light Communications. *Photonics* **2023**, *10*, 648. https://doi.org/10.3390/photonics10060648.
20. Fang, C.; Li, S.; Wang, Y.; Wang, K. High-Speed Underwater Optical Wireless Communication with Advanced Signal Processing Methods Survey. *Photonics* **2023**, *10*, 811. https://doi.org/10.3390/photonics10070811.
21. Huang, R.; Yamazato, T. A Review on Image Sensor Communication and Its Applications to Vehicles. *Photonics* **2023**, *10*, 617. https://doi.org/10.3390/photonics10060617.

References

1. Haas, H.; Yin, L.; Wang, Y.; Chen, C. What is lifi? *J. Light. Technol.* **2015**, *34*, 1533–1544. [CrossRef]
2. Chi, N.; Zhou, Y.; Wei, Y.; Hu, F. Visible light communication in 6G: Advances, challenges, and prospects. *IEEE Veh. Technol. Mag.* **2020**, *15*, 93–102. [CrossRef]
3. Ferreira, R.X.; Xie, E.; McKendry, J.J.; Rajbhandari, S.; Chun, H.; Faulkner, G.; Watson, S.; Kelly, A.E.; Gu, E.; Penty, R.V.; et al. High bandwidth GaN-based micro-LEDs for multi-Gb/s visible light communications. *IEEE Photonics Technol. Lett.* **2016**, *28*, 2023–2026. [CrossRef]
4. Guo, Y.; Alkhazragi, O.; Kang, C.H.; Shen, C.; Mao, Y.; Sun, X.; Ng, T.K.; Ooi, B.S. A tutorial on laser-based lighting and visible light communications: Device and technology. *Chin. Opt. Lett.* **2019**, *17*, 040601.
5. O'Brien, D.; Rajbhandari, S.; Chun, H. Transmitter and receiver technologies for optical wireless. *Philos. Trans. R. Soc. A* **2020**, *378*, 20190182. [CrossRef] [PubMed]
6. Elgala, H.; Mesleh, R.; Haas, H. Indoor optical wireless communication: Potential and state-of-the-art. *IEEE Commun. Mag.* **2011**, *49*, 56–62. [CrossRef]
7. Wang, Z.; Wang, Q.; Huang, W.; Xu, Z. *Visible Light Communications: Modulation and Signal Processing*; John Wiley & Sons: Hoboken, NJ, USA, 2017.
8. Shi, J.; Niu, W.; Ha, Y.; Xu, Z.; Li, Z.; Yu, S.; Chi, N. AI-enabled intelligent visible light communications: Challenges, progress, and future. *Photonics* **2022**, *9*, 529. [CrossRef]
9. Sun, X.; Kang, C.H.; Kong, M.; Alkhazragi, O.; Guo, Y.; Ouhssain, M.; Weng, Y.; Jones, B.H.; Ng, T.K.; Ooi, B.S. A review on practical considerations and solutions in underwater wireless optical communication. *J. Light. Technol.* **2020**, *38*, 421–431. [CrossRef]
10. Yamazato, T.; Takai, I.; Okada, H.; Fujii, T.; Yendo, T.; Arai, S.; Andoh, M.; Harada, T.; Yasutomi, K.; Kagawa, K.; et al. Image-sensor-based visible light communication for automotive applications. *IEEE Commun. Mag.* **2014**, *52*, 88–97. [CrossRef]
11. Armstrong, J.; Sekercioglu, Y.A.; Neild, A. Visible light positioning: A roadmap for international standardization. *IEEE Commun. Mag.* **2013**, *51*, 68–73. [CrossRef]
12. Sadat, H.; Abaza, M.; Mansour, A.; Alfalou, A. A survey of NOMA for VLC systems: Research challenges and future trends. *Sensors* **2022**, *22*, 1395. [CrossRef] [PubMed]
13. He, C.; Wang, T.Q.; Armstrong, J. Performance of optical receivers using photodetectors with different fields of view in a MIMO ACO-OFDM system. *J. Light. Technol.* **2015**, *33*, 4957–4967. [CrossRef]
14. Matthews, W.; Collins, S. A Roadmap for Gigabit to Terabit Optical Wireless Communications Receivers. *Sensors* **2023**, *23*, 1101. [CrossRef] [PubMed]
15. Do, T.H.; Yoo, M. An in-depth survey of visible light communication based positioning systems. *Sensors* **2016**, *16*, 678. [CrossRef] [PubMed]

Article

Decoding-Order-Based Power Allocation (DOPA) Scheme for Non-Orthogonal Multiple Access (NOMA) Visible Light Communication Systems

Ali H. Alqahtani [1,*], Esam M. Almohimmah [2], Mohammed T. Alresheedi [2], Ahmad Fauzi Abas [2], Ahmad-Adnan Qidan [3] and Jaafar Elmirghani [3]

[1] College of Applied Engineering, King Saud University, Riyadh 11451, Saudi Arabia
[2] Department of Electrical Engineering, King Saud University, Riyadh 11421, Saudi Arabia
[3] School of Electronic and Electrical Engineering, University of Leeds, Leeds LS2 9JT, UK
* Correspondence: ahqahtani@ksu.edu.sa

Abstract: Non-orthogonal multiple access (NOMA) is an effective multiple access scheme that can be used to improve considerably the spectral efficiency of indoor downlink visible light communication (VLC) systems. However, NOMA suffers from inevitable multi-user interference which degrades the system performance. In this paper, a NOMA scheme is applied in a downlink VLC system and the impact of the multi-user interference on the system performance is studied. A closed-form expression for the user symbol error rate (SER) is derived and a decoding-order-based power allocation (DOPA) method is proposed to reduce the multi-user interference and find the optimal power allocation that minimizes the SER. The significance of the proposed schemes is demonstrated by simulation. The results show that the proposed DOPA method is able to reduce effectively the multi-user interference and provide more sum rate in comparison with benchmarking schemes such as the gain ratio PA (GRPA) and the normalized gain difference PA (NGDPA) methods.

Keywords: non-orthogonal multiple access (NOMA); visible light communication (VLC); successive interference cancellation (SIC); interference management

Citation: Alqahtani, A.H.; Almohimmah, E.M.; Alresheedi, M.T.; Abas, A.F.; Qidan, A.-A.; Elmirghani, J. Decoding-Order-Based Power Allocation (DOPA) Scheme for Non-Orthogonal Multiple Access (NOMA) Visible Light Communication Systems. *Photonics* **2022**, *9*, 718. https://doi.org/10.3390/photonics9100718

Received: 17 August 2022
Accepted: 27 September 2022
Published: 2 October 2022

Publisher's Note: MDPI stays neutral with regard to jurisdictional claims in published maps and institutional affiliations.

1. Introduction

Visible light communication (VLC) is one of the most promising technologies that is attracting more attention as the demand for wireless data communication continues to increase. As the radio frequency (RF) spectrum is becoming more and more saturated, RF-based technologies, such as cellular networks and Wi-Fi, will no longer keep up with the growing demand for more data rate [1]. In contrast, VLC systems occupy the license-free light spectrum which is out of the RF spectrum and can provide a much wider bandwidth spanning approximately from 400 THz to 800 THz [2–4]. VLC is an alternative technology for indoor wireless communications which is expected to be an integral part of the future wireless networks, such as networks used beyond 5G, due to its energy efficiency and its ability to achieve high data rates [5,6]. Although the most potential applications of VLC are in indoor scenarios, VLC has many promising applications in outdoor environments such as vehicular VLC (V-VLC) as an alternative vehicular access solution to RF-based vehicular communications [7–9].

In order to realize a multi-user VLC system, a multiple access (MA) scheme is required. Many MA schemes have been proposed for VLC such as time division multiple access (TDMA) [10], space division multiple access (SDMA) [11], code division multiple access (CDMA) [12], and orthogonal frequency division multiple access (OFDMA) [13]. One of the recently proposed MA schemes is the non-orthogonal multiple access (NOMA) which is characterized by its high spectral efficiency and its ability to support more users [14–17].

The essence of NOMA is to exploit the power domain to squeeze more users in the same time-frequency (TF) resources. Hence, a significant increase in the system capacity can be achieved but at the cost of increasing the multi-user interference and the complexity of the receiver. NOMA employs a multi-user detection (MUD) technique such as successive interference cancellation (SIC) to eliminate the multi-user interference and decode the desired signal.

Power allocation (PA) plays an important role in the performance of NOMA. Hence, a suitable PA method with low complexity is required to exploit the power domain optimally. Recently, several PA methods for NOMA have been proposed. For example, one of the low-complexity PA methods is the fractional transmission PA (FTPA) [18]. In FTPA, the powers are allocated to the users based on their channel gains.

1.1. Related Work and Motivation

Recently, NOMA has been proposed for VLC to enhance the spectral efficiency of the system. In [19], NOMA has been applied in a VLC system and a gain ratio PA (GRPA) strategy has been proposed. It was demonstrated that the performance of the system can be improved by tuning the transmission angle of the light emitting diode (LED) and the field of view (FOV) of the receiver. In [20], the performance of NOMA in an indoor 2×2 multiple input multiple output (MIMO) VLC system has been investigated, and a normalized gain difference PA (NGDPA) method has been proposed to improve the achievable sum rate of the system. The authors in [21] proposed an enhanced PA (EPA) method for maximizing the sum rate of an OFDM-NOMA VLC system with an arbitrary number of multiplexed users. In [22], the authors propose a low-complexity PA scheme, called simplified gain ratio power allocation (S-GRPA), for NOMA-based indoor VLC systems. In S-GRPA, the users' channel gains are obtained by a lookup-table method to reduce the complexity and GRPA is used as the PA method. In [23], a low-complexity PA called a simple fair power allocation strategy (SFPA) was proposed to ensure a fair distribution of transmission capacity in a multi-user scenario. SFPA shows robustness to channel estimation errors. The performance of SFPA was compared to other PA methods. Although SFPA offers the highest fairness among the other methods, NGDPA outperforms SFPA in terms of the average sum rate.

The bit error rate (BER) performance of a downlink VLC with NOMA has been analyzed in [24] considering both perfect and imperfect channel state information (CSI). NOMA with OFDM has been applied to an indoor VLC system and its superior performance over OFDMA was demonstrated in [25]. In [26], the authors have derived closed-form expressions for the system coverage probability and the ergodic sum rate. Moreover, the probability that NOMA has higher individual rates than OMA has also been derived. In [27], it has been shown that the LED driving power, which ensures the same quality of service for each user, is lower for NOMA than OMA. In [28], a simple user pairing scheme for NOMA-based VLC systems has been proposed. The superiority of NOMA over conventional OMA has been shown. A flexible-rate SIC-free NOMA technique is considered in [29] for downlink VLC systems, using uneven constellation demapping (UCD) and constellation partitioning coding (CPC).

In [30], a convex optimization is applied to NOMA-based VLC system for downlink transmission in terms of the bit error rate (BER) and the sum-rate. The authors in [31] have proposed and demonstrated a real-time software reconfigurable dynamic power-and-subcarrier allocation scheme for OFDM-NOMA in VLC systems. In [32], an offset QAM/OFDM combined with NOMA (OQAM/OFDM-NOMA) modulation scheme is employed in an asynchronous multi-user multi-cell VLC system with experimental demonstratation. In [33], inter-cell interference mitigation is considered in a dimming-aware way for multi-cell NOMA-VLC systems through efficient time-scheduling, scaling, and coordination of NOMA transmissions at the access points. In [34], error analysis is presented for a downlink power-domain NOMA-based VLC system with high-order square quadrature amplitude modulation (QAM) schemes where imperfect SIC is considered.

Most of the existing NOMA VLC studies assume perfect SIC and ignore the serious effect of the multi-user interference on the system performance. This could lead to an inaccurate and impractical analysis on such systems, as multi-user interference is inevitable in NOMA-based VLC systems. Reducing the multi-user interference can be achieved by a careful PA. In this paper, we provide a theoretical analysis of the multi-user interference and user symbol error rate (SER) in NOMA-based VLC systems. Furthermore, we propose a low-complexity PA method that meets NOMA requirements for VLC systems and reduces the multi-user interference.

1.2. Contributions

In this paper, a NOMA scheme is applied in a downlink VLC system. The performance of the system is analyzed and evaluated in terms of the user SER. The main contributions of this paper are summarized as follows:

(1) *SER Analysis*: An exact closed-form expression for the user SER in a NOMA-based VLC is derived with imperfect SIC. The pulse amplitude modulation (PAM) is considered. The derived expressions are validated through computer simulations.
(2) *Power Allocation*: A decoding-order-based PA (DOPA) method is proposed which reduces the multi-user interference and satisfies all the PA constraints of NOMA-based VLC. In addition, the optimal PA, which mitigates the multi-user interference and minimizes the error during SIC, is found. The performance of DOPA is compared with GRPA and NGDPA.
(3) *Multi-user interference Effect*: The impact of the multi-user interference on the performance of NOMA-based systems and the role of PA in controlling the multi-user interference are shown.
(4) *Sum-rate Improvement*: The contribution of DOPA in the sum rate performance is illustrated compared with GRPA and NGDPA.

1.3. Paper Organization

The rest of this paper is organized as follows. In Section 2, the system model of VLC propagation is described. Section 3 discusses management of inter-group interference for multiple users in NOMA VLC system. In Section 4, the concept of NOMA-based VLC system is illustrated. Then, Section 5 introduces the SER performance analysis. The proposed DOPA is introduced in Section 6. The performance evaluation of the proposed scheme with simulation results are presented and discussed in Section 7. Finally, the conclusion is given in Section 8.

2. System Model

We consider a downlink VLC network composed of multiple optical APs given by $L, l = \{1, \ldots, L\}$, serving $K, k = \{1, \ldots, K\}$, users distributed on the receiving plane as shown in Figure 1. In this paper, NOMA is proposed as a multiple access scheme, and therefore, the K users must be divided into multiple groups given by $G, g = \{1, \ldots, G\}$, each with multiple users given by $K_g, k_g = \{1, \ldots, K_g\}$, grouped based on the difference in the channel gain. In this context, the received signal of user k_g served by L optical APs regardless of the transmission scheme, i.e., NOMA, is

$$
y^{[k_g, g]} = \mathbf{h}^{[k_g, g]} \mathbf{x}^{[k_g, g]} + \underbrace{\sum_{k'_g = 1, k'_g \neq k_g}^{K_g} \mathbf{h}^{[k_g, g]} \mathbf{x}^{[k'_g, g]}}_{\text{Intra-group-interference}} +
$$

$$
\underbrace{\sum_{g' = 1, g' \neq g}^{G} \mathbf{h}^{[k_g, g]} \sum_{k_{g'} = 1,}^{K_{g'}} \mathbf{x}^{[k_{g'}, g']} + \mathbf{z}^{[k_g, g]}}_{\text{Inter-group-interference}},
$$

(1)

where $\mathbf{h}^{[k_g,g]} \in \mathbb{R}_+^{L\times 1}$ is the channel vector between the L optical APs and user k_g belonging to group $g, g \in G$, and $\mathbf{x}^{[k_g,g]}$ is the signal transmitted to user k_g. Moreover, $\mathbf{x}^{[k'_g,g]}$ and $\mathbf{x}^{[k_{g'},g']}$ are interference signals received due to transmission to user k'_g belonging to the same group g and to user $k_{g'}$ belonging to the adjacent group g', respectively. Finally, $\mathbf{z}^{[k_g,g]}$ is defined as background noise given by the sum of thermal and shot noise [35].

In the work, we propose the implementation of NOMA to manage the interference among users of the same group, while inter-group interference is avoided by considering orthogonal or blind interference alignment (BIA) transmission schemes. It is worth mentioning that, the distribution of users and the channel coherence time are known to the VLC network through the uplink transmission, which can be established using frequency other than the optical one. In the following the optical channel is derived.

Indoor VLC Propagation Model

The main parameters of the indoor VLC attocell with a single LED are shown in Figure 1. The vertical distance from the LED to the receiving plane is denoted by L, the horizontal distance from the center point to user k is denoted by r_k, and the direct distance from the LED to user k is denoted by d_k. The semi-angle of the LED is $\Phi_{1/2}$ and the FOV of user k is Ψ_k. The irradiance and incidence angles are denoted by ϕ_k and ψ_k, respectively.

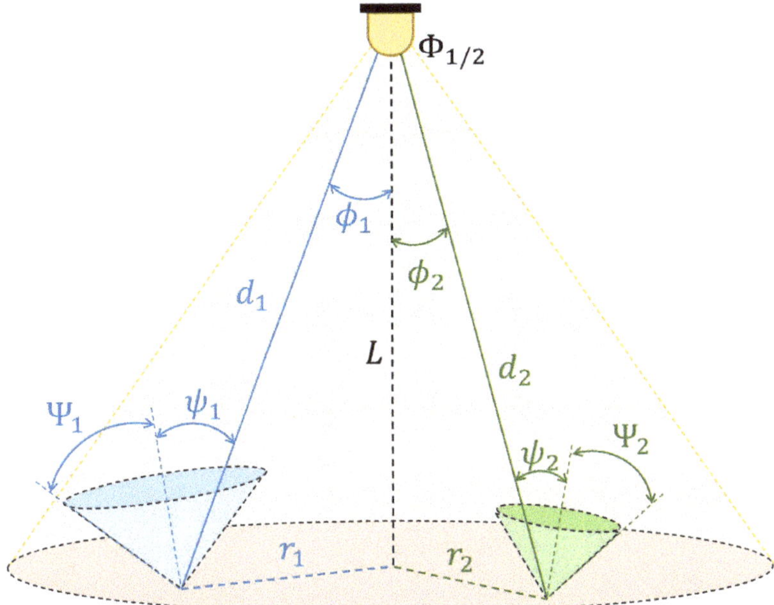

Figure 1. A VLC attocell with a single LED and two users.

The optical channel gain for user k is given by [35]:

$$h_k = \frac{(m+1)\, A_k\, T(\psi_k)\, g(\psi_k)}{2\pi d_k^2}\, cos^m(\phi_k)\, cos(\psi_k), \tag{2}$$

where $m = -1/log_2(cos(\Phi_{1/2}))$ is the order of Lambertian emission; A_k is the effective area of the photodetector (PD); $T(\psi_k)$ is the optical filter gain; and $g(\psi_k) = n^2/sin^2(\Psi_k)$ is the gain of the optical concentrator with a refractive index n. Since we assume that the users directly face the ceiling of the room, the irradiance angle and the incidence angle are equal. Therefore, the optical channel gain of the line-of-sight (LOS) link can be expressed as a function of r_k as:

$$h_k = \frac{\zeta_k}{\sqrt{(L^2 + r_k^2)^{m+3}}},$$

(3)

where $\zeta_k = (m+1)A_k T(\psi_k)g(\psi_k)L^{m+1}/2\pi$.

3. Inter-Group Interference Management

The performance of the NOMA systems can be improved by user pairing into groups based on their channel conditions [36]. Considering the implementation of NOMA, multiple users are divided into multiple groups as in [37]. Basically, an algorithm referred to as next-largest-difference user pairing (NLUPA) is implemented in which the users are sorted in ascending order from lower to higher channel gains as $h^{[1]} \leq \ldots h^{[k]} \leq \cdots \leq h^{[K]}$. After that, each user with lower channel gain is grouped with a strong user. It it worth mentioning that, the efficiency of the groping algorithm is determined based on the channel gain difference among the users of the same group.

In this work, we assume that K users are divided into G groups, each is composed of $K_g = 2$, $g \in G$, users as in [37]. The aim of this section is addressing interference management among multiple groups. Orthogonal transmission schemes can be easily implemented where different time or frequency slots are assigned to different groups, and therefore, inter-group interference can be managed as noise. However, this way leads to limit the capacity of the VLC network. In [38], BIA was implemented for RF networks to manage the interference among multiple users maximizing the multiplexing gain. In [39,40], BIA is considered for VLC networks to minimize CSI at transmitters, which is difficult to provide in RF or VLC networks. It is shown that BIA provides high performance for VLC networks compared with zero forcing (ZF) or maximum ratio combining (MRC) schemes. In [41], BIA is adopted to manage the inter-group interference in a MISO RF scenario avoiding the limitations of orthogonal transmission schemes.

In the following, the methodology of BIA is defined first for a simple VLC scenario for the sake of easy understanding, and then, the VLC general case is considered.

3.1. Motivational Example

For illustrative purposes, we consider an example comprises $L = 2$ optical APs serving $K = 6$ users. The users are divided into $G = 3$ groups, each with $K_g = 2$ users. Notice that, the purpose of this example is managing the interference among multiple groups considering BIA, and therefore, the following notations and equations are derived regardless of NOMA, which is presented in Section 4 managing the intra-group interference. Basically, the transmission based on BIA occurs over a block named supersymbol, which comprises multiple symbol extensions, i.e., time slots. The supersymbol of BIA for this particular example is composed of 4 times slots as shown in Figure 2. That is, the transmitted signal can be expressed as

$$\mathbf{X} = \begin{bmatrix} \mathbf{x}[1] \\ \mathbf{x}[2] \\ \mathbf{x}[3] \\ \mathbf{x}[4] \end{bmatrix} = \underbrace{\begin{bmatrix} \mathbf{I}_2 \\ \mathbf{I}_2 \\ \mathbf{0}_2 \\ \mathbf{0}_2 \end{bmatrix}}_{\mathbf{W}^{[1]}} \mathbf{s}^{[1]}\sqrt{P_1} + \underbrace{\begin{bmatrix} \mathbf{I}_2 \\ \mathbf{0}_2 \\ \mathbf{I}_2 \\ \mathbf{0}_2 \end{bmatrix}}_{\mathbf{W}^{[2]}} \mathbf{s}^{[2]}\sqrt{P_2} + \underbrace{\begin{bmatrix} \mathbf{I}_2 \\ \mathbf{0}_2 \\ \mathbf{0}_2 \\ \mathbf{I}_2 \end{bmatrix}}_{\mathbf{W}^{[3]}} \mathbf{s}^{[3]}\sqrt{P_3},$$

(4)

where $\mathbf{x}[n] \in \mathbb{R}_+^{L \times 1}$ is the signal transmitted over the time slot n, $\mathbf{I}$ and $\mathbf{0}$ are 2×2 identity and zero matrices, respectively. Moreover, $\mathbf{s}^{[g]} = (\mathbf{s}^{[1,g]} + \mathbf{s}^{[2,g]})$ is the symbol transmitted to $K_g = 2$ users belonging to group g, $g \in G$. Furthermore, P_g and $\mathbf{W}^{[g]}$ are the power allocated and the precoding matrix of group g, respectively. Notice that, over the supersympol that comprised 4 time slots, a resource block composed of two time slots, i.e., each time slot corresponds to one transmitter, allocated to each group. Focussing on group 1, i.e., $g = 1$, the first and second time slots form the resource block selecting $\mathbf{h}(1)$ and $\mathbf{h}(2)$, respectively. Therefore, the received signal, for example, by user 1 is given by

$$
\begin{bmatrix} y^{[1,1]}[1] \\ y^{[1,1]}[2] \\ y^{[1,1]}[3] \\ y^{[1,1]}[4] \end{bmatrix} = \underbrace{\begin{bmatrix} \mathbf{h}^{[1,1]^T}(1) \\ \mathbf{h}^{[1,1]^T}(2) \\ \mathbf{0}_{2,1}{}^T \\ \mathbf{0}_{2,1}{}^T \end{bmatrix}}_{\text{rank}=2} (\mathbf{s}^{[1,1]} + \mathbf{s}^{[2,1]}) \sqrt{P_1} +
$$

$$
\underbrace{\begin{bmatrix} \mathbf{h}^{[1,1]^T}(1) \\ \mathbf{0}_{2,1}{}^T \\ \mathbf{h}^{[1,1]^T}(1) \\ \mathbf{0}_{2,1}{}^T \end{bmatrix}}_{\text{rank}=2} (\mathbf{s}^{[1,2]} + \mathbf{s}^{[2,2]}) \sqrt{P_2} + \underbrace{\begin{bmatrix} \mathbf{h}^{[1,1]^T}(1) \\ \mathbf{0}_{2,1}{}^T \\ \mathbf{0}_{2,1}{}^T \\ \mathbf{h}^{[1,1]^T}(1) \end{bmatrix}}_{\text{rank}=2}
$$

$$
(\mathbf{s}^{[1,3]} + \mathbf{s}^{[2,3]}) \sqrt{P_3} + \begin{bmatrix} z^{[1,1]}[1] \\ z^{[1,1]}[2] \\ z^{[1,1]}[3] \\ z^{[1,1]}[4] \end{bmatrix}. \tag{5}
$$

Notice that, the users of group 1 receive their information over the first and second time slots, and therefore, the interference between them must be managed based on NOMA. Moreover, they receive interference over the first time slot due to transmission to adjacent groups 2 and 3. Based on BIA, orthogonal transmission is carried out over second, third and fourth time slots as in (5). Therefore, the users of group 1 can dedicate the third and fourth time slots to measure the interference received over the first time slot. That is, the signal received by user 1 after interference subtraction is given by

$$
\mathbf{y}^{[1,1]} = \begin{bmatrix} \mathbf{h}^{[1,1]^T}(1) \\ \mathbf{h}^{[1,1]^T}(2) \end{bmatrix} (\mathbf{s}^{[1,1]} + \mathbf{s}^{[2,1]}) \sqrt{P_1} +
$$

$$
\begin{bmatrix} z^{[1,1]}[1] - z^{[1,1]}[3] - z^{[1,1]}[4] \\ z^{[1,1]}[2] \end{bmatrix}. \tag{6}
$$

Notice that, user 2 of group 1 follows the same procedure. As a result, the symbols transmitted to the users of group 1 can be decoded where the interference among groups is aligned and subtracted. It is worth mentioning that, over a supersymbol comprised 4 time slots, 2 DoF can be achieved by group 1. The groups 2 and 3, following the same methodology above, can achieve 2 DoF each. Therefore, for the considered example, the sum DoF equals $\frac{12}{3}$ considering the number of users that form each group. However, applying NOMA, there is a certain level of interference among the users of the same group that must be taken into consideration, which is the aim of this work, in order to further enhance the performance of the VLC network.

Time slot	Group 1	Group 2	Group 3
1	$\mathbf{h}^{[1]}(1)$	$\mathbf{h}^{[2]}(1)$	$\mathbf{h}^{[3]}(1)$
2	$\mathbf{h}^{[1]}(2)$	$\mathbf{h}^{[2]}(1)$	$\mathbf{h}^{[3]}(1)$
3	$\mathbf{h}^{[1]}(1)$	$\mathbf{h}^{[2]}(2)$	$\mathbf{h}^{[3]}(1)$
4	$\mathbf{h}^{[1]}(1)$	$\mathbf{h}^{[2]}(1)$	$\mathbf{h}^{[3]}(2)$

Figure 2. The supersymbol of BIA for a VLC network with $L = 2$ transmitters and $G = 3$ groups.

3.2. General Case

According to [38], the supersymbol of BIA is divided into two blocks referred to as Block 1 and Block 2. In the example above, Block 1 is given by the first time slot, while Block 2 is formed by the second, third and fourth time slots.

Let us consider a general case in which L optical APs serving K users divided into G groups. In contrast to BIA in [38], in this work, the supersymbol of BIA is given by the number of transmitters and groups as in [41]. In Block 1, all groups receive their information simultaneously generating sever inter-group interference. Therefore, Block 1 comprises $(L-1)^G$ time slots. In Block 2, orthogonal transmission is carried out in order to give the users of each group enough dimensions to measure the interference received over Block 1. Therefore, Block 2 comprises $G(L-1)^{G-1}$ time slots. As a result, $(L-1)^{G-1}$ resource blocks are allocated for each group. Notice that, the first $L-1$ time slots of each resource block allocated for each group are provided over Block 1, while the last time slot is provided over Block 2. To conclude, the length of the supersymbol of BIA implemented for inter-group interference mitigation is given by

$$\Gamma_{BIA} = (L-1)^G + G(L-1)^{G-1}. \tag{7}$$

The methodology of BIA is explained in more details in [38].

4. NOMA-Based VLC

As in [37], the users are divided into multiple groups, each with strong and weak users. In NOMA, two main processes must be considered. Firstly, superposition coding (SC) at transmitters where two different levels of power are assigned to the signals of the strong and weak users through an appropriate PA method. These signals are then superposed into a single signal representing the instantaneous electrical current that drives the transmitter.

Secondly, performing SIC at the users side giving each user the ability to cancel the interference received due to the transmission to users with lower channel gains. In SIC, the strong users decode the signals with high power and subtract them from the received signal unit the desired signal is decoded. Both SC and SIC are conducted in the electrical domain. In Figure 3, the concept of NOMA in a VLC system is summarized.

The instantaneous electrical current that drives optical transmitters is given by:

$$\mathbf{X_e} = \sum_{g=1}^{G} \mathbf{W}^{[g]} \sum_{k_g=1}^{K_g} \sqrt{a_{k_g} P_g}\, \mathbf{s}^{[k_g,g]} + I_{DC}, \tag{8}$$

where a_{k_g} is the power allocation factor and I_{DC} is a DC current to ensure that the total current is unipolar. Here, $\mathbf{s}^{[k_g,g]}$ is assumed to be a vector that contains the symbols transmitted to the group over resource blocks allocated from BIA implemented to align the interference among multiple groups.

The optical transmitted signal is given by [26]:

$$\mathbf{X_o} = \eta \mathbf{X_e},\tag{9}$$

where η is the efficiency of the transmitter to convert the electrical current to optical power (W/A). The received signal by user k_g belonging to group g after canceling the inter-group interference is given by :

$$\mathbf{y}^{[k_g,g]} = \mathbf{h}^{[k_g,g]} \mathbf{W}^{[g]} \sum_{k_g=1}^{K_g} \sqrt{a_{k_g} P_g}\, \mathbf{s}_\ell^{[k_g,g]} + \mathbf{z}^{[k_g,g]},\tag{10}$$

where $\mathbf{s}_\ell^{[k_g,g]}$ contains the symbols transmitted to group g over the resource block ℓ. Notice that, $(L-1)^{G-1}$ resource blocks allocated to each group due to the implementation of BIA. Although, in (10) the inter-group interference is canceled, user k_g is subject to multi-user interference resulted from applying NOMA. In the following, this issue is addressed in more details aiming to enhance the performance of NOMA.

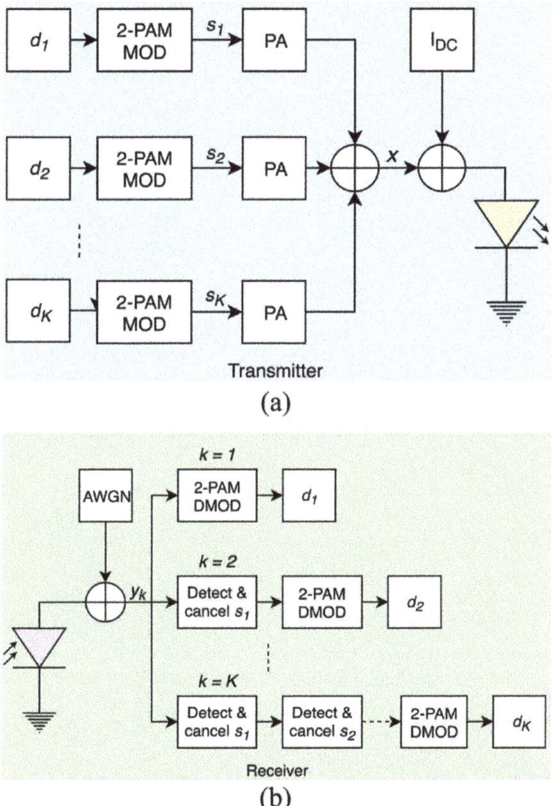

Figure 3. The concept of a NOMA-based VLC system: (a) Transmitter (b) Receiver.

Multi-User Interference in NOMA

From now on, without loss of generality, we focus on a generic group and eliminate the index of the group for the sake of simplicity. To explain the multi-user interference in NOMA, the received signal in (10) is rewritten as:

$$
y_k = c_k \left(\underbrace{\sum_{i=1}^{k-1} \sqrt{a_i} s_i}_{\text{interference}} + \underbrace{\sqrt{a_k} s_k}_{\text{signal}} + \underbrace{\sum_{j=k+1}^{K} \sqrt{a_j} s_j}_{\text{noise}} \right) + z_k ,
\tag{11}
$$

where $c_k = v h_k \eta \sqrt{P_T}$ and v is the PD responsivity. There are two types of user interference in NOMA systems in (11); the interference from users with lower decoding order, i.e., users with higher powers, and the interference from users with higher decoding order, i.e., users with lower powers. The first type should be successfully canceled during SIC process, while the second type can be considered as noise. The interference can be controlled through the PA process. To ensure the success of SIC, the power of the signal to be decoded should be higher than the power of the second-type interference plus noise. Otherwise, the interference could become severe and leads to unsuccessful SIC, which results in a potential performance degradation. On the other hand, the powers of the signals with higher decoding orders should be high enough to be distinguished from the noise. Consequently, there is a trade-off between reducing the interference to decode the first-user signal and reducing the effect of the noise to decode the second-user signal. In other words, the higher the power allocated to the first user the less interference from the second user. However, the effect of AWGN on the second-user signal will be increased.

5. SER Analysis

Typically, VLC can be realized as an intensity modulation/direct detection (IM/DD) system [42], which requires that the modulated signal has to be both real-valued and non-negative (unipolar). Hence, more efficient modulation schemes, such as QAM cannot be directly implemented for IM/DD because they are complex-valued and bipolar in nature. Optical orthogonal frequency division multiplexing (O-OFDM) systems are the most commonly utilized solution [43]. In O-OFDM, a real-valued time-domain signal is obtained at the expense of a 50% reduction of the spectral efficiency due to Hermitian symmetry constraint. A unipolar time-domain signal can then be obtained by adding a positive DC to the OFDM signal, such as in the DC-biased O-OFDM (DCO-OFDM) [44], or using techniques that obtain a unipolar time-domain signal such as in asymmetrically clipped O-OFDM (ACO-OFDM) with a further 50% reduction of the data rate [44].

In this work, PAM with DC-biasing is adopted as a modulation scheme to avoid the reduction in spectral efficiency imposed by the requirements of using higher-order modulation schemes in VLC systems. In this section, we derive SER expressions for the users in NOMA-based VLC systems. First, we derive the SER expressions when the modulation scheme for both users is 2-PAM. Then, we generalize the derivation for M-PAM.

5.1. Derivation of First User SER

In NOMA systems, the signal constellation corresponding to the signal of user k can be expressed by the vector $\mathbf{s}_k = (s_{k1}, \ldots, s_{kM})$, where s_{km} is the symbol m for user k and M is the total number of symbols. Figure 4 shows the constellations of both users and the superposed signal in a 2-user NOMA system with 2-PAM signaling. $\mathbf{s}_1 = (-\sqrt{a_1 E_s}, \sqrt{a_1 E_s})$ and $\mathbf{s}_2 = (-\sqrt{a_2 E_s}, \sqrt{a_2 E_s})$, where a_k and E_s are the power allocation factor and the maximum energy per symbol before the PA, respectively. The signal constellation corresponding to the superposed transmitted signal (x) consists of four constellation points, i.e., $\mathbf{s}_x = \{s_{x1}, s_{x2}, s_{x3}, s_{x4}\}$, where $s_{x1} = -\sqrt{a_1 E_s} - \sqrt{a_2 E_s}$, $s_{x2} = -\sqrt{a_1 E_s} + \sqrt{a_2 E_s}$, $s_{x3} = \sqrt{a_1 E_s} - \sqrt{a_2 E_s}$, and $s_{x4} = \sqrt{a_1 E_s} + \sqrt{a_2 E_s}$. The symbols s_{x1}, s_{x2}, s_{x3}, and s_{x4} are assumed to be equally probable, i.e., $Pr\{s_{x1}\} = Pr\{s_{x2}\} = Pr\{s_{x3}\} = Pr\{s_{x4}\} = \frac{1}{4}$.

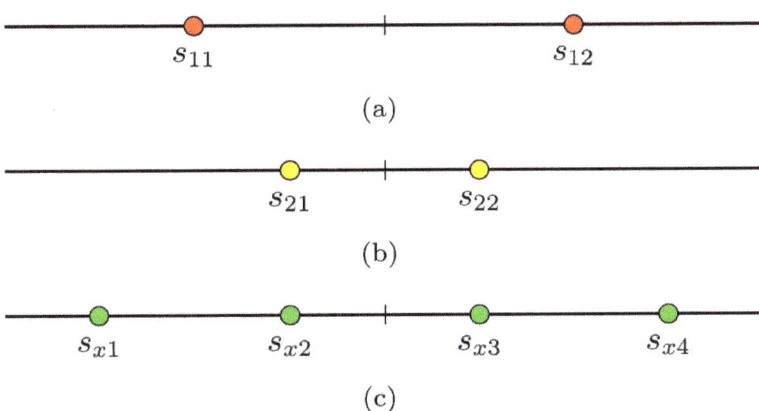

Figure 4. The signal constellation of: (**a**) weak user (u_1), (**b**) strong user (u_2), and (**c**) superposed transmitted signal (x).

The received symbol can be expressed as:

$$\tilde{y}_k = g_k s_x + \tilde{n}_k , \tag{12}$$

where $g_k = \upsilon \eta h_k$ is the total channel gain and the signal $s_x \in \{s_{x1}, s_{x2}, s_{x3}, s_{x4}\}$. Since the noise n is assumed to be zero-mean AWGN with N_0 variance, i.e., $\tilde{n}_k \sim \{0, N_0\}$, the probability density function (PDF) of $\tilde{n}_k$ is given by:

$$Pr(\tilde{n}_k) = \frac{1}{\sqrt{2\pi\sigma^2}} \, exp\left(-\frac{(\tilde{n}_k - \mu)^2}{2\sigma^2}\right), \tag{13}$$

where $\mu = 0$ is the mean and $\sigma^2 = N_0$ is the variance. We assume that g_k is previously known at the receiver. Hence, equalization is performed by dividing the received symbol $\tilde{y}_k$ by g_k, then:

$$y_k = s_x + n_k, \tag{14}$$

where $n_k = \tilde{n}_k / g_k$ is the additive Gaussian noise scaled by the total channel gain. The conditional PDF of the equalized received signal y_k given s_{x3} was sent is given by:

$$Pr(y_k|s_{x3}) = \frac{1}{\sqrt{2\pi N_k}} \, exp\left(-\frac{(y_k - s_{x3})^2}{2N_k}\right), \tag{15}$$

where $N_k = N_0 / g_k^2$. Similarly,

$$Pr(y_k|s_{x4}) = \frac{1}{\sqrt{2\pi N_k}} \, exp\left(-\frac{(y_k - s_{x4})^2}{2N_k}\right). \tag{16}$$

The optimal decision threshold for the maximum likelihood (ML) receiver of the first user is $\lambda_{th} = 0$. Therefore, the ML receiver should detect the symbol $s_{11} = -\sqrt{a_1 E_s}$ if either s_{x1} or s_{x2} was sent. Similarly, it should detect the symbol $s_{12} = +\sqrt{a_1 E_s}$ if either s_{x3} or s_{x4} was sent.

Figure 5 shows the conditional PDF of the received signal y_k given s_{x3} was sent and the conditional PDF of the received signal y_k given s_{x4} was sent. The dark region represents the probability of error.

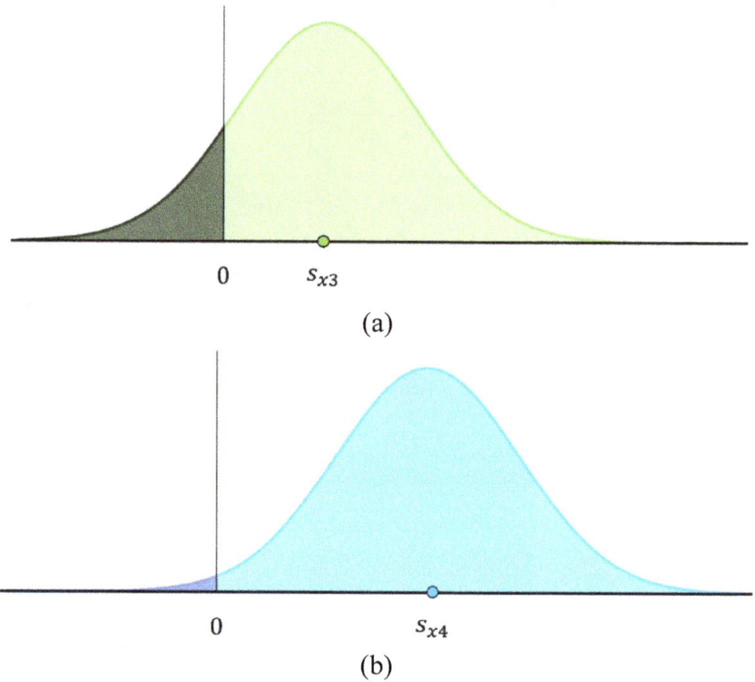

(a)

(b)

Figure 5. The conditional PDF (**a**) $Pr(y_k|s_{x3})$. (**b**) $Pr(y_k|s_{x4})$. The dark region represents the probability of error.

Therefore, the probability of error given s_{x3} was sent is given by:

$$
\begin{aligned}
Pr\{e|s_{x3}\} &= \frac{1}{\sqrt{2\pi N_k}} \int_{-\infty}^{0} exp\left(-\frac{(y_k - s_{x3})^2}{2N_k}\right) dy_k \\
&= \frac{1}{\sqrt{2\pi}} \int_{\frac{s_{x3}}{\sqrt{N_k}}}^{\infty} exp\left(-\frac{z^2}{2}\right) dz \\
&= Q\left(\frac{s_{x3}}{\sqrt{N_k}}\right) \\
&= Q(\sqrt{\gamma_k}\,\Gamma_1) ,
\end{aligned}
\tag{17}
$$

where $\gamma_k = \frac{g_k^2 E_s}{N_0}$, $\Gamma_1 = \sqrt{a_1} - \sqrt{a_2}$, and $Q(\cdot)$ is the Q-function defined by $Q(x) = \frac{1}{\sqrt{2\pi}} \int_x^{\infty} exp(-\frac{u^2}{2})\, du$. Similarly, the probability of error given s_{x4} was sent is given by:

$$
Pr\{e|s_{x4}\} = Q\left(\frac{s_{x4}}{\sqrt{N_k}}\right) = Q(\sqrt{\gamma_k}\,\Gamma_2) ,
\tag{18}
$$

where $\Gamma_2 = \sqrt{a_1} + \sqrt{a_2}$.

The SER for the first user can be expressed as:

$$
\begin{aligned}
SER_1 = {}& Pr\{s_{x1}\}Pr\{e|s_{x1}\} + Pr\{s_{x2}\}Pr\{e|s_{x2}\} \\
&+ Pr\{s_{x3}\}Pr\{e|s_{x3}\} + Pr\{s_{x4}\}Pr\{e|s_{x4}\} .
\end{aligned}
\tag{19}
$$

Since the constellation of x is symmetric, then $Pr\{e|s_{x1}\} = Pr\{e|s_{x4}\}$, and $Pr\{e|s_{x2}\} = Pr\{e|s_{x3}\}$. Therefore, (19) can be reduced to:

$$\text{SER}_1 = \frac{1}{2}(Pr\{e|s_{x3}\} + Pr\{e|s_{x4}\}) \,. \tag{20}$$

By substituting (17) and (18) in (20), we obtain:

$$\text{SER}_1 = \frac{1}{2}Q(\sqrt{\gamma_1}\,\Gamma_1) + \frac{1}{2}Q(\sqrt{\gamma_1}\,\Gamma_2). \tag{21}$$

5.2. Derivation of Second User SER

For the sake of the second user, since it is required to cancel the interference comming from the signal of the first user, the SER of the second user depends on whether the first user symbol was successfully detected or not. Therefore, the SER for the second user can be calculated as:

$$\text{SER}_2 = (1 - \text{SER}_{2\to1})\text{SER}_{2\to2} + \text{SER}_{2\to1} \,, \tag{22}$$

where $\text{SER}_{2\to1}$ is the SER for the second user to detect the signal of the first user and $\text{SER}_{2\to2}$ is the SER for the second user to detect its own signal after detecting the signal of the first user successfully. In (22), we consider the worst-case scenario in which the second user cannot decode its signal successfully whenever the signal of the first user was unsuccessfully detected. $\text{SER}_{2\to1}$ can be calculated in a similar way to (21). Therefore:

$$\text{SER}_{2\to1} = \frac{1}{2}Q(\sqrt{\gamma_2}\,\Gamma_1) + \frac{1}{2}Q(\sqrt{\gamma_2}\,\Gamma_2). \tag{23}$$

After canceling the interference from the signal of the first user successfully, the second user will detect its signal as if there is no interference. Therefore, $\text{SER}_{2\to2}$ can be calculated easily as:

$$\text{SER}_{2\to2} = Q(\sqrt{a_2\gamma_2}). \tag{24}$$

Combining (22) and (24), the SER for the second user can be calculated as:

$$\text{SER}_2 = Q(\sqrt{a_2\gamma_2}) + \text{SER}_{2\to1}(1 - Q(\sqrt{a_2\gamma_2})). \tag{25}$$

Hence, the SER of the second user can be calculated using (23) and (25).

5.3. Derivation of General SER Expression

The general case for high order PAM schemes can be derived following the same fashion proposed in Sections 5.1 and 5.2. Let M_1 and M_2 be the modulation orders for user 1 and user 2, respectively. The symbols alphabet for user 1 could be represented as

$$A_{1m_1} = 2m_1 - 1 - M_1, \qquad m_1 = 1, 2, \cdots, M_1 \tag{26}$$

Similarly, The symbols alphabet for user 2 could be represented as

$$A_{2m_2} = 2m_2 - 1 - M_2, \qquad m_2 = 1, 2, \cdots, M_2 \tag{27}$$

Therefore, user 1 and user 2 symbols can be expressed as

$$s_{km_k} = A_{km_k}\varsigma_k\sqrt{a_k E_s}, \qquad k = 1, 2 \tag{28}$$

where $\varsigma_k = \sqrt{\frac{3}{M_k^2 - 1}}$ is a normalization factor. The signal constellation corresponding to the superposed signal (x) consists of $M_1 M_2$ constellation points. The superimposed symbol is expressed as follows:

$$s_x = \sqrt{E_s}\left(A_{1m_1}\varsigma_1\sqrt{a_1} + A_{2m_2}\varsigma_2\sqrt{a_2}\right). \tag{29}$$

Figure 6 shows the constellations of both users and the superposed signal when both users have 4-PAM signaling. The vertical lines represent the boundaries of the decision regions for user 1.

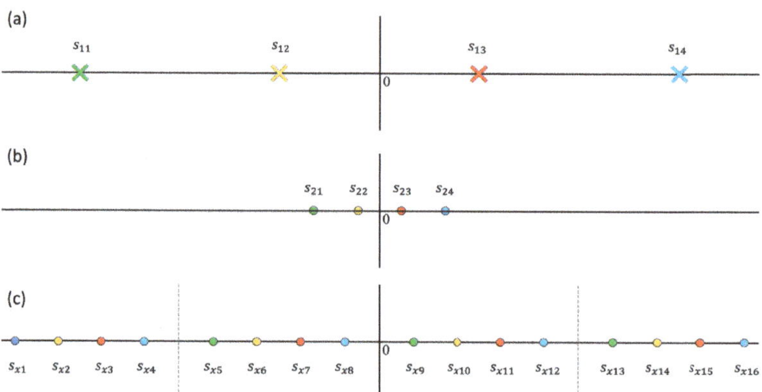

Figure 6. Signal constellation when both users have 4-PAM. (**a**) user 1, (**b**) user 2, and (**c**) superposed signal (x).

To calculate the SER for user 1, we follow the same procedures in Section 5.1 to find $Pr\{e|s_{x_k}\}$ and exploit the constellation symmetry around the y-axis to reduce the calculations. Thus, we need only to calculate the error probability for the symbols on the right-hand side or left-hand side. To further reduce the calculations, the symmetry of the error probability of some symbols in the decision region can be exploited. For example, in 4-PAM case $Pr\{e|s_{x_9}\} = Pr\{e|s_{x_{12}}\}$ and $Pr\{e|s_{x_{10}}\} = Pr\{e|s_{x_{11}}\}$. Therefor, the general expression for SER for user 1 can be expressed as:

$$\text{SER}_1 = \frac{2(M_1 - 1)}{M_1 M_2} \sum_{k=1}^{M_2/2} (Q(\sqrt{\gamma_1}\,\Gamma_1) + Q(\sqrt{\gamma_1}\,\Gamma_2)), \tag{30}$$

where $\Gamma_1 = \xi_1\sqrt{a_1} - (2k - 1)\xi_2\sqrt{a_2}$ and $\Gamma_2 = \xi_1\sqrt{a_1} + (2k - 1)\xi_2\sqrt{a_2}$.

The SER for user 2 in the general case can be calculated following same procedures in Section 5.2. Equation (22) is still applicable for determining SER_2 in the general case, but the terms $\text{SER}_{2\to1}$ and $\text{SER}_{2\to2}$ need to be updated. $\text{SER}_{2\to1}$ can be calculated in a similar way to (30). Therefore:

$$\text{SER}_{2\to1} = \frac{2(M_1 - 1)}{M_1 M_2} \sum_{k=1}^{M_2/2} (Q(\sqrt{\gamma_2}\,\Gamma_1) + Q(\sqrt{\gamma_2}\,\Gamma_2)). \tag{31}$$

After canceling the interference from user 1 signal, the second user will decode its signal as if there is no interference. Therefore, $\text{SER}_{2\to2}$ can be calculated easily as:

$$\text{SER}_{2\to2} = 2\left(1 - \frac{1}{M_2}\right)Q(\sqrt{a_2\gamma_2}). \tag{32}$$

Combining (22) and (32), the SER for the second user can be calculated as:

$$\text{SER}_2 = 2\left(1 - \frac{1}{M_2}\right)Q(\sqrt{a_2\gamma_2}) + \text{SER}_{2\to1}\left(\left(\frac{2}{M_2} - 1\right)Q(\sqrt{a_2\gamma_2})\right). \tag{33}$$

Hence, the SER of the second user can be calculated using (31) and (33).

6. Power Allocation in NOMA

In NOMA systems, the users are multiplexed in the power domain, hence, the power allocated to each user should be carefully determined. For PA in NOMA-based VLC systems, two main power constraints should be satisfied: the first constraint is requires that the total power should be kept constant since the light sources in VLC systems are used for both illumination and data transmission, i.e., $\sum_{i=1}^{K} p_i = P_T$. The second constraint is that the powers allocated to the users with lower decoding order, i.e., lower channel gains, should be greater than the powers assigned to the users with higher decoding order, i.e., higher channel gains. In other words, NOMA users are allocated with power levels inversely proportional to their channel gain, i.e., $p_1 > p_2 > \cdots > p_k$. This constraint is to ensure successful decoding during SIC [45].

Since the indoor VLC attocells have small coverage area and due to the fact that the users may have the same horizontal distance, the users in an attocell typically have close or similar channel gains [46]. Although the performance of NOMA is degraded when the channel gains are close or similar, it is shown that NOMA can always increase the sum rate of VLC systems even if the users have similar channel gains. However, using NOMA for pairing users with close or similar channel gains may increase the effect of multi-user interference on decoding the signals of the users.

To solve the previous issue, we propose a low-complexity PA method that guarantees different powers allocated to the users in all cases even if the users have similar channel gains. The proposed PA method is a decoding-order-based PA (DOPA) in which the powers are allocated to the users based on their indices in the decoding order and their channel gains. In DOPA, the power allocated to user k is given by:

$$p_k = \frac{(K-k+1)^\alpha h_{K-k+1}}{\sum_{i=1}^{K}(K-i+1)^\alpha h_{K-i+1}} P_T \, , \tag{34}$$

where α is an optimization parameter that should be predetermined to maximize the target performance metric such as the system sum rate, the fairness among the users or the energy efficiency. The powers allocated to the users with DOPA satisfy the following constraints:

- $\sum_{i=1}^{K} p_i = P_T$
- $p_k > p_{k+1}$, $\forall k \in \{1,2,\ldots,K-1\}$ and $\alpha > 0$

The first point represents the first PA constraint in NOMA-based VLC. The sum of all powers allocated to the users using DOPA can be calculated as:

$$\sum_{k=1}^{K} p_k = \sum_{k=1}^{K} \frac{(K-k+1)^\alpha h_{K-k+1}}{\sum_{i=1}^{K}(K-i+1)^\alpha h_{K-i+1}} P_T \, . \tag{35}$$

Let $u = K - i + 1$ and $v = K - k + 1$, then:

$$\sum_{k=1}^{K} p_k = \sum_{v=1}^{K} \frac{v^\alpha h_v}{\sum_{u=1}^{K} u^\alpha h_u} P_T \tag{36}$$

$$= \frac{\sum_{v=1}^{K} v^\alpha h_v}{\sum_{u=1}^{K} u^\alpha h_u} P_T = P_T \, .$$

Therefore, the total power of DOPA is always constant. The second point states that the powers allocated to the users by DOPA are distinctive and follow the optimal decoding order, i.e., $p_1 > p_2 > \cdots > p_k$. Using (34), p_{k+1} can be calculated by:

$$p_{k+1} = \frac{(K-k)^\alpha h_{K-k}}{\sum_{i=1}^{K}(K-i+1)^\alpha h_{K-i+1}} P_T \, . \tag{37}$$

Let $m = K - k$ and substitute in (34) and (37), then:

$$p_k = \frac{(m+1)^\alpha h_{m+1}}{\sum_{i=1}^{K}(K-i+1)^\alpha h_{K-i+1}} P_T .$$

(38)

and

$$p_{k+1} = \frac{(m)^\alpha h_m}{\sum_{i=1}^{K}(K-i+1)^\alpha h_{K-i+1}} P_T .$$

(39)

From (38) and (39), we obtain

$$p_k = \left(\frac{m+1}{m}\right)^\alpha \left(\frac{h_{m+1}}{h_m}\right) p_{k+1} .$$

(40)

Since $h_m \leq h_{m+1}$ is given and $m + 1 > m$, it is clear from (40) that p_k is always greater than p_{k+1} for all possible values of h_m and h_{m+1} as long as $\alpha > 0$. Therefore, the proposed DOPA satisfies the PA constraints of NOMA-based VLC.

Figure 7 shows the PA coefficients of the first user (u_1) and the second user (u_2) using DOPA with different values of the optimization parameter (α). Whether the users have similar or different horizontal distances, i.e., similar or different channel gains, the power allocated to u_1 is always higher than the power allocated to u_2 except when both users have similar horizontal distances and $\alpha = 0$. As α increases, the powers allocated to the users become more distinct. However, for high values of α, the power of u_2 becomes very low. Consequently, u_2 becomes more susceptible to the AWGN. Therefore, the value of α should be determined carefully.

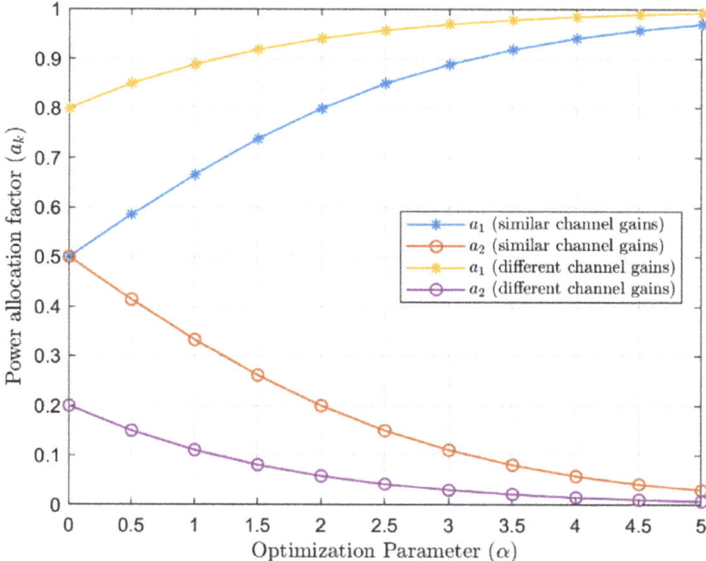

Figure 7. PA coefficients using DOPA with the optimization parameter.

The optimal value of α that minimizes SER$_1$ is infinity since SER$_1$ reduces as the power allocated for the first user is increased. On the other hand, to get the optimal value of the optimization parameter α that minimizes the SER of the second user in a 2-user NOMA, the previous SER analysis is used. The problem can be formulated as follows:

$$\hat{a} = \arg \min_\alpha (\text{SER}_2) ,$$

(41)

where SER_2 is given by (25). Using $K = 2$ in (34), we obtain:

$$a_1 = \frac{2^\alpha h_2}{h_1 + 2^\alpha h_2}$$
$$a_2 = \frac{h_1}{h_1 + 2^\alpha h_2}$$

(42)

Substituting (25) in (41), the problem becomes:

$$\hat{\alpha} = \arg\min_\alpha (Q(\sqrt{\gamma_2 a_2}) + (\frac{1}{2}Q(\sqrt{\gamma_2}\,\Gamma_1) + \frac{1}{2}Q(\sqrt{\gamma_2}\,\Gamma_2))\Theta),$$

(43)

where $\Theta = 1 - Q(\sqrt{a_2\gamma_2})$.

Although the problem (43) cannot be solved easily by setting the first derivative (with respect to α) of (43) equal to zero and finding the solution of the resulting equation, it is convex and can be solved numerically or using optimization software. Figure 8, presents the optimal values of α that minimize SER_2. The Golden Section Search and Parabolic Interpolation algorithm [47] are used to find $\hat{\alpha}$ that satisfies (43). As can be seen, the optimal value of α for high γ_2 is almost fixed. For example, when $h_2 = 2h_1$, $\hat{\alpha} = 1$, and when $h_2 = 3h_1$, $\hat{\alpha} \approx 0.4$.

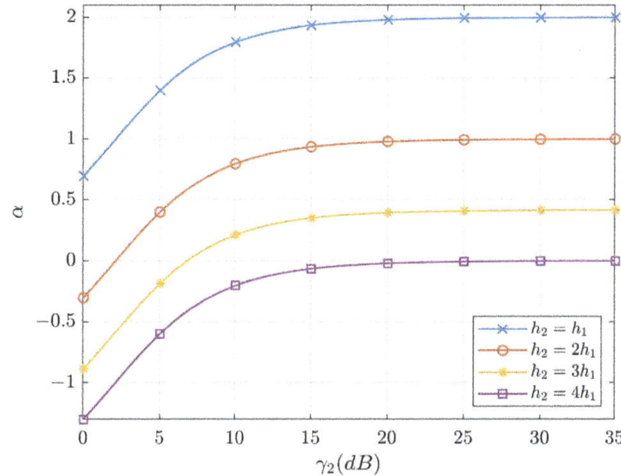

Figure 8. Optimal values of the optimization parameter α that minimize SER_2.

It is worth noting that both GRPA, NGDPA, and proposed DOPA depend on the channel gains of the users, which can be obtained through channel estimation. In the proposed DOPA, the decoding order of the users is included to guarantee different powers allocated to the users. Moreover, the proposed DOPA is more flexible than GRPA and NGDPA since it has the optimization parameter α which could be predetermined for minimizing the symbol error rate or maximizing the target performance metric, such as the system sum rate, the fairness, or the energy efficiency.

In terms of complexity, both GRPA, NGDPA, and proposed DOPA with fixed α have nearly the same complexity. For the optimal α, the complexity of the proposed DOPA increases depending on the applied optimization algorithm to find optimal α. Nonetheless, the complexity of the proposed DOPA, in this case, can be significantly reduced by predetermining the $\hat{\alpha}$ values for different SNR values and channel gain differences. Then, the predetermined values can be stored in lookup tables instead of optimizing during runtime.

7. Performance Evaluation

In this section, analytical and simulation results are presented for the SER performance of a NOMA-based VLC system. DOPA is compared with GRPA and NGDPA. Table 1 displays the simulation parameters.

The simulation for SERs computation is performed as follows. First, a frame of 1000 uniformly-distributed random symbols is generated for each user. Then, the generated symbols are modulated using M-PAM, depending on the modulation order of each user. After that, the modulated signals are normalized and the PA is applied to the normalized signals. Then, the signals are superimposed into one signal and the channel effect is added to the superimposed signal depending on the signal-to-noise ratio (SNR). User 1 signal is demodulated directly from the received signal. In the case of User 2, SIC is performed on the received signal by decoding User 1 signal and then subtracting it from the received signal. Then, User 2 signal is demodulated from the resulting signal. The demodulated signals are compared to the corresponding generated signals and the errors are computed. Finally, the above steps are repeated for 1000 iterations and the cumulative errors are used to find the SER for each user.

Table 1. Simulation Parameters.

Parameter	Notation	Value
Vertical distance	L	2 m
LED input power	P_T	1 W
LED semi-angle	$\Phi_{1/2}$	60°
LED efficiency	η	1 W/A
PD FOV	Ψ	60°
PD area	A	10^{-4} m^2
PD responsivity	v	0.6 A/W
Optical filter gain	T	1
Refractive index	n	1.5
Noise power spectral density	N_0	10^{-15} A^2/Hz

7.1. SER Evaluation

In Figure 9, theoretical and simulated SER of two users, u_1 and u_2, in a NOMA-based VLC system are plotted in two cases. In the first case, the power allocation coefficients of u_1 and u_2 are $a_1 = 0.85$ and $a_2 = 0.15$, respectively, while in the second case $a_1 = 0.9$ and $a_1 = 0.1$. From Figure 6, it is obevious that simulation outcomes match well with theoretical analysis, which validates and confirms the derived SER expressions. Moreover, it can be observed that the SER of u_2 is higher than the SER of u_1 at the same SNR value because the power allocated to u_2 is lower and u_2 needs to decode u_1 signal first before decoding its own signal.

Figures 10 and 11 show the comparison of DOPA, GRPA, and NGDPA in terms of SER$_1$ and SER$_2$, respectively. When both users have similar channel gains, GRPA allocates equal power level to both users. Hence, the interference generated by the second user on the first user is very high and cannot be neglected. Similarly, the second user cannot successfully cancel the interference inserted by the first user. This explains the bad performance of GRPA in this case for both users. On the other hand, NGDPA allocates zero power to the second user when both users have similar channel gains. In this case, there is no interference at the first user which explains its good performance. However, the second signal is lost. Unlike

GRPA and NGDPA, DOPA offers better performance for both users because it allocates different non-zero power levels to the users.

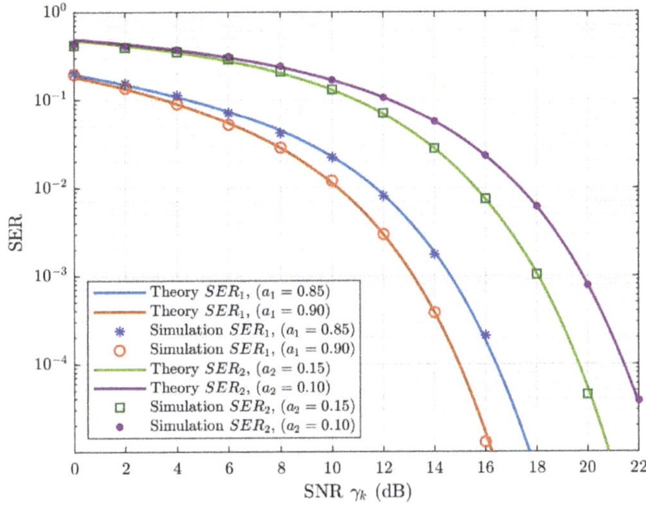

Figure 9. Theoretical and simulated SER of two users in a NOMA−based VLC system.

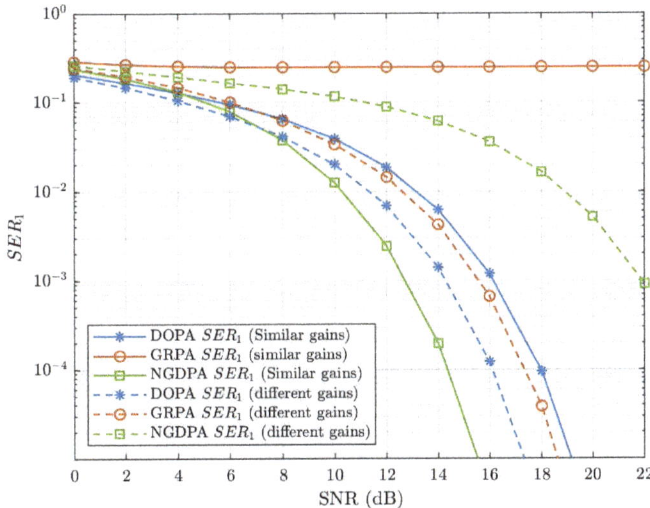

Figure 10. SER$_1$ using DOPA, GRPA and NGDPA. $\alpha = 2$ for similar channel gains and $\alpha = 1$ for different channel gains.

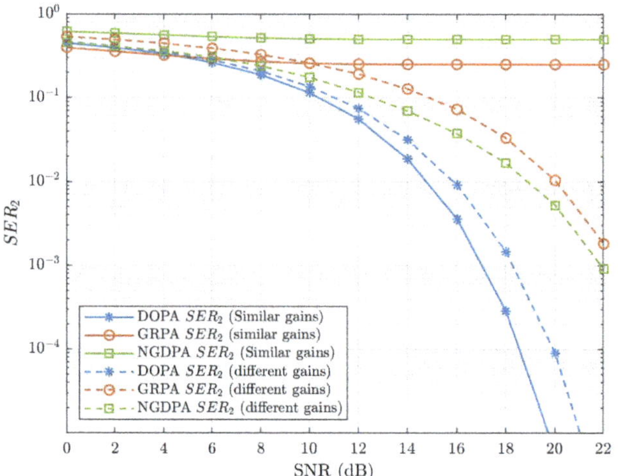

Figure 11. SER_2 using DOPA, GRPA and NGDPA. $\alpha = 2$ for similar channel gains and $\alpha = 1$ for different channel gains.

Figure 12 shows how the SER of u_1 varies with α at different SNR values. Without loss of generality, the horizontal distances of the users are assumed to be $r_1 = r_2 = 1.5$ m (i.e., both users are on the circumference of a circle with a radius r of 1.5). From Figure 12, as α increases, SER_1 is decreased. This is because of that the higher the value of α, the higher the power allocated to u_1 and the lower the power allocated to u_2, which means the multi-user interference is decreased. Therefore, SER_1 is also decreased. Moreover, SER_1 decreases as the SNR increases and the value of α that gives a specific SER_1 is also decreased. For example, when the SNR is 15 dB, the value of α is about 3.2 to achieve SER_1 of 10^{-4}, whereas the value of α is about 2 when the SNR is 18 dB to achieve the same SER_1.

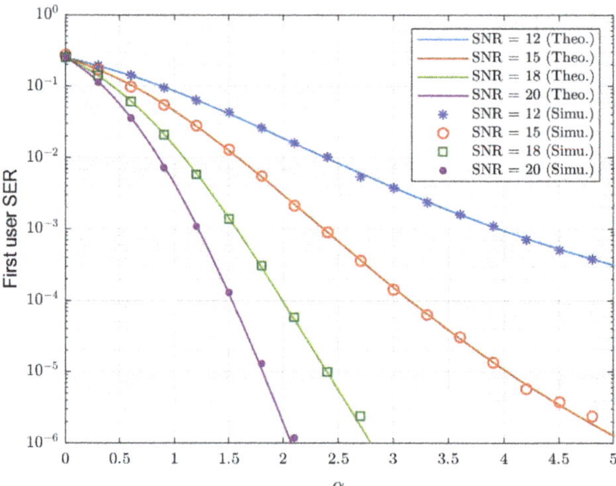

Figure 12. SER_1 with α at different SNR values.

Figure 13, shows how the SER of u_2 varies with α at different SNR values. As the value of α increases, the SER_2 starts to decrease until it reaches a minimum value and then increases again. When the SNR value increases, the SER_2 decreases faster and reaches a lower minimum value and increases faster again. The SER_2 initially decreases because the multi-user interference is reduced when α increases. Hence, u_2 can more accurately cancel the interference from u_1 during the SIC process. However, as α increases, the power allocated to u_2 decreases. Thus, the AWGN added to the signal of u_2 increases. Therefore, the SER_2 finally increases again. From Figure 13, we can also notice that the optimal value of α that gives the minimum value of the SER_2 is roughly the same regardless of the SNR value.

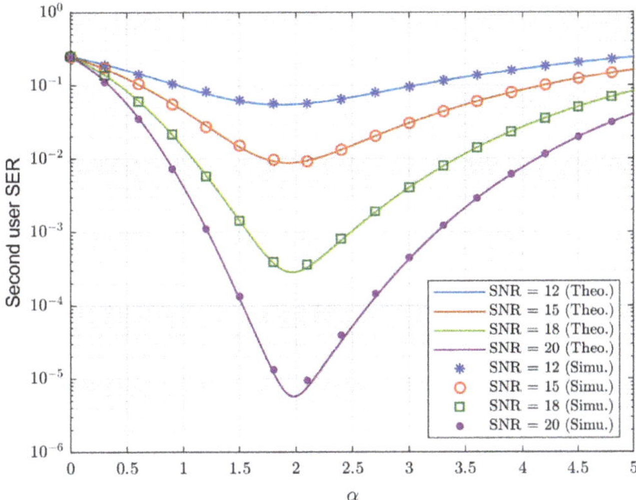

Figure 13. SER_2 with α at different SNR values.

7.2. Achievable Rate

The sum rate is an important metric to show the performance of the VLC network. In Figure 14, the achievable sum rate of the proposed DOPA against different values of the SNR ranging from 10 dB to 20 dB is depicted. The aim of this figure is showing the effectiveness of formulating the optimization problem considering the SER of the second user in comparison with benchmarking schemes, i.e., GRPA and NGDPA. It is shown that the DOPA scheme provides a greater sum rate, which is increased considerably as the value of the SNR increases, due to allocating the power among the users of each group, i.e., the first and second users, considering the minimization of the interference between them. It is worth mentioning that the interference among groups is avoided by implementing BIA. On the other hand, the GRPA scheme achieves a poor sum rate compared with the DOPA, due to the fact that the power is allocated equally between the first and second users, and therefore, the users of each group generate severe interference between them resulting in the degradation of the sum rate. The worst scenario is shown by the NGDPA scheme, it is easy to see that the sum rate is slightly increased as the SNR increases. Where for the highest value of the SNR, which equals to 20 dB, a low sum-rate around 2.5 [bits/s/Hz] is achieved. This is because, the NGDPA allocates a zero power level for the second user belonging to each group contradicting the concept of NOMA, i.e, the second users in the network achieve zero rates. As a result, allocating the power based on the objective function of the proposed DOPA scheme makes NOMA more suitable for the VLC network achieving high rates.

The achievable user rate of the proposed DOPA scheme is derived after considering the approach used for inter-group interference management in Section 3 power allocation approach in Section 6. Note that, after the formation of multiple groups, the bandwidth is divided according to the transmission block of BIA aiming to dedicate resources for each group, while inter-group interference is aligned into less dimensions than the useful information (please see the motivational example and general case in Section 3). It is worth mentioning that inter-group interference in the proposed scheme is aligned at the cost of increasing the noise that results from interference cancellation. In other words, a given user in a certain group must cancel the signals intended to other groups. Following this methodology, the overall noise in our work is given by inter-group interference cancellation noise, background noise and successive interference cancellation (SIC) noise. Subsequently, the useful signal intended to each user belonging to a certain group is determined after performing power allocation according to our approach reported in detail in Section 6. Finally, Shannon capacity is used to calculate the data rate received by each user. Note that the user rates of all the counterparts schemes considered in the paper are calculated following the same fashion, while they differ in power allocation between the weak and strong users.

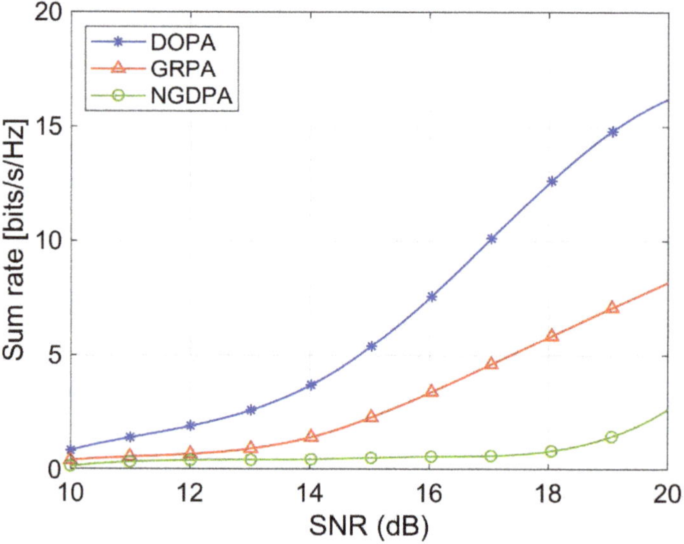

Figure 14. The sum rate of the proposed DOPA against the SNR for $\alpha = 1$ and $K = 10$ users, in comparison with the benchmark GRPA and NGDPA schemes.

7.3. Transmitting an Image

To further show the impact of the multi-user interference on the system performance, we assume that an image is sent for each user simultaneously. These images have the same size and there is no data or channel encryption. Furthermore, it is assumed that the users have similar channel gains and have SNR = 16 dB. The impact of the multi-user interference on the received images is shown in Figure 15. The top row images are for u_1 and the second-row images are for u_2. The transmitted images for u_1 and u_2 are shown in Figure 15a. The received images are shown in different cases. In Figure 15b, the received images are shown when GRPA is used. In this case, the multi-user interference can be clearly observed in both images, which means that the interference from u_2 is too high to be ignored by u_1 and the interference from u_1 can also be observed in u_2 image because of the unsuccessful SIC. The high multi-user interference, in this case, is due to the equal powers allocated to the users by GRPA.

Figure 15. The impact of multi-user interference on the received images. The top row images are for u_1 and the second-row images are for u_2. (**a**) Transmitted images. (**b–f**) Received images with: (**b**) GRPA, (**c**) NGDPA, (**d**) DOPA ($\alpha = 1$), (**e**) DOPA ($\alpha = 2$), and (**f**) DOPA ($\alpha = 4$).

In Figure 15c, the received images are shown when NGDPA is used. In this case, the multi-user interference has been totally eliminated from u_1 image; however, u_2 image has become just noise since NGDPA allocates zero power to the second user when the users' channel gains are equal. Figure 15d–f show the received images when the proposed DOPA is used with $\alpha = 1$, $\alpha = 2$, and $\alpha = 4$, respectively. When $\alpha = 1$, the multi-user interference has been reduced compared with GRPA case because DOPA allocates different powers to the users even if they have similar channel gains. When $\alpha = 2$, which is the optimal value in this case as shown in Figure 13, the multi-user interference has been significantly reduced in both images. When $\alpha = 4$, the interference in u_1 image has been reduced further, however, the effect of the AWGN on u_2 image has been increased, which is expected because the power allocated to u_2 decreases as α is increased.

8. Conclusions

In this paper, the effect of multi-user interference on the performance of a downlink NOMA-besed VLC system has been presented. A closed-form expression for the user symbol error rate has been derived and verified by computer simulation. A low-complexity power allocation (PA) scheme has been proposed to reduce the inevitable multi-user interference. The proposed PA method is a decoding-order-based PA (DOPA) scheme in which the powers are allocated to the users based on their indices in the decoding order and their channel gains. It has been demonistrated that DOPA is able to reduce the interference in NOMA-based VLC systems, and achieve better sum rate compared to GRPA and NGDPA schemes.

Author Contributions: Conceptualization, A.H.A., M.T.A. and J.E.; methodology, E.M.A. and A.-A.Q.; software, E.M.A. and A.-A.Q.; validation, J.E., M.T.A., A.F.A. and A.H.A.; formal analysis, A.H.A., A.-A.Q. and M.T.A.; investigation, E.M.A., A.H.A. and A.-A.Q.; resources, E.M.A. and A.-A.Q.; data curation, E.M.A. and A.-A.Q.; writing—original draft preparation, A.H.A. and E.M.A.; writing—review and editing, A.H.A. and E.M.A.; visualization, M.T.A., J.E., and A.H.A.; supervision, J.E., A.H.A. and M.T.A.; project administration, M.T.A.; funding acquisition, M.T.A. All authors have read and agreed to the published version of the manuscript.

Funding: This work was supported by King Saud University, Kingdom of Saudi Arabia under Researchers Supporting Project number (RSP-2021/336).

Institutional Review Board Statement: Not applicable.

Informed Consent Statement: Not applicable.

Data Availability Statement: All data are provided in full in the results section of this paper.

Acknowledgments: The authors would like to thank King Saud University for supporting this research under the Researchers Supporting Project number (RSP-2021/336).

Conflicts of Interest: The authors declare no conflict of interest.

References

1. Pathak, P.H.; Feng, X.; Hu, P.; Mohapatra, P. Visible Light Communication, Networking, and Sensing: A Survey, Potential and Challenges. *IEEE Commun. Surv. Tutor.* **2015**, *17*, 2047–2077. [CrossRef]
2. Al Hammadi, A.; Sofotasios, P.C.; Muhaidat, S.; Al-Qutayri, M.; Elgala, H. Non-Orthogonal Multiple Access for Hybrid VLC-RF Networks with Imperfect Channel State Information. *IEEE Trans. Veh. Technol.* **2021**, *70*, 398–411. [CrossRef]
3. Chowdhury, M.Z.; Hossan, M.T.; Islam, A.; Jang, Y.M. A Comparative Survey of Optical Wireless Technologies: Architectures and Applications. *IEEE Access* **2018**, *6*, 9819–9840. [CrossRef]
4. Chen, J.; Wang, Z.; Jiang, R. Downlink Interference Management in Cell-Free VLC Network. *IEEE Trans. Veh. Technol.* **2019**, *68*, 9007–9017. [CrossRef]
5. Albraheem, L.I.; Alhudaithy, L.H.; Aljaser, A.A.; Aldhafian, M.R.; Bahliwah, G.M. Toward Designing a Li-Fi-Based Hierarchical IoT Architecture. *IEEE Access* **2018**, *6*, 40811–40825. [CrossRef]
6. Hussein, A.T.; Alresheedi, M.T.; Elmirghani, J.M.H. 20 Gb/s Mobile Indoor Visible Light Communication System Employing Beam Steering and Computer Generated Holograms. *J. Light. Technol.* **2015**, *33*, 5242–5260. [CrossRef]
7. Karbalayghareh, M.; Miramirkhani, F.; Eldeeb, H.B.; Kizilirmak, R.C.; Sait, S.M.; Uysal, M. Channel Modelling and Performance Limits of Vehicular Visible Light Communication Systems. *IEEE Trans. Veh. Technol.* **2020**, *69*, 6891–6901. [CrossRef]
8. Amjad, M.S.; Tebruegge, C.; Memedi, A.; Kruse, S.; Kress, C.; Scheytt, J.C.; Dressler, F. Towards an IEEE 802.11 Compliant System for Outdoor Vehicular Visible Light Communications. *IEEE Trans. Veh. Technol.* **2021**, *70*, 5749–5761. [CrossRef]
9. Memedi, A.; Dressler, F. Vehicular Visible Light Communications: A Survey. *IEEE Commun. Surv. Tutor.* **2021**, *23*, 161–181. [CrossRef]
10. Abdelhady, A.M.; Amin, O.; Chaaban, A.; Shihada, B.; Alouini, M. Downlink Resource Allocation for Dynamic TDMA-Based VLC Systems. *IEEE Trans. Wirel. Commun.* **2019**, *18*, 108–120. [CrossRef]
11. Chen, Z.; Haas, H. Space division multiple access in visible light communications. In Proceedings of the 2015 IEEE International Conference on Communications (ICC), London, UK, 8–12 June 2015; pp. 5115–5119. [CrossRef]
12. Qiu, Y.; Chen, S.; Chen, H.; Meng, W. Visible Light Communications Based on CDMA Technology. *IEEE Wirel. Commun.* **2018**, *25*, 178–185. [CrossRef]
13. Bawazir, S.S.; Sofotasios, P.C.; Muhaidat, S.; Al-Hammadi, Y.; Karagiannidis, G.K. Multiple Access for Visible Light Communications: Research Challenges and Future Trends. *IEEE Access* **2018**, *6*, 26167–26174. [CrossRef]

14. Saito, Y.; Kishiyama, Y.; Benjebbour, A.; Nakamura, T.; Li, A.; Higuchi, K. Non-Orthogonal Multiple Access (NOMA) for Cellular Future Radio Access. In Proceedings of the 2013 IEEE 77th Vehicular Technology Conference (VTC Spring), Dresden, Germany, 2–5 June 2013; pp. 1–5. [CrossRef]
15. Zhang, Y.; Wang, H.M.; Zheng, T.X.; Yang, Q. Energy-Efficient Transmission Design in Non-orthogonal Multiple Access. *IEEE Trans. Veh. Technol.* **2017**, *66*, 2852–2857. [CrossRef]
16. Wu, Q.; Chen, W.; Ng, D.W.K.; Schober, R. Spectral and Energy-Efficient Wireless Powered IoT Networks: NOMA or TDMA? *IEEE Trans. Veh. Technol.* **2018**, *67*, 6663–6667. [CrossRef]
17. Liu, X.; Chen, Z.; Wang, Y.; Zhou, F.; Luo, Y.; Hu, R.Q. BER Analysis of NOMA-Enabled Visible Light Communication Systems with Different Modulations. *IEEE Trans. Veh. Technol.* **2019**, *68*, 10807–10821. [CrossRef]
18. Benjebbour, A.; Saito, Y.; Kishiyama, Y.; Li, A.; Harada, A.; Nakamura, T. Concept and practical considerations of non-orthogonal multiple access (NOMA) for future radio access. In Proceedings of the 2013 International Symposium on Intelligent Signal Processing and Communication Systems, Naha, Japan, 12–15 November 2013; pp. 770–774. [CrossRef]
19. Marshoud, H.; Kapinas, V.M.; Karagiannidis, G.K.; Muhaidat, S. Non-Orthogonal Multiple Access for Visible Light Communications. *IEEE Photonics Technol. Lett.* **2016**, *28*, 51–54. [CrossRef]
20. Chen, C.; Zhong, W.; Yang, H.; Du, P. On the Performance of MIMO-NOMA-Based Visible Light Communication Systems. *IEEE Photonics Technol. Lett.* **2018**, *30*, 307–310. [CrossRef]
21. Fu, Y.; Hong, Y.; Chen, L.; Sung, C.W. Enhanced Power Allocation for Sum Rate Maximization in FDM-NOMA VLC Systems. *IEEE Photonics Technol. Lett.* **2018**, *30*, 1218–1221. [CrossRef]
22. Zhao, Q.; Jiang, J.; Wang, Y.; Du, J. A low complexity power allocation scheme for NOMA-based indoor VLC systems. *Opt. Commun.* **2020**, *463*, 125383. [CrossRef]
23. Lopez, C.A.; Reguera, V.A. Simple Fair Power Allocation for NOMA-Based Visible Light Communication (VLC) Systems. *arXiv* **2021**, arXiv:2112.04410.
24. Marshoud, H.; Sofotasios, P.C.; Muhaidat, S.; Karagiannidis, G.K.; Sharif, B.S. On the Performance of Visible Light Communication Systems with Non-Orthogonal Multiple Access. *IEEE Trans. Wirel. Commun.* **2017**, *16*, 6350–6364. [CrossRef]
25. Kizilirmak, R.C.; Rowell, C.R.; Uysal, M. Non-orthogonal multiple access (NOMA) for indoor visible light communications. In Proceedings of the 2015 4th International Workshop on Optical Wireless Communications (IWOW), Istanbul, Turkey, 7–8 September 2015; pp. 98–101. [CrossRef]
26. Yin, L.; Popoola, W.O.; Wu, X.; Haas, H. Performance Evaluation of Non-Orthogonal Multiple Access in Visible Light Communication. *IEEE Trans. Commun.* **2016**, *64*, 5162–5175. [CrossRef]
27. Kafafy, M.; Fahmy, Y.; Khairy, M. Minimizing the driving power for Non Orthogonal Multiple Access in Indoor Visible Light Communication. *Opt. Switch. Netw.* **2019**, *33*, 169–176. [CrossRef]
28. Almohimmah, E.M.; Alresheedi, M.T.; Abas, A.F.; Elmirghani, J. A Simple User Grouping and Pairing Scheme for Non-Orthogonal Multiple Access in VLC System. In Proceedings of the 2018 20th International Conference on Transparent Optical Networks (ICTON), Bucharest, Romania, 1–5 July 2018; pp. 1–4. [CrossRef]
29. Chen, C.; Zhong, W.; Yang, H.; Du, P.; Yang, Y. Flexible-Rate SIC-Free NOMA for Downlink VLC Based on Constellation Partitioning Coding. *IEEE Wirel. Commun. Lett.* **2019**, *8*, 568–571. [CrossRef]
30. Tahira, Z.; Asif, H.M.; Khan, A.A.; Baig, S.; Mumtaz, S.; Al-Rubaye, S. Optimization of Non-Orthogonal Multiple Access Based Visible Light Communication Systems. *IEEE Commun. Lett.* **2019**, *23*, 1365–1368. [CrossRef]
31. Shi, J.; Hong, Y.; Deng, R.; He, J.; Chen, L.K.; Chang, G.K. Demonstration of Real-Time Software Reconfigurable Dynamic Power-and-Subcarrier Allocation Scheme for OFDM-NOMA-Based Multi-User Visible Light Communications. *J. Light. Technol.* **2019**, *37*, 4401–4409. [CrossRef]
32. Shi, J.; He, J.; Wu, K.; Ma, J. Enhanced Performance of Asynchronous Multi-Cell VLC System Using OQAM/OFDM-NOMA. *J. Light. Technol.* **2019**, *37*, 5212–5220. [CrossRef]
33. Eltokhey, M.W.; Khalighi, M.A.; Ghassemlooy, Z. Dimming-Aware Interference Mitigation for NOMA-Based Multi-Cell VLC Networks. *IEEE Commun. Lett.* **2020**, *24*, 2541–2545. [CrossRef]
34. Almohimmah, E.M.; Alresheedi, M.T. Error Analysis of NOMA-Based VLC Systems with Higher Order Modulation Schemes. *IEEE Access* **2020**, *8*, 2792–2803. [CrossRef]
35. Komine, T.; Nakagawa, M. Fundamental analysis for visible-light communication system using LED lights. *IEEE Trans. Consum. Electron.* **2004**, *50*, 100–107. [CrossRef]
36. Ding, Z.; Fan, P.; Poor, H.V. Impact of User Pairing on 5G Nonorthogonal Multiple-Access Downlink Transmissions. *IEEE Trans. Veh. Technol.* **2016**, *65*, 6010–6023. [CrossRef]
37. Liang, W.; Ding, Z.; Li, Y.; Son, L. User Pairing for Downlink Non-Orthogonal Multiple Access Networks Using Matching Algorithm. *IEEE Trans. Commun.* **2017**, *65*, 5319–5332. [CrossRef]
38. Gou, T.; Wang, C.; Jafar, S.A. Aiming Perfectly in the Dark-Blind Interference Alignment through Staggered Antenna Switching. *IEEE Trans. Signal Process.* **2011**, *59*, 2734–2744. [CrossRef]
39. Adnan-Qidan, A.; Cespedes, M.M.; Armada, A.G. User-Centric Blind Interference Alignment Design for Visible Light Communications. *IEEE Access* **2019**, *7*, 21220–21234. [CrossRef]
40. Adnan-Qidan, A.; Cespedes, M.M.; Armada, A.G. Load Balancing in Hybrid VLC and RF Networks Based on Blind Interference Alignment. *IEEE Access* **2020**, *8*, 72512–72527. [CrossRef]

41. Cespedes, M.M.; Dobre, O.A.; Armada, A.G. Semi-Blind Interference Aligned NOMA for Downlink MU-MISO Systems. *IEEE Trans. Commun.* **2020**, *68*, 1852–1865. [CrossRef]
42. Dimitrov, S.; Haas, H. *Principles of LED Light Communications: Towards Networked Li-Fi*; Cambridge University Press: Cambridge, UK, 2015.
43. Armstrong, J. OFDM for optical communications. *J. Light. Technol.* **2009**, *27*, 189–204. [CrossRef]
44. Dissanayake, S.D.; Armstrong, J. Comparison of aco-ofdm, dco-ofdm and ado-ofdm in im/dd systems. *J. Light. Technol.* **2013**, *31*, 1063–1072. [CrossRef]
45. Yang, Z.; Xu, W.; Pan, C.; Pan, Y.; Chen, M. On the Optimality of Power Allocation for NOMA Downlinks with Individual QoS Constraints. *IEEE Commun. Lett.* **2017**, *21*, 1649–1652. [CrossRef]
46. Marshoud, H.; Muhaidat, S.; Sofotasios, P.C.; Hussain, S.; Imran, M.A.; Sharif, B.S. Optical Non-Orthogonal Multiple Access for Visible Light Communication. *IEEE Wirel. Commun.* **2018**, *25*, 82–88. [CrossRef]
47. Brent, R.P. *Algorithms for Minimization without Derivatives*; Courier Corporation: North Chelmsford, MA, USA, 2013.

 *photonics*

Article

Reconstructed Hybrid Optical OFDM-NOMA for Multiuser VLC Systems

Baolong Li [1], Jianfeng Shi [1,2] and Simeng Feng [3]*

[1] School of Electronic and Information Engineering, Nanjing University of Information Science and Technology, Nanjing 210044, China
[2] National Mobile Communications Research Laboratory, Southeast University, Nanjing 210096, China
[3] Key Laboratory of Dynamic Cognitive System of Electromagnetic Spectrum Space, Ministry of Industry and Information Technology, Nanjing University of Aeronautics and Astronautics, Nanjing 210016, China
* Correspondence: simeng-feng@nuaa.edu.cn

Abstract: Non-orthogonal multiple access (NOMA) is deemed to be a prospective multiple access technology of the next generation. However, in visible light communication (VLC), when advanced hybrid optical orthogonal frequency division multiplexing (O-OFDM), such as hybrid asymmetrically clipped O-OFDM (HACO-OFDM), is combined with NOMA, error propagation is induced, which degrades the system performance. Therefore, a novel reconstructed hybrid O-OFDM-NOMA (RHO-OFDM-NOMA) scheme is conceived in this paper. In order to eliminate the error propagation, the users in RHO-OFDM-NOMA opt for the ACO-OFDM or clipping-free O-OFDM signals according to their channel qualities, which are subsequently superimposed on pulse-amplitude-modulated discrete multitone (PAM-DMT) to yield the spectrum-efficient hybrid O-OFDM signal. Furthermore, a reconstruction process is designed to ensure the non-negativity. Compared with HACO-OFDM, the proposed RHO-OFDM can retain the error propagation in NOMA-VLC, whilst maintaining the superiorities of high spectral and power efficiency. It is demonstrated by simulation results that RHO-OFDM-NOMA can support a notably higher data rate than the NOMA schemes using conventional O-OFDM.

Keywords: visible light communication (VLC); non-orthogonal multiple access (NOMA); optical orthogonal frequency division multiplexing (O-OFDM)

Citation: Li, B.; Shi, J.; Feng, S. Reconstructed Hybrid Optical OFDM-NOMA for Multiuser VLC Systems. *Photonics* **2022**, *9*, 857. https://doi.org/10.3390/photonics9110857

Received: 29 August 2022
Accepted: 8 November 2022
Published: 13 November 2022

Publisher's Note: MDPI stays neutral with regard to jurisdictional claims in published maps and institutional affiliations.

1. Introduction

Visible light communication (VLC) exploits light rays emitted by light-emitting diode (LED) to transmit data [1,2]. Acting as a burgeoning wireless communication technology, VLC possesses many remarkable advantages, including low cost, license-free optical spectrum, high-speed data transmission, no electromagnetic contamination, etc. Given these remarkable advantages, VLC has gained great attention from both academia and the industry. It has also been recognized as a potential technology of the sixth generation (6G) of wireless communication [3,4].

In wireless communications, the booming development of information technology has led to explosively increased mobile data and smart devices [5]. Therefore, how to significantly boost the data rate and enhance the ability of user connectivity has become an urgent problem of 6G communication, which is also one of the main ongoing research efforts of VLC. On the other hand, one of the main drawback lies in the low communication bandwidth of the existing commercial LEDs in VLC, which makes this urgent problem more challenging for VLC [6,7]. In order to tackle the problem, multiple access (MA) technology plays a significant role.

In conventional orthogonal MA (OMA) schemes, strict orthogonality in the time or frequency domains is required to eliminate the multiuser interference. However, due to this strict orthogonality of the supported users, the OMA schemes cannot accommodate requirements of the explosively increased mobile data and smart devices well. Against this background, non-orthogonal MA (NOMA) has gained tremendous attention and is deemed to be a potential MA technology of the next generation. Superior to OMA, NOMA can serve the simultaneous communication of multiple users in the same time and frequency resource, and notably enhances spectral efficiency and ability of the user connectivity in comparison with OMA [8]. Therefore, NOMA has numerous applications in VLC, and is regarded as one of the research directions [7,9]. NOMA should be used in combination with modulation. Relying on high spectral efficiency, orthogonal frequency division multiplexing (OFDM) constitutes competitive modulation in VLC. Furthermore, more ambitious system performance can be expected by combing NOMA with OFDM, which constitutes the concept of OFDM-NOMA [10].

In VLC, a real and non-negative signal is required due to the intensity-modulated direct detection [11,12]. Therefore, a variety of optical OFDM (O-OFDM) strategies have been conceived for VLC. With the aid of frequency-domain Hermitian symmetry, a real O-OFDM signal can be produced. For the sake of the non-negativity, numerous strategies have been conceived. A simple strategy is direct-current-biased O-OFDM (DCO-OFDM), in which a direct-current bias is used [13]. However, DCO-OFDM has to employ a comparatively large DC bias to avert non-linear distortion, resulting in a poor performance in terms of power efficiency. To address it, asymmetrically clipped O-OFDM (ACO-OFDM) has been designed, which generates a unipolar signal through direct clip of the negative part [14,15]. The same philosophy has been extended to discrete multitone modulation (DMT), yielding pulse-amplitude-modulated DMT (PAM-DMT). Nevertheless, these O-OFDM schemes waste half of the subcarrier resources, thus leading to spectral inefficiency [16].

To address the problem of conventional O-OFDM, more advanced hybrid O-OFDM schemes have been conceived [17]. In VLC, hybrid ACO-OFDM (HACO-OFDM) is deemed to be one of the widely used hybrid schemes, which exploits an amalgam of the ACO-OFDM and PAM-DMT techniques [18]. To expound further, the time-domain ACO-OFDM and PAM-DMT signals occupying different subcarriers are transmitted in parallel through direct superimposition. An iterative receiver can be utilized in HACO-OFDM to successively detect the two signal components. Given that the clipping operation of ACO-OFDM induces the interference with PAM-DMT, additional operation of eliminating the clipping distortion is required before detecting the PAM-DMT signal [19]. Compared with conventional schemes, HACO-OFDM can achieve substantial improvement of the spectral efficiency, whist maintaining high power efficiency. Therefore, HACO-OFDM has been widely applied in various scenarios of VLC [20–22].

Recently, there have been several important contributions to OFDM-NOMA in VLC. In [23], the performance of a multiple-input multiple-output-based, multiuser VLC system using DCO-OFDM-NOMA was investigated. By using the real-time software reconfigurable technique, Shi et al. [24] demonstrated an OFDM-NOMA VLC system with dynamic resource allocation. Furthermore, an OFDM-NOMA VLC with the aid of offset quadrature amplitude modulation (QAM) was experimentally demonstrated in [25]. For the sake of the improvement in both user fairness and throughput, a resource allocation method was investigated for the OFDM-NOMA VLC in [26]. In [27], joint power allocation and user pairing was studied for ACO-OFDM-NOMA system to achieve massive connectivity and energy saving. However, these exciting works are mainly developed based on ACO-OFDM and DCO-OFDM, which suffer from spectral and power inefficiency, respectively. OFDM-NOMA using the more advanced hybrid O-OFDM has not been deeply investigated. Moreover, users with worse channel quality can only partially decode the transmitted symbols. Although, when the existing hybrid O-OFDM scheme is combined with NOMA, the clipping distortion elimination should depend on the transmitted symbols of all users. Therefore, error propagation can be induced for users with the worse channel, leading to performance degradation.

In the paper, a novel reconstructed hybrid O-OFDM-NOMA (RHO-OFDM-NOMA) is designed for VLC. In RHO-OFDM-NOMA, the ACO-OFDM and clipping-free O-OFDM signals using the odd-indexed subcarrier for transmission are adopted for different NOMA users according to their channel quality in the proposed scheme to avoid error propagation. Furthermore, these O-OFDM signals are combined with PAM-DMT to yield the spectrum-efficient RHO-OFDM signal, which is subsequently made non-negative by introducing a reconstruction process. The novelty and contributions are summarized as follows:

1. The proposed RHO-OFDM exploits the simultaneous transmission of multiple O-OFDM signals, which effectively enhances the spectral efficiency compared with ACO-OFDM. Meanwhile, no direct-current bias is added in RHO-OFDM, thus leading to high power efficiency.
2. Moreover, compared with conventional HACO-OFDM, RHO-OFDM can eliminate error propagation in NOMA-VLC systems, whilst maintaining both high spectral and power efficiency.
3. Thanks to no error propagation, the proposed RHO-OFDM-NOMA achieves better BER performance than HACO-OFDM-NOMA for users with worse channel quality. Moreover, a significantly high data rate is achieved by RHO-OFDM-NOMA compared with the NOMA schemes using conventional O-OFDM.

2. System Model and VLC Channel

2.1. System Model

In the paper, a typical VLC system is considered, in which a single LED-based transmitter installed on the ceiling is used to support the downlink communication, as shown in Figure 1 [28]. On the receiver plane, M users each equipped with a photo-diode (PD) are simultaneously served by the NOMA transmission in the VLC system. The transmitted data for the M users are modulated by the light intensity of LED. The light signal is transformed to an electrical signal by the photodiode of each user, and the direct detection is subsequently performed to extract the transmitted data.

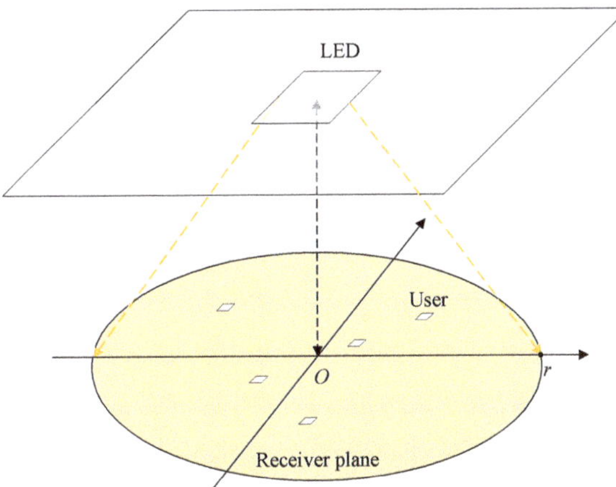

Figure 1. VLC system model.

2.2. VLC Channel

In VLC, the channel attenuation from the LED-based transmitter to the m-th user is given by

$$h_m = \begin{cases} \frac{\rho A_m(d+1)}{2\pi l_m^2}\cos^d(\phi_m)T_s(\psi_m)g(\psi_m)\cos(\psi_m), & 0 \leq \psi_m \leq \psi_c \\ 0, & \psi_m > \psi_c \end{cases} \tag{1}$$

where $m = 1, 2, \cdots, M$, ρ is the O/E conversion efficiency; l_m denotes the distance from the transmitter to the PD; A_m represents the detector area; d denotes the order of Lambertian emission and is a function of the semi-angle at half power $\Phi_{1/2}$ via $d = -\ln 2 / \ln(\cos(\Phi_{1/2}))$; ϕ represents the angle of irradiance; ϕ_m is the angle of incidence; ψ_c is the field of view (FOV) of the receiver; and $T_s(\psi_m)$ denotes the gain of the optical concentrator, which is calculated as

$$g(\psi_m) = \begin{cases} \dfrac{n_c^2}{\sin^2(\psi_c)}, & 0 \leq \psi_m \leq \psi_c \\ 0, & \psi_m > \psi_c \end{cases} \tag{2}$$

where n_c is the refractive index. Furthermore, the received signal is contaminated by noise, which can be modeled as the additive Gaussian noise with zero mean and variance of

$$\sigma_m^2 = \sigma_{\text{sh},m}^2 + \sigma_{\text{th},m}^2, \tag{3}$$

Here, σ_m^2 is the variance of the m-th user, $\sigma_{\text{sh},m}^2$ and $\sigma_{\text{th},m}^2$ are the variances of the shot noise and thermal noise, which can be calculated according to [28].

3. Transmitter Design of RHO-OFDM-NOMA

The transmitted symbols of the NOMA users are superimposed in different power levels. Without loss of generality, we sort the M users in ascending order on the basis of channel qualities. In the NOMA system, successive interference cancellation (SIC) is employed to decode the transmitted symbols. To be specific, the m-th user decodes the transmitted symbols of the first $m - 1$ users, and subsequently removes the interference of the decoded symbols from the received signal. Furthermore, the symbol of the m-th user can be decoded by treating the transmitted symbols of the remaining $M - m$ users as noise. When HACO-OFDM is directly combined with NOMA, the clipping distortion is determined by the transmitted symbols of all users. However, the users with worse channel qualities can only decode the transmitted symbols of part users. Therefore, the clipping distortion cannot be successfully removed for the users with worse channel quality, leading to unavoidable error propagation.

To address the problem of error propagation, a novel hybrid OFDM scheme, termed RHO-OFDM, is conceived for NOMA-VLC. In the proposed hybrid methodology, different OFDM schemes are adopted for the transmission of the NOMA users according to their channel qualities at the odd-indexed subcarriers. Since the transmitted symbol of the first user can be decoded by all users and consumes most of the transmitted power, the power-efficient ACO-OFDM scheme is adopted for the first user. Let $Q_{m,k}$ denote the QAM symbol loaded at the k-th subcarrier for the m-th user. The frequency-domain signal of the first user is written as

$$X_k = \begin{cases} \sqrt{p_{1,k}} Q_{1,k}, & k = 2i + 1, \\ \sqrt{p_{1,k}} Q_{1,N-k}^*, & k = N - (2i + 1), \\ 0, & \text{otherwise,} \end{cases} \tag{4}$$

where $i = 0, 1, \cdots, N/4 - 1$, $p_{1,k}$ is the power allocated to $Q_{1,k}$, and N represents the number of subcarriers. After the IFFT operation is performed on X_k, the signal x_n is obtained as

$$x_n = \frac{1}{\sqrt{N}} \sum_{k=0}^{N-1} X_k e^{j \frac{2\pi nk}{N}}, n = 0, 1, \cdots, N - 1. \tag{5}$$

By directly removing the negative part of x_n, the non-negative ACO-OFDM signal can be generated in a power-efficient manner. Furthermore, the transmitted symbols of the remaining users cannot be decoded by all users. Therefore, the clipping-free O-OFDM signal is used for the transmitted symbols of the remaining users, in which the clipping operation is not used to avoid the clipping distortion. The frequency-domain signal for the remaining users is written as

$$
X_k^{\text{free}} = \begin{cases} \sum\limits_{m=2}^{M} \sqrt{p_{m,k}} Q_{m,k}, & k = 2i+1, \\ \sum\limits_{m=2}^{M} \sqrt{p_{m,k}} Q_{m,k}^*, & k = N-(2i+1), \\ 0, & \text{otherwise}, \end{cases} \tag{6}
$$

where $p_{m,k}$ denotes the power allocated to $Q_{m,k}$. The time-domain clipping-free signal, denoted by x_n^{free}, is generated through the IFFT operation, which is expressed as

$$
x_n^{\text{free}} = \frac{1}{\sqrt{N}} \sum_{k=0}^{N-1} X_k^{\text{free}} e^{j\frac{2\pi nk}{N}}, n = 0, 1, \cdots, N-1. \tag{7}
$$

Since the clipping distortion is avoided, the error propagation can be eliminated.

In order to effectively exploit the subcarrier resource, the M NOMA users further share the real parts of the even-indexed subcarriers through PAM-DMT. Let $P_{m,k}$ represent the PAM symbol loaded at the k-th subcarrier for the m-th user. Furthermore, the frequency-domain signal of PAM-DMT-NOMA is expressed as

$$
Y_k = \begin{cases} j \sum\limits_{m=1}^{M} \sqrt{p_{m,k}} P_{m,k}, & k = 2q, \\ -j \sum\limits_{m=1}^{M} \sqrt{p_{m,k}} P_{m,k}, & k = N-2q, \\ 0, & \text{otherwise}, \end{cases} \tag{8}
$$

where $q = 1, 2, \cdots, N/4 - 1$. The signal Y_k is input into the IFFT module to yield the signal y_n, given by

$$
y_n = \frac{1}{\sqrt{N}} \sum_{k=0}^{N-1} Y_k e^{j\frac{2\pi nk}{N}}, n = 0, 1, \cdots, N-1, \tag{9}
$$

which is subsequently clipped for the sake of the non-negativity. Furthermore, the ACO-OFDM, clipping-free O-OFDM, and PAM-DMT signals are superimposed to realize high spectral efficiency, which is given by

$$
z_n = \lfloor x_n \rfloor_c + x_n^{\text{free}} + \lfloor y_n \rfloor_c. \tag{10}
$$

where $\lfloor \cdot \rfloor_c$ denotes the clipping operation. Note that z_n is bipolar due to the signal component x_n^{free}. Therefore, a reconstruction signal is further introduced to guarantee the non-negativity. In this way, the RHO-OFDM signal is written as

$$
z_n^{\text{RHO}} = z_n + b_n, \tag{11}
$$

where b_n is the reconstruction signal. For the sake of the non-negativity and no interference with the transmitted symbols, the reconstruction signal can be calculated as

$$
b_n = -\min\left\{ z_n, z_{\text{mod}(\frac{N}{2}-n,N)}, z_{\text{mod}(\frac{N}{2}+n,N)}, z_{\text{mod}(N-n,N)} \right\}, \tag{12}
$$

where $\min\{\cdot\}$ denotes the minimum of the sequence, and $\text{mod}\{\}$ is the operation for calculating the remainder. It can be proved that the reconstruction signal in (12) can guarantee the non-negativity, and is only loaded at the real part of the even-indexed subcarrier, which implies that no interference is imposed on the transmitted symbols. The detailed proof is provided in Appendix A.

The architecture of the RHO-OFDM-NOMA transmitter is provided in Figure 2. The transmitted symbols of the NOMA users are modulated by ACO-OFDM, the clipping-free

O-OFDM, and PAM-DMT. Subsequently, the three O-OFDM signal components are combined for hybrid transmission, and the reconstruction signal is further added to generate the non-negative RHO-OFDM signal.

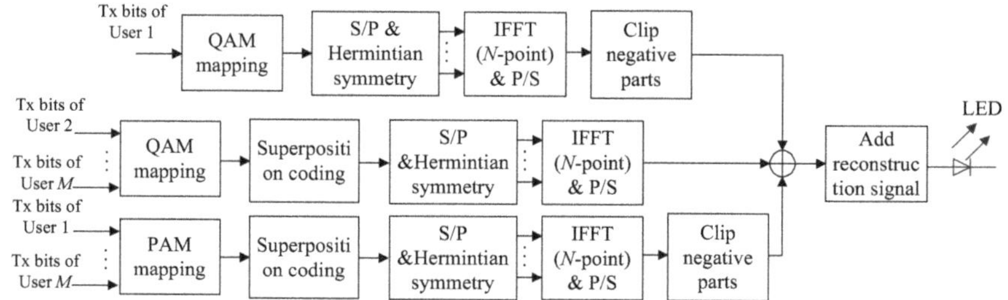

Figure 2. Transmitter architecture of the proposed RHO-OFDM-NOMA.

4. Receiver of RHO-OFDM-NOMA

With the aid of the photodiode, the light signal is detected and converted into the electrical signal at the receiver, which is expressed as

$$z_{m,n} = h_{m,n} * z_n^{\text{RHO}} + w_{m,n}, \tag{13}$$

where $*$ denotes the convolution operation, $h_{m,n}$ is the channel impulse response, and $w_{m,n}$ denotes the noise with variance of σ_m^2. Since the multipath delays are negligible in VLC, $h_{m,n} = h_m \delta(n)$ is considered in the simulation, where $\delta(n)$ is the Dirac Delta function. The received signal is subsequently processed by an FFT block. The frequency-domain signal for the m-th user is given by

$$Z_{m,k} = H_{m,k} Z_k^{\text{RHO}} + W_{m,k}, \tag{14}$$

where $H_{m,k}$, Z_k^{RHO}, and $W_{m,k}$ are the FFT output of $h_{m,n}$, z_n^{RHO}, and $w_{m,n}$, respectively. After one-tap equalization, the equalized signal of $Z_{m,k}$ is written as

$$\widehat{Z}_{m,k} = Z_k^{\text{RHO}} + W_{m,k}/H_{m,k}. \tag{15}$$

Observing the odd-indexed subcarriers that convey the transmitted symbols, we have

$$\widehat{Z}_{m,2i+1} = \sum_{m=1}^{M} \sqrt{p_{m,2i+1}} Q_{m,2i+1} + W_{m,2i+1}/H_{m,2i+1}. \tag{16}$$

The user can decode the transmitted symbol through the SIC process. Furthermore, at the imaginary part of the even-indexed subcarrier, the frequency-domain signal is written as

$$\text{Im}\left\{\widehat{R}_{m,2q}\right\} = \sum_{m=1}^{M} \sqrt{p_{m,2q}} P_{2q} + X_{2q}^{\text{clip}} + \text{Im}\left\{\widehat{W}_{m,2q}\right\}, \tag{17}$$

where X_{2q}^{clip} is the clipping distortion caused by ACO-OFDM, and $\text{Im}\left\{\widehat{R}_{m,2q}\right\}$ and $\text{Im}\left\{\widehat{W}_{m,2q}\right\}$ denote the imaginary part of $\widehat{R}_{m,2q}$ and $\widehat{W}_{m,2q}$, respectively. Note that the clipping distortion is only determined by the transmitted symbol of the first user, which can be decoded by all users. Therefore, the clipping distortion is first regenerated and removed from the received signal $\text{Im}\left\{\widehat{R}_{m,2q}\right\}$ based on the decoded symbol of the first user. Subsequently, the transmitted PAM symbol can be decoded through the SIC process.

The block diagram of the RHO-OFDM-NOMA receiver is provided in Figure 3. By executing the FFT operation on the received signal, the frequency-domain signal is obtained, which can be used to decode the transmitted QAM symbol of the user through the

SIC process. Furthermore, the clipping distortion caused by the ACO-OFDM branch is reproduced and removed from the received signal based on the decoded QAM symbol of the first user. Subsequently, the transmitted PAM symbol of the user can be decoded through the SIC process.

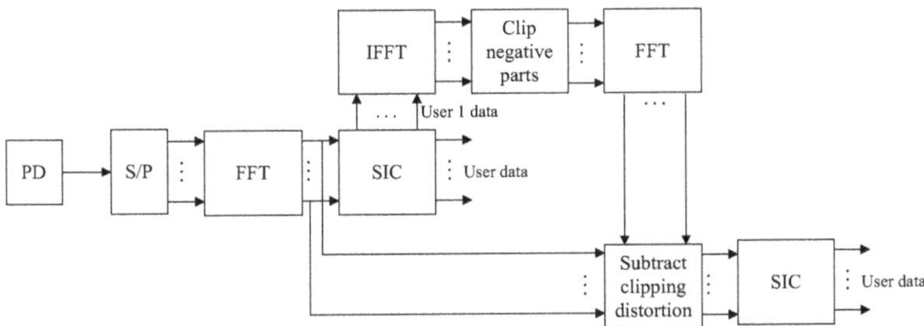

Figure 3. Receiver architecture of the proposed RHO-OFDM-NOMA.

5. Theoretical Performance Analysis

5.1. Sum Rate of RHO-OFDM-NOMA

The sum rate of the proposed RHO-OFDM-NOMA is theoretically analyzed. Since the m-th user can decode the transmitted symbols of the first $m - 1$ users and treat the remaining ones as noise, the signal to interference plus noise ratio (SINR) for the m-th user at the $(2i + 1)$-th subcarrier is computed as

$$\text{SINR}_{m,2i+1} = \frac{p_{m,2i+1}}{\sum\limits_{l=m+1}^{M} p_{l,2i+1} + \sigma_m^2 / H_{m,2i+1}^2}. \tag{18}$$

In RHO-OFDM-NOMA, the clipping distortion is only determined by the transmitted symbol of first user, which can be decoded by all users according to the principle of NOMA. On the other hand, when the transmitted symbols of all users are modulated by ACO-OFDM in HACO-OFDM, the clipping distortion is determined by all users. Therefore, in order to successfully eliminate the clipping distortion, the user is required to decode the transmitted symbols of all users, which restricts the data rate to the minimum one of all users. By contrast, the proposed RHO-OFDM-NOMA removes the restriction on the data rate by the well-designed hybrid transmission architecture. In this way, the data rate of the $(2i + 1)$-th subcarrier achieved by the m-th user in RHO-OFDM-NOMA can be computed as

$$R_{m,2i+1} = \log_2(1 + \text{SINR}_{2i+1})$$

$$= \log_2\left(1 + \frac{p_{m,2i+1}}{\sum\limits_{l=m+1}^{M} p_{l,2i+1} + \sigma_m^2 / H_{m,2i+1}^2}\right). \tag{19}$$

Furthermore, the SINR for the m-th user at the imaginary part of the $2q$-th subcarrier is expressed as

$$\text{SINR}_{m,2q} = \frac{p_{m,2q}}{\sum\limits_{l=m+1}^{M} p_{l,2q} + \sigma_m^2 / H_{m,2q}^2}. \tag{20}$$

Therefore, the data rate at the imaginary part of the $2q$-th subcarrier is given by

$$R_{m,2q} = \log_2\left(1 + \mathrm{SINR}_{m,2q}\right)$$

$$= \log_2\left(1 + \frac{p_{m,2q}}{\sum\limits_{l=m+1}^{M} p_{l,2q} + \sigma_{m,2q}^2 / \left(2H_{m,2q}^2\right)}\right). \tag{21}$$

The sum rate achieved by RHO-OFDM-NOMA is expressed as

$$R_{\mathrm{sum}} = \sum_{m=1}^{M} \sum_{i=0}^{N/4-1} R_{m,2i+1} + \sum_{m=1}^{M} \sum_{q=1}^{N/4-1} R_{m,2q}. \tag{22}$$

5.2. Complexity Analysis

The complexity of the OFDM-based system is dominated by IFFT/FFT, while the complexity of other operations can be negligible [21]. Therefore, the computational complexity of the IFFT/FFT operations is analyzed to characterize the complexity of the proposed scheme. When the radix-2 algorithm is adopted, we can use $2\mathcal{O}(N\log_2 N)$ to characterize the computational complexity of the N-point IFFT/FFT operation executed on the complex-valued frame.

At the transmitter, three IFFT operations are required in RHO-OFDM-NOMA. To expound further, the ACO-OFDM and clipping-free O-OFDM signals exploit one IFFT operation performed on the complex-valued QAM symbols, respectively. Since only the odd-indexed subcarriers are used, half of the computational complexity can be reduced. Therefore, the computational complexity of ACO-OFDM and clipping-free O-OFDM can be characterized as $\mathcal{O}(N\log_2 N)$. Additionally, one IFFT operation performed on the imaginary-valued PAM symbols is required in PAM-DMT, which leads to $\frac{1}{2}\mathcal{O}(N\log_2 N)$ since only even-indexed subcarriers are used. Therefore, the total computational complexity of the RHO-OFDM-NOMA transmitter is $\frac{5}{2}\mathcal{O}(N\log_2 N)$, which is relatively higher than that of the HACO-OFDM-NOMA transmitter, i.e., $\frac{3}{2}\mathcal{O}(N\log_2 N)$. Nevertheless, the increased complexity of the RHO-OFDM-NOMA transmitter is not a problem since the transmitter generally possesses strong ability on signal processing.

At the receiver, the proposed RHO-OFDM-NOMA uses the same detection process as HACO-OFDM-NOMA, in which two FFT operations and one IFFT operation are required. The two FFT operations are performed on the real-valued frame, leading to the total complexity of $2\mathcal{O}(N\log_2 N)$. Additionally, due to the complex-valued frame, the IFFT operation has the complexity of $\mathcal{O}(N\log_2 N)$. Therefore, the total complexity of the RHO-OFDM-NOMA receiver is $\mathcal{O}(N\log_2 N)$, which is the same as that of the HACO-OFDM-NOMA receiver.

6. Simulation Results and Discussion

Simulations are performed to characterize the performance of different NOMA schemes. In the simulation, one LED array installed on the ceiling is used as a transmitter, which simultaneously serves two users distributed on the receiver plane. The receiver plane is located 2 m away from the ceiling. The relevant system parameters are summarized in Table 1 [29,30].

Table 1. Parameters of the LED and receiver in NOMA-VLC.

Parameter	Value	Parameter	Value
Field of view of the receiver	80 (deg.)	Detector area	1 (cm^2)
Gain of the optical concentrator	1.0	Refractive index	1.5
O/E conversion efficiency	0.23 (A/W)	Communication bandwidth	100 (MHz)
Number of LEDs per array	60×60	LED interval	0.01 (m)
Semi-angle at half power	70 (deg.)		

The BER curves of two users in the hybrid OFDM-NOMA system are illustrated in Figures 4 and 5, respectively. In the simulation, 64-QAM and 8-PAM are adopted, and two scenarios are considered. In Scenario 1, the locations of the two users are $(3.0, 2.9)$ and $(0.8, 0.1)$, respectively. In Scenario 2, the locations of the two users are $(3.3, 3.3)$ and $(-1.3, -1.3)$, respectively. The two users in Scenario 1 are located close to each other while the users are located far from each other in Scenario 2. In both scenarios, User 1 has worse channel quality than User 2 since it is located farther from the LED array. Therefore, most of the power is allocated to User 1 according to the philosophy of NOMA. In the simulation, the proportion of the power of User 1 is set to 99.5%.

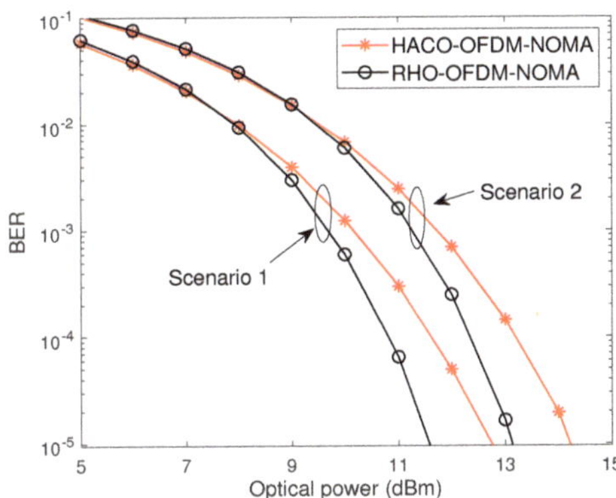

Figure 4. BER performance of User 1 using HACO-OFDM and RHO-OFDM-NOMA.

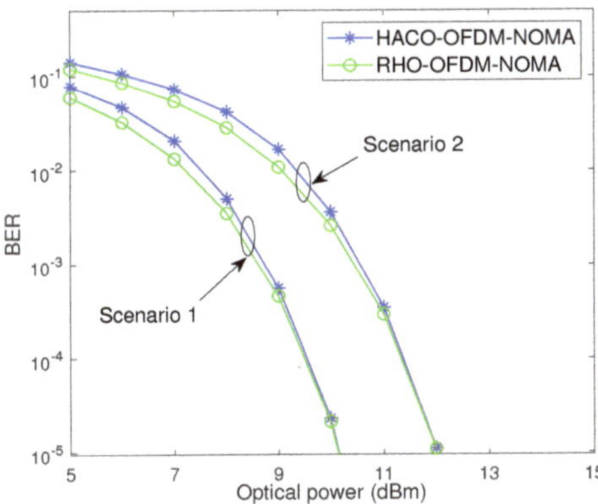

Figure 5. BER performance of User 2 using HACO-OFDM and RHO-OFDM-NOMA.

In the NOMA-based system, successive interference cancellation (SIC) is employed for NOMA users at the receiver. Since the first user has worse channel quality, the first user can only decode its own transmitted symbol, and the transmitted symbol of User 2 is treated as noise. However, when HACO-OFDM is directly combined with NOMA, the clipping noise is determined by the transmitted symbols of all users [18]. Therefore, the clipping distortion cannot be successfully regenerated and eliminated for User 1, leading to error propagation. By contrast, the error propagation can be eliminated in the proposed RHO-OFDM-NOMA. Therefore, it is observed from Figure 4 that RHO-OFDM-NOMA achieves better BER performance than HACO-OFDM-NOMA for User 1. On the other hand, User 2 can decode the transmitted symbols of two users. Therefore, the clipping distortion can be eliminated in HACO-OFDM-NOMA. Meanwhile, in contrast to HACO-OFDM, the reconstruction process is introduced to guarantee the non-negativity. It is clearly seen from Figure 5 that HACO-OFDM-NOMA and RHO-OFDM-NOMA achieve similar BER performance for User 2, which implies that the proposed RHO-OFDM still has the superiority of high power efficiency, regardless of using the reconstruction process or not.

Furthermore, the sum rates of RHO-NOMA-VLC for Scenario 1 and Scenario 2 are illustrated in Figures 6 and 7, respectively. The sum rates of HACO-OFDM-NOMA, DCO-OFDM-NOMA, and ACO-OFDM-NOMA are provided for comparison [26,31]. The DC bias of 10 dB is taken into consideration in DCO-OFDM-NOMA. Compared with conventional DCO-OFDM-NOMA and ACO-OFDM-NOMA, the proposed RHO-OFDM-NOMA has the advantage of high power efficiency, whist improving spectral efficiency. Meanwhile, the error propagation in HACO-OFDM-NOMA induces the sum rate reduction. Therefore, it is seen from Figures 6 and 7 that the proposed RHO-OFDM-NOMA substantially outperforms other NOMA schemes in terms of the sum rate.

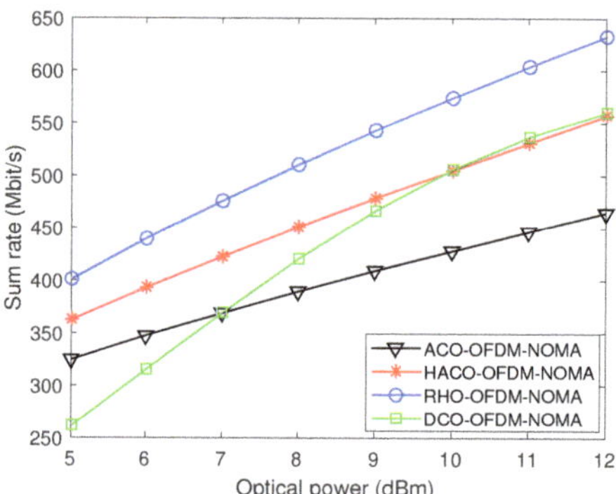

Figure 6. Sum rate of the NOMA-VLC system using different OFDM schemes for Scenario 1.

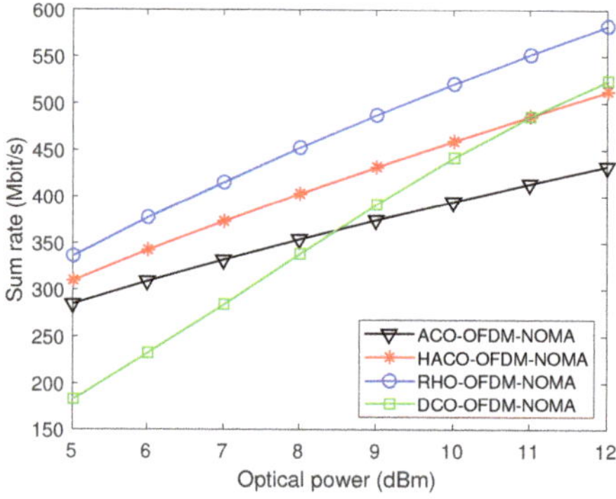

Figure 7. Sum rate of the NOMA-VLC system using different OFDM schemes for Scenario 2.

In order to make a comprehensive evaluation, the average sum rate of different OFDM-NOMA schemes for VLC system are provided in Figure 8. A circular VLC cell with a radius of $r = 3$ m is considered. The average sum rate is obtained by assuming that the two users are uniformly distributed in the VLC cell. It is seen from Figure 8 that the performance of RHO-OFDM-NOMA is much better than that of the conventional scheme. Furthermore, the spectral efficiency is defined as the ratio of the data rate to the bandwidth. Since the same bandwidth is adopted for different schemes, the higher data rate indicates higher spectral efficiency. Therefore, the proposed RHO-OFDM-NOMA has higher spectral efficiency than other schemes, which makes it a competitive transmission scheme for multiuser VLC.

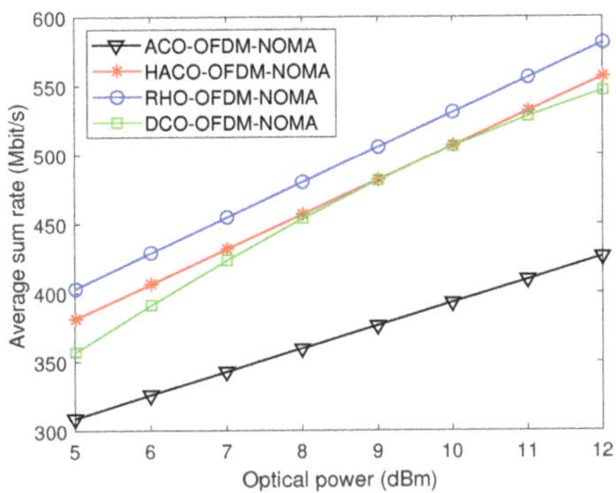

Figure 8. Average sum rate of the NOMA-VLC system using different OFDM schemes.

7. Conclusions

A novel RHO-OFDM-NOMA is conceived for VLC in this paper. In RHO-OFDM-NOMA, ACO-OFDM and clipping-free O-OFDM are used for different users based on their channel qualities, which are further combined with PAM-DMT to enhance the spectral efficiency. The well-designed hybrid transmission architecture in RHO-OFDM-NOMA can eliminate the error propagation that arises in the NOMA-VLC system using conventional hybrid O-OFDM. Owing to having no error propagation, the proposed RHO-OFDM-NOMA can support better BER performance compared with HACO-OFDM-NOMA for the users with worse channel quality. Moreover, a much higher data rate is obtained by RHO-OFDM-NOMA than the NOMA schemes using DCO-OFDM and ACO-OFDM, which makes it a prospective technology of the multiuser transmission for VLC.

Author Contributions: Conceptualization, B.L. and S.F.; methodology, B.L.; software, J.S.; validation, B.L., J.S., and S.F.; formal analysis, B.L.; investigation, J.S.; resources, B.L.; data curation, J.S.; writing—original draft preparation, B.L.; writing—review and editing, J.S.; visualization, J.S.; supervision, S.F.; project administration, S.F.; funding acquisition, S.F. All authors have read and agreed to the published version of the manuscript.

Funding: This research was funded by the National Natural Science Foundation of China under Grants 62001219, 62201275 and 62201274, the Natural Science Foundation of Jiangsu Province under Grants BK20190582 and BK20210641, the open research fund of National Mobile Communications Research Laboratory, Southeast University under Grant 2021D11, and the Startup Foundation for Introducing Talent of NUIST.

Institutional Review Board Statement: Not applicable.

Informed Consent Statement: Not applicable.

Data Availability Statement: The data that support the findings of this study are available from the corresponding author upon reasonable request.

Conflicts of Interest: The authors declare no conflict of interest.

Appendix A

We first prove that the reconstruction signal in (12) can generate the non-negative RHO-OFDM signal. Based on (12), the following inequality holds:

$$z_n \geq \min\left\{z_n, z_{\mathrm{mod}(\frac{N}{2}-n,N)}, z_{\mathrm{mod}(\frac{N}{2}+n,N)}, z_{\mathrm{mod}(N-n,N)}\right\}. \tag{A1}$$

Therefore, we have

$$z_n^{\mathrm{RHO}} = z_n + b_n = z_n - \min\left\{z_n, z_{\mathrm{mod}(\frac{N}{2}-n,n)}, z_{\mathrm{mod}(\frac{N}{2}+n,n)}, z_{\mathrm{mod}(N-n,n)}\right\} \geq 0. \tag{A2}$$

It is observed that the introduced reconstruction signal can make the RHO-OFDM signal non-negative.

Furthermore, we will prove that the reconstruction signal is loaded at the real part of the even-indexed subcarrier. According to (12), the reconstruction signal has the following characteristics:

$$\begin{cases} b_n = b_{n+\frac{N}{2}}, & n = 0, \frac{N}{4}, \\ b_n = b_{\frac{N}{2}-n} = b_{n+\frac{N}{2}} = b_{N-n}, & n = 1, 2, \cdots, \frac{N}{4} - 1. \end{cases} \tag{A3}$$

By performing FFT operation on b_n, we have

$$\begin{aligned}
B_k &= \frac{1}{\sqrt{N}} \sum_{n=0}^{N-1} b_n e^{-j\frac{2\pi nk}{N}} = \frac{1}{\sqrt{N}} \sum_{n=1}^{\frac{N}{4}-1} b_n \left(e^{-j\frac{2\pi nk}{N}} + e^{-j\pi k + j\frac{2\pi nk}{N}} + e^{-j\frac{2\pi nk}{N} - j\pi k} + e^{j\frac{2\pi nk}{N}}\right) \\
&\quad + \frac{1}{\sqrt{N}}\left[b_0\left(1 + e^{-j\pi k}\right) + b_{\frac{N}{4}} e^{-j\frac{\pi}{2}k}\left(1 + e^{-j\pi k}\right)\right] \\
&= \frac{1}{\sqrt{N}} \sum_{n=1}^{\frac{N}{4}-1} b_n \left(e^{-j\frac{2\pi nk}{N}} + e^{j\frac{2\pi nk}{N}}\right)\left(1 + e^{-j\pi k}\right) + \frac{1}{\sqrt{N}}\left(b_0 + b_{\frac{N}{4}} e^{-j\frac{\pi}{2}k}\right)\left(1 + e^{-j\pi k}\right) \\
&= \frac{1}{\sqrt{N}} \sum_{n=1}^{\frac{N}{4}-1} 2b_n \cos\left(\frac{2\pi nk}{N}\right)\left(1 + e^{-j\pi k}\right) + \frac{1}{\sqrt{N}}\left(b_0 + b_{\frac{N}{4}} e^{-j\frac{\pi}{2}k}\right)\left(1 + e^{-j\pi k}\right)
\end{aligned} \tag{A4}$$

For the odd-indexed subcarrier, i.e., $k = 2i + 1$, we have

$$1 + e^{-j\pi k} = 1 + e^{-j2\pi i - j\pi} = 0. \tag{A5}$$

Therefore, B_k at the odd-indexed subcarrier is calculated as

$$B_k = 0, \ k = 2i + 1, \tag{A6}$$

which implies that the reconstruction signal does not induce any interference with the transmitted symbol at the odd-indexed subcarrier. Furthermore, for the even-indexed subcarrier, i.e., $k = 2q$, we have

$$1 + e^{-j\pi k} = 2, \ 1 + e^{-j\frac{\pi}{2}k} = (-1)^q. \tag{A7}$$

Therefore, B_k at the even-indexed subcarrier is written as

$$B_k = \frac{1}{\sqrt{N}} \sum_{n=1}^{\frac{N}{4}-1} 4b_n \cos\left(\frac{2\pi nk}{N}\right) + \frac{2}{\sqrt{N}}\left[b_0 + (-1)^q b_{\frac{N}{4}}\right], \ k = 2q. \tag{A8}$$

It is found that B_k is real for $k = 2q$, which indicates that the reconstruction signal is only loaded at the real parts of the even-indexed subcarrier. Hence, the introduced reconstruction signal does not contaminate the QAM and PAM symbols at the corresponding subcarriers.

References

1. Shi, J.; Niu, W.; Ha, Y.; Xu, Z.; Li, Z.; Yu, S.; Chi, N. AI-enabled intelligent visible light communications: Challenges, progress, and future. *Photonics* **2022**, *9*, 529. [CrossRef]
2. Abdelhady, A.M.; Amin, O.; Salem, A.K.S.; Alouini, M.-S.; Shihada, B. Channel characterization of IRS-based visible light communication systems. *IEEE Trans. Commun.* **2022**, *70*, 1913–1926.
3. Ndjiongue, A.R.; Ngatched, T.M.N.; Dobre O.A.; Armada, A.G. VLC-based networking: Feasibility and challenges. *IEEE Netw.* **2020**, *34*, 158–165. [CrossRef]
4. Wang, J.; Ge, H.; Lin, M.; Wang, J.; Dai, J.; Alouini, M.S. On the secrecy rate of spatial modulation-based indoor visible light communications. *IEEE J. Sel. Areas Commun.* **2019**, *37*, 2087–2101. [CrossRef]
5. Liu, Y.; Mu, X.; Liu, X.; Renzo, M.D.; Ding, Z.; Schober, R. Reconfigurable intelligent surface-aided multi-user networks: Interplay between NOMA and RIS. *IEEE Wireless Commun.* **2022**, *29*, 169–176.
6. Arfaoui, M.A.; Ghrayeb, A.; Assi, C.; Qaraqe, M. CoMP-assisted NOMA and cooperative NOMA in indoor VLC cellular systems. *IEEE Trans. Commun.* **2022**, *70*, 6020–6034. [CrossRef]
7. Marshoud, H.; Muhaidat, S.; Sofotasios, P.C.; Hussain, S.; Imran, M.A.; Sharif, B.S. Optical non-orthogonal multiple access for visible light communication. *IEEE Wireless Commun.* **2018**, *25*, 82–88. [CrossRef]
8. Liu, Y.; Qin, Z.; Elkashlan, M.; Ding, Z.; Nallanathan A.; Hanzo, L. Nonorthogonal multiple access for 5G and beyond. *Proc. IEEE* **2017**, *105*, 2347–2381.
9. Obeed, M.; Salhab, A.M.; Alouini M.-S.;Zummo, S.A. On optimizing VLC networks for downlink multi-user transmission: A survey. *IEEE Commun. Surveys Tuts.* **2019**, *21*, 2947–2976.
10. Xu, W.; Li, X.; Lee, C.-H.; Pan M.; Feng, Z. Joint sensing duration adaptation, user matching, and power allocation for cognitive OFDM-NOMA systems. *IEEE Trans. Wireless Commun.* **2018**, *17*, 1269–1282. [CrossRef]
11. Hong, H.; Li, Z. Hybrid adaptive bias OFDM-based IM/DD visible light communication system. *Photonics* **2021**, *8*, 257.
12. Cao, B.; Yuan, K.; Li, H.; Duan, S.; Li, Y.; Ouyang, Y. The performance improvement of VLC-OFDM system based on reservoir computing. *Photonics* **2022**, *9*, 185. [CrossRef]
13. Carruthers, J.B.; Kahn, J.M. Multiple-subcarrier modulation for nondirected wireless infrared communication. *IEEE J. Sel. Areas Commun.* **1996**, *14*, 538–546.
14. Armstrong, J.; Lowery, A.J. Power efficient optical OFDM. *Electron. Lett.* **2006**, *42*, 370–372. [CrossRef]
15. Lee, S.; Randel, S.; Breyer, F.; Koonen, A. PAM-DMT for intensity-modulated and direct-detection optical communication systems. *IEEE Photon. J.* **2009**, *21*, 1749–1751.
16. Dissanayake, S.D.; Armstrong, J. Comparison of ACO-OFDM, DCO-OFDM and ADO-OFDM in IM/DD systems. *J. Lightw. Technol.* **2013**, *31*, 1063–1072.
17. Zhang, X.; Babar, Z.; Petropoulos, P.; Haas, H.; Hanzo, L. The evolution of optical OFDM. *IEEE Commun. Surveys Tuts.* **2021**, *23*, 1430–1457. [CrossRef]
18. Ranjha, B.; Kavehrad, M. Hybrid asymmetrically clipped OFDM based IM/DD optical wireless system. *J. Opt. Commun. Netw.* **2014**, *6*, 387–396. [CrossRef]
19. Li, B.; Xu, W.; Zhang, H.; Zhao, C.; Hanzo, L. PAPR reduction for hybrid ACO-OFDM aided IM/DD optical wireless vehicular communications. *IEEE Trans. Veh. Technol.* **2017**, *66*, 9561–9566. [CrossRef]
20. Yang, F.; Gao, J. Dimming control scheme with high power and spectrum efficiency for visible light communications. *IEEE Photon. J.* **2017**, *9*, 1–13.
21. Hu, W.-W. Design of cyclic shifted PAM-DMT signals in HACO-OFDM visible light communication. *IEEE Commun. Lett.* **2020**, *24*, 2834–2838.
22. Wang, Q.; Wang, Z.; Dai, L. Asymmetrical hybrid optical OFDM for visible light communications with dimming control. *IEEE Photon. Technol. Lett.* **2015**, *27*, 974–977. [CrossRef]
23. Chen, C.; Zhong, W.; Yang, H.; Du, P. On the performance of MIMO-NOMA-based visible light communication systems. *IEEE Photon. Technol. Lett.* **2017**, *30*, 307–310. [CrossRef]
24. Shi, J.; Hong, Y.; Deng, R.; He, J.; Chen, L.-K.; Chang, G.-K. Demonstration of real-time software reconfigurable dynamic power-and-subcarrier allocation scheme for OFDM-NOMA-based multi-user visible light communications. *J. Lightw. Technol.* **2019**, *37*, 4401–4409. [CrossRef]
25. Shi, J.; He, J.; Wu, K.; Ma, J. Enhanced performance of asynchronous multi-cell VLC system using OQAM/OFDM-NOMA. *J. Lightw. Technol.* **2019**, *37*, 5212–5220.
26. Wang, G.; Shao, Y.; Chen, L. -K.; Zhao, J. Subcarrier and power allocation in OFDM-NOMA VLC systems. *IEEE Photon. Technol. Lett.* **2021**, *33*, 189–192.
27. Jiang, R.; Sun, C.; Tang, X.; Zhang, L.; Wang, H.; Zhang, A. Joint user-subcarrier pairing and power allocation for uplink ACO-OFDM-NOMA underwater visible light communication systems. *J. Lightw. Technol.* **2020**, *39*, 1997–2007.
28. Marshoud, H.; Sofotasios, P.C.; Muhaidat, S.; Karagiannidis, G.K.; Sharif, S.B. On the performance of visible light communication systems with non-orthogonal multiple access. *IEEE Trans. Wireless Commun.* **2017**, *16*, 6350–6364. [CrossRef]

29.	Li, B.; Wang, J.; Zhang, R.; Shen, H.; Zhao, C.; Hanzo, L. Multiuser MISO transceiver design for indoor downlink visible light communication under per-LED optical power constraints. *IEEE Photon. J.* **2015**, *7*, 1–13.
30.	Komine, T.; Nakagawa, M. Fundamental analysis for visible-light communication system using LED lights. *IEEE Trans. Consum. Electron.* **2004**, *50*, 100–107.
31.	Jayashree, P.; Holey, P.; Kappala, V.K.; Das, S.K. Performance analysis of ACO-OFDM NOMA for VLC communication. *Opt. Quant. Electron.* **2022**, *54*, 1–16.

Article

DeepGOMIMO: Deep Learning-Aided Generalized Optical MIMO with CSI-Free Detection

Xin Zhong [1], Chen Chen [1,*], Shu Fu [1], Zhihong Zeng [2] and Min Liu [1]

[1] School of Microelectronics and Communication Engineering, Chongqing University, Chongqing 400044, China
[2] LiFi Research and Development Centre, Institute for Digital Communications, The University of Edinburgh, Edinburgh EH9 3JL, UK
* Correspondence: c.chen@cqu.edu.cn

Abstract: Generalized optical multiple-input multiple-output (GOMIMO) techniques have been recently shown to be promising for high-speed optical wireless communication (OWC) systems. In this paper, we propose a novel deep learning-aided GOMIMO (DeepGOMIMO) framework for GOMIMO systems, wherein channel state information (CSI)-free detection can be enabled by employing a specially designed deep neural network (DNN)-based MIMO detector. The CSI-free DNN detector mainly consists of two modules: one is the preprocessing module, which is designed to address both the path loss and channel crosstalk issues caused by MIMO transmission, and the other is the feedforward DNN module, which is used for joint detection of spatial and constellation information by learning the statistics of both the input signal and the additive noise. Our simulation results clearly verify that, in a typical indoor 4×4 MIMO-OWC system using both generalized optical spatial modulation (GOSM) and generalized optical spatial multiplexing (GOSMP) with unipolar nonzero 4-level pulse-amplitude modulation (4-PAM) modulation, the proposed CSI-free DNN detector achieves near the same bit error rate (BER) performance as the optimal joint maximum-likelihood (ML) detector, but with much-reduced computational complexity. Moreover, because the CSI-free DNN detector does not require instantaneous channel estimation to obtain accurate CSI, it enjoys the unique advantages of improved achievable data rate and reduced communication time delay in comparison to the CSI-based zero-forcing DNN (ZF-DNN) detector.

Keywords: optical wireless communication; multiple-input multiple-output; deep learning

Citation: Zhong, X.; Chen, C.; Fu, S.; Zeng, Z; Liu, M. DeepGOMIMO: Deep Learning-Aided Generalized Optical MIMO with CSI-Free Detection. *Photonics* **2022**, *9*, 940. https://doi.org/10.3390/photonics9120940

Received: 7 November 2022
Accepted: 2 December 2022
Published: 5 December 2022

Publisher's Note: MDPI stays neutral with regard to jurisdictional claims in published maps and institutional affiliations.

1. Introduction

Due to the exhaustion of radio frequency (RF) spectrum resources, optical wireless communication (OWC) which explores the infrared, visible light, or ultraviolet spectrum has been envisioned as a promising candidate to satisfy the ever-increasing data demand in future indoor environments [1]. In recent years, bidirectional OWC, which is also named light fidelity (LiFi), has been widely considered as one of the key enabling technologies for 5G/6G and Internet of things (IoT) communications [2–5]. Although OWC systems have many inherent advantages such as abundant license-free spectrum resources, no electromagnetic interference (EMI) and enhanced physical-layer security, the practically achievable capacity of OWC systems is largely limited by the small modulation bandwidth of commercial off-the-shelf (COTS) optical elements, especially for illumination light-emitting diodes (LEDs) [6].

As a very natural way to efficiently improve the achievable capacity of indoor OWC systems that use LEDs, multiple-input multiple-output (MIMO) transmission has attracted great attention recently, which fully exploits the existing LED fixtures in the ceiling of a typical room to harvest substantial diversity or multiplexing gain [7–9]. So far, various optical MIMO techniques have been introduced for OWC systems, among which optical spatial multiplexing (OSMP) and optical spatial modulation (OSM) are two of the most popular. Specifically, OSMP can achieve a full multiplexing gain and hence a relative high spectral efficiency, but suffers from severe interchannel interference (ICI) [10]. In contrast, OSM can remove ICI by activating a single LED to transmit signal at each time slot. Although OSM can transmit additional index bits, only one constellation symbol can be transmitted at each time slot, and hence it is challenging for OSM systems to achieve high spectral efficiency [11]. Lately, generalized optical MIMO (GOMIMO) techniques, including generalized OSM (GOSM) and generalized OSMP (GOSMP), have been further proposed to boost the capacity of MIMO-OWC systems [12–15]. In GOSM systems, multiple LEDs are activated to transmit the same signal, and therefore more index bits can be transmitted and the diversity gain can also be increased. In GOSMP systems, only a subset of LEDs are activated to transmit different signals, resulting in reduced multiplexing gain. However, additional index bits can be transmitted, and the ICI can also been reduced in GOSMP systems.

In order to successfully implement GOMIMO systems, an efficient MIMO detection scheme should be adopted. Generally, the joint maximum-likelihood (ML) detector serves as the optimal detector for GOMIMO systems [16]. Nevertheless, the ML detector usually has high computational complexity, making it infeasible in practical applications. Instead, the combination of zero-forcing (ZF) equalization and ML detection can be a practical low-complexity detection scheme for GOMIMO systems [16]. However, ZF equalization inevitably leads to noise amplification due to high channel correlation in typical indoor MIMO-OWC systems. Moreover, the ZF-ML detector also suffers from the adverse effect of error propagation, because the detection error of spatial symbols might propagate to the estimation of constellation symbols.

With the rapid development of machine learning technology, machine learning has revealed its great potential in wireless communication systems [17,18]. Moreover, machine learning techniques have also been widely applied in optical communication systems. In [19], a distributed collaborative learning approach was proposed for cognitive and autonomous multidomain elastic optical networking. In [20], two machine learning algorithms were proposed for bit error rate (BER) degradation detection and failure identification in elastic optical networks. In [21], a machine learning method was proposed for quality of transmission prediction of unestablished lightpaths. In [22], a convolutional neural networks-based error vector magnitude estimation scheme was proposed for fast and accurate signal quality monitoring in coherent optical communications. In [23], a long-short-term-memory (LSTM) algorithm was proposed to mitigate transmission impairments of 4-level pulse-amplitude modulation (PAM4) produced by silicon-microring modulator. Most recently, deep learning techniques have been further introduced in OWC systems for binary signaling design [24], mitigation of both linear and nonlinear impairments [25], energy-efficient resource management [26], and so on. More specifically, a ZF-based deep neural network (DNN)-detection scheme has been proposed for MIMO detection in GOMIMO systems [27]. The obtained results in [27] show that the ZF-DNN detector can achieve comparable BER performance as the optimal joint ML detector with greatly reduced computational complexity. Nevertheless, the ZF-DNN detector takes the ZF equalized signal as its input, which requires accurate channel state information (CSI), i.e., the MIMO channel matrix, to successfully perform ZF equalization. Although CSI can be estimated by using training symbols [28], training-based instantaneous channel estimation inevitably causes both the loss of achievable data rate and the increase of communication time delay.

In this paper, to address the disadvantages of CSI-based ZF-DNN detection due to the requirement of instantaneous CSI for ZF equalization, we for the first time propose a DeepGOMIMO framework for GOMIMO systems where CSI-free MIMO detection is achieved by a novel DNN detection scheme. By adding a specially designed preprocessing module before the feedforward DNN module, CSI-free detection can be successfully enabled for GOMIMO systems. The key difference between our previous work [15] and this current work can be described as follows: our previous work [15] mainly proposed four OFDM-based GOMIMO schemes, whereas this current work proposes a CSI-free DNN detection scheme for PAM-based GOMIMO systems. Numerical simulations are extensively conducted to evaluate the performance of the proposed CSI-free DNN detector, which is also compared with other three benchmark schemes including the joint ML detector, the

ZF-ML detector and the ZF-DNN detector. Our simulation results verify the advantages of the proposed CSI-free DNN detector in comparison to other benchmark schemes in GOMIMO systems. To the best of our knowledge, it is the first time that a CSI-free DNN detection scheme is proposed and evaluated in detail for PAM-based GOMIMO systems.

The rest of this paper is organized as follows. In Section 2, we describe the mathematical model of a general GOMIMO system. In Section 3, we introduce four detection schemes for GOMIMO systems. Detailed simulation setup and results are presented in Section 4. Finally, Section 5 concludes the paper.

2. System Model

In this section, we introduce the mathematical model of a general GOMIMO system equipped with N_t LEDs and N_r photodetectors (PDs). The channel model is first described, and then the basic principle of GOMIMO is further reviewed.

2.1. Channel Model

Letting $\mathbf{x} = [x_1, x_2, \cdots, x_{N_t}]^T$ be the transmitted signal vector, $\mathbf{H}$ represent the $N_r \times N_t$ MIMO channel matrix and $\mathbf{n} = [n_1, n_2, \cdots, n_{N_r}]^T$ denote the additive noise vector, the received signal vector $\mathbf{y} = [y_1, y_2, \cdots, y_{N_r}]^T$ is obtained by

$$\mathbf{y} = \mathbf{Hx} + \mathbf{n}, \tag{1}$$

and the corresponding channel matrix $\mathbf{H}$ can be expressed by

$$\mathbf{H} = \begin{bmatrix} h_{11} & \cdots & h_{1N_t} \\ \vdots & \ddots & \vdots \\ h_{N_r1} & \cdots & h_{N_rN_t} \end{bmatrix}, \tag{2}$$

where h_{rt} ($r = 1, 2, \cdots, N_r; t = 1, 2, \cdots, N_t$) denotes the direct current (DC) channel gain between the r-th PD and the t-th LED. Assuming that each LED follows the general Lambertian radiation pattern and only the line-of-sight (LOS) transmission is considered, h_{rt} is calculated by [29]

$$h_{rt} = \frac{(l+1)\rho A}{2\pi d_{rt}^2} \cos^m(\varphi_{rt}) T_s(\theta_{rt}) g(\theta_{rt}) \cos(\theta_{rt}). \tag{3}$$

In Equation (3), $l = -\ln2/\ln(\cos(\Psi))$ denotes the Lambertian emission order, with Ψ being the semiangle at half power of the LED; ρ and A represent the responsivity and the physical area of the PD, respectively; d_{rt} is the distance between the r-th PD and the t-th LED; φ_{rt} and θ_{rt} are the emission angle and the incident angle, respectively. $T_s(\theta_{rt})$ is the gain of optical filter, and $g(\theta_{rt}) = \frac{n^2}{\sin^2\Phi}$ is the gain of optical lens, where n and Φ are the refractive index and the half-angle field-of-view (FOV) of the optical lens, respectively.

Moreover, the additive noise in typical OWC systems consists of both shot and thermal noises, and it is reasonable to model the additive noise as a real-valued zero-mean additive white Gaussian noise (AWGN) [8]. Letting N_0 denote the noise power spectral density (PSD) and B be the signal bandwidth, the power of the additive noise is given by $P_n = N_0 B$.

2.2. Principle of GOMIMO

The concept of GOMIMO was first proposed in [15], which aims to fully explore the potential of MIMO transmission for spectral efficiency enhancement of bandlimited OWC systems. Specifically, GOMIMO techniques can be generally divided into two main categories: one is GOSM, in which all the activated LED transmitters transmit the same signal, and the other is GOSMP, in which the activated LED transmitters transmit different signals. For more details about the GOMIMO techniques, please refer to our previous work [15].

Figure 1 illustrates the block diagram of a general $N_r \times N_t$ GOMIMO system, where N_a ($1 \leq N_a \leq N_t$) LEDs are activated for signal transmission during GOMIMO mapping. As we can see, the input bits are first divided into two streams: one is fed into the constellation mapper which converts the binary bits into constellation symbols, and the other is sent into the LED index selector which selects the desired LEDs to transmit the generated constellation symbols accordingly. Based on the obtained constellation symbol vector **c** and spatial index vector **v**, GOMIMO (GOSM or GOSMP) mapping is performed to generate the transmitted signal vector **x**. The mapping tables for GOSM and GOSMP with $N_t = 4$ and $N_a = 2$ are given in insets (a) and (b) of Figure 1, respectively. On the receiver side, the received signal vector **y** is fed into the GOMIMO detector which finally yields the output bits. The detailed GOMIMO detection schemes will be discussed in the following section.

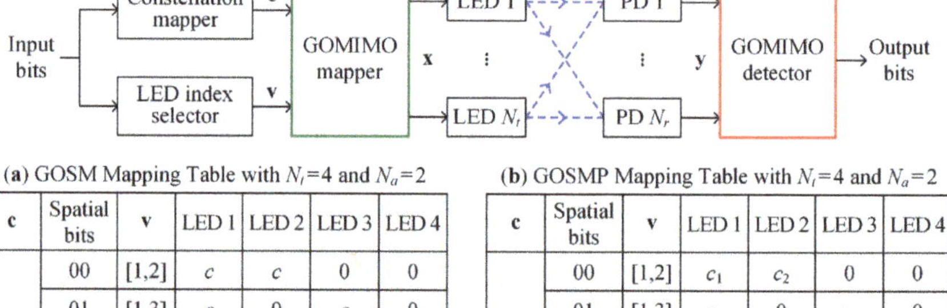

(a) GOSM Mapping Table with N_t=4 and N_a=2

c	Spatial bits	**v**	LED 1	LED 2	LED 3	LED 4
[c,c]	00	[1,2]	c	c	0	0
	01	[1,3]	c	0	c	0
	11	[2,4]	0	c	0	c
	10	[3,4]	0	0	c	c

(b) GOSMP Mapping Table with N_t=4 and N_a=2

c	Spatial bits	**v**	LED 1	LED 2	LED 3	LED 4
[c_1,c_2]	00	[1,2]	c_1	c_2	0	0
	01	[1,3]	c_1	0	c_2	0
	11	[2,4]	0	c_1	0	c_2
	10	[3,4]	0	0	c_1	c_2

Figure 1. Block diagram of a general $N_r \times N_t$ GOMIMO system. Insets (**a**) and (**b**) show the mapping tables of GOSM and GOSMP, respectively.

In typical LED-based OWC systems, intensity modulation with direct detection (IM/DD) is generally applied due to the noncoherence nature of LEDs. As a result, only real-valued nonnegative signals can be successfully transmitted in the IM/DD OWC systems [29]. In this work, unipolar M-ary PAM (M-PAM) is adopted as the modulation format for GOMIMO systems. In order to avoid the loss of spatial information when performing GOMIMO mapping, the M-PAM symbols cannot have zero values [15]. Therefore, unipolar nonzero M-PAM modulation is utilized here and the corresponding intensity levels are given by

$$I_m = \frac{2I_{\mathrm{av}}}{M+1}m, \quad m = 1, \cdots, M,$$ (4)

where I_{av} denotes the average optical power emitted [8]. By using M-PAM modulation, the spectral efficiencies (bits/s/Hz) of the $N_r \times N_t$ GOMIMO system with N_a activated LEDs applying GOSM and GOSMP mappings are respectively given by

$$\eta_{\mathrm{GOSM}} = \log_2(M) + \lfloor \log_2(C(N_t, N_a)) \rfloor,$$ (5)

$$\eta_{\mathrm{GOSMP}} = N_a \log_2(M) + \lfloor \log_2(C(N_t, N_a)) \rfloor,$$ (6)

where $\lfloor \cdot \rfloor$ denotes the floor operator, which outputs an integer smaller or equal to its input value and $C(\cdot, \cdot)$ represents the binomial coefficient.

3. Detection Schemes for GOMIMO Systems

In this section, we first introduce two conventional detection schemes for GOMIMO systems by utilizing M-PAM modulation, including the optimal joint ML detection and the ZF-ML detection. After that, we further present two deep learning-aided detection schemes, including the CSI-based ZF-DNN detection and our newly proposed CSI-free DNN detection.

3.1. Joint ML Detection

Assuming perfect CSI, joint ML detection is the optimal detection scheme for GOMIMO systems with M-PAM modulation. More specifically, the joint ML detector estimates the transmitted constellation and spatial information simultaneously in a joint manner. By applying the joint ML detector, the transmitted signal vector $\mathbf{x}$ can be estimated by

$$\hat{\mathbf{x}}_{JML} = \arg\min_{\mathbf{x} \in \mathbb{X}} \|\mathbf{y} - \mathbf{H}\mathbf{x}\|^2, \tag{7}$$

where $\|\cdot\|_2$ denotes the modulus operator and $\mathbb{X}$ represents the set of all the considered transmitted signal vectors.

Although the joint ML detection can achieve optimal performance, it suffers from high computational complexity. Therefore, it is usually not feasible to apply the joint ML detector in practical GOMIMO systems.

3.2. ZF-ML Detection

In order to avoid the high computational complexity of joint ML detection, a low-complexity ZF-ML detection scheme can be applied in GOMIMO systems, which is basically a three-step detection scheme [15,30]. In the first step, ZF equalization is performed for MIMO demultiplexing. The estimate of the transmitted signal vector $\mathbf{x}$ after ZF equalization can be obtained by

$$\hat{\mathbf{x}}_{ZF} = \mathbf{H}^{\dagger}\mathbf{y} = \mathbf{x} + \mathbf{H}^{\dagger}\mathbf{n}, \tag{8}$$

where $\mathbf{H}^{\dagger}$ denotes the pseudoinverse of $\mathbf{H}$ [10].

In the second step, ML detection is executed to obtain the estimate of the spatial index vector according to $\hat{\mathbf{x}}_{ZF}$. Finally, in the third step, the estimate of the constellation symbol vector can be obtained accordingly by using $\hat{\mathbf{x}}_{ZF}$ and the estimate of the spatial index vector. For more details about the principle of ZF-ML detection for GOMIMO systems, please refer to our previous work [15].

Compared with joint ML detection, the computational complexity of ZF-ML detection is significantly reduced. Nevertheless, the performance of ZF-ML detection is also largely degraded in comparison to that of joint ML detection, which can be explained as follows. On the one hand, ZF equalization inevitably causes severe noise amplification due to the high channel correlation in typical MIMO-OWC systems [8], which might greatly degrade the performance of GOMIMO systems. On the other hand, the detection error of spatial symbols might propagate to the estimation of the constellation symbols [31], which leads to further substantial performance degradation of GOMIMO systems.

3.3. CSI-Based ZF-DNN Detection

To efficiently address both the high computational complexity issue of joint ML detection and the noise-amplification and error-propagation issues of ZF-ML detection, a ZF-DNN detection has been proposed for GOSMP systems in [27]. The key idea of the ZF-DNN detection scheme is to employ a feedforward DNN module to directly and simultaneously estimate the transmitted spatial and constellation bits by taking the ZF equalized signal vector $\hat{\mathbf{x}}_{ZF}$ as input. For more details about the implementation of the ZF-DNN detector, please refer to [31,32]. In a word, the feedforward DNN module can fulfill the tasks of spatial index vector estimation, constellation symbol vector estimation, spatial symbol demodulation, and constellation symbol demodulation at the same time. Simulation results in [27] clearly show that, by selecting a proper training signal-to-noise ratio (SNR), the ZF-DNN detector can achieve nearly the same BER performance as the optimal joint ML detector, but with a significantly reduced computational complexity.

Despite the near-optimal BER performance and low computational complexity of the ZF-DNN detector, it takes the ZF equalized signal vector $\hat{\mathbf{x}}_{ZF}$ as the input of the feedforward DNN module. As per (8), $\hat{\mathbf{x}}_{ZF}$ is obtained by multiplying the received signal vector $\mathbf{y}$ with $\mathbf{H}^{\dagger}$, i.e., the pseudoinverse of the channel matrix $\mathbf{H}$. Generally, the CSI (i.e., the channel matrix) can be efficiently estimated by transmitting training symbols [28]. Nevertheless, the use of training symbols for accurate CSI estimation inevitably reduces the achievable data rate of GOMIMO systems, especially for low SNR scenarios. Furthermore, because the channel matrix is highly related to the specific location of the MIMO receiver, i.e., the PD array, channel estimation needs to be executed instantaneously with the change of receiver location. In consequence, instantaneous channel estimation inevitably introduces additional communication time delay and computational complexity in practical GOMIMO systems.

3.4. Proposed CSI-Free DNN Detection

Considering the many disadvantages of CSI-based ZF-DNN detection due to the requirement of instantaneous CSI for ZF equalization, in this work, we for the first time propose a novel CSI-free DNN detection scheme for GOMIMO systems. Figure 2 depicts the schematic diagram of the proposed CSI-free DNN detector, which consists of a pre-processing module and a feedforward DNN module. It can be seen that a preprocessing module is placed in front of the feedforward DNN module in the proposed CSI-free DNN detector, which is the key to dealing with the impact of MIMO transmission through free-space channels and hence realize detection without the need of CSI. Specifically, as can be found from (1), the impact of MIMO transmission on the transmitted signal vector $\mathbf{x}$ can be characterized from the following two aspects. First, because the channel coefficients in typical MIMO-OWC systems are within the region from 10^{-6} to 10^{-4} [8,33], the electrical path loss caused by MIMO transmission is about 80 to 120 dB. Secondly, MIMO transmission also inevitably leads to channel crosstalk, which might cause severe ICI, especially for GOSMP systems. As a result, the designed preprocessing module should be able to address both the path loss issue and the channel crosstalk issue caused by MIMO transmission.

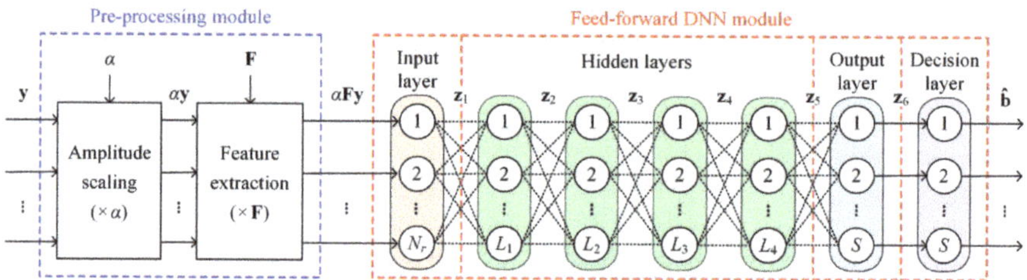

Figure 2. Schematic diagram of the proposed CSI-free DNN detector consisting of a preprocessing module and a feedforward DNN module.

As shown in Figure 2, our specially designed preprocessing module mainly contains two parts: one is the amplitude scaling part and the other is the feature-extraction part. Specifically, the amplitude scaling part is adopted to address the path loss issue by multiplying the received signal vector $\mathbf{y}$ with a scaling factor α. Note that a proper α value is determined in advance for each receiver location in the GOMIMO system, and hence no instantaneous CSI is needed to achieve amplitude scaling. Moreover, the feature-extraction part is used to address the channel crosstalk issue, which multiplies the scaled received signal vector $\alpha\mathbf{y}$ by a feature matrix $\mathbf{F}$. Hence, the output signal vector of the preprocessing module in the CSI-free DNN detector, i.e., $\hat{\mathbf{y}} = [\hat{y}_1, \hat{y}_2, \cdots, \hat{y}_{N_r}]^T$, can be obtained by

$$\hat{\mathbf{y}} = \alpha\mathbf{F}\mathbf{y}. \tag{9}$$

In order to provide enough information for the following feedforward DNN module to efficiently learn and remove the impact of channel crosstalk caused by MIMO transmission, the feature matrix $\mathbf{F}$ should be able to reflect all the potential signal superposition cases at the receiver side. Consequently, according to the mapping tables of both GOSM and GOSMP in Figure 1, we adopt the corresponding unified mapping matrix as the feature matrix, i.e.,

$$\mathbf{F} = \begin{bmatrix} 1 & 1 & 0 & 0 \\ 1 & 0 & 1 & 0 \\ 0 & 1 & 0 & 1 \\ 0 & 0 & 1 & 1 \end{bmatrix}. \tag{10}$$

Subsequently, the preprocessed signal vector $\hat{\mathbf{y}}$ is fed into a feedforward DNN module, which mainly consists of an input layer, multiple hidden layers, an output layer, and a decision layer. Because $\hat{\mathbf{y}}$ is a vector with N_r elements, the input layer contains N_r neurons accordingly. Moreover, we set totally four fully connected hidden layers in the feedforward DNN module, which are used to learn the statistical characteristics of both the input signal and the additive noise. The number of neurons in the i-th ($1 \leq i \leq 4$) hidden layer is denoted by L_i, and the rectified linear unit (ReLU) function, i.e., $f_{\mathrm{ReLU}}(\alpha) = \max(0, \alpha)$, is adopted as the activation function of the hidden layers. For the output layer, it adopts the Sigmoid function, i.e., $f_{\mathrm{Sigmoid}}(\alpha) = 1/(1 + \exp^{-\alpha})$, as the activation function to generate a fuzzy bit information, so as to map the output of each neuron within the range $[0, 1]$. Because the DNN detector takes the input binary bits corresponding to a transmitted signal vector as the output, both the output layer and the decision layer have the same number of neurons, which is equal to the spectral efficiency of the GOMIMO system, i.e., $S = \eta_{\mathrm{GOMIMO}}$. Therefore, letting $\mathbf{z}_k$ denote the output of the k-th ($1 \leq k \leq 6$) layer of the feedforward DNN module, the corresponding input–output relationship can be described by

$$\mathbf{z}_k = \begin{cases} \alpha \mathbf{F} \mathbf{y}, & k = 1 \\ f_{\mathrm{ReLU}}(\mathbf{W}_{k-1}\mathbf{z}_{k-1} + \mathbf{b}_{k-1}), & 2 \leq k \leq 5 \\ f_{\mathrm{Sigmoid}}(\mathbf{W}_{k-1}\mathbf{z}_{k-1} + \mathbf{b}_{k-1}), & k = 6 \end{cases}, \tag{11}$$

where $\mathbf{W}_p$ and $\mathbf{b}_p$ with $1 \leq p \leq 5$ represent the corresponding weight matrix and the bias vector, respectively. According to (11), the mean-square error (MSE) loss can be calculated as follows:

$$e_{\mathrm{MSE}} = \frac{1}{S}\|\mathbf{z}_6 - \mathbf{b}\|^2, \tag{12}$$

where $\mathbf{b}$ denotes the corresponding transmitted bit vector ans S is the length of $\mathbf{b}$.

Finally, the decision layer is utilized to determine the noninteger output of each neuron in the output layer to be 0 or 1. Letting $\hat{\mathbf{b}} = [\hat{b}_1, \hat{b}_2, \cdots, \hat{b}_S]^T$ denote the final output binary bit vector, the q-th ($q = 1, 2, \cdots, S$) binary bit in $\hat{\mathbf{b}}$ can be estimated by

$$\hat{b}_q = \begin{cases} 0, & z_{6,q} < 0.5 \\ 1, & z_{6,q} \geq 0.5 \end{cases}. \tag{13}$$

4. Simulation Results

In this section, we evaluate and compare the performance of four different detection schemes in a typical indoor GOMIMO system through numerical simulations.

4.1. Simulation Setup

In our simulations, we consider a 4×4 ($N_r = N_t = 4$) GOMIMO system configured in a typical 5 m $\times$ 5 m $\times$ 3 m room. The 2×2 square LED array is placed at the center of the ceiling and the spacing between two adjacent LEDs is 2 m. The height of the receiving plane is 0.85 m, and two receiver locations over the receiving plane, i.e., the center (2.5 m, 2.5 m, 0.85 m) and the corner (0 m, 0 m, 0.85 m), are considered for performance evaluation. The receiver consists of a 2×2 square PD array, where the spacing between two adjacent PDs is 10 cm. For both GOSM and GOSMP mappings, two out of four LEDs are activated for signal transmission, i.e., $N_a = 2$. Moreover, unipolar nonzero 4-PAM modulation is adopted in the GOMIMO system, and hence the corresponding spectral efficiencies for GOSM and GOSMP mappings are 4 and 6 bits/s/Hz, respectively. In addition, we adopt transmitted SNR as the measure to evaluate the BER performance of the GOMIMO system, which is defined as the ratio of the transmitted electrical signal power to the additive noise power [8,15]. The other simulation parameters of the GOMIMO system can be found in Table 1.

Table 1. Simulation parameters.

Parameter	Value
Room dimension	$5\,\text{m} \times 5\,\text{m} \times 3\,\text{m}$
Height of receiving plane	0.85 m
Number of LEDs	4
Semi-angle at half power of LED	60°
LED spacing	2.5 m
Gain of optical filter	0.9
Refractive index of optical lens	1.5
Half-angle FOV of optical lens	72°
Number of PDs	4
Responsivity of PD	1 A/W
Active area of PD	$1\,\text{cm}^2$
PD spacing	10 cm
Number of activated LEDs, N_a	2
PAM levels, M	4

The detailed parameters of the CSI-free DNN detectors for GOSM and GOSMP are given in Table 2. For GOSM, the number of neurons of four hidden layers is 128, 64, 32, and 16, respectively. The learning rate is 0.01 when the receiver is located at the center of the receiving plane, and it is reduced to 0.001 when the receiver is moved to the corner. Moreover, the scaling factors are set to 1×10^5 and 2×10^5 when the receiver is located at the center and the corner, respectively. For GOSMP, every hidden layer contains 64 neurons, and the learning rates are 0.01 and 0.005 when the receiver is located at the center and the corner of the receiving plane, respectively. In addition, the scaling factors of 1×10^5 and 1×10^6 are used for center and corner received locations, respectively. For both GOSM and GOSMP, the lengths of training set and validation set are assumed to be 150,000 and 50,000, respectively. In order to accelerate the convergence speed, we use the minibatch technique in training, and each minibatch contains 100 transmitted signal vectors.

Table 2. Parameters of the DNN detector for GOSM and GOSMP.

Parameter	GOSM	GOSMP
Receiver locations	(2.5 m, 2.5 m, 0.85), (0 m, 0 m, 0.85)	
Number of input nodes	4	
Number of hidden layers	4	
Number of neurons	$128 \times 64 \times 32 \times 16$	$64 \times 64 \times 64 \times 64$
Number of output nodes	4	6
Hidden layer activation	ReLU	
Output layer activation	Sigmoid	
Loss function	MSE	
Optimizer	Adamax	
Learning rate	0.01 | 0.001	0.01 | 0.005
Length of training set	150,000	
Length of validation set	50,000	
Scaling factor	1×10^5 | 2×10^5	1×10^5 | 1×10^6

4.2. MSE Loss

We first analyze the MSE loss of the proposed CSI-free DNN detector in the 4×4 GOMIMO system. Figure 3a,b show the MSE losses versus the number of epochs for GOSM and GOSMP, respectively, where the receiver is located at the center of the receiving plane. As we can see, the MSE loss decreases rapidly with the increase of training epochs for both GOSM and GOSMP. Moreover, the MSE loss is much reduced when a higher training SNR is used, especially for GOSMP. It can be seen from Figure 3a that the MSE loss for GOSM converges quickly with only a few epochs. For GOSMP, as shown in Figure 3b, about 20 epochs are required for the MSE loss to converge. Hence, owing to the use of the minibatch technique, the CSI-free DNN detector only requires a very limited number of epochs for efficient training, indicating that it can be deployed rapidly in practical applications.

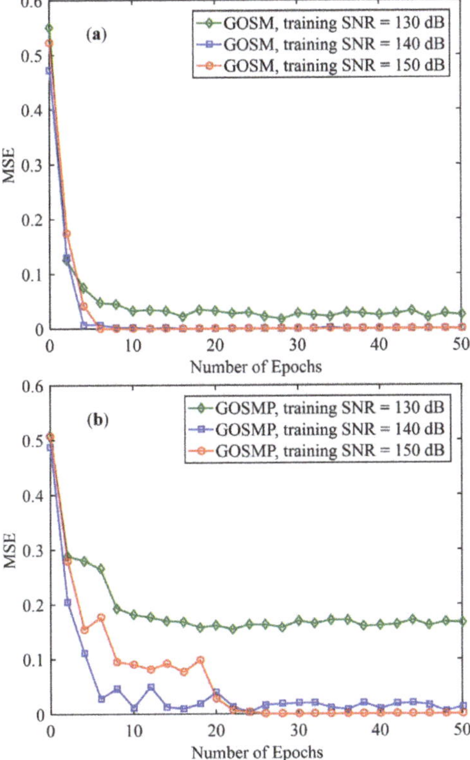

Figure 3. MSE training loss of the proposed CSI-free DNN detector with receiver located at the center of the receiving plane for (**a**) GOSM and (**b**) GOSMP.

4.3. BER Performance

We further evaluate and compare the BER performance of the proposed CSI-free DNN detector with the other three benchmark detectors in the 4×4 GOMIMO system. Figure 4a,b compare the BER performance of four detectors for GOSM with the receiver located at the center and the corner of the receiving plane, respectively. When the receiver is located at the center of the receiving plane, as shown in Figure 4a, the ZF-ML detector requires a high transmitted SNR of 163.4 dB to achieve the target BER of 10^{-3}. However, the required SNR to reach BER = 10^{-3} is reduced to 138.9 dB for the joint ML detector. As a result, a substantial 24.5-dB SNR gain can be obtained by the joint ML detector in comparison to the ZF-ML detector, which is mainly because the ZF-ML detector suffers from severe noise amplification and error propagation. Moreover, it can be further seen that the ZF-DNN detector with an optimal 140-dB training SNR can achieve comparable BER performance as the joint ML detector in the high SNR region, suggesting the excellent error performance of the ZF-DNN detector under the condition of accurate CSI for ZF equalization. Finally, for our proposed CSI-free DNN detector, we investigate the impact of training SNR on its error performance and three different training SNRs of 130, 140, and 150 dB are considered. It is clearly shown that the CSI-free DNN detector with 140-dB training SNR achieves nearly the same BER performance as the joint ML detector across the whole SNR region, which slightly outperforms the ZF-DNN detector in the low SNR region. However, the joint ML detector outperforms the CSI-free DNN detector when a lower training SNR of 130 dB or a higher training SNR of 150 dB is adopted, and the reasons can be explained as follows. The DNN module can better learn the statistics of the noise with a relatively small training SNR, whereas the statistics of the data symbols can be more accurately learned when the training SNR is relatively large. As a result, there exists an optimal training SNR which can make a tradeoff for the DNN module to learn the statistics of both the noise and the data symbols and hence lead to a minimum overall BER. When the receiver is moved to the corner of the receiving plane, as shown in Figure 4b, we can observe that the joint ML detector outperforms the ZF-ML detector by an SNR gain of more than 40 dB at BER = 10^{-3}, whereas the ZF-DNN detector with an optimal training SNR of 160 dB obtains near-optimal BER performance as the joint ML detector only for relatively low BERs. Furthermore, the CSI-free DNN detector achieves comparable BER performance as the joint ML detector in the high SNR region, which outperforms the ZF-DNN detector in the low SNR region. It should be noted that an error floor occurs for the CSI-free DNN detector with a lower training SNR of 140 dB, which is mainly due to the insufficient learning of the statistics of the data symbols in a very noisy environment.

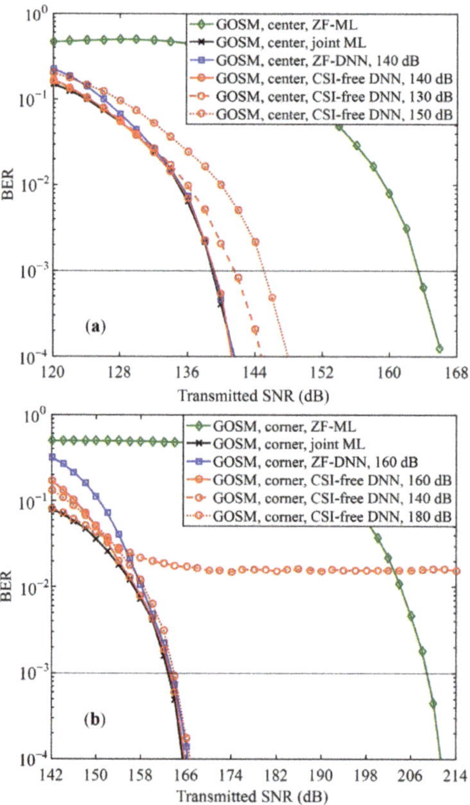

Figure 4. BER comparison of the proposed CSI-free DNN detector and three benchmark detectors for GOSM at (**a**) the center and (**b**) the corner.

The BER versus transmitted SNR for GOSMP is plotted in Figure 5. As we can see, the ZF-DNN detector with an optimal training SNR can achieve very close performance as the joint ML detector when the receiver is located at the center of the receiving plane, but it performs worse than the joint ML detector when the receiver is moved to the corner, especially in the low SNR region. In contrast, the proposed CSI-free DNN detector can achieve comparable BER performance as the joint ML detector for both center and corner receiver locations. Moreover, error floors occur for the CSI-free DNN detector when the adopted training SNR is too small or too large. It can be further observed from Figures 4 and 5 that the optimal training SNRs for the ZF-DNN detector and the CSI-free DNN detector at the same receiver location are generally the same in GOMIMO systems.

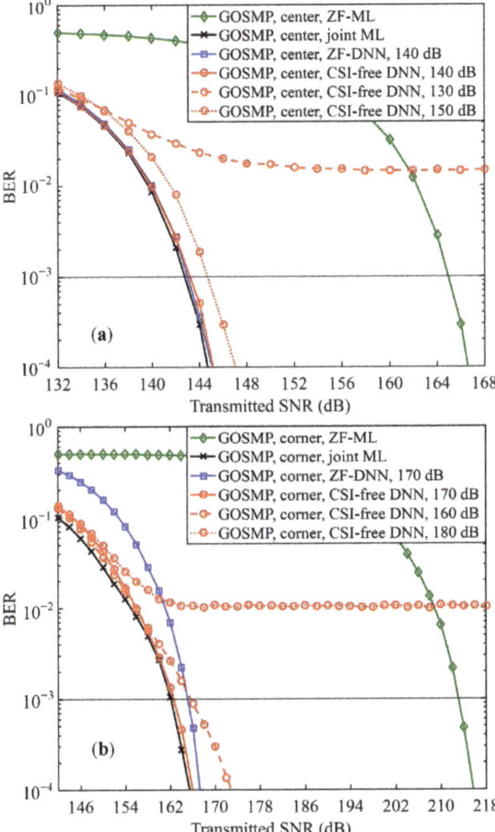

Figure 5. BER comparison of the proposed CSI-free DNN detector and three benchmark detectors for GOSMP at (**a**) the center and (**b**) the corner.

4.4. Impact of Input Pre-Processing

It can be seen from Figure 2 that the preprocessing module, which preprocesses the input of the feedforward DNN module, plays a vital role to guarantee that the proposed CSI-free DNN detector can successfully perform MIMO detection without the need of CSI. In the next, we evaluate the impact of input preprocessing on the performance of the CSI-free DNN detector. Here, two different inputs of the feedforward DNN module are considered: one is $\alpha\mathbf{y}$, i.e., the preprocessing module only performs amplitude scaling, and the other is $\alpha\mathbf{Fy}$, i.e., the preprocessing module performs both amplitude scaling and feature extraction. Figure 6 compares the BER performance of the proposed CSI-free DNN detector where the feedforward DNN module having different inputs for both GOSM and GOSMP with the receiver located at the center of the receiving plane. As we can see, for GOSM, the BER performance is only slightly improved when the input is changed from $\alpha\mathbf{y}$ to $\alpha\mathbf{Fy}$, and the SNR gain at BER = 10^{-3} is only 0.8 dB. In contrast, for GOSMP, a noticeable BER improvement can be obtained by replacing the input $\alpha\mathbf{y}$ with $\alpha\mathbf{Fy}$, and the corresponding SNR gain at BER = 10^{-3} is increased to 2.4 dB. The difference between BER improvements for GOSM and GOSMP can be explained as follows. As discussed in Section 3.4, because the feature matrix $\mathbf{F}$ contains the spatial mapping information of GOMIMO systems, the feedforward DNN module can use the spatial mapping information to remove the channel crosstalk. As a result, the feedforward DNN module with input $\alpha\mathbf{Fy}$ can efficiently mitigate the adverse effect of error propagation. However, in GOSM systems, the activated LEDs are used to transmit the same signal and hence error propagation only leads to reduced diversity gain, which might not significantly degrade the BER performance. In contrast, because the activated LEDs transmit different signals in GOSMP systems, error propagation leads to the missing of constellation information and hence results in significant BER degradation.

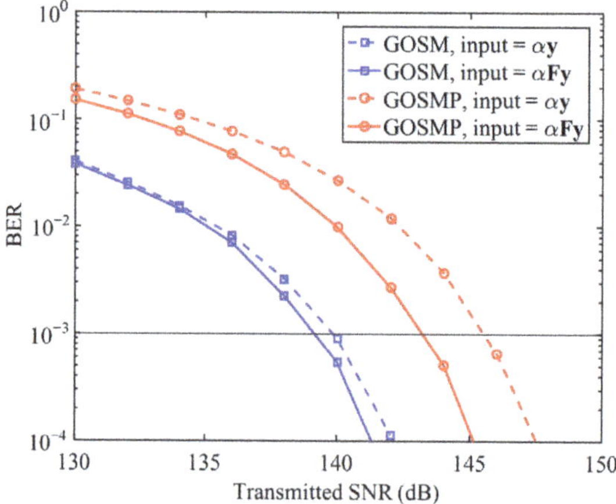

Figure 6. BER comparison of the proposed CSI-free DNN detector where the feedforward DNN module having different inputs for both GOSM and GOSMP at the center of the receiving plane.

Due to the substantial path loss during MIMO transmission, the received signal needs to be properly amplified before it can be fed into the feedforward DNN module. Figure 7a,b show the BER versus $\mathrm{Log}_{10}\alpha$ with different transmitted SNRs for GOSM and GOSMP, respectively. For GOSM, as shown in Figure 7a, we can observe that a feasible range of α is around $[10^5, 10^7]$ when the receiver is located at the center of the receiving plane. Moreover, the feasible range of α keeps the same for different transmitted SNR values. When the receiver is moved to the corner, the feasible range of α is $[10^5, 10^8]$. For GOSMP, as shown in Figure 7b, the same feasible range of α is obtained as that of GOSM when the receiver is located at the center of the receiving plane. However, the feasible range of α for GOSMP is only around 10^6 when the receiver is moved to the corner. To successfully implement the proposed CSI-free DNN detector, the proper α value with respect to each receiver location is determined in advance.

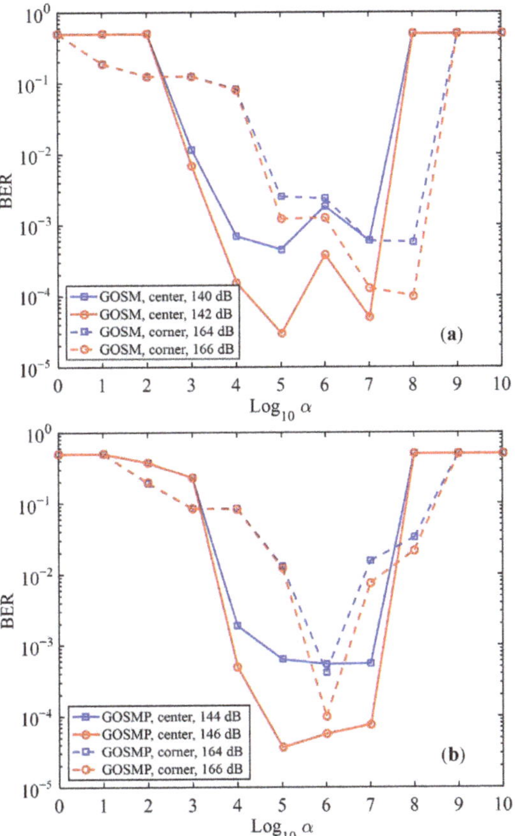

Figure 7. BER vs. $\log_{10}\alpha$ of the proposed CSI-free DNN detector for (**a**) GOSM and (**b**) GOSMP.

4.5. Computational Complexity

Finally, we evaluate the computational complexity of the proposed CSI-free DNN detector and compare it with other benchmark detectors. For both the CSI-free DNN detector and the ZF-DNN detector, once the detector has been successfully trained, it can be used for MIMO detection for a long period of time without further retraining, unless the system parameters such as receiver location have been changed [27]. Hence, only the computational complexity of the online detection process is considered for the CSI-free DNN detector and the ZF-DNN detector, whereas the complexity of the offline training process is not taken into account. Moreover, the computational complexity of the proposed DNN detectors and the other three benchmark detectors is evaluated and compared in terms of computation time, which is a common way for computational complexity evaluation in the literature [34]. Figure 8a,b compare the computation time of the proposed DNN detectors and the other three benchmark detectors for GOSM and GOSMP, respectively. As we can see, for GOSM, the CSI-free DNN detector, the ZF-DNN detector and the ZF-ML detector require nearly the same computation time which is less than 3 s. However, the joint ML detector requires totally 48.42 s to finish the computation, which is significantly longer than that of the other three detectors. It is the same for GOSMP that the CSI-free DNN detector, the ZF-DNN detector, and the ZF-ML detector require comparable computation time, which is much shorter than that required by the joint ML detector. Therefore, the proposed CSI-free DNN detector achieves near-optimal BER performance as the joint ML detector, but with substantially lower computational complexity.

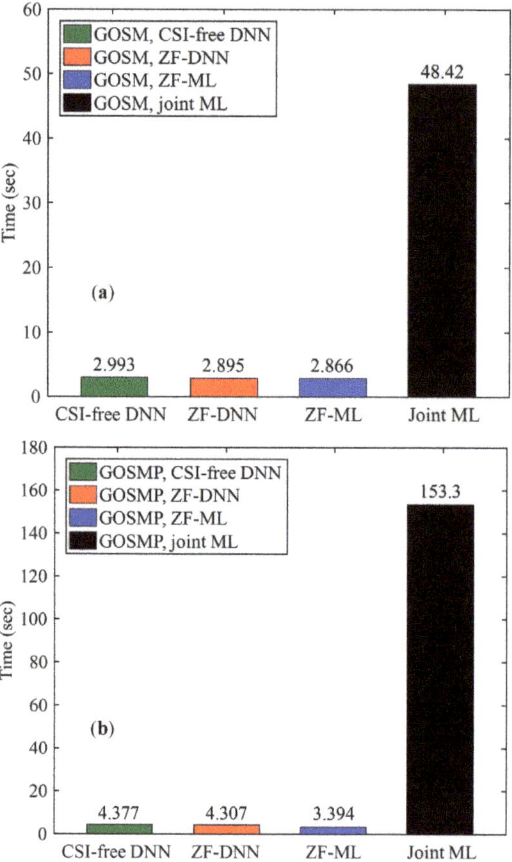

Figure 8. Computation time comparison of the proposed CSI-free DNN detector and three benchmark detectors for (**a**) GOSM and (**b**) GOSMP at the center of the receiving plane.

5. Conclusions

In this paper, we have for the first time proposed a novel DeepGOMIMO framework for GOMIMO systems, where a DNN-based detector is specially designed to realize CSI-free detection of the received MIMO signals. The proposed CSI-free DNN detector contains a preprocessing module and a feedforward DNN module, which are used to address the adverse effects of MIMO transmission and to perform joint detection of spatial and constellation information, respectively. It is shown by our simulation results that, in a typical indoor 4×4 MIMO-OWC system adopting both GOSM and GOSMP with unipolar nonzero 4-PAM modulation, the CSI-free DNN detector achieves comparable BER performance as the optimal joint ML detector, which greatly outperforms the ZF-ML detector. Moreover, the CSI-free DNN detector, the ZF-DNN detector and the ZF-ML detector require nearly the same computation time to perform detection, which is significantly shorter than that required by the joint ML detector. In addition, compared with the ZF-DNN detector, the CSI-free DNN detector can achieve an improved achievable data rate and reduced communication time delay because it does not require instantaneous channel estimation to obtain accurate CSI for ZF equalization. In conclusion, our proposed DeepGOMIMO can be a potential candidate for the implementation of practical high-speed and low-complexity OWC systems.

Author Contributions: Formal analysis, C.C., Z.Z. and M.L.; funding acquisition, C.C.; investigation, X.Z. and S.F.; methodology, S.F.; project administration, C.C.; resources, Z.Z.; software, S.F.; supervision, C.C. and M.L.; validation, X.Z.; writing—original draft, X.Z.; writing—review & editing, C.C., S.F., Z.Z. and M.L. All authors have read and agreed to the published version of the manuscript.

Funding: This research was funded by the National Natural Science Foundation of China (62271091 and 61901065) and the Natural Science Foundation of Chongqing (cstc2021jcyj-msxmX0480).

Institutional Review Board Statement: Not applicable.

Informed Consent Statement: Not applicable.

Data Availability Statement: Not applicable.

Acknowledgments: The authors would like to thank the anonymous reviewers for their valuable comments and suggestions.

Conflicts of Interest: The authors declare no conflict of interest.

References

1. Ghassemlooy, Z.; Arnon, S.; Uysal, M.; Xu, Z.; Cheng, J. Emerging optical wireless communications-advances and challenges. *IEEE J. Sel. Areas Commun.* **2015**, *33*, 1738–1749. [CrossRef]
2. Cogalan, T.; Haas, H. Why would 5G need optical wireless communications? In Proceedings of the IEEE International Symposium on Personal, Indoor and Mobile Radio Communications, Montreal, QC, Canada, 8–13 October 2017; pp. 1–6.
3. Chi, N.; Zhou, Y.; Wei, Y.; Hu, F. Visible light communication in 6G: Advances, challenges, and prospects. *IEEE Veh. Technol. Mag.* **2020**, *15*, 93–102. [CrossRef]
4. Demirkol, I.; Camps-Mur, D.; Paradells, J.; Combalia, M.; Popoola, W.; Haas, H. Powering the Internet of Things through light communication. *IEEE Commun. Mag.* **2019**, *57*, 107–113. [CrossRef]
5. Chen, C.; Fu, S.; Jian, X.; Liu, M.; Deng, X.; Ding, Z. NOMA for energy-efficient LiFi-enabled bidirectional IoT communication. *IEEE Trans. Commun.* **2021**, *69*, 1693–1706. [CrossRef]
6. Le Minh, H.; O'Brien, D.; Faulkner, G.; Zeng, L.; Lee, K.; Jung, D.; Oh, Y.; Won, E.T. 100-Mb/s NRZ visible light communications using a postequalized white LED. *IEEE Photonics Technol. Lett.* **2009**, *21*, 1063–1065. [CrossRef]
7. Zeng, L.; O'Brien, D.C.; Le Minh, H.; Faulkner, G.E.; Lee, K.; Jung, D.; Oh, Y.; Won, E.T. High data rate multiple input multiple output (MIMO) optical wireless communications using white LED lighting. *IEEE J. Sel. Areas Commun.* **2009**, *27*, 1654–1662. [CrossRef]
8. Fath, T.; Haas, H. Performance comparison of MIMO techniques for optical wireless communications in indoor environments. *IEEE Trans. Commun.* **2013**, *61*, 733–742. [CrossRef]
9. Chen, C.; Zhong, W.D.; Wu, D. On the coverage of multiple-input multiple-output visible light communications [Invited]. *J. Opt. Commun. Netw.* **2017**, *9*, D31–D41. [CrossRef]
10. Chen, C.; Yang, H.; Du, P.; Zhong, W.D.; Alphones, A.; Yang, Y.; Deng, X. User-centric MIMO techniques for indoor visible light communication systems. *IEEE Syst. J.* **2020**, *14*, 3202–3213. [CrossRef]
11. Mesleh, R.; Elgala, H.; Haas, H. Optical spatial modulation. *J. Opt. Commun. Netw.* **2011**, *3*, 234–244. [CrossRef]
12. Alaka, S.; Narasimhan, T.L.; Chockalingam, A. Generalized spatial modulation in indoor wireless visible light communication. In Proceedings of the IEEE Global Communications Conference (GLOBECOM), San Diego, CA, USA, 6–10 December 2015, pp. 1–7.
13. Wang, F.; Yang, F.; Song, J. Constellation optimization under the ergodic VLC channel based on generalized spatial modulation. *Opt. Exp.* **2020**, *28*, 21202–21209. [CrossRef] [PubMed]
14. Wang, K. Indoor optical wireless communication system with filters-enhanced generalized spatial modulation and carrierless amplitude and phase (CAP) modulation. *Opt. Lett.* **2020**, *45*, 4980–4983. [CrossRef] [PubMed]
15. Chen, C.; Zhong, X.; Fu, S.; Jian, X.; Liu, M.; Yang, H.; Alphones, A.; Fu, H.Y. OFDM-based generalized optical MIMO. *J. Lightw. Technol.* **2021**, *39*, 6063–6075. [CrossRef]
16. Özbilgin, T.; Koca, M. Optical spatial modulation over atmospheric turbulence channels. *J. Lightw. Technol.* **2015**, *33*, 2313–2323. [CrossRef]
17. LeCun, Y.; Bengio, Y.; Hinton, G. Deep learning. *Nature* **2015**, *521*, 436–444. [CrossRef]
18. Wang, T.; Wen, C.K.; Wang, H.; Gao, F.; Jiang, T.; Jin, S. Deep learning for wireless physical layer: Opportunities and challenges. *Chin. Commun.* **2017**, *14*, 92–111. [CrossRef]
19. Chen, X.; Li, B.; Proietti, R.; Liu, C.Y.; Zhu, Z.; Yoo, S.B. Demonstration of distributed collaborative learning with end-to-end QoT estimation in multi-domain elastic optical networks. *Opt. Exp.* **2019**, *27*, 35700–35709. [CrossRef]
20. Vela, A.P.; Ruiz, M.; Fresi, F.; Sambo, N.; Cugini, F.; Meloni, G.; Potì, L.; Velasco, L.; Castoldi, P. BER degradation detection and failure identification in elastic optical networks. *J. Lightw. Technol.* **2017**, *35*, 4595–4604. [CrossRef]
21. Rottondi, C.; Barletta, L.; Giusti, A.; Tornatore, M. Machine-learning method for quality of transmission prediction of unestablished lightpaths. *J. Opt. Commun. Netw.* **2018**, *10*, A286–A297. [CrossRef]

22. Fan, Y.; Udalcovs, A.; Pang, X.; Natalino, C.; Furdek, M.; Popov, S.; Ozolins, O. Fast signal quality monitoring for coherent communications enabled by CNN-based EVM estimation. *J. Opt. Commun. Netw.* **2021**, *13*, B12–B20. [CrossRef]
23. Peng, C.W.; Chan, D.W.; Tong, Y.; Chow, C.W.; Liu, Y.; Yeh, C.H.; Tsang, H.K. Long short-term memory neural network for mitigating transmission impairments of 160 Gbit/s PAM4 microring modulation. In Proceedings of the Optical Fiber Communication Conference (OFC), Optica Publishing Group, Washington, DC, USA, 6–11 June 2021; paper Tu5D.3.
24. Lee, H.; Lee, I.; Quek, T.Q.; Lee, S.H. Binary signaling design for visible light communication: A deep learning framework. *Opt. Exp.* **2018**, *26*, 18131–18142. [CrossRef]
25. Lu, X.; Lu, C.; Yu, W.; Qiao, L.; Liang, S.; Lau, A.P.T.; Chi, N. Memory-controlled deep LSTM neural network post-equalizer used in high-speed PAM VLC system. *Opt. Exp.* **2019**, *27*, 7822–7833. [CrossRef]
26. Yang, H.; Alphones, A.; Zhong, W.D.; Chen, C.; Xie, X. Learning-based energy-efficient resource management by heterogeneous RF/VLC for ultra-reliable low-latency industrial IoT networks. *IEEE Trans. Ind. Inform.* **2019**, *16*, 5565–5576. [CrossRef]
27. Wang, T.; Yang, F.; Song, J. Deep learning-based detection scheme for visible light communication with generalized spatial modulation. *Opt. Exp.* **2020**, *28*, 28906–28915. [CrossRef]
28. Wang, Y.; Chi, N. Demonstration of high-speed 2 × 2 non-imaging MIMO Nyquist single carrier visible light communication with frequency domain equalization. *J. Lightw. Technol.* **2014**, *32*, 2087–2093. [CrossRef]
29. Komine, T.; Nakagawa, M. Fundamental analysis for visible-light communication system using LED lights. *IEEE Trans. Consum. Electron.* **2004**, *50*, 100–107. [CrossRef]
30. Tavakkolnia, I.; Yesilkaya, A.; Haas, H. OFDM-based spatial modulation for optical wireless communications. In Proceedings of the IEEE Globecom Workshops (GC Wkshps), Abu Dhabi, United Arab Emirates, 9–13 December 2018; pp. 1–6.
31. Chen, C.; Zeng, L.; Zhong, X.; Fu, S.; Liu, M.; Du, P. Deep learning-aided OFDM-based generalized optical quadrature spatial modulation. *IEEE Photonics J.* **2022**, *14*, 7302306. [CrossRef]
32. Zhong, X.; Chen, C.; Zeng, L.; Zhang, R.; Tang, Y.; Nie, Y.; Liu, M. Joint detection for generalized optical MIMO: A deep learning approach. In Proceedings of the IEEE Conference on Industrial Electronics and Applications (ICIEA), Chengdu, China, 1–4 August 2021; pp. 1317–1321.
33. Ying, K.; Qian, H.; Baxley, R.J.; Yao, S. Joint optimization of precoder and equalizer in MIMO VLC systems. *IEEE J. Sel. Areas Commun.* **2015**, *33*, 1949–1958. [CrossRef]
34. Albinsaid, H.; Singh, K.; Biswas, S.; Li, C.P.; Alouini, M.S. Block deep neural network-based signal detector for generalized spatial modulation. *IEEE Commun. Lett.* **2020**, *24*, 2775–2779. [CrossRef]

Article

Asymptotic Capacity Maximization for MISO Visible Light Communication Systems with a Liquid Crystal RIS-Based Receiver

Qi Wu, Jian Zhang *, Yanyu Zhang, Gang Xin and Dongqin Guo

National Digital Switching System Engineering and Technological Research Center, Zhengzhou 450000, China
* Correspondence: zhang_xinda@126.com

Abstract: Combined with reconfigurable intelligent surfaces (RISs), visible light communications (VLCs) can increase the communication performance to a great degree. However, the research into RIS-aided VLC systems has mainly focused on mirror array-based RISs deployed on the wall while neglecting liquid crystal (LC)-based RISs in VLC receivers. With the development of advanced materials, the LC RIS has been gradually attracting attention from researchers. Inspired by the current research into the LC RIS, the applications of the LC RIS in multiple-input single-output (MISO) VLC systems are investigated in this paper. We formulate an optimization problem with asymptotic capacity maximization as the objective function and the refractive index of the LC RIS as the independent variable. As for this nonconvex optimization problem, we propose the particle swarm optimization (PSO) algorithm to determine the configuration of parameters for the LC RIS. The simulation results indicate that the employment of the LC RIS in VLC receivers can raise the communication performance of the MISO-VLC systems; meanwhile, the proposed algorithm is an effective way to deal with the optimization problems for LC RIS-based MISO-VLC systems when compared with the exhaustive search method and a baseline scheme. The LC RIS is also expected to solve the dead zone problem in traditional VLC systems.

Keywords: visible light communication (VLC); reconfigurable intelligent surface (RIS); asymptotic capacity; liquid crystals (LC); optimization algorithm

Citation: Wu, Q.; Zhang, J.; Zhang, Y.; Xin, G.; Guo, D. Asymptotic Capacity Maximization for MISO Visible Light Communication Systems with a Liquid Crystal RIS-Based Receiver. *Photonics* **2023**, *10*, 128. https://doi.org/10.3390/photonics10020128

Received: 1 December 2022
Revised: 30 December 2022
Accepted: 6 January 2023
Published: 27 January 2023

1. Introduction

As candidate technologies for the upcoming sixth generation (6G) wireless communication, visible light communication (VLC) and reconfigurable intelligent surface (RIS) have been developed rapidly in recent years to address the shortcomings of the fifth generation (5G) wireless communication [1]. Specifically, VLC is able to provide a large amount of unlicensed bandwidth offering abundant communication resources to complement the scarcity of spectrum resources for radio frequency (RF) communications [2]. Meanwhile, VLC is considered an environmental and green communication technology due to its ability to achieve a communication process when the illumination need is satisfied. On the other hand, RIS is regarded as a prospective technology to enhance communication performance by its characteristic of manipulating the wireless propagation environment in an intelligent way in real time [3]. This is a prospective research direction, and the employment of RIS in RF communication has been investigated recently [4–6].

An RIS can be defined as a metasurface comprised of artificial meta-atoms or a mirror array composed of low-cost passive reflective elements. Simultaneously, an RIS controller is integrated into the technology to sense the environment, judge the changes in circumstance, and make adaptive decisions on the metasurface or mirror array in order to achieve the changes in the transmission environment intelligently [7,8]. In consideration of the potential of VLC and the ability of the RIS to enhance communication performance, combined with the technical bottleneck faced by VLC technology (e.g., high path loss, alignment issues with the transmitters and receivers, and high penetration loss [9]), RIS has been widely applied in VLC systems to obtain a performance gain in the communication process [10–14] and overcome the shortcomings of VLC [15,16]. The area of VLC RIS mentioned above has concentrated mainly on the part between the transmitter and the receiver, and the process is achieved by the mirror array-based VLC RIS deployed on walls, since the VLC RIS comprised of a mirror array exceeds a metasurface-based VLC RIS [17]. However, the research into liquid crystal (LC) RISs in VLC receivers for beam steering and light amplification is sparse and still in the initial stages. Based on the current research on LC RIS, this paper concentrates on the employment of an LC RIS in MISO-VLC systems to investigate the benefits LC RIS can bring.

In [18], the authors outlined the effectiveness of LC RIS in increasing the signal-to-noise ratio (SNR) and enhancing the field-of-view (FoV) in the receiver. With the existence of etendue reducers composed of convex lenses in traditional VLC receivers, combined with the reflection at the upper surface of the lens, the transmission process from the incident light to the photodetector (PD) can produce up to 30% optical intensity losses [19]. The application of convex lenses in VLC receivers limits the transmission capabilities of VLC systems owing to the reflection process on the surfaces of lenses; consequently, some methods were examined in [20] to try to steer the incident light dynamically and amplify the incident light intensity simultaneously. Comparing various methods, the LC RIS is considered as a low-cost and robust method to overcome the drawbacks faced by traditional VLC receivers. Specifically, an external voltage is used to reorient the LC molecules to control the refractive index of the LC RIS. Consequently, the LC RIS is able to steer the incident light in the VLC receivers, tuning the refractive index dynamically. The work on realizing LC RIS-based VLC receivers in practical life was conducted in [21–24], and a practical design for an LC RIS was proposed in [25]. Inspired by the work in [25], we propose an LC RIS-based MISO-VLC system and formulate an optimization problem with asymptotic capacity maximization as the utility function. Our main contributions are as follows.

- An LC RIS-aided MISO-VLC system is proposed with an LC RIS in the VLC receiver to steer the incident light dynamically; meanwhile, the corresponding asymptotic capacity in the MISO-VLC system is enhanced after applying the LC RIS;
- For the LC RIS-aided MISO-VLC system, the asymptotic capacity in high SNR with peak-constrained inputs is derived. Additionally, we formulate an optimization problem with the asymptotic capacity derived by us as the objective function and the refractive index of LC RIS as the independent variable. For this nonconvex optimization problem, we propose a metaheuristic optimization algorithm (particle swarm optimization algorithm) to determine the optimal refractive index of the LC RIS according to the environmental changes;
- The simulation results demonstrate that the employment of an LC RIS in VLC receivers can raise the communication performance of MISO-VLC systems to a greater degree. Simultaneously, compared with the exhaustive search method, the PSO algorithm is an effective method to deal with the optimization problems for LC RIS-based MISO-VLC systems. Meanwhile, we found that there was a significant performance gain compared with a benchmark scheme (BSch) (randomly selecting a refractive index of LC RIS) or the MISO-VLC systems with receivers without the LC RIS. In addition, the LC RIS is expected to solve the dead zone problem in traditional VLC systems by analyzing the growth rate of communication performance for each location on the floor.

2. System Model

In this section, an LC RIS-aided MISO-VLC system is modeled, and accordingly, the channel gain is formulated according to the content in [25]. As shown in Figure 1, an LC RIS-aided MISO-VLC system is modeled with N transmitters and one PD. In particular, there is an LC RIS deployed in front of the PD for steering the incident light and amplifying the

SNR in the receiver. For simplicity, we assume that the transmitters are fixed to the ceiling, and similarly, the receiver is assumed to be fixed to the floor. Consequently, according to the locations of the transmitters and receivers, the refractive index of LC RIS can be confirmed. The cartesian coordinate system is established in the figure, and for simplicity, the norm vectors of the transmitters and the receiver are perpendicular to the ground. It is assumed that the room size is $x_{max} \times y_{max} \times z_{max}$ (m^3). This model of the LC RIS-aided MISO-VLC system is first proposed in this research, and the details of the model are introduced.

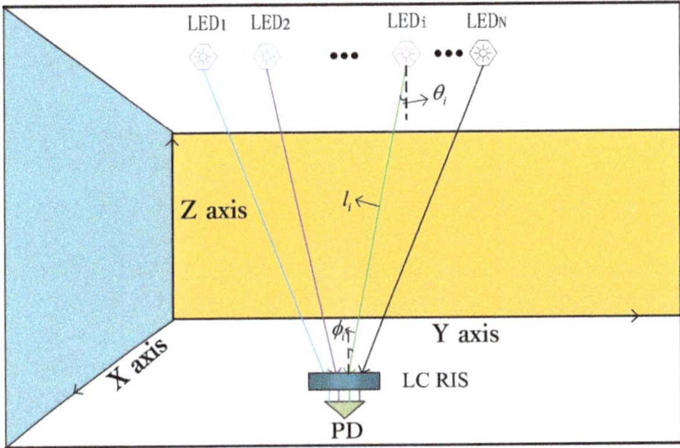

Figure 1. A MISO-VLC system with an LC RIS in the receiver.

Before introducing the derivation of the channel gain for the LC RIS-aided MISO-VLC system, we need to focus on the central part in the transmission process that contributes considerably to the channel gain to facilitate the analysis of the influence of the LC RISs on the MISO-VLC systems. To be specific, the transmission process of VLC systems mainly concentrates on line-of-sight (LoS) transmission links due to the distinct modulation scheme adopted in the VLC. The application of intensity modulation and direct detection (IM/DD) guarantees the nonnegativity of input signals, and accordingly, the diffuse reflections from walls, ceiling, and floor can only produce low responses, which have little influence on the transmission process [26]. Consequently, we concentrate on the LoS channels and ignore the diffused reflections for the LC RIS-aided MISO-VLC system in the following content. The system model of the LC RIS-aided MISO-VLC is characterized by

$$Y = \mathbf{h}^\top \mathbf{X} + Z, \tag{1}$$

where Y denotes the received signals, $\mathbf{h} = (h_1, h_2, \ldots, h_N)^\top$ represents the channel gain, $\mathbf{X} = (X_1, X_2, \ldots, X_N)^\top$ is the channel input, and $Z \sim \mathcal{N}(0, \sigma^2)$ denotes the additive white gaussian noise (AWGN) with σ^2 as the variance of the noise. The application of the LC RIS can reconfigure the channel $\mathbf{h}$ by changing the refractive index of the LC RIS. This process can be realized by applying external voltage to the LC cell. This paper endeavors to investigate the optimal refractive index of the LC RIS to enhance the communication performance of LC RIS-aided MISO-VLC systems.

2.1. Channel Gain

2.1.1. Channel Gain through the Air

According to [27], the LoS channel can be modeled as a Lambertian model in traditional VLC systems. For the LC RIS-aided MISO-VLC system, the i-th LoS transmission path contains two parts: the propagation via the air and the LC RIS, respectively. The former can be derived according to the Lambertian model in traditional VLC systems, and the latter can be characterized on the basis of the analysis in [25]. Consequently, the i-th channel gain for the LC RIS-aided MISO-VLC system is expressed as

$$g_i = G_i \times \alpha_i, \quad 1 \leq i \leq N, \tag{2}$$

where G_i represents the i-th channel gain through the air, and α_i denotes the transition coefficient of the i-th channel through the LC RIS. Meanwhile, α_i is correlated with the angle of incidence from the i-th transmission path and the tunable refractive index of LC RIS. The former is labeled as ψ_i, and the latter is denoted by η_c. Finally, the number of transmitters in the system is represented by N. The channel gain through the air for the i-th LoS link can be characterized by [27].

$$G_i = \frac{(m+1)A_{\mathrm{PD}}}{2\pi l_i^2} \cos^m(\theta_i) \cos(\phi_i) T_{of}(\phi_i) T_{oc}(\phi_i), \tag{3}$$

where m represents the Lambertian order calculated by the equation $m = -1/\log_2(\cos(\psi_{1/2}))$, with $\psi_{1/2}$ being the light-emitting diode (LED) half-power semiangle. The physical area of the PD is denoted by A_{PD}, the distance between the i-th LED and the PD is denoted by l_i, the angle of the irradiance for the i-th LED is represented by θ_i, ϕ_i represents the angle of the incidence for the incident light from the i-th LED, and the gain of the optical filter is labeled as $T_{of}(\phi_i)$, which is usually set as a constant. $T_{oc}(\phi_i)$ denotes the optical concentrator gain correlated with the FoV labeled as Φ_{FoV} and the internal refractive index in the PD labeled as a. Specifically, the gain of the optical concentrator can be expressed as

$$T_{oc}(\phi_i) = \begin{cases} a^2/\sin^2 \Phi_{\mathrm{FoV}}, & 0 \leq \phi_i \leq \Phi_{\mathrm{FoV}}, \\ 0, & \text{otherwise}, \end{cases} \tag{4}$$

where Φ_{FoV} is satisfied by $\Phi_{\mathrm{FoV}} \leq \frac{\pi}{2}$. The transition coefficient of the i-th channel through the LC RIS α_i is discussed in the following according to the work in [25].

2.1.2. The Transition Coefficient through the LC RIS

As shown in Figure 2, the propagation process of light through the LC RIS is displayed. According to Fresnel's law, the quantification of the reflected light at the upper surface of the LC RIS for the i-th channel can be expressed as [28]

$$R_{in}(\phi_i, \delta_i) = \frac{1}{2}\left(\frac{\eta \cos \phi_i - \cos \delta_i}{\eta \cos \phi_i + \cos \delta_i}\right)^2 + \frac{1}{2}\left(\frac{\cos \phi_i - \eta \cos \delta_i}{\cos \phi_i + \eta \cos \delta_i}\right)^2, \tag{5}$$

where the relative refractive index of the air with respect to the LC RIS is represented by $\eta = \eta_c/\eta_a$. The refractive index of the air is denoted by η_a, and the LC RIS is labeled as η_c. According to Snell's law, we can obtain $\eta_a \sin \phi_i = \eta_c \sin \delta_i$, and the amount of the reflected light can be further derived as

$$R_{in}(\phi_i) = \frac{1}{2}\left(\frac{\eta^2 \cos \phi_i - \sqrt{\eta^2 - \sin^2 \phi_i}}{\eta^2 \cos \phi_i + \sqrt{\eta^2 - \sin^2 \phi_i}}\right)^2 + \frac{1}{2}\left(\frac{\cos \phi_i - \sqrt{\eta^2 - \sin^2 \phi_i}}{\cos \phi_i + \sqrt{\eta^2 - \sin^2 \phi_i}}\right)^2. \tag{6}$$

We assume there is no light absorbed on the surface of the LC RIS, and consequently, the amount of the refractive light through the LC RIS can be characterized by $T_{in}(\phi_i) = 1 - R_{in}(\phi_i)$. Meanwhile, the transition coefficient for the refracted process from the air to the LC RIS is given by

$$\alpha_i^{in} = (\eta)^2 T_{in}(\phi_i). \tag{7}$$

Similarly, the transition coefficient for the refracted process from the LC cell to the air is given by

$$\alpha_i^{out} = (\eta_1)^2 T_{out}(\delta_i),\tag{8}$$

where $\eta_1 = \eta_a/\eta_c$ is the relative refractive index, $T_{out}(\delta_i) = 1 - R_{out}(\delta_i)$ represents the amount of the refractive light from the LC cell to the air, $R_{out}(\delta_i)$ denotes the the amount of the reflected light expressed as

$$R_{out}(\delta_i) = \frac{1}{2}\left(\frac{\eta_1^2\cos\delta_i - \sqrt{\eta_1^2 - \sin^2\delta_i}}{\eta_1^2\cos\delta_i + \sqrt{\eta_1^2 - \sin^2\delta_i}}\right)^2 + \frac{1}{2}\left(\frac{\cos\delta_i - \sqrt{\eta_1^2 - \sin^2\delta_i}}{\cos\delta_i + \sqrt{\eta_1^2 - \sin^2\delta_i}}\right)^2;\tag{9}$$

finally, the PD detects the refracted light existed from the LC RIS, and the final transition coefficient of the i-th channel through the LC RIS is

$$\alpha_i = \alpha_i^{in} \times \alpha_i^{out} = T_{in}(\phi_i) \times T_{out}(\delta_i).\tag{10}$$

From the derivation mentioned above, we can know that the overall transition coefficient for the i-th incident light is correlated with the refractive index of LC RIS and the angle of the incidence for the incident light from the i-th LED. The angle of the incidence is usually related to the location of the receiver in the VLC after fixing the locations of the transmitters. In addition, the LC RIS can change the refractive index dynamically according to the environmental changes by applying an external voltage v_e, as shown in Figure 2. The applied voltage can change the tilt angle specifying the molecular orientations of the LC RIS and, accordingly, influence the refractive index. Figures 3 and 4 indicate the molecular orientations of the LC RIS cell before and after the external voltage is applied, respectively. Specifically, the tilt angle can be characterized by

$$\epsilon = \begin{cases} \frac{\pi}{2} - 2\arctan[\exp(-\frac{v_e - v_{th}}{v_0})], & v_e > v_{th}, \\ 0, & v_e \le v_{th}, \end{cases}\tag{11}$$

where v_{th} denotes the threshold voltage that allows the tilt process of the molecule to begin, v_0 denotes a constant, and ϵ represents the tilt angle. Meanwhile, the relationship between the tilt angle ϵ and the refractive index of LC RIS η_c is given by [29].

$$\frac{1}{\eta_c^2(\epsilon)} = \frac{\cos^2\epsilon}{\eta_{up}^2} + \frac{\sin^2\epsilon}{\eta_{low}^2},\tag{12}$$

where η_{up} and η_{low} are the upper and lower limit of the tunable refractive index for the LC RIS.

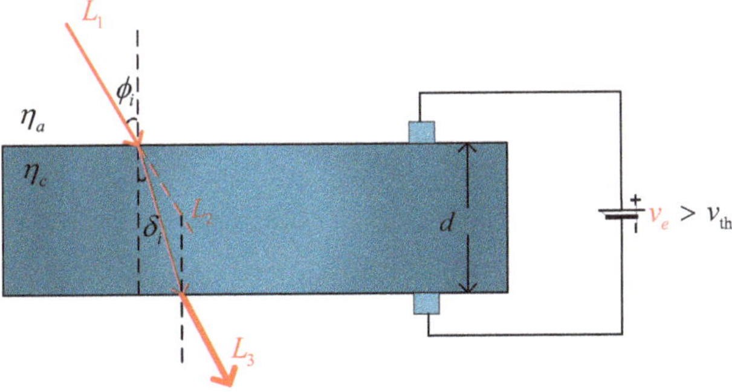

Figure 2. The detail of the propagation through the LC RIS.

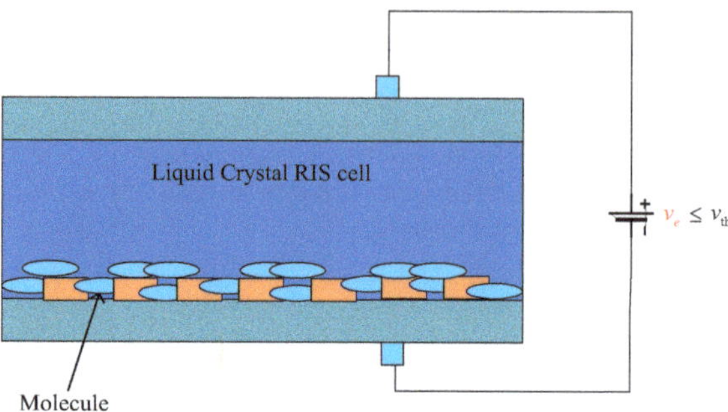

Figure 3. The molecular orientation of the LC RIS before applying the voltage.

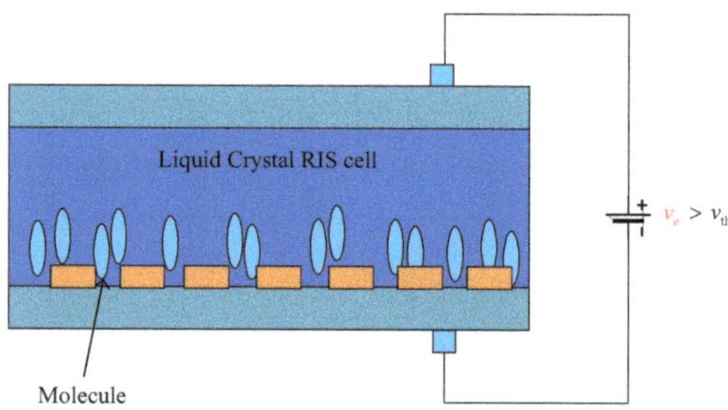

Figure 4. The molecular orientation of the LC RIS after applying the voltage.

2.2. Amplification Coefficient

The light traveling through the LC RIS can be amplified with the application of external voltage [25], as shown in Figure 2. From the figure, we observe that the light intensity from the LC RIS, labeled as L_3, was greater than the incident light intensity entering the LC RIS, labeled as L_1. The amplification gain for the i-th channel can be expressed as [25]

$$\beta_i = \alpha_i \times \exp(Y_i d), \tag{13}$$

where d is the depth of the LC cell, and Y_i is the negative absorption coefficient of the i-th channel for the LC RIS expressed as

$$Y_i = \frac{2\pi v_e \eta_c^3}{\lambda d \cos \phi_i} \gamma, \tag{14}$$

where the wavelength of the incident light is denoted by λ, and the electro-optical conversion coefficient is represented by γ.

Consequently, the overall transmission gain can be characterized by

$$h_i = g_i \times \beta_i = G_i \times \alpha_i \times \exp(Y_i d), \quad 1 \le i \le N. \tag{15}$$

Consequently, we see that the dynamic change in the refractive index can influence the overall transmission channel and enhance the communication performance from the derivation above.

3. Asymptotic Capacity Optimization

The asymptotic capacity with peak-constrained inputs for the LC RIS-aided MISO-VLC system is formulated in this section. Meanwhile, the tunable refractive index of the LC RIS is considered as an independent variable to optimize the asymptotic capacity. This section formulates the optimization problem, and accordingly, a metaheuristic algorithm is proposed to confirm the optimal refractive index.

3.1. Asymptotic Capacity Maximization Problem

The asymptotic capacity for the LC RIS-aided MISO-VLC system shaped like Figure 1 can be given by [30].

$$R(\eta_c) = \lim_{\mathcal{A} \to \infty} \frac{1}{2} \log(\frac{(\sum_{i=1}^{N} h_i)^2 \mathcal{A}^2}{2\pi \exp(1)\sigma^2}), \tag{16}$$

where the peak-power constraint for the input signal is denoted by $\mathcal{A}$. From the derivation in (15), we see that the tunable refractive index of the LC RIS can only influence the overall transmission gain h_i, and consequently, we can formulate the optimization problem as

$$\max_{\eta_c} \sum_{i=1}^{N} h_i \tag{17}$$

$$\text{s.t.} \quad \eta_{\text{low}} \leq \eta_c \leq \eta_{\text{up}}.$$

The problem is a highly nonconvex optimization problem, and consequently, it is difficult to solve the problem by adopting traditional algorithms. Hence, the particle swarm optimization (PSO) algorithm is proposed to solve the problem in our work. The PSO algorithm is a metaheuristic optimization algorithm leveraged to solve nonconvex problems. The principle of this algorithm is to simulate the foraging behavior of birds in a constrained search area [31]. In the following, a description of this algorithm's specifics is provided.

3.2. Proposed Solution Algorithm

This paper adopts the PSO algorithm to deal with the optimization problems. For the practical problem proposed above, the location of each particle represents one possible value of the tunable refractive index. The searching space of the particle swarm is the set of all the possible values for the refractive index. The optimality of each particle is measured by the calculated value according to the objective function in (17). The location and velocity of one particle is updated iteratively to search for the optimal solution of the problem; meanwhile, the updating process should obey the laws characterized by

$$\mathbf{v}_i^{t+1} = \omega \mathbf{v}_i^t + c_1 r_1(\mathbf{x}_{i,pbest}^t - \mathbf{x}_i^t) + c_2 r_2(\mathbf{x}_{i,gbest}^t - \mathbf{x}_i^t), \tag{18}$$

$$\mathbf{x}_i^{t+1} = \mathbf{x}_i^t + \mathbf{v}_i^{t+1}, \tag{19}$$

where the velocity of the i-th particle in the swarm, at the t-th iteration, is represented by $\mathbf{v}_i^t$, and the best-recorded location of the i-th particle, until the t-th iteration evolution, is denoted by $\mathbf{x}_{i,pbest}^t$, and $\mathbf{x}_{i,gbest}^t$ denotes the best-recorded location of the entire particle swarm until the t-th iteration evolution. Furthermore, ω is considered as the inertia weight, c_1 and c_2 are the acceleration constants (also known as "learning factors"), and r_1 and r_2 denote randomly selected constants subject to a uniform distribution in $[0, 1]$. Meanwhile, the constraint on velocity needs to be imposed to prevent missing the ideal solution due to excessive velocity or a delay in achieving the final solution caused by the stagnant updating velocity for particles. Consequently, the velocity of the updating process meets $\mathbf{v}_i^t \subseteq [-\mathbf{v}_{\max}, \mathbf{v}_{\max}]$, where $\mathbf{v}_{\max}$ is the maximum updating velocity set for each particle. All the vectors mentioned above have K elements, and K is the dimension of the independent variable for the optimization problems. According to (17), we see that the dimension of the vector is reduced to a one-dimensional variable (i.e., $K = 1$). Algorithm 1 indicates the pseudocode of the PSO algorithm.

Algorithm 1 The PSO Algorithm for Asymptotic Capacity Maximization

---

Input: Size of the particle swarm $\mathcal{I}$; maximum iterations $t_{\max}$.

1: **Set:** Constraints on velocity and location $\eta_{\mathrm{low}}, \eta_{\mathrm{up}}, -v_{\max}, v_{\max}$; objective function in (17); inertia weight ω; acceleration constraints c_1, c_2.

2: **Initialization:** Location of each particle $x_i = \eta_c^i \subseteq [\eta_{\mathrm{low}}, \eta_{\mathrm{up}}], 1 \leq i \leq \mathcal{I}$; velocity of each particle: $v_i \subseteq [-v_{\max}, v_{\max}], 1 \leq i \leq \mathcal{I}$; the best-recorded location for the i-th particle: $x_{i,pbest}, 1 \leq i \leq \mathcal{I}$; the best-recorded location for the particle swarm: $x_{i,gbest}$; all the initial values are satisfied with the constraints mentioned above.

3:   **While** $t \leq t_{\max}$

4:   **While** $1 \leq i \leq \mathcal{I}$

5:     **Refresh:** The velocity and location of the i-th particle based on (18) and (19);

6:     **Calculate:** The corresponding fitness according to the objective function in (17);

7:     **If** the fitness of x_i^{t+1} is larger than the fitness of $X_{i,pbest}^t$

8:       **Choose:** x_i^{t+1} as the the i-th particle's best-recorded location;

9:     **End If**

10:    **If** the fitness of $x_{i,pbest}^t$ is larger than that of x_{gbest}^t

11:      **Choose:** $x_{i,pbest}^t$ as the particle swarm's best-recorded location;

12:    **End If**

13:   **End While**

14:   **End While**

15: **return** $\eta_c^* = x_{gbest}^{t_{\max}}$ and the corresponding optimal fitness of the particle swarm.

Output: The best location of the particle swarm $\eta_c^* = x_{gbest}^{t_{\max}}$; the optimum value for the objection function.

3.3. Complexity Analysis

We consider the computational complexity of the proposed algorithm in this subsection. Firstly, the same number of operations is needed for the process of generating the initial particle and the initial velocity, characterized by $\mathcal{O}(K\mathcal{I})$, where $\mathcal{I}$ represents the size of the particle swarm, and K denotes the dimension of the independent variable. Hence, the process of initialization needs $\mathcal{O}(K\mathcal{I}) + \mathcal{O}(K\mathcal{I}) = \mathcal{O}(K\mathcal{I})$ operations. There are $\mathcal{I}$ operations required to calculate the best-recorded location for each particle; meanwhile, the same number of operations is demanded to select the best-recorded location of the particle swarm. Secondly, the updating process for locations and velocities of the particle swarm needs $\mathcal{O}(K\mathcal{I}t_{\max})$ operations at most, where $t_{\max}$ denotes the maximum iterations. It requires $\mathcal{O}(\mathcal{I}t_{\max})$ operations at most to calculate the historically optimal fitness of each particle; meanwhile, the same number of operations is needed to select the best-recorded position of the particle swarm according to the updated particles. Consequently, the computational complexity for the whole updating process in the worst-case scenario is $\mathcal{O}(K\mathcal{I}t_{\max})$. Finally, the overall computational complexity of the proposed algorithm is $\mathcal{O}(K\mathcal{I}) + \mathcal{O}(K\mathcal{I}t_{\max}) \approx \mathcal{O}(K\mathcal{I}t_{\max})$.

4. Simulation Results and Analysis

Some simulations are detailed in this section to demonstrate the availability of the proposed algorithm in searching for the optimal value of the tunable refractive index for the LC RIS in MISO-VLC systems; meanwhile, we visualize the improvement provided by the LC RIS in the communication performance for the MISO-VLC systems.

4.1. Simulation Parameters

A $5 \times 5 \times 3$ (m^3) room and 3×1 LC RIS-aided MISO-VLC system were considered in our simulations. We assumed the location of the PD was randomly chosen and satisfied with uniform distribution on the floor to guarantee the generality. Meanwhile, we chose the most probable locations in our daily life as the positions of the LEDs. The parameters in Algorithm 1 were set as $\mathcal{I} = 10$ and $t_{\max} = 30$. The remaining values of the simulation parameters are summarized in Table 1.

Table 1. Simulation Parameters.

Name of Parameter	Value
$\psi_{1/2}$	70°
A_{PD}	1 cm^2
a	1.5
Φ_{FoV}	80°
$T_{of}(\phi_i)$	1.0
γ	12 pm/V
η_{up}	1.7
η_{low}	1.5
η_a	1.0
v_0	0.8 V
v_{th}	1.2 V
d	0.80 mm
σ^2	1×10^{-14}
$x_{\max} \times y_{\max} \times z_{\max}$ (m^3)	$5 \times 5 \times 3$ (m^3)

4.2. Simulation Results

4.2.1. Convergence Analysis for the Proposed Algorithm

The convergence process of the proposed algorithm with different wavelengths of the transmission lights is displayed in Figure 5. The maximization of the asymptotic capacity for different wavelengths was achieved through the PSO algorithm with several iterations. By analyzing the results displayed in the figure, we see that the convergence rate of the proposed algorithm was fast enough to search for the optimal solution for the optimization problem proposed in the paper.

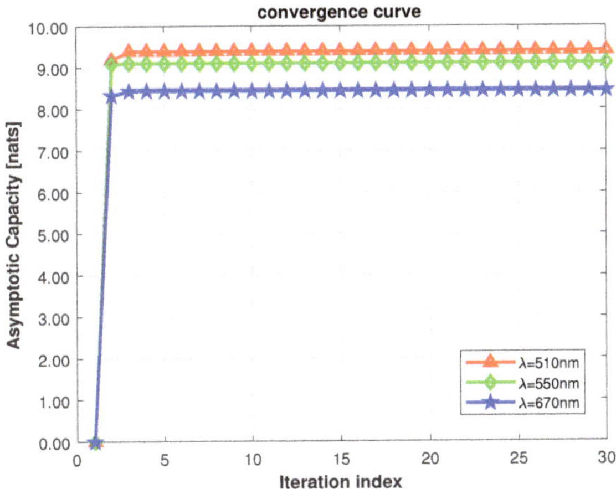

Figure 5. The convergence process of Algorithm 1 with different wavelengths of the light beams.

4.2.2. Asymptotic Capacity Performance Gain for the Optimal Design of LC RIS versus a Baseline Scheme

Figure 6 shows the asymptotic capacity performance gain for the optimal refractive index of LC RIS versus a randomly selected refractive index of LC RIS. The latter was considered as the baseline scheme for comparison. We chose the location of the PD randomly to guarantee generality. Meanwhile, the wavelength of the transmission light signal and the assignment scheme of the tuning refractive index for the LC RIS were changed to investigate the influence of the optimization algorithm and the wavelengths of the light signals on the communication performance for the LC RIS-aided MISO-VLC system. In this figure, "no-LC RIS" means that the receiver of the system was organized by an ordinary receiver containing a convex lens, "LC RIS-aided BSch" indicates that the LC RIS was tuned according to a baseline scheme, and the refractive index in this scheme was selected randomly from all the feasible values. "LC RIS-aided: PSO" represents that the refractive index was calculated by the proposed algorithm (the PSO algorithm). The difference among the wavelengths of transmission light signals is distinguished by different colors.

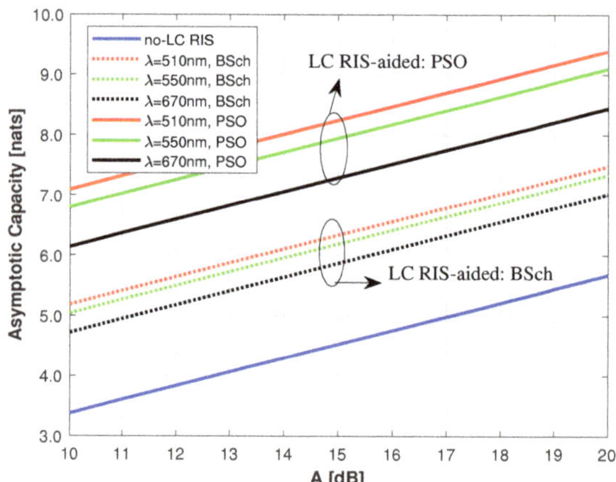

Figure 6. Asymptotic capacity versus SNR with different schemes.

By comparing the "no-LC RIS" and "LC RIS-aided" curves from the figure, we see that the communication performance for the MISO-VLC systems was improved with the application of the LC RIS. Meanwhile, for the problems shaped as (17), the figure demonstrates that the PSO algorithm was an effective method. The proposed algorithm solved the nonconvex problem with several iterations, and the results provided by the PSO algorithm were much better than the BSch algorithm.

4.2.3. Growth Rate of the Performance versus the Position of the Receiver

Figure 7 shows the growth rate of the communication performance versus the position of the receiver after fixing the locations of the transmitters. We assumed that the wavelengths of the transmission lights were constant; meanwhile, the tunable refractive index of the LC RIS was calculated by the proposed algorithm according to the position of the receiver to confirm the optimal performance of the RIS. The growth rate in the corner of the room was higher than other locations, as shown the figure. This is a meaningful result, since the corner in indoor VLC systems is usually a dead zone (the area that the light cannot illuminate), and the quality of service in the corner is usually weak. Consequently, the application of the LC RIS improved the communication quality and solved the dead zone problems of the traditional VLC systems.

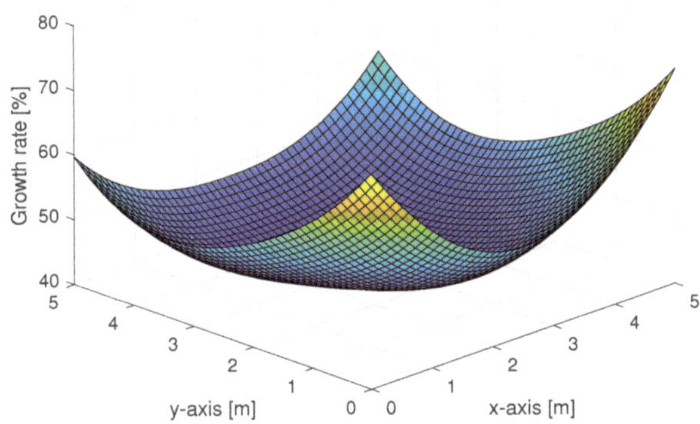

Figure 7. Growth rate versus the position of receiver.

5. Conclusions and Future Research Directions

In this paper, an LC RIS-aided MISO-VLC system was first proposed, and an optimization problem with the tunable refractive index of LC RIS as the optimization variable was formulated accordingly. We considered the asymptotic capacity as the criterion characterizing the communication performance of the newly-established system model. As for the nonconvex optimization problem given in this paper, instead of adopting the exhaustive search method, the PSO algorithm was considered as the proper algorithm to solve the optimization problem. The simulation results demonstrated that the proposed algorithm was an effective approach to solve the proposed optimization problems, and the optimal refractive index of the LC RIS was found with several iterations. Meanwhile, compared with other assignment schemes for the LC RIS (random selection or no LC RIS), the refractive index calculated by the PSO algorithm helped the LC RIS-aided MISO-VLC system to maximize the communication performance improvement. Finally, we conducted simulations, which indicated that the application of the LC RIS improved the growth rate of the communication performance in the corner of the room moreso than other locations, which is an excellent result that contributes to solving the dead zone problems of traditional VLC systems.

The application of the RIS in VLC systems usually focuses on one type of RIS such as an LC RIS or a mirror-array-based RIS. However, the combination of different types of RIS may enable better performance improvement for traditional VLC systems. In the future, we would like to combine the mirror-array-based RIS deployed on walls and the LC RIS deployed in the receiver to achieve better communication performance for traditional VLC systems.

Author Contributions: Conceptualization, Q.W.; methodology, D.G.; software, Q.W.; validation, Q.W.; formal analysis, Q.W.; investigation, Q.W.; resources, Q.W.; data curation, G.X.; writing—original draft preparation, Q.W.; writing—review and editing, J.Z.; visualization, Q.W.; supervision, Y.Z.; project administration, J.Z.; funding acquisition, J.Z. All authors have read and agreed to the published version of the manuscript.

Funding: This research was supported by the National Natural Science Foundation of China (NSFC) under Grant (62071489) and the National Key Research and Development Project (2018YFB1801903).

Institutional Review Board Statement: Not applicable.

Informed Consent Statement: Not applicable.

Data Availability Statement: Not applicable.

Conflicts of Interest: The authors declare no conflict of interest.

References

1. Samsung Research. 6G: The Next Hyper Connected Experience for All. 2020. Available online: https://research.samsung.com/next-generation-communications (accessed on 24 July 2022).
2. Karunatilaka, D.; Zafar, F.; Kalavally, V.; Parthiban, R. LED based indoor visible light communications: State of the art. *IEEE Commun. Surv. Tutorials* **2015**, *17*, 1649–1678. [CrossRef]
3. Luo, S.; Hao, J.; Ye, F.; Li, J.; Ruan, Y.; Cui, H.; Liu, W.; Chen, L. Evolution of the electromagnetic manipulation: From tunable to programmable and intelligent metasurfaces. *Micromachines* **2021**, *12*, 988. [CrossRef] [PubMed]
4. ElMossallamy, M.A.; Zhang, H.; Song, L.; Seddik, K.G.; Han, Z.; Li, G.Y. Reconfigurable intelligent surfaces for wireless communications: Principles, challenges, and opportunities. *IEEE Trans. Cogn. Commun. Netw.* **2020**, *6*, 990–1002. [CrossRef]
5. Basar, E.; Di Renzo, M.; De Rosny, J.; Debbah, M.; Alouini, M.S.; Zhang, R. Wireless communications through reconfigurable intelligent surfaces. *IEEE Access* **2019**, *7*, 116753–116773. [CrossRef]
6. Guo, H.; Liang, Y.C.; Chen, J.; Larsson, E.G. Weighted sum-rate maximization for reconfigurable intelligent surface aided wireless networks. *IEEE Trans. Wirel. Commun.* **2020**, *19*, 3064–3076. [CrossRef]
7. Ma, Q.; Bai, G.D.; Jing, H.B.; Yang, C.; Li, L.; Cui, T.J. Smart metasurface with self-adaptively reprogrammable functions. *Light Sci. Appl.* **2019**, *8*, 98. [CrossRef]
8. Renzo, M.D.; Debbah, M.; Phan-Huy, D.T.; Zappone, A.; Alouini, M.S.; Yuen, C.; Sciancalepore, V.; Alexandropoulos, G.C.; Hoydis, J.; Gacanin, H.; et al. Smart radio environments empowered by reconfigurable AI meta-surfaces: An idea whose time has come. *EURASIP J. Wirel. Commun. Netw.* **2019**, *2019*, 129. [CrossRef]
9. Feng, L.; Yang, H.; Hu, R.Q.; Wang, J. MmWave and VLC-based indoor channel models in 5G wireless networks. *IEEE Wirel. Commun.* **2018**, *25*, 70–77. [CrossRef]
10. Aboagye, S.; Ngatched, T.M.; Dobre, O.A.; Ndjiongue, A.R. Intelligent reflecting surface-aided indoor visible light communication systems. *IEEE Commun. Lett.* **2021**, *25*, 3913–3917. [CrossRef]
11. Wu, Q.; Zhang, J.; Guo, J.N. Position Design for Reconfigurable Intelligent-Surface-Aided Indoor Visible Light Communication Systems. *Electronics* **2022**, *11*, 3076. [CrossRef]
12. Wu, Q.; Zhang, J.; Guo, J. Capacity Maximization for Reconfigurable Intelligent Surface-Aided MISO Visible Light Communications. *Photonics* **2022**, *9*, 487. [CrossRef]
13. Qian, L.; Chi, X.; Zhao, L.; Chaaban, A. Secure visible light communications via intelligent reflecting surfaces. In Proceedings of the ICC 2021-IEEE International Conference on Communications, Montreal, QC, Canada, 14–23 June 2021; pp. 1–6.
14. Sun, S.; Yang, F.; Song, J.; Han, Z. Joint resource management for intelligent reflecting surface–aided visible light communications. *IEEE Trans. Wirel. Commun.* **2022**, *21*, 6508–6522. [CrossRef]
15. Sun, S.; Yang, F.; Song, J.; Zhang, R. Intelligent Reflecting Surface for MIMO VLC: Joint Design of Surface Configuration and Transceiver Signal Processing. *arXiv* **2022**, arXiv:2206.14465.
16. Han, Y.; Xiao, Y.; Zhang, X.; Gao, Y.; Zhu, Q.; Dong, B. Design of dynamic active-passive beamforming for reconfigurable intelligent surfaces assisted hybrid VLC/RF communications. *IET Commun.* **2022**, *16*, 1531–1544. [CrossRef]
17. Abdelhady, A.M.; Salem, A.K.S.; Amin, O.; Shihada, B.; Alouini, M.S. Visible light communications via intelligent reflecting surfaces: Metasurfaces vs. mirror arrays. *IEEE Open J. Commun. Soc.* **2020**, *2*, 1–20. [CrossRef]
18. Aboagye, S.; Ndjiongue, A.R.; Ngatched, T.; Dobre, O.; Poor, H.V. RIS-Assisted Visible Light Communication Systems: A Tutorial. *arXiv* **2022**, arXiv:2204.07198.
19. Ndjiongue, A.R.; Ngatched, T.M.; Dobre, O.A.; Haas, H. Re-configurable intelligent surface-based VLC receivers using tunable liquid-crystals: The concept. *J. Light. Technol.* **2021**, *39*, 3193–3200. [CrossRef]
20. Ndjiongue, A.R.; Ngatched, T.M.; Dobre, O.A.; Haas, H. Toward the use of re-configurable intelligent surfaces in VLC systems: Beam steering. *IEEE Wirel. Commun.* **2021**, *28*, 156–162. [CrossRef]
21. Forkel, G.J.; Krohn, A.; Hoeher, P.A. Optical interference suppression based on LCD-filtering. *Appl. Sci.* **2019**, *9*, 3134. [CrossRef]
22. Krohn, A.; Forkel, G.J.; Hoeher, P.A.; Pachnicke, S. LCD-based optical filtering suitable for non-imaging channel decorrelation in VLC applications. *J. Light. Technol.* **2019**, *37*, 5892–5898. [CrossRef]
23. Zhang, X.G.; Sun, Y.L.; Zhu, B.; Jiang, W.X.; Yu, Q.; Tian, H.W.; Qiu, C.W.; Zhang, Z.; Cui, T.J. A metasurface-based light-to-microwave transmitter for hybrid wireless communications. *Light Sci. Appl.* **2022**, *11*, 126. [CrossRef]
24. Krohn, A.; Pachnicke, S.; Hoeher, P.A. Genetic optimization of liquid crystal matrix based interference suppression for VLC MIMO transmissions. *IEEE Photonics J.* **2021**, *14*, 1–5. [CrossRef]
25. Aboagye, S.; Ndjiongue, A.R.; Ngatched, T.M.; Dobre, O.A. Design and Optimization of Liquid Crystal RIS-Based Visible Light Communication Receivers. *IEEE Photonics J.* **2022**, *14*, 1–7. [CrossRef]
26. Lee, K.; Park, H.; Barry, J.R. Indoor channel characteristics for visible light communications. *IEEE Commun. Lett.* **2011**, *15*, 217–219. [CrossRef]
27. Komine, T.; Nakagawa, M. Fundamental analysis for visible-light communication system using LED lights. *IEEE Trans. Consum. Electron.* **2004**, *50*, 100–107. [CrossRef]

28. Hébert, M.; Hersch, R.D.; Emmel, P. Fundamentals of optics and radiometry for color reproduction. In *Handbook of Digital Imaging*; John Wiley & Sons: Hoboken, NJ, USA, 2015; pp. 1–57.
29. Saleh, B.E.; Teich, M.C. *Fundamentals of Photonics*; John Wiley & Sons: Hoboken, NJ, USA, 2019.
30. Moser, S.M.; Wang, L.; Wigger, M. Capacity results on multiple-input single-output wireless optical channels. *IEEE Trans. Inf. Theory* **2018**, *64*, 6954–6966. [CrossRef]
31. Marini, F.; Walczak, B. Particle swarm optimization (PSO). A tutorial. *Chemom. Intell. Lab. Syst.* **2015**, *149*, 153–165. [CrossRef]

 *photonics* 

Article

LiNEV: Visible Light Networking for Connected Vehicles

Osama Saied [1], Omprakash Kaiwartya [1,*], Mohammad Aljaidi [2], Sushil Kumar [3], Mufti Mahmud [1], Rupak Kharel [4], Farah Al-Sallami [5] and Charalampos C. Tsimenidis [6]

[1] Department of Computer Science, Nottingham Trent University, Nottingham NG11 8NS, UK; osama.saied@ntu.ac.uk (O.S.); mufti.mahmud@ntu.ac.uk (M.M.)

[2] Computer Science Department, Faculty of Information Technology, Zarqa University, Zarqa 13110, Jordan; mjaidi@zu.edu.jo

[3] School of Computer and Systems Sciences, Jawaharlal Nehru University, New Delhi 110067, India; skdohare@mail.jnu.ac.in

[4] School of Psychology and Computer Science, University of Central Lancashire, Preston PR1 2HE, UK; rkharel1@uclan.ac.uk

[5] School of Future Transport Engineering, Coventry University, Coventry CV1 2TU, UK; ad9051@coventry.ac.uk

[6] Department of Engineering, Nottingham Trent University, Nottingham NG11 8NS, UK; charalampos.tsimenidis@ntu.ac.uk

* Correspondence: omprakash.kaiwartya@ntu.ac.uk

Abstract: DC-biased optical orthogonal frequency division multiplexing (DCO-OFDM) has been introduced to visible light networking framework for connected vehicles (LiNEV) systems as a modulation and multiplexing scheme. This is to overcome the light-emitting diode (LED) bandwidth limitation, as well as to reduce the inter-symbol interference caused by the multipath road fading. Due to the implementation of the inverse fast Fourier transform, DC-OFDM suffers from its large peak-to-average power ratio (PAPR), which degrades the performance in LiNEV systems, as the LEDs used in the vehicles' headlights have a limited optical power-current linear range. To tackle this issue, discrete Fourier transform spread-optical pulse amplitude modulation (DFTS-OPAM) has been proposed as an alternative modulation scheme for LiNEV systems instead of DCO-OFDM. In this paper, we investigate the system performance of both schemes considering the light-emitting diode linear dynamic range and LED 3 dB modulation bandwidth limitations. The simulation results indicate that DCO-OFDM has a 9 dB higher PAPR value compared with DFTS-OPAM. Additionally, it is demonstrated that DCO-OFDM requires an LED with a linear range that is twice the one required by DFTS-OPAM for the same high quadrature amplitude modulation (QAM) order. Furthermore, the findings illustrate that when the signal bandwidth of both schemes significantly exceeds the LED modulation bandwidth, DCO-OFDM outperforms DFTS-OPAM, as it requires a lower signal-to-noise ratio at a high QAM order.

Keywords: DFT spread-optical pulse amplitude modulation; DC-biased optical orthogonal frequency division multiplexing; peak-to-average power ratio; light-emitting diode dynamic range; light-emitting diode limited bandwidth

Citation: Saied, O.; Kaiwartya, O.; Aljaidi, M.; Kumar, S.; Mahmud, M.; Kharel, R.; Al-Sallami, F.; Tsimenidis, C.C. LiNEV: Visible Light Networking for Connected Vehicles. *Photonics* 2023, 10, 925. https://doi.org/10.3390/photonics10080925

Received: 28 May 2023
Revised: 7 July 2023
Accepted: 7 August 2023
Published: 11 August 2023

1. Introduction

The constant increase in the use of the radio frequency (RF) spectrum leads to RF wavelength interference which limits the required speed of wireless communication applications [1]. To alleviate the RF spectrum crunch, the huge unlicensed visible light spectrum ranging from 380 to 780 nm (i.e., offers a bandwidth of up to 300 THz) has been extensively investigated to be used in current and next wireless communication generations (i.e., the fifth and sixth wireless communication generations) [2]. As such, VLC is now playing a significant role as a complimentary technology to most of the indoor RF applications (i.e., museums, general offices, shopping centers, railways, airports, and hospitals). In addition to its indoor applications, VLC is also now being considered in some outdoor applications, particularly in some congested outdoor environment applications such as in automated and connected vehicle network applications (i.e., vehicle to everything communication (V2X)) where RF systems face major challenges to fulfill the V2X latency, reliability, scalability, and capacity requirements in such a congested environment [3–5].

VLC-V2X system performance is mainly affected by being interfered with by the optical natural source's light and the reflected vehicle's light, where the former interference introduces a variation amount of background noise up to 20 dB at noon daytime and the latter one causes inter-symbol interference (ISI). The background noise can be reduced by implementing a diversity receiver with a selective combining technique, which results in a 5 dB signal to noise ratio (SNR) improvement, as shown in [6]. Furthermore, a 6.47 dB improvement in the SNR can also occur by implementing optical filtering at the receiver (Rx) side, as illustrated in [7].

On the other hand, the ISI issue can be addressed by letting the transmitted signal bandwidth be less than the coherence bandwidth of the VLC-V2X channel. This has been achieved by introducing the attractive orthogonal frequency division multiplexing (OFDM) signal scheme for VLC-VTX systems. In addition to reducing the ISI, OFDM can also overcome the light-emitting diode 3 dB modulation bandwidth (LED_{3dbBW}) limitation, which is only a few MHz. This is achieved by investigating the bit and power OFDM loading feature. As such, an OFDM-based VLC system achieved a transmission data rate of 15 Gbps [8]. Furthermore, the utilization of Turbo coding in conjunction with OFDM can effectively mitigate the adverse impacts of channel impairments, thereby significantly enhancing the overall system performance [9].

However, implementing OFDM in VLC systems involves two challenges including intensity modulation (IM) constraints and the limited linear dynamic range of LEDs. According to the IM requirements, the OFDM signal must be real and positive before being passed to the LED. The real constraint was addressed by applying Hermitian Symmetry (HS) to the OFDM symbols at the cost of halving the available electrical bandwidth. The positive constraint was tackled by adding a DC bias to the OFDM signal at the cost of the power consumption, known as DC-biased optical OFDM (DCO-OFDM). Alternatively, asymmetric clipped optical OFDM (ACO-OFDM) was adopted to meet the positive signal by modulating only the odd subcarriers of the OFDM signal at the expense of halving the spectrum efficiency compared with DCO-OFDM [10]. In addition to the IM constraints challenge, OFDM-based VLC systems suffer from the OFDM high peak-to-average power ratio (PAPR) time domain signal [11]. This is because the LEDs have a limited linear dynamic range (LED-DR), where any signal beyond or above this linear range must be clipped before being passed to the LED [12]. To address the OFDM nonlinear signal distortion and clipping challenge, the complex interleaved frequency division multiple access (IFDMA) signal scheme was modified to be used in VLC systems instead of OFDM schemes [13–19].

1.1. Related Work and the Problem Identification

ACO-single-carrier frequency domain equalization (ACO-SCFDE) and unipolar-pulse amplitude modulation frequency division multiplexing (U-PAM-FDM) are two IFDMA-modified schemes introduced by [14,15] to address the PAPR of the ACO-OFDM signal. The only difference between ACO-OFDM and ACO-SCFDE is the addition of FFT and IFFT blocks at the transmitter (Tx) and Rx sides of the ACO-SCFDE, respectively. In U-PAM-FDM Tx, the quadrature amplitude modulation (QAM) mapping block in ACO-SCFDE Tx is replaced by the PAM block, while the interleaving mapping and HS blocks are replaced by the symmetrically conjugate (SCG) block.

Although the simulation results show that implementing ACO-SCFDE and U-PAM-FDM in VLC systems can improve the ACO-OFDM PAPR value by 2.1 dB and 3.6 dB, respectively, the PAPR values of these modified IFDMA schemes (ACO-SCFDE and U-PAM-FDM) still remain high compared with the PAPR value of the RF-IFDMA scheme.

This is because the implementation of HS and SCG blocks changes the subcarrier orders of the RF-IFDMA, as was justified in [16].

In addition to ACO-SCFDE- and UPAM-FDM-modified IFDMA schemes, the optical single-carrier-interleaved frequency division multiplexing (OSC-IFDM) scheme, introduced in [17,18], aims to achieve a low PAPR for optical IFDMA comparable to that of RF-IFDMA. This is accomplished by setting the mapping factor (Q) of the RF-IFDMA scheme to two. Consequently, the IFDMA time domain vector is doubled with the first half transmitting real samples and the second half transmitting complex samples. Simulation results show that implementing the OSC-IFDM scheme in VLC systems can reduce the PAPR by 10 dB compared with DCO-OFDM. However, these results also reveal that OSC-IFDM requires an SNR of more than 3 dB compared with DCO-OFDM to achieve the same bit error rate (BER) level. This is due to the fact that the first OSC-IFDM sub-carrier (DC-subcarrier) must be a modulated subcarrier, which can be affected by the DC bias and introduce distortion noise in all time domain samples, making this scheme impractical.

In contrast to other modified RF-IFDMA schemes [13–18], ref. [19] introduced a novel scheme known as discrete Fourier transform spread-optical pulse amplitude modulation (DFTS-OPAM) to make RF-IFDMA signals suitable for VLC without increasing the SNR or the PAPR. In the DFTS-OPAM Tx, the PAM and the repeating mapping (RM) blocks were used instead of the QAM and the interleaved mapping blocks at the RF-IFDMA Tx. Since the DFTS-OPAM transmitted symbols were PAM symbols (real symbols), the output FFT subcarriers at the DFTS-OPMA Tx were symmetrically conjugated, except for the first and the middle subcarriers, therefore passing these sub-carriers through the RM block before IFFT implementation, resulting in a version copy of the transmitted PAM symbol at the even samples of the IFFT output time domain and zeros at the odd samples. Please note that in the RM block the output FFT vector of DFTS-OPAM was repeated to ensure a real time domain signal with as low a PAPR as that of RF-IFDMA.

As a result of this significant PAPR reduction, DFTS-OPAM offers a 2.5 dB improvement in power consumption compared with the traditional DCO-FDM, as demonstrated in practical experiments [19]. Furthermore, the practical results in [19] show a 33% improvement in the distance between the Tx and the Rx when DFTS-OPAM is implemented compared with DCO-OFDM, thanks to the low PAPR chrematistics of DFTS-OPAM in the time domain.

Although DFTS-OPAM using PAM symbols results in half the spectral efficiency compared with DCO-OFDM, it offers a 2.5 dB lower power consumption. However, if both schemes are considered as multiple access schemes based on time division multiple access (TDMA) techniques, as in VLC-V2X and other systems as illustrated in Figure 1 (i.e., indoor ceiling access point and outdoor flying access point applications), they would provide the same spectral efficiency. This is because the odd time domain samples in DFTS-OPAM do not carry any data. In addition, these unused samples can also be utilized for various vehicular traffic environment applications such as illumination, security, positioning, localization, and time domain equalization. Furthermore, due to the RM process, any affected DFTS-OPAM subcarrier can be easily compensated [19].

1.2. Contributions of this Paper

In the context of using VLC for V2X communications under dense vehicular environments, a visible Light Networking framework for Connected Vehicles (LiNEV) is presented in this paper. The major contributions of this paper are as follows:

- A system model for a novel multiple access scheme to enable VLC-V2X traffic use cases is developed.
- The workflow of the model is mathematically derived for highlighting the scientific novelty of the model.
- An extensive performance evaluation of the proposed framework is carried out under the influence of different QAM orders with a range of LED-DR values and limited LED bandwidth.

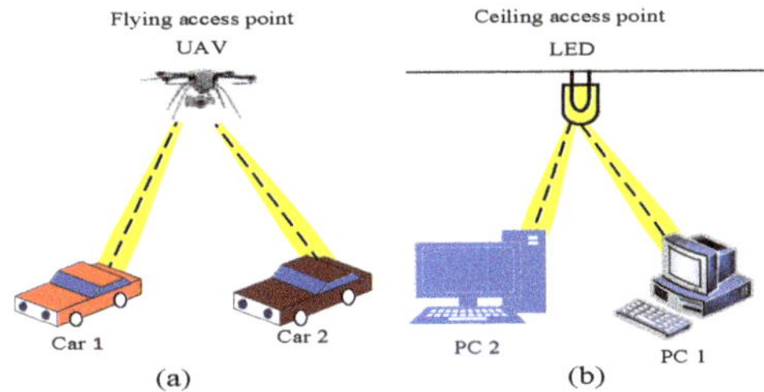

Figure 1. TDMA-VLC applications where (**a**) uses UAV as flying access point to assist vehicular communication while an LED is used in (**b**) as a ceiling access point to provide indoor multiple access.

However, to visually illustrate these contributions, we have included Figure 2, which presents a detailed contribution map of our paper.

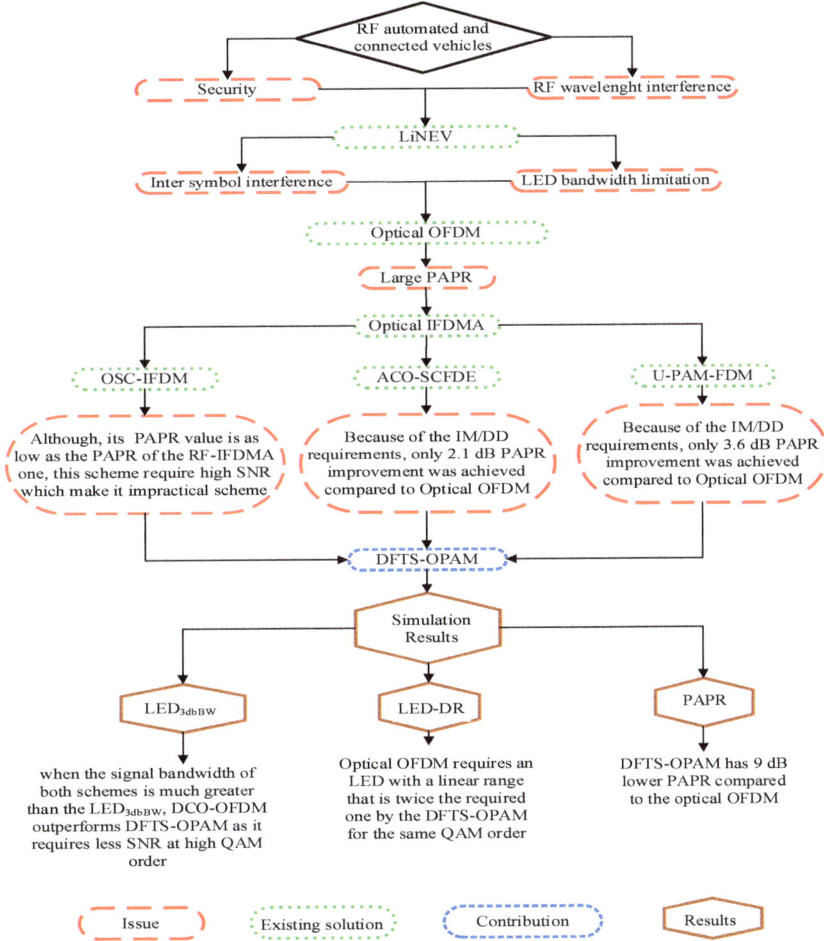

Figure 2. Paper contribution map.

The rest of the paper is structured as follows. Section 2 presents the details of the proposed DFTS-OPAM for connected vehicles. The simulation results are critically discussed in Section 3, followed by the conclusions, presented in Section 4.

2. LiNEV: DFTS-OPAM for Connected Vehicles

2.1. System Model

Figure 3 illustrates a block diagram of the DFTS-OPAM transceiver. The only difference between the DFTS-OPAM Tx and the traditional DCO-OFDM Tx is that the HS block at the DCO-OFDM is replaced by FFT, and RM blocks. Also, the PAM is used in DFTS-OPAM as a transmitted symbol instead of QAM. In [19], we mathematically and practically proved that the output of the DFTS-OPAM IFFT x is a real signal with similar DCO-OFDM features (i.e., reducing the ISI and bit and power loading features) and with as low a PAPR as the single-carrier modulation. Information and/or security data (i.e., text, image, or video message) are firstly converted to a stream of binary bits (i.e., A converted to 01000001) and input to the transmitter side of Figure 3 to be processed before being intensity-modulated and transmitted to the Rx by the LED headlight. In order to increase the transmitted data rate, these binary bits are converted to parallel bits and mapped to PAM symbols, where the order of PAM depends on the SNR level (i.e., a low SNR requires a low PAM order, and vice versa). For example, $[0, 1, 0, 0, 0, 0, 0, 0, 1]$ is mapped to $[-1, 1, -1, -1, -1, -1, -1, 1]$ or $[-1, -3, -3, -1]$ for 2- and 4-PAM mapping symbols, respectively. To reduce the ISI as well as to transmit symbols even beyond the 3 dB LED modulation bandwidth, these PAM symbols pass to the IFFT operation. As such, the symbol time duration is now greater than the maximum time delay spread duration of the VLC-V2X channel (i.e., the signal bandwidth is divided into several sub-bands, where each sub-band is less than the VLC-V2X channel coherence bandwidth).

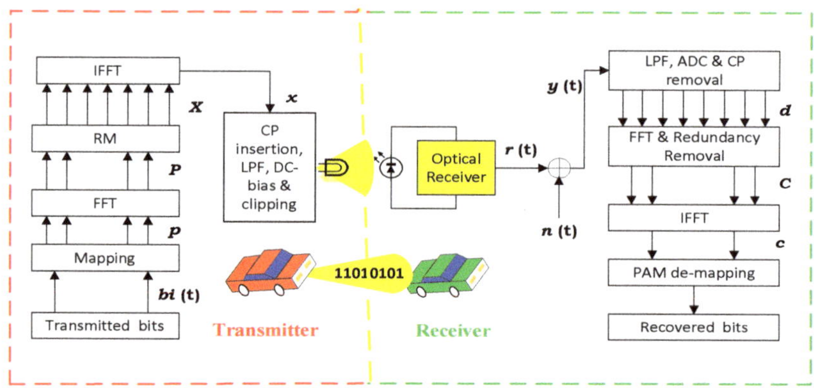

Figure 3. DFTS-OPAM transceiver block diagram.

However, implementing IFFT operation to a number of PAM symbols results in a complex and high PAPR time domain signal, while the LEDs used in cars' headlights have a limited linear dynamic range and only modulate the real time domain signals. To make the IFFT output time domain signal real with low a PAPR value, we inserted FFT and RM blocks before the implementation of the IFFT operation, as will be explained in more detail in the following subsection. As such, we introduced a real time domain signal with a low PAPR value that can overcome the ISI issue and transmit data even beyond the LED 3 dB modulation bandwidth. Finally, this sampled signal was converted to an analog signal before being intensity-modulated and transmitted as a light signal by the LED.

2.2. Workflow of DFTS-OPAM

The signal processing steps at the Tx are described as follows. First, the serial binary bits $bi\,(t)$ are converted into parallel data streams and mapped onto a group of real PAM symbols P as given by $p = [p_0,\ p_1,\ p_2 \ldots\ p_{M-1}]$, where M is the number of data symbols. The real symbols are then transformed to the frequency domain by being passed to the FFT implementation, as given by:

$$P_k = \sum_{m=0}^{M-1} p_m e^{\frac{-j2\pi mk}{M}},$$

(1)

where P_k is the data frequency domain at the kth subcarrier, $P = [P_0,\ P_1,\ P_2 \ldots\ P_{K-1}]$, and $K = M$ is the number of used subcarriers. Because the FFT inputs are PAM symbols, the output FFT data subcarriers are symmetrically conjugated around $P_{\left(\frac{K}{2}+1\right)}$, except the P_0, as was mathematically proved in [19] and is illustrated in Figure 4.

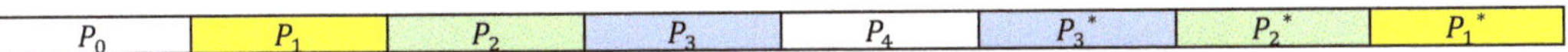

Figure 4. The subcarriers for eight real input samples at the output of the FFT module.

P is passed to the RM block, where the output vector X is a double of P, as illustrated in (2) and shown in Figure 5.

$$X = \{p_0,\ p_1,\ p_2,\ \ldots\ p_{M-1},\ p_0,\ p_1,\ p_2,\ \ldots\ p_{M-1}\},$$
$$X = \{X_0,\ X_1,\ X_2, X_3,\ X_4,\ ,,,,,,,,,,,,\ \ldots\ X_{N-1}\}.$$

(2)

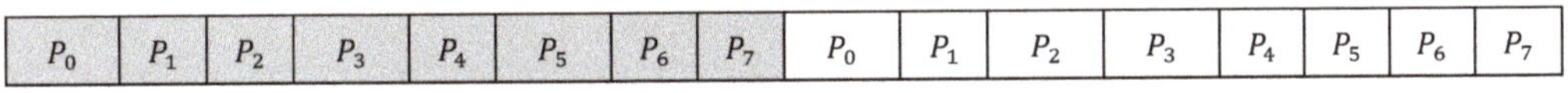

Figure 5. The subcarriers for eight real input samples at the RM module.

Then, X is converted back to the time domain vector x by being implemented in the IFFT operation. However, as was mathematically proved in [19], because of the FFT and RM processes, the even samples of x are a version of p, and the odd ones are zeros, as illustrated in (3). Indeed, x has the characteristics of a single carrier with a low PAPR.

$$x_n = \frac{1}{N} \sum_{l=0}^{N-1} X_l\, e^{\frac{j2\pi ln}{N}}$$

$$x_n = \frac{1}{N} \left| \sum_{l=0}^{\frac{N}{2}-1} X_l\, e^{\frac{j2\pi ln}{N}} + \sum_{l=\frac{N}{2}}^{N-1} X_l\, e^{\frac{j2\pi ln}{N}} \right|$$

(3a)

From Equation (3a), the following equation can be deduced:

$$x_n = \frac{1}{N} \left| \sum_{k=0}^{M-1} P_l\, e^{\frac{j2\pi nk}{N}} + \sum_{k=0}^{M-1} P_k\, e^{\frac{j2\pi n(M+k)}{N}} \right|$$

$$x_n = \frac{1}{N} \left| \sum_{k=0}^{M-1} P_l\, e^{\frac{j2\pi nk}{N}} + \sum_{k=0}^{M-1} P_k\, e^{j\left(\pi n + \frac{2\pi nk}{N}\right)} \right|$$

(3b)

Therefore, the odd and even samples of x can be, respectively, defined as:

$$x_{n\_Odd} = \frac{1}{N} \left| \sum_{k=0}^{M-1} P_l\, e^{\frac{j2\pi nk}{N}} - \sum_{k=0}^{M-1} P_l\, e^{\frac{j2\pi nk}{N}} \right| = 0$$

$$x_{n\_Even} = \frac{1}{2M} \left| \sum_{k=0}^{M-1} P_l\, e^{\frac{j2\pi nk}{M}} + \sum_{k=0}^{M-1} P_l\, e^{\frac{j2\pi nk}{M}} \right|$$

(3c)

$$x_{n\_Even} = \frac{1}{M} \sum_{k=0}^{M-1} P_l\, e^{\frac{j2\pi nk}{M}} \tag{3d}$$

where n is the nth sample of x, l is the lth subcarrier of X, and $N = 2M$ is the number of OFDM samples after the RM process.

It is important to note that in DFTS-OPAM, the FFT operation initially spreads the symbols across the subcarriers, and if the IFFT operation is directly applied without reordering any subcarriers, it would effectively undo the spreading, leading to a signal similar to conventional single-carrier modulation. To address this issue while ensuring a consistent PAPR and preserving the real-time requirements of VLC systems, the RM block was implemented between the FFT and IFFT blocks at the DFTS-OPAM Tx (please see Equation (3) and Figure 3).

Finally, x is passed through parallel to serial (P/S) converter, cyclic prefix (CP) insertion, digital to analog converter (DAC), low-pass filter (LPF), DC bias, and clipping processes before being intensity-modulated and transmitted by the LED.

Following optical detection, the received electrical signal is $y(t) = r(t) + n\,(t)$, where $r(t) = Rs(t) * h(t)$, R is the photodiode responsivity, $s\,(t)$ is the transmitted optical signal, the symbol $*$ denotes the linear convolution operation, $h\,(t)$ is the impulse response of the system, and $n\,(t)$ is the additive white Gaussian noise (AWGN). Note that, for the purpose of simplicity and without the loss of generality, we assume that $h\,(t) = R = 1$. Then, $y\,(t)$ is passed to LPF, analog to digital converter (ADC), CP removal, and serial to parallel (S/P) converter processes before being converted to the frequency domain by the FFT process. Finally, redundant subcarriers are removed and the result signal $C \approx P + n\,(t)$ is passed to the IFFT and PAM de-mapping blocks to recover the transmitted bits.

3. Simulation Results

In this study, we have focused on evaluating the performance of the proposed system using well-established simulation models. While we acknowledge that additional experimental results could provide more specific insights into the system's performance under different conditions, we believe that the simulation-based approach provides valuable and representative findings.

In these simulations, there were 256 IFFT points and 4-, 16-, 64- and 256-QAM constellation points for both DCO-OFDM and DFTS-OPAM. Note that for DFTS-OPAM, the PAM symbols were created by separating the real and the imaginary parts of QAM symbols (i.e., the QAM symbol $(a + ib)$ was separated into 'a' and 'b' PAM symbols, where these symbols were recovered and combined at the Rx to reconstruct the QAM). The LED_{3dbBW} was 10 MHz, the channel was considered as AWGN, and, to avoid the ISI, the CP duration (T_{CP}) as well as the subcarrier bandwidth (Sub_{BW}) of both schemes were chosen, as defined in (4) [20].

$$T_{CP} \geq T_{RMS} \geq \frac{1}{5\,(Sub_{BW})}, \tag{4}$$

where $T_{RMS} = 206.1$ ns is the root mean square time delay spread for the VLC-V2X multipath channel at an 18 m Inter-vehicular distance [21]. Finally, regarding the third generation partnership project (3GPP) standards, the EVM of 4-, 16-, 64- and 256-QAM should be less or equal to 17.5%, 12.5%, 8%, and 3.5%, respectively [22,23]. In this study, we defined these threshold values as EVM_{opt}.

Figure 6 shows that the probability of the PAPR values of both schemes is higher than a certain threshold level (i.e., $PAPR_0$), for which a complementary cumulative distribution function (CCDF) value of 10^{-4}, i.e., pr $\{PAPR > PAPR_0\} = 0.0001$, is considered [24,25]. In this figure, it is illustrated that DCO-OFDM has a 9 dB higher PAPR value compared with DFTS-OPAM. This PAPR improvement of the DFTS-OPAM scheme is due to the insertion of the FFT and the RM blocks prior to IFFT at the OFDM Tx.

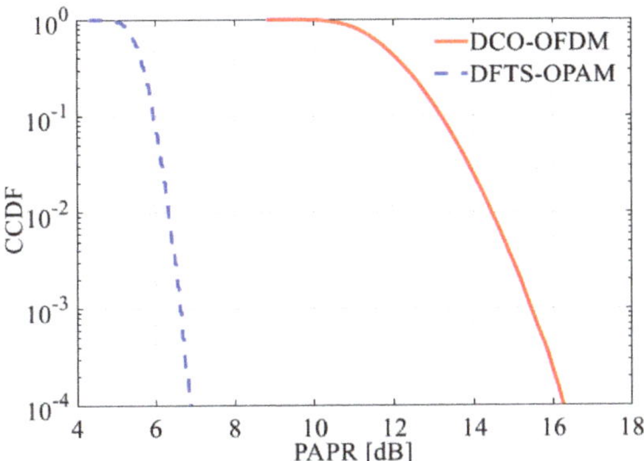

Figure 6. CCDF vs. PAPR for DCO-OFDM and DFTS-OPAM, where 256 IFFT points were considered.

The minimum achievable EVM% (EVM_{min}) of DCO-OFDM and DFTS-OPAM at different values of the LED-DR and QAM orders is investigated in Figure 7. As in [15], the average power of the AWGN (P_{awgn}) was set to -10 dBm and the average transmitted power (P_{ave}) of both schemes varied from 0 dBm to 30 dBm (i.e., 10 dB $\leq$ SNR $\leq$ 40 dB), where the EVM_{min} was achieved when the P_{ave} reached the maximum linear range of the LED, as the clipping noise occurred just after this value and, consequently, the EVM% started increasing again. The figure illustrates that implementing 256-, 64-, 16- and 4-QAM for DCO-OFDM or DFTS-OPAM requires an LDE with a linear dynamic range greater or equal to 3, 1.3, 0.8, and 0.5 V for the former and 1.5, 0.65, 0.5, and 0.3 V for the later.

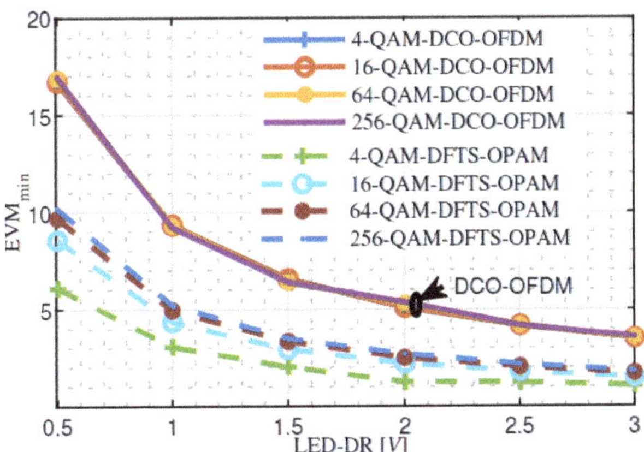

Figure 7. EVM_{min} versus LED-DR for DCO-OFDM and DFTS-OPAM.

The figure also depicts that changing the QAM orders of the DCO-OFDM scheme has an unnoticeable impact on the EVM_{min}, while it causes variations in EVM_{min} values in DFTS-OPAM. This can be justified in Figure 8, which illustrates the probability density function of the normal time domain signal amplitude for both techniques at modulation orders of 4 and 256. The figure shows that changing the QAM orders of DFTS-OPAM varies the standard deviation of distribution, while it remains constant regardless of the QAM orders in DCO-OFDM.

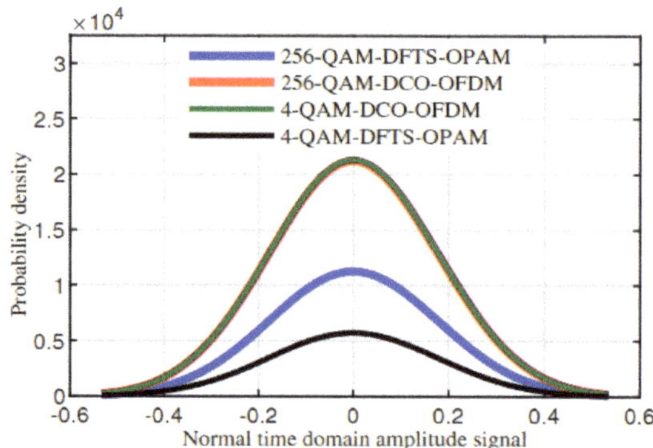

Figure 8. Probability density function of the normal time domain signal amplitude for DCO-OFDM and DFTS-OPAM at QAM modulation orders of 4 and 256.

As the main purpose of the headlight LED is to provide illumination, the diming control parameter should be considered in VLC systems. Diming control can be achieved by adjusting the DC bias above and beyond the middle point of the LED-DR, which is limited by the maximum and minimum values of the P_{ave} (P_{max} and P_{min}). Values of P_{max} and P_{min} for different QAM orders of both schemes are provided in Table 1. As such, the DC bias of both schemes can only be varied from P_{min} to P_{max}, as increasing P_{ave} above or below these values will introduce upper or below clipping noise, respectively.

Table 1. P_{max} and P_{min} of DCO-OFDM and DFTS-OPAM schemes.

		DCO-OFDM QAM Orders				DFTS-OPAM QAM Orders			
		4	**16**	**64**	**256**	**4**	**16**	**64**	**256**
LED-DR = 0.5	P_{max}	11.1	0	0	0	16.25	13	0	0
	P_{min}	8.75	0	0	0	5.7	8.68	0	0
LED-DR = 1	P_{max}	18.15	17	0	0	22.5	19.25	17.75	0
	P_{min}	8.75	11.65	0	0	5.7	8.68	12.25	0
LED-DR = 1.5	P_{max}	21.7	20.85	19.5	0	26.1	22.7	21.4	20.05
	P_{min}	8.75	11.65	15.6	0	5.7	8.68	12.25	20
LED-DR = 2	P_{max}	24.3	23.4	22.3	0	28.6	25.3	23.85	22.8
	P_{min}	8.75	11.65	15.6	0	5.7	8.68	12.25	20
LED-DR = 2.5	P_{max}	26.3	25.18	24.36	0	30.45	27.2	25.8	24.72
	P_{min}	8.75	11.65	15.6	0	5.7	8.68	12.25	20
LED-DR = 3	P_{max}	27.84	27	26	23.21	32	28.75	27.4	26.4
	P_{min}	8.75	11.65	15.6	23.2	5.7	8.68	12.25	20

The adjusted available power values ($P_{Just} = P_{max} - P_{min}$) of both schemes are illustrated in Figure 9. From the figure, it can be clearly noticed that increasing the LED-DR value as well as decreasing the QAM order of both schemes provides wider diming control. However, from the same figure, it can also be recognized that DFTS-OPAM outperforms DCO-OFDM in terms of supporting diming control. For example, the average power of DFTS-OPAM can be justified by 10 dBm around the LED-DR middle point without causing a clipping error when the 4-QAM order and the 0.5 V LED-DR are considered, while the average power of the DCO-OFDM can only be justified by 2 dBm around the LED-DR middle points for the same given assumptions, resulting in DFTS-OPAM providing wider dimming control compared with DCO-OFDM.

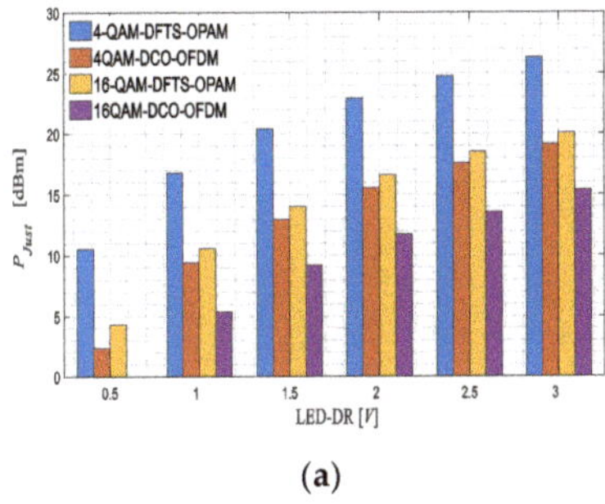

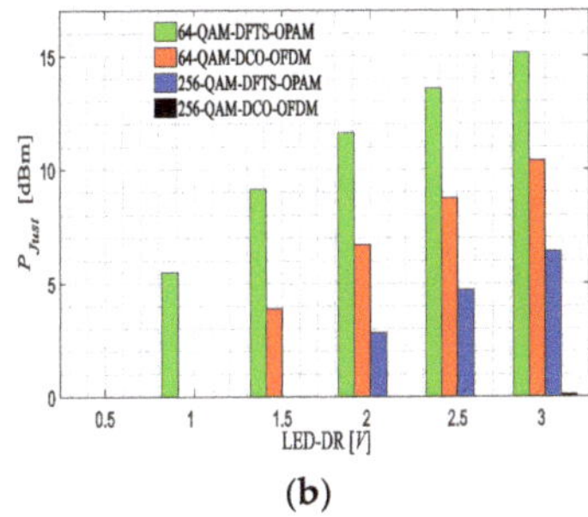

(a)   (b)

Figure 9. P_{Just} for DCO-OFDM and DFTS-OPAM, where different QAM orders and different LED-DR values are considered, (**a**) 4–16 QAM, (**b**) 64–256 QAM.

Figure 10 provides an example of how P_{min}, P_{max}, and EVM_{min} were measured in this paper. The figure illustrates EVM% against P_{ave} for 4 QAM DFTS-OPAM, where LED-DR = 1 V, P_{awgn} = -10 dBm, and P_{ave} varied from 0 dBm to 30 dBm. The figure shows that the EVM decays with increasing P_{ave} until it reaches EVM_{min}. After this turning point, EVM increases again as P_{ave} reaches the maximum value of the LED-DR, and hence, clipping noise occurs. P_{min} and P_{max} were achieved when EVM = EVM_{opt} = 17.5% before and after EVM_{min}, respectively.

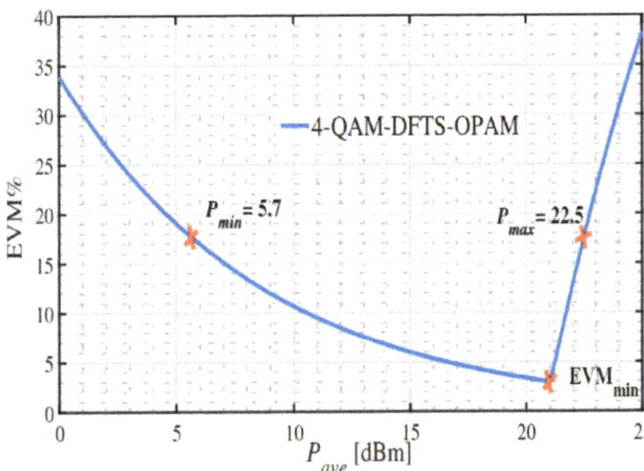

Figure 10. EVM versus P_{ave} for 4-QAM DFTS-OPAM, where 1 V LED-DR is considered.

Finally, Figure 11 illustrates the minimum required SNR to achieve EVM_{opt} versus signal bandwidth (S_{BW}) for different QAM orders of DCO-OFDM and DFTS-OPAM schemes where an LED_{3dbBW} of 10 MHz is considered, while the signal bandwidth of both schemes varied from 5 MHz to 30 MHz and the minimum required SNR to obtain EVM_{opt} for both schemes was measured at 5, 10, 15, 20, 25 and 30 MHz. The figure shows that when S_{BW} $\leq LED_{3dbBW}$ (i.e., $S_{BW} \leq 10$ MHz), DFTS-OPAM outperforms DCO-OFDM for all QAM orders, as it requires less SNR to achieve EVM_{opt} and that is because DFTS-OPAM has a lower PAPR.

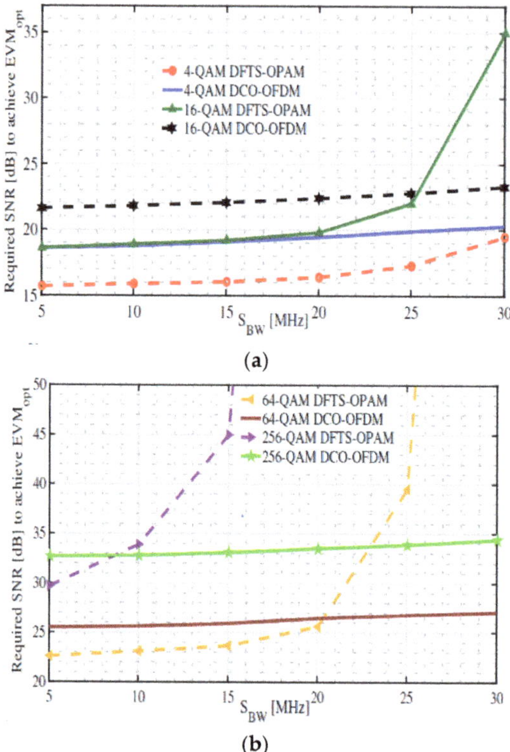

(a)

(b)

Figure 11. Required SNR to achieve EVM_{Opt} versus S_{BW} for different QAM orders of DCO-OFDM and DFTS-OPAM schemes where LED_{3dbBW} of 10 MHz is considered, (**a**) 4-16 QAM, (**b**) 64-256 QAM.

However, transmitting data beyond LED_{3dbBW} (i.e., $S_{BW} \geq 10$ MHz) more severely impacted the performance of DFTS-OPAM than DCO-OFDM, particularly at high QAM orders and high S_{BW} values. This is due to the FFT implementation at the DFTS-OPAM, as the errors that occurred from the subcarriers located beyond the LED_{3dbBW} will spread across all transmitted symbols in DFTS-OPAM, while they will only affect these subcarriers in DCO-OFDM. For instance, the 16-, 64- and 256-QAM DCO-OFDM outperformed the 16-, 64- and 256-QAM DFTS-OPAM when S_{BW} was greater or equal to 2.5 LED_{3dbBW}, 2 LED_{3dbBW}, and LED_{3dbBW}, respectively. However, as already illustrated, the P_{ave} (i.e., the SNR) of both schemes is limited by the limited LED-DR. For example, to obtain a 30 dB SNR value for $P_{awgn} = -10$ dBm, an LED with a 3 V and 1.5 V linear dynamic range is required for the DCO-OFDM and DFTS-OPAM schemes, respectively.

4. Conclusions and Future Work

In this paper, we introduced the DFTS-OPAM scheme as a multiple access scheme for visible light networking in connected vehicle systems, replacing the traditional DCO-OFDM. The decision to adopt DFTS-OPAM was motivated by its significant lower PAPR value in the time domain compared with that of DCO-OFDM. The system performance of the DCO-OFDM and DFTS-OPAM schemes under the influence of limitations such as the LED dynamic range and the 3 dB LED bandwidth were compared.

Simulation results demonstrated that DFTS-OPAM outperforms DCO-OFDM when considering a narrow LED-DR. Specifically, DFTS-OPAM requires an LED with a linear range that is half of what is needed for DCO-OFDM. Additionally, DFTS-OPAM showed superior performance in terms of supporting dimming control compared with DCO-OFDM, which struggled to support dimming control, especially with a narrow LED-DR or a high order of QAM. Furthermore, we observed that when both schemes transmitted signals below the 3 dB LED modulation bandwidth, DCO-OFDM outperformed DFTS-OPAM at high QAM orders, while DFTS-OPAM surpassed DCO-OFDM at low QAM orders.

In future research, the bit and power loading feature of both schemes will be investigated using artificial intelligence techniques. By leveraging the achieved diversity of this feature, we aim to enhance the physical layer security of the LiNEV framework systems. The utilization of artificial intelligence will enable us to optimize the allocation of bits and power across the system, thereby improving the overall security performance. This research will contribute to the development of more robust and secure LiNEV communication systems.

Author Contributions: Conceptualization O.S.; Formal analysis O.S.; Investigation, O.S.; Methodology, O.S.; Resources, O.K.; Supervision, O.K. and S.K.; Validation, O.K.; Writing, O.S.; Review and Editing, O.K., M.A., M.M., R.K., F.A.-S. and C.C.T.; All authors have read and agreed to the published version of the manuscript.

Funding: This research is funded by the QR-Fund of the B11 Unit of Assessment, Computing and Informatics Research Center, Department of Computer Science, Nottingham Trent University, UK.

Data Availability Statement: Research data will be available on individual requests to the corresponding author considering collaboration possibilities with the researcher or research team and with restrictions that the data will be used only for further research in the related literature progress.

Conflicts of Interest: The authors declare no conflict of interest.

References

1. Matheus, L.E.M.; Vieira, A.B.; Vieira, L.F.M.; Vieira, M.A.M.; Gnawali, O. Visible Light Communication: Concepts, Applications and Challenges. *IEEE Commun. Surv. Tutor.* **2019**, *21*, 3204–3237. [CrossRef]
2. Chi, N.; Wei, Y.; Hu, F. Visible Light Communication in 6G: Advances, Challenges, and Prospects. *IEEE Veh. Technol. Mag.* **2020**, *15*, 93–102. [CrossRef]
3. Cervinka, D.; Ahmad, O.S.; Rajbhandari, S. A Study of Yearly Sunlight Variance Effect on Vehicular Visible Light Communication for Emergency Service Vehicles. In Proceedings of the IEEE 12th International Symposium on Communication Systems, Networks and Digital Signal Processing (CSNDSP), Porto, Portugal, 20–22 July 2020.
4. Kaiwartya, O.; Kumar, S. Guaranteed geocast routing protocol for vehicular adhoc networks in highway traffic environment. *Wirel. Pers. Commun.* **2015**, *83*, 2657–2682. [CrossRef]
5. Khasawneh, A.M.; Helou, M.A.; Khatri, A.; Aggarwal, G.; Kaiwartya, O.; Altalhi, M.; Abu-Ulbeh, W.; AlShboul, R. Service-centric heterogeneous vehicular network modeling for connected traffic environments. *Sensors* **2022**, *22*, 1247. [CrossRef] [PubMed]
6. Lee, I.E.; Sim, M.L.; Kung, F.W.L. Performance enhancement of outdoor visible-light communication system using selective combining receiver. *IET Optoelectron.* **2009**, *3*, 30–39. [CrossRef]
7. Islim, M.S.; Videv, S.; Safari, M.; Xie, E.; McKendry, J.J.D.; Gu, E.; Dawson, M.D.; Haas, H. The impact of solar irradiance on visible light communications. *J. Light. Technol.* **2018**, *36*, 2376–2386. [CrossRef]
8. Bian, R.; Tavakkolnia, I.; Haas, H. 15.73 Gb/s Visible Light Communication with Off-the-Shelf LEDs. *J. Light. Technol.* **2019**, *37*, 2418–2424. [CrossRef]
9. Chronopoulos, S.K.; Vasilis, C.; Giorgos, T.; Tatsis, G. Performance of turbo coded OFDM under the presence of various noise types. *J. Wirel. Pers. Commun.* **2016**, *87*, 1319–1336. [CrossRef]
10. Dissanayake, S.D.; Armstrong, J. Comparison of ACO-OFDM, DCO-OFDM and ADO-OFDM in IM/DD Systems. *J. Light. Technol.* **2013**, *31*, 1063–1072. [CrossRef]
11. Chronopoulos, S.K.; Tatsis, G.; Vasilis, R.; Kostarakis, P. Enhanced PAPR in OFDM without deteriorating BER performance. *J. Wirel. Pers. Commun.* **2011**, *4*, 164. [CrossRef]
12. Zhang, X.; Han, L.; Wang, J.; Ling, X.; Gao, X. PAPR Reduction under EVM Constraint in DCO-OFDM Systems. In Proceedings of the IEEE 21st International Conference on Communication Technology (ICCT), Tianjin, China, 13–16 October 2021.
13. Lin, B.; Tang, X.; Yang, H.; Ghassemlooy, Z.; Zhang, S.; Li, Y.; Lin, C. Experimental Demonstration of IFDMA for Uplink Visible Light Communication. *EEE Photonics Technol. Lett.* **2016**, *28*, 2218–2220. [CrossRef]
14. Mesleh, R.; Elgala, H.; Haas, H. LED nonlinearity mitigation techniques in optical wireless OFDM communication systems. *IEEE OSA J. Opt. Commun. Netw.* **2012**, *4*, 865–875. [CrossRef]
15. Saied, O.; Ghassemlooy, Z.; Zvanovec, S.; Kizilirmak, R.C.; Lin, B. Unipolar-pulse amplitude modulation frequency division multiplexing for visible light communication systems. *Opt. Eng.* **2020**, *59*, 096108. [CrossRef]

16. Wu, C.; Zhang, H.; Xu, W. On visible light communication using LED array with DFT-Spread OFDM. In Proceedings of the IEEE International Conference on Communications (ICC), Sydney, NSW, Australia, 10–14 June 2014.
17. Saied, O.; Ghassemlooy, Z.; Kizilirmak, R.C.; Dai, X.; Ribeiro, C.; Zhang, M.; Rajbhandari, S. Single carrier optical FDM in visible light communication. In Proceedings of the IEEE 10th International Symposium on Communication Systems, Networks and Digital Signal Processing (CSNDSP), Prague, Czech Republic, 20–22 July 2016.
18. Saied, O.; Ghassemlooy, Z.; Rajbhandari, S.; Burton, A. Optical single carrier-interleaved frequency division multiplexing for visible light communication systems. *Optik* **2019**, *194*, 162910. [CrossRef]
19. Saied, O.; Li, X.; Rabie, K.M. DFT Spread-Optical Pulse Amplitude Modulation for Visible Light Communication Systems. *IEEE Access* **2022**, *10*, 15956–15967. [CrossRef]
20. Turan, B.; Narmanlioglu, O.; Koc, O.N.; Kar, E.; Coleri, S.; Uysal, M. Measurement Based Non-Line-of-Sight Vehicular Visible Light Communication Channel Characterization. *IEEE Trans. Veh. Technol.* **2022**, *71*, 10110–10114. [CrossRef]
21. Turan, B.; Gurbilek, G.; Uyrus, A.; Ergen, S.C. Vehicular VLC Frequency Domain Channel Sounding and Characterization. In Proceedings of the IEEE Vehicular Networking Conference (VNC), Taipei, Taiwan, 5–7 December 2018.
22. Li, H.; Hu, R.; Yang, Q.; Luo, M.; He, Z.; Jiang, P.; Liu, Y.; Li, X.; Yu, S. Improving performance of mobile fronthaul architecture employing high order delta-sigma modulator with PAM-4 format. *Opt. Exp.* **2017**, *25*, 1–9. [CrossRef] [PubMed]
23. *ETSI TS 136 104*; Evolved Universal Terrestrial Radio Access (E-UTRA). Base Station (BS) Radio Transmission and Reception (3GPP TS 36.104 Version 8.6. 0 Release 8). ETSI: Sophia Antipolis, France, 2009. Available online: https://www.etsi.org/deliver/etsi_ts/136100_136199/136104/14.03.00_60/ts_136104v140300p.pdf (accessed on 1 May 2023).
24. Wang, M.; Jiang, Y.; Zhu, X.; Li, H.; Wang, T. Multi-Layer Superimposed PAPR Reduction for ACO-OFDM VLC Systems. In Proceedings of the IEEE International Conference on Communications (ICC), Seoul, Republic of Korea, 16–20 May 2022.
25. Zhang, T.; Ji, H.; Ghassemlooy, Z.; Tang, X.; Lin, B.; Qiao, S. Spectrum-Efficient Triple-Layer Hybrid Optical OFDM for IM/DD-Based Optical Wireless Communications. *IEEE Access* **2020**, *8*, 10352–10362. [CrossRef]

Article

An Experimental and Numerical Study of the Impact of Ambient Light of SiPMs in VLC Receivers

William Matthews and Steve Collins *

Department of Engineering Science, University of Oxford, Parks Road, Oxford OX1 3PJ, UK
* Correspondence: steve.collins@eng.ox.ac.uk

Abstract: Silicon photomultiplier's relatively large area and ability to detect single photons makes them attractive as receivers for visible light communications. However, their non-linear response has a negative impact on the receiver performance, including making them particularly sensitive to ambient light. Experiments and Monte Carlo simulations have been used to study this non-linearity. The resulting detailed understanding of the origins of the non-linear response leads to concerns over the accuracy of some previous simulations of SiPMs. In addition, it leads to simple methods to determine the maximum rate at which an SiPM can count photons and of determining the impact of a SiPMs non-linearity on its performance of a receiver. Finally, a method of determining which filters should be used to protect an SiPM from ambient light is proposed.

Keywords: visible light communications; SiPM; Monte Carlo

Citation: Matthews, W.; Collins, S. An Experimental and Numerical Study of the Impact of Ambient Light of SiPMs in VLC Receivers. *Photonics* **2022**, *9*, 888. https://doi.org/10.3390/photonics9120888

Received: 18 October 2022
Accepted: 18 November 2022
Published: 22 November 2022

Publisher's Note: MDPI stays neutral with regard to jurisdictional claims in published maps and institutional affiliations.

1. Introduction

Visible light communications (VLC) and optical wireless communications (OWC) have been proposed as approaches to increasing local wireless communications capacity using visible or optical wavelengths [1]. An important parameter for any communications system is the rate at which it makes errors. This is characterized by the bit error rate (BER), which depends upon the signal to noise ratio (SNR) at the output of the receiver. One approach to increasing the SNR of a VLC or OWC systems that are designed to operate at data rates of more than 100 Mbps is to use a silicon photomultiplier (SiPM) as a receiver [2–16]. These devices are arrays of microcells, containing a single photon avalanche diode (SPAD), and each microcell is designed so that an output pulse is generated whenever a photon initiates an avalanche event. It is the resulting ability to detect single photons which allows SiPM receivers to operate within a few photons per bit of the noise floor determined by Poisson statistics [5]. However, an intrinsic part of the microcell's photon detection mechanism is the quenching of the avalanche process by reducing the bias voltage across its avalanche photodiode (APD). After the avalanche process has been quenched, the microcell has to be recharged so that another photon can be detected. Unfortunately, this means that the SiPM has a non-linear response [4].

SiPMs are commercially available with different characteristics, including area, numbers of microcells, photon detection efficiencies (PDEs), recharge times and output bandwidths, which are expected to impact their performance in receivers. The performance of receivers containing SiPMs can be determined experimentally [3–7,9–16]. However, these experiments should be performed very carefully and are time consuming. In addition, other parts of the system, particularly the transmitter, can have a significant impact on the performance of a system. Furthermore, even when this does not happen, it can be difficult to separate the impact of different SiPM characteristics. Finally, it is not always possible to test a receiver in some environments, for example, outside. These issues mean that a model or simulation of a SiPM receiver can complement experimental results.

Previously, SiPMs have been modelled using equivalent circuits or Monte Carlo simulations [17]. However, numerical methods, including Monte Carlo simulations, have been preferred when the performance of SiPMs in receivers is modelled. Some Monte Carlo simulations have focussed on the impact of the SiPM's non-linear response on their ability to count photons [18]. However, this means that it is not necessary to take the finite width of the SiPMs output pulses into account. Furthermore, it was assumed that a microcell cannot detect photons whilst it is recovering [18]. Alternatively, the performance of SiPM receivers has been studied by evaluating relevant equations [19–22]. It should be possible to evaluate a series of equations in less time than it takes to perform Monte Carlo simulations. Unfortunately, sometimes these equations assume a feature, such as a digital output, which are not relevant to commercial SiPMs [19]. Alternatively, they are relevant to OFDM [22], which is not as energy efficient as on-off keying (OOK) [23] and currently gives a lower data rate than OOK at eye safe irradiances [14]. In other cases, the equations assume that, since the microcells are passively quenched, they are paralysable [18,20]. This assumption is correct when the microcells have a digital output [19], but, commercial SiPMs have analog outputs and they are therefore not necessarily paralysable. To create a simulation that is based upon the fewest possible assumptions, a Monte Carlo simulation of the physical processes in a SiPM has been created.

The parameters in the simulation are obtained from either the relevant data sheet or the experimental results. The results of the simulations are then validated by comparing them to the results of the experiments. In particular, they are compared to the measurements of the bias current needed to sustain an over-voltage on the SiPM and the impact of ambient light on the performance of receivers containing SiPMs. The simulation results are then used to show that microcells are able to detect photons whilst they are recharging. The simulation results also lead to a new simple method of predicting the impact of the non-linear response of the SiPM on receiver performance in ambient light and a method for selecting optical filters that should be used in receivers. In the future, it should be possible to use the Monte Carlo simulation to devise an efficient means of compensation for any SiPM non-linear caused by the transmitted data or to predict the performance of receivers containing existing or future SiPMs in a wide variety of situations.

This paper is organized as follows. Section 2 contains descriptions of the operation of a SiPM, the experimental procedure used to test receivers containing SiPMs and the Monte Carlo simulation. This is followed in Section 3 by the results of the experiments to determine the voltage dependence of the microcells PDE. This section also contains the results of the experiments to determine either the irradiance dependence current needed to sustain an over-voltage or the impact of ambient light on the performance of receivers containing SiPMs. In both cases, these experimental results are compared to the results of Monte Carlo simulations of the same experiments. Finally, Section 4 contains results which show that microcells can detect photons before they are fully charged. Results are also presented which show that despite this behavior, the maximum count rate of an SiPM can be determined using an equation that was derived assuming that there was a minimum time between photons that could be detected, a time previously known as the dead time. The non-linear response of the SiPM is then shown to arise from a combination of changes to the average PDE and microcell charge when photons are detected. This leads to a simple method of predicting the impact of the SiPM's non-linearity on the performance of a receiver in ambient light. Finally, this section includes a suggested method for selecting optical filters to use with SiPMs in receivers and a discussion of the possible future uses of the Monte Carlo simulation.

2. Experimental Procedure and Monte Carlo Simulation of SiPMs

2.1. Description of SiPMs and Their Response to Light

A SiPM is an array of microcells that is connected in parallel. Each microcell contains an APD which is biased above its breakdown voltage, $V_{\text{breakdown}}$, by an amount known as the over-voltage, V_{ov}. The probability that an avalanche will occur [24] means that the PDE of a microcell can be calculated using

$$PDE(\lambda, t) = PDE_{\max}(\lambda) \times (1 - \exp(-V_{\mathrm{ov}}(t)/V_{\mathrm{char}})) \qquad (1)$$

where $V_{\mathrm{ov}}(t)$ is the instantaneous over-voltage, $PDE_{\max}(\lambda)$ is the maximum possible PDE at a particular wavelength and V_{char} is a characteristic voltage at this wavelength for the APD.

If the over-voltage is positive and the microcell only contained an APD, then a photon could initiate a self-sustained avalanche event. This means that only one photon could be detected. This avalanche event therefore has to be quenched so that other photons can be detected. In the case of the commercially available SiPMs manufactured by Broadcomm, Hamamatsu and Onsemi, a resistor is placed in series with the APD within each microcell. Consequently, the current caused by an avalanche process results in a voltage drop across the resistor, which reduces the voltage across the APD. Once this voltage equals the APDs breakdown voltage, the self-sustained avalanche process is quenched. The capacitance in the microcell is then recharged via this resistor and the resistance between the microcell and the source of the SiPM bias voltage. This means that the recharging process can be represented by the equation

$$V_{\mathrm{ov}}(t) = V_{\mathrm{ov}}(1 - \exp(-t/\tau_{RC})) \qquad (2)$$

where t is the time since the avalanche process was quenched and τ_{RC} is the time constant for the recharging process. This time constant can be determined from individual pulses that occur when photons are detected and is typically tens of nanoseconds. If the capacitance of the microcell is C_{cell}, then the additional charge stored in the microcell will be $C_{\mathrm{cell}}V_{\mathrm{ov}}$. The results in Figure 1 show that the sensitivity of the PDE to the over-voltage means that it recovers more quickly than the over-voltage. Since the additional charge stored on the microcell is proportional to the over-voltage, the PDE also recovers more quickly than the additional charge stored in the microcell.

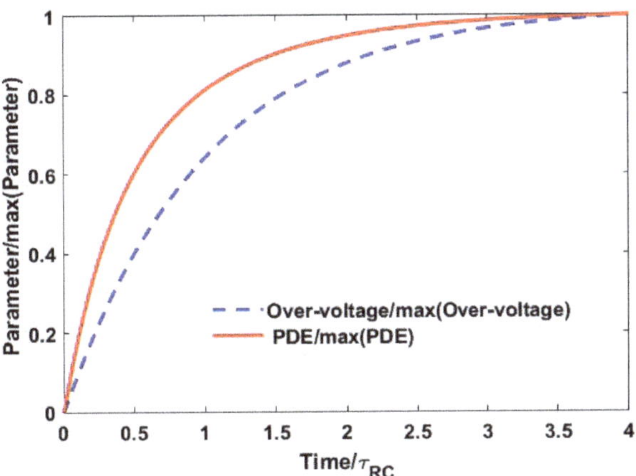

Figure 1. The recovery of the over-voltage and the photon detection efficiency determined using Equations (1) and (2).

Photons can be detected by monitoring the bias current that flows into the SiPM to recharge each microcell. This current flows because the microcell is discharged when it detects a photon, and since the amount of charge on the microcell is independent of the photon wavelength, the current is independent of the wavelength of the detected photon. If the interval between photons being detected by the SiPM is significantly longer than

τ_{RC}, then each detected photon results in a pulse with a fast rising edge, followed by the exponential decay expected from Equation (2). This mechanism can be used to detect and count photons using any of the commercially available SiPMs.

Figure 2 is a schematic diagram showing how a bias voltage was applied to a SiPM manufactured by Onsemi and how a digital multimeter was connected to measure the current flowing to sustain this voltage. This figure also shows that these particular SiPMs have an output known as the fast output. In addition, a second output can be created by placing a resistor between the SiPM's anode and ground. This output is equivalent to the output of SiPMs manufactured by other companies and it is possible to detect individual photons using this output. However, the width of the voltage pulses on this output is determined by the recharge time constant of the microcells. Since this time is longer than the fast output pulse width, this output is referred to as the slow output.

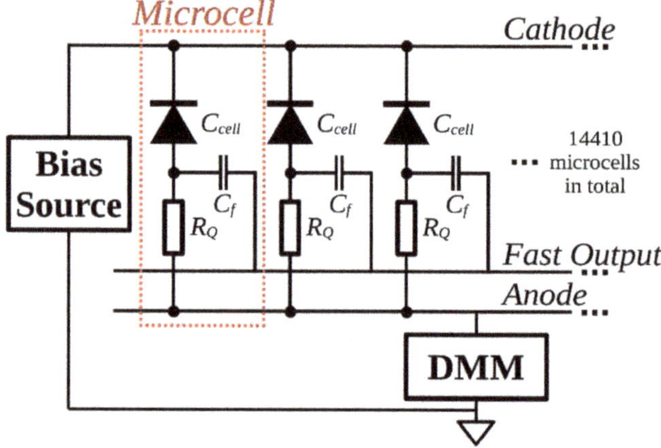

Figure 2. A schematic diagram showing three microcells in a representative Onsemi SiPM. The diagram also shows how these are connected the source of the bias voltage and a digital multimeter that is used to measure the bias current needed to sustain the bias voltage.

As shown in Figure 2, the fast output is created by capacitively coupling a common output to the connection between the APD and the quenching resistor in each microcell [25]. This capacitive coupling means that the signal on this fast output line is proportional to the rate of change of the voltage across the APD. The charging of the node between the APD and the resistor form the slow output pulses. This capacitance therefore means that the pulses on the fast output are a high pass filtered version of the slow output pulses. This removes the dc level component of the signal and explains why the fast output pulses are at least an order to magnitude narrower than the pulses on the slow output.

At low irradiances, each microcell has time to recover before the next photon is detected and the SiPM has a linear response. However, increasing the irradiance falling on the microcells reduces the average time between successive photons passing through each microcell. Eventually, photons arrive at microcells whilst they are still recharging. The result is that the SiPM has a non-linear response. Previously, this non-linear response has been observed by measuring the bias current needed to sustain the over-voltage on the SiPM as the irradiance falling on the SiPM is increased [10].

2.2. Experimental Procedure

A schematic diagram of the equipment used to characterise a SiPM and determine its performance as a VLC receiver is shown in Figure 3. Previously, experiments were performed with J series SiPMs mounted on SMA evaluation boards. These boards are convenient to use. However, they contain a resistor in series with the SiPM so that slow output pulses can be detected. Unfortunately, this resistor both increases the time needed for each microcell to recharge and decreases the effective over-voltage, and hence PDE, at high irradiances [16]. Since the fast output is used for data transmission experiments, this resistor is not needed. More recently, experiments have therefore been performed using a J series 30020 SiPM mounted on an SMPTA board, whose key characteristics are listed in Table 1. Without a resistor in series with the SiPM, the SMTPA boards have a shorter recharge time, and their PDE is not degraded at high ambient light levels. This means a SiPM on an SMTPA board is both easier to model and, more importantly, is a better receiver. As shown in Figure 2, in the absence of a resistor in series with the SiPM on the SMPTA board, the current needed to sustain the over-voltage was measured with a Keithley 196 digital multimeter.

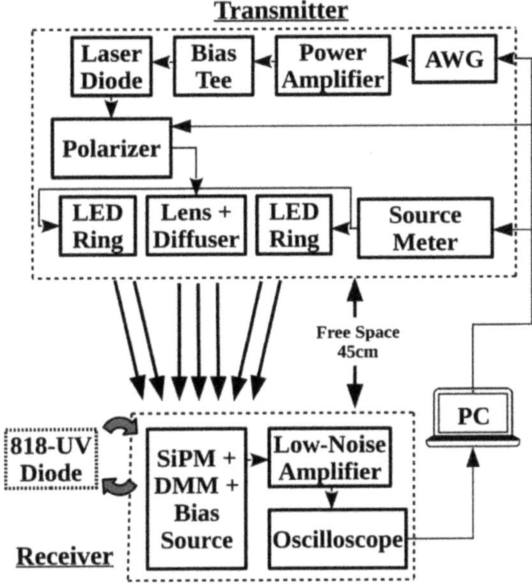

Figure 3. System block diagram describes the experimental setup used to evaluate the ambient light performance of the SiPM. The AWG was a 25 GS/s AWG70002A, the Power Amplifier is a Fairview FMAM3269 10 MHz to 6 GHz Amplifier, which feeds a Bias Tee (Thorlabs ZFBT-4R2GW+) and the Laser Diode a ThorLabs L405P20. The LED ring includes eight UV3TZ-405-15 LEDs, and is driven by a Keithley 224 Source Meter. During some experiments, the bias voltage applied to these 405 nm LEDs was varied to control the effective ambient light level. On the receiver side, the SiPM is coupled to a ZX60-43-S+ 4 GHz Low Noise Amplifier, which feeds a Keysight MSO64 (4 GHz, 25 GS/s) oscilloscope. The polarizer, source meter, AWG and oscilloscope are computer-controlled by MATLAB®.

Table 1. Key parameters obtained from the manufactures data sheet for a j series 30020 [26].

Parameter	30020
Number of microcells	14,410
Microcells active area diameter (μm)	20
Fill factor (%)	62
Recharge/recovery time constant (ns)	15
Dark Count Rate (MHz)	1.2 (@ 5 V_{ov})
Fast output pulse width (ns)	1.4

To obtain reproducible results from data transmission experiments, particular care had to be taken to minimize the impact of RF interference. When the beam from the transmitter to the receiver was blocked, a 5 mV$_{\mathrm{pp}}$ interference signal was initially observed. Since the signal when a photon was detected was 15 mV$_{\mathrm{pp}}$, this level of interference was unacceptable. A near field probe was therefore used to determine that the source of the interference was the transmitter. The optical cage system containing the transmitter and the SMA cable connecting the transmitter to the AWG was therefore covered with a metallized cloth and the probe was then used to confirm that this cloth prevented this type of interference.

Even with this precaution, it was sometimes impossible to obtain reproducible BER measurement results consistently. By watching the oscilloscope as it captured data, it became clear that a 20 mV$_{\mathrm{pp}}$ signal occurred frequently enough to explain the difficulties in reproducing results. A subsequent investigation showed that the frequency spectrum of this intermittent interference was consistent with it being caused by Wi-Fi and other RF signals transmitted by colleagues' electronic devices. The experimental procedure was therefore changed so that any data captured when there was a significant level of this interference was discarded and the data was transmitted again. However, there was so much interference during normal working hours that most results were captured overnight.

2.3. Monte Carlo Simulation of an SiPM

Results from experiments with a 30020 on an SMPTA board have been compared to results from a Monte Carlo simulation of this SiPM. These simulations were performed with a time variable that increased by the minimum of one twentieth of a nanosecond and one twentieth of the bit time of the OOK data. Since the charge on the microcell and the microcell's recovery are independent of the detected photons wavelength, all the simulated photons are assumed to have the same wavelength as the transmitters' output. This means that the impact of ambient light is represented by the irradiance at this wavelength, which gives rise to the same count rate.

In some simulations, the irradiance on the SiPM was assumed to be constant. However, when simulating data transmission experiments, the irradiance was modulated to represent OOK data. At each time, the instantaneous irradiance, the bit time and the Poisson probability density function

$$Poisson(n) = m^n e^{-m} / n! \tag{3}$$

were used to determine the number of photon incidents on the SiPM in a bit time, n, where m is the mean of the distribution. At a time, t, this mean was calculated using

$$m(t) = (L_{TX}(t) + L_{amb}(t)) A_{SiPM}.dt / E_p \tag{4}$$

where $L_{TX}(t)$ is the irradiance from the transmitter at time t and $L_{amb}(t)$ is the irradiance representing ambient light at the same time. In addition, dt is the time step used in the simulation, E_p is the energy of a photon from the transmitter and A_{SiPM} is the area of the SiPM. The n photons calculated using (3) and (4) were then randomly distributed in the bit time. This was done using a random number with a Poisson distribution so that the time between photons had the required exponential distribution.

Once the photon stream had been generated, an event-driven Monte Carlo simulation was started and the following quantities were calculated:

(i) The total charge on all microcells.

$$Q_{total}(t) = \sum_{n=1}^{Ncells} C_{cell} V_{ov}(n, t) \tag{5}$$

where C_{cell} is the capacitance of a microcell and $V_{ov}(n,t)$ is the over-voltage on the nth microcell at time t.

(ii) The average charge on the microcells that have detected a photon at this time. In this case, (5) is evaluated, but only the microcells that have detected a photon at this time are included in the summation. This sum is then divided by the number of microcells that have detected a photon at this time.

(iii) The instantaneous current needed to recharge each microcell was calculated by multiplying the increase in the over-voltage for each microcell since the previous time by the microcell capacitance and dividing the result by dt. The total current was then calculated by adding all these contributions; hence

$$I_{bias}(t) = \sum_{n=1}^{Ncells} \begin{cases} \frac{C_{cell}}{dt}(V_{ov}(n,t) - V_{ov}(n,t-dt)) & if\ V_{ov}(n,t) > V_{ov}(n,t-dt) \\ 0 & otherwise \end{cases} \qquad (6)$$

(iv) The proportion of microcells that are fully charged was calculated by determining the proportion of microcells whose over-voltage was more than 99% of the maximum over-voltage.

(v) The average PDE of all the microcells was determined using (1) the instantaneous PDE of each microcell and then calculating the average value.

The simulation started by initiating the microcells in the SiPM into a state that is consistent with the initial irradiance. The simulation was then evolved by up-dating the over-voltage and PDE of each microcell using Equations (1) and (2) until the time at which the next photon or photons are incident on the SiPM. At each of these times, the first step was to use a uniformly distributed random number to determine which microcell might detect the photon. The instantaneous PDE of the selected microcell and a second random number were then used to determine if the photon was detected. If the photon was detected, the over-voltage and PDE of the microcell were both instantaneously set to zero. In addition, the charge on this microcell was added to the sum of the charge on microcells that had detected a photon at this time. This process was then repeated for all photons incident on the SiPM at the same time. Once the process of detecting photons at a particular time had been completed, all the quantities of interest were calculated. The simulation was then evolved until the time when the next photon or photons were incident on the SiPM. This process was then repeated until the end of the simulated time.

At the end of the simulation, the sum of the charge on microcells that detected photons at each time was convolved with a Gaussian kernel, which represented the fast output pulses. The result was a fast output pulse whose integral is proportional to the charge discharged by all the photons detected at a particular simulated time. If the incident irradiance was modulated to represent OOK data, the resulting simulated fast output was processed in the same way as the fast output from a SiPM in an experiment.

When writing the simulation, a decision was made not to include three non-ideal behaviors of SiPMs, specifically dark counts, after-pulsing and optical cross-talk. Dark counts are spontaneous avalanche events that occur in the dark and in the 30020 they occur at a rate of 1 MHz [26]. This is much smaller than the anticipated rate at which ambient light photons are detected and so it was not included in the simulation. After-pulsing occurs when a charge carrier initiates an avalanche event in the same microcell after being temporarily trapped in the high field region of the microcell [17]. Similarly, cross-talk occurs when a secondary photon produced by an avalanche event initiates an avalanche in another microcell either immediately or after a delay [17]. In a 30020, the cross-talk occurs after less than 7.5% of avalanche events and after-pulsing after less than 5% of avalanche events. It is not clear from this data if these effects needed to be included to achieve the required modelling accuracy. Furthermore, the data required to model the delays in these effects is not provided by the manufacturer. The pragmatic decision was therefore taken to create a numerical model that excluded these effects and then reconsider this decision once its results had been compared to experimental data. The results in Sections 3.3 and 3.4 suggest that it is not necessary to include these effects in the Monte Caro simulation.

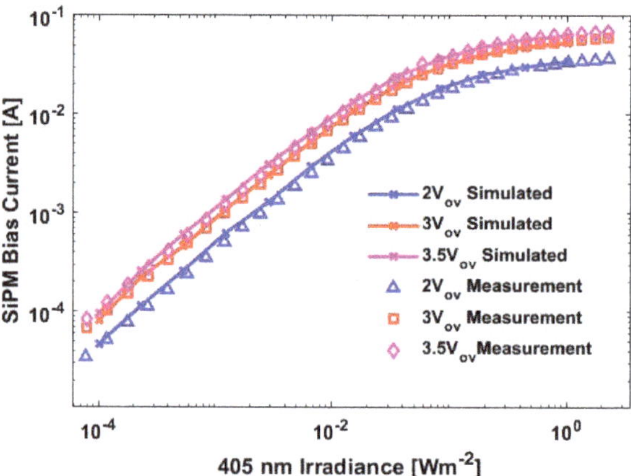

Figure 6. A comparison of the measured and simulated currents needed to sustain three different over-voltages on a 30020 SiPM.

3.4. Data Transmission Experiments in Ambient Light

Figure 7 shows the results of experiments to determine the irradiance from the transmitter required to achieve a BER of 3.8×10^{-3} when the ambient light irradiance increases. Eye safe transmitters have been described providing a radius of horizontal coverage in a typical office of 2 m and which provide a minimum transmitter irradiance at 405 nm of 2 mWm^{-2} [11]. Figure 7 shows that with this transmitter irradiance, it is possible to support data rates up to 1.5 Gbps with a BER of 3.8×10^{-3}. However, as the data rate increases, the ambient light irradiance which may be tolerated decreases. In particular, with a transmitter irradiance of 2 mWm^{-2}, ambient irradiances of up to the equivalent of 1 mWm^{-2}, 3 mWm^{-2} and 5 mWm^{-2} of 405 nm light are tolerated at 1.5 Gbps, 1 Gbps and 500 Mbps.

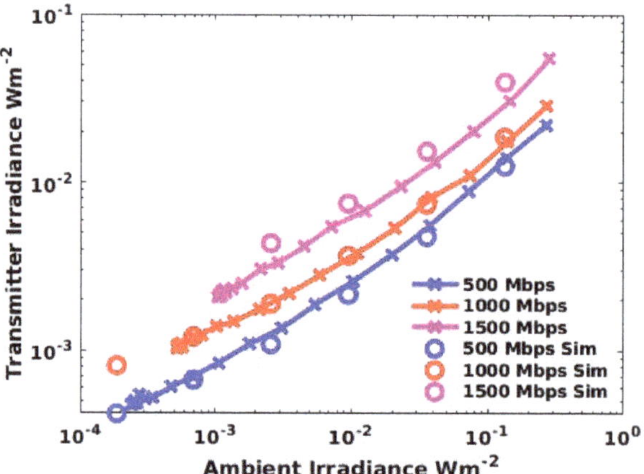

Figure 7. A comparison of the measured and simulated irradiances needed to support three data rates as the incident ambient light irradiance increased. The *x* axis is the equivalent 405 nm irradiance that generates the same count rate and hence bias current as the incident ambient light.

The dominant noise source in a SiPM receiver is expected to be Poisson noise. If this is the case, the BER when an on-off keyed signal is transmitted can be calculated using [5]

$$\text{BER} = \frac{1}{2}\left[\sum_{k=0}^{n_T} \frac{(N_{Tx} + N_b)^k}{k!} . e^{-(N_{Tx} + N_b)} + \sum_{k=n_T}^{\infty} \frac{(N_b)^k}{k!} . e^{-N_b} \right] \tag{11}$$

where N_b is the average number of photons detected per bit time when a zero is received, N_{Tx} is the number of additional detected photons per bit time needed from the transmitter when one is received and n_T is the threshold used to differentiate a one from a zero. The value of n_T that minimizes the BER has to be determined for particular combinations of N_b and N_{Tx}.

Equation (11) shows that the important parameters are the numbers of detected photons per bit when a zero and a one are received. These parameters have therefore been used as the axes in Figure 8 to show the results of experiments during which the ambient light level, and hence the number of photons detected when a zero is transmitted, was varied. As expected from (11), using this x-axis, the results for 500 Mbps and 1000 Mbps fall on the same curve. However, the results for 1500 Mbps suggest that there is a relatively small, but noticeable, power penalty for this data rate. This may be caused by the width of the SiPM fast pulses or the limited bandwidth of another part of the link.

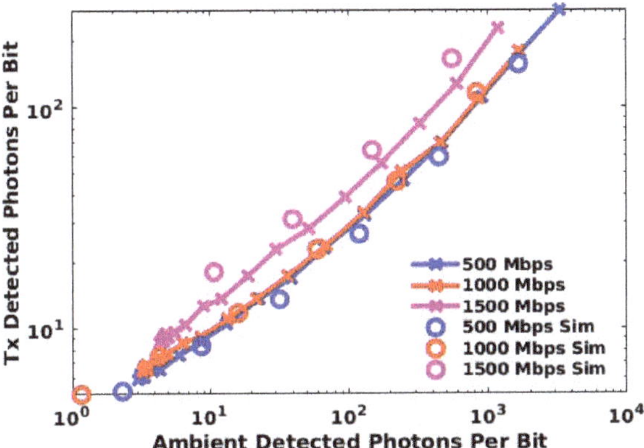

Figure 8. A comparison of the measured and simulated irradiances needed to support three data rates as the number of detected ambient light photons per bit time is increased.

In addition to the experimental results, Figure 8 also shows the results of Monte Carlo simulations of these experiments. Excellent agreement is obtained for data rates of 500 Mbps and 1000 Mbps. However, the agreement is not as good for 1500 Mbps. The simulation included the width of the fast output pulses and the difference between the simulated 1000 Mbps and 1500 Mbps results. These results show that the width of the output pulses is starting to have an effect at data rates above 1000 Mbps. The difference between the results from experiments and the simulations at 1500 Mbps must therefore be due to something that has not been included in the simulations, for example the bandwidth of the transmitter. More importantly, the results in Figure 8 confirm that, if the links performance is determined by the SiPM, then its performance can be predicted using this Monte Carlo simulation.

4. Discussion

4.1. The Origins of the SiPMs Non-Linearity

The count rate for a SiPM such as the 30020 can be related to the irradiance of monochromatic light falling on the SiPM, L, by [10]

$$C_{rate} = N_{cells}\alpha\, L / (1 + \alpha\, \tau_p\, L)$$

(12)

where N_{cells} is the number of microcells and τ_p is a characteristic time. In addition, the parameter α is

$$\alpha = \frac{\eta(V_{ov}, \lambda) A_\mu}{E_p}$$

(13)

where $\eta(V_{ov}, \lambda)$ is the photon detection efficiency of the SiPM at a particular over-voltage and wavelength and V_{ov}, is the over-voltage, E_p is the energy of each photon and A_μ is the active area of a microcell.

Equation (12) was suggested as a function which is consistent with the SiPM having a linear response at low irradiances and a saturated response at high irradiances. Furthermore, when (12) was suggested, it was assumed that each microcell cannot detect a photon whilst it was being recharged [4]. The latter assumption meant that previously the parameter τ_p was referred to as the dead-time for the microcell [4].

The assumption that a microcell cannot detect a photon until it is fully recharged means that a charge $C_{cell} V_{ov}$ is discharged when a photon is detected. Consequently, the bias current needed to sustain the over-voltage is

$$I_{bias} = C_{cell}\, V_{ov}\, N_{cells}\alpha\, L / \left(1 + \alpha\, \tau_p\, L\right) \tag{14}$$

Previously, this equation has been shown to agree with experimental results [10]. It therefore appears that the assumption that a microcell cannot detect a photon whilst it is recharging is correct and this assumption has been used to simulate SiPMs in receivers [19,22,27].

One advantage of developing a detailed Monte Carlo simulation is that it allows users to understand the physical processes occurring in microcells in detail. Figure 9 shows the behavior of a microcell when the average time between detected photons is longer than the time that the microcell needs to fully recharge. As expected, in these circumstances, the microcell is usually fully recharged before it detects a photon. However, the results in Figure 9 show a photon being detected when the microcell is only partly recharged. This event clearly shows that, despite the concept of dead time leading to an equation, Equation (14), that fits the measured bias current data, microcells can detect photons when only partially recharged. Some conclusions arising from any simulations which assume that microcells are unable to detect photons whilst they are recharging will therefore not be reliable. In addition, it should be possible to improve on any methods to compensate for the impact of the non-linearity which arises from these simulations.

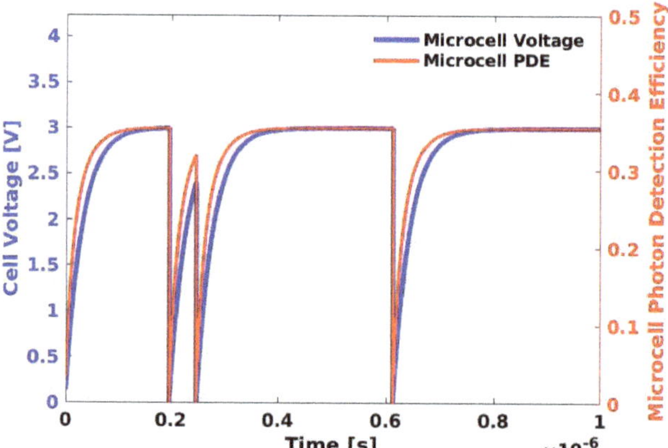

Figure 9. A representative microsecond of a simulation of one microcell showing the recovery of the over-voltage and PDE. In addition, these results show an example, at approximately 0.25 μs, of a photon being detected before the microcell is fully recharged.

4.2. A Simple Method of Estimating the Maximum Count Rate

Although the concept of dead time, which was part of the derivation of (14), is not accurate, this equation has been shown to agree with the measured bias current data. An important aspect of the derivation of Equation (14) [4] was that it assumed that there was a minimum time between photons that a microcell could detect, τ_p. However, this parameter was not related to the recharge time of the microcell and it was therefore used to fit (14) to a particular set of experimental data.

The reason why it has previously been possible to show agreement between (14) and the experimental results can be understood by considering the current flowing when the SiPM response is saturated. Saturation occurs when the denominator of (14) is dominated by the second term and the resulting current when the SiPM saturates is

$$I_{sat} = N_{cells} C_{cell} V_{ov} / \tau_p \tag{15}$$

This means that

$$(I_{sat}(V_{ov1}) / V_{ov1}) / (I_{sat}(V_{ov2}) / V_{ov2}) = \tau_p(V_{ov2}) / \tau_p(V_{ov1}) \tag{16}$$

Consequently, the ratio of characteristic times needed to fit (14) to bias currents measured at different over-voltages can be determined from (16). This ratio of characteristic times has been determined for a wide range of over-voltages. The results in Figure 10 show that, once the over-voltage is more than 1.5 V, this characteristic time is almost constant. This means that, for the range of over-voltages that are typically used, the maximum count rate of a SiPM can be estimated using

$$C_{max} = N_{cells} / \tau_p \tag{17}$$

where τ_p is approximately 2.2 times the RC time constant of the microcells [10].

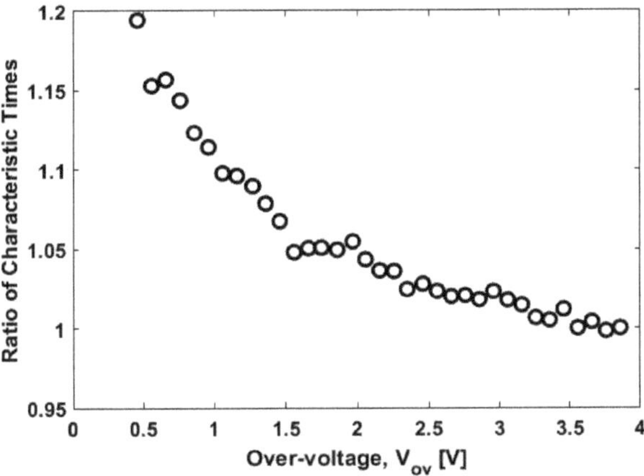

Figure 10. The ratio between the characteristic time obtained from the saturated current at each over-voltage to this time for an over-voltage of 3.85 V.

4.3. A Simple Method of Predicting the Impact of Ambient Light

The results in Figures 7 and 8 show that the results of the Monte Carlo simulations can be used to predict the results of data transmission over a wide range of ambient light conditions. However, each simulation can take an inconvenient time. An even simpler prediction method would therefore be advantageous. The experimental results for the two data rates, 500 Mbps and 1000 Mbps, for which the VLC systems performance is determined by the SiPM alone are shown in Figure 11. This figure also includes the performance of these systems predicted using the SiPM parameters and (11). The results in this figure show that the performance of the SiPM receiver at 500 Mbps and 1000 Mbps can be predicted using Poisson statistics until approximately 100 detected ambient light photons per bit. However, by 1000 detected ambient light photons per bit, there is an error of a factor of approximately two in the prediction. If the photon detection efficiency is 0.35, then 1000 detected photons per bit corresponds to an irradiance of 78 mWm^{-2}.

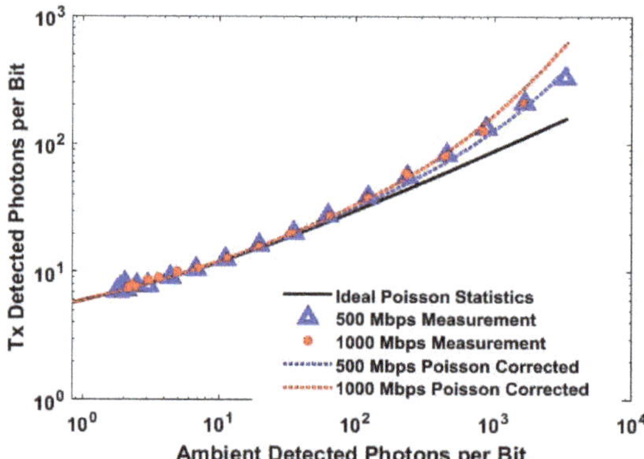

Figure 11. Experimental results for 500 Mbps and 1000 Mbps compared to the results expected from Poisson Theory and these results are combined with the correction, Equation (18). The x axis is the equivalent 405 nm irradiance that generates the same count rate and hence bias current as the incident ambient light.

Figure 9 shows that photons can be detected before a microcell is fully charged and, hence, whilst the microcells' PDE is less than its maximum value. Furthermore, as the irradiance increases, more microcells will detect photons whilst their PDE is less than the maximum. The average PDE of the array at times when photons are detected has been calculated for different simulated irradiances. The results in Figure 12 show that, as expected, when the irradiance is high enough, this array average PDE when any photon is detected decreases. This change in the array average PDE alone might explain the non-linear response of the SiPM. However, the irradiance at which the array average PDE falls to half its maximum value is 193 mWm^{-2}. In contrast, the current falls to half the value expected from its linear response when the irradiance is 73.6 mWm^{-2}. The change in the array average PDE when photons are detected cannot therefore be the only mechanism contributing to the SiPMs non-linear response.

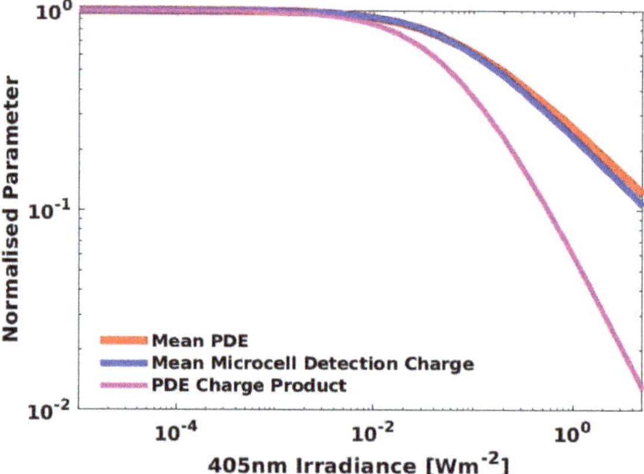

Figure 12. The mean array PDE when photons are detected and the mean charge on the microcells that have detected a photons.

An important assumption in the Monte Carlo simulation is that the height of the fast output pulse generated when a photon is detected is proportional to the charge on the microcell when that photon is detected. This means that the smaller charge stored on a microcell when it detects a photon before it is fully recharged may contribute to the non-linear response of both the bias current and the fast output used when the SiPM is a receiver. This may explain why the irradiance at which the transmitters sensitivity is half the expected value, 78 mWm^{-2}, is similar to the irradiance at which the measured bias current is half the expected value.

It appears that the charge stored when a photon is detected contributes to the SiPMs' non-linearity. The average charge stored on a microcell when it detects a photon has therefore been calculated at different irradiances. The results in Figure 12 show that this effect is as significant as the change in the array average PDE when a photon is detected. Consequently, when the two processes are taken into account, the average signal per incident photon falls to half its maximum value at an irradiance of 65 mWm^{-2}, which is much closer to the irradiance at which the bias current is half the value expected from its linear response.

The origin of the fast output pulses and the results in Figure 12 suggest that the non-linearity in the bias current should also have an impact on the performance of the SiPM as a receiver. In this case, the impact of the non-linear SiPM response on the performance of a VLC system can be predicted by multiplying the predictions from Poisson statistics by a correction factor

$$1 + \alpha\, \tau_p\, L \tag{18}$$

The results in Figure 11 show that with this correction, the experimental results for 500 Mbps and 1000 Mbps can be predicted accurately under a wide range of ambient light conditions.

4.4. Selecting Optical Filters for Operation in Ambient Light

Results such as those in Figure 7 show that even in the presence of a significant amount of ambient light, data rates up to at least 1500 Mbps can be received. However, the noise added by the ambient light increases the irradiance from the transmitter required to achieve a particular combination of BER and data rate. In addition, at high ambient light irradiances, the non-linear response of the SiPM can cause an additional increase in the required transmitter irradiance. This means that the SiPM should be protected from ambient light using optical filters.

In the past, optical filters with narrow pass-bands have been used to protect SiPMs from ambient light [5,6,10]. However, they restrict the receiver's field-of-view. Consequently, optical filters which absorb light and which support wider fields of view are preferred [11]. The first priority when selecting filters should be to limit the impact of the SiPMs non-linearity. Equation (18) is valid for monochromatic light and the equivalent equation for ambient light would need to take into account the spectrum of the ambient light and the wavelength dependence of the SiPMs PDE. However, this non-linearity affects the bias current. Consequently, the effectiveness of filters can be determined by measuring the bias current for a particular SiPM and ambient light source when different filters, or combinations of filters, are placed in front of the SiPM. If the ambient light is strong enough to force the SiPM into its non-linear region, the first priority is to use filters that reduce its impact so that the impact of the SiPM's non-linearity is reduced. The non-linearity will double the required transmitter irradiance when

$$\alpha\, \tau_p\, L_{\text{eff}} = 1 \tag{19}$$

where L_{eff} is the 405 nm irradiance that gives the same bias current as the ambient light. At this irradiance, the bias current is half the maximum bias current. The first aim should be to ensure that the non-linearity increases the required transmitter irradiance by a factor of two or less. This means reducing the measured bias current to less than half its maximum value. However, if the bias current can be reduced to less than one tenth of its maximum value, then the non-linearity is only increasing the required irradiance by approximately 10%, and may therefore be considered negligible.

A potential problem with aiming to reduce the bias current using filters is that it may require filters that also attenuate the wavelength used to transmit data. Even when filters are used in high levels of ambient light, the number of detected photons per bit will probably be high enough for the Poisson distribution to be approximated by a normal distribution. If this is the case, the noise caused by the ambient light will be proportional to the square-root of the rate at which ambient light photons are detected. This means that if using a filter reduces the bias current by a factor of $1/n$, then the signal to noise ratio, and hence bit error rate, will be maintained if the filter also reduces the bias current from the transmitter alone by a factor of $1/\sqrt{n}$. This means that it is not always necessary to use optical filters which transmit all of the photons from the transmitter.

4.5. Future Work

In the future understanding of the origins of the SiPMs, non-linear responses obtained from Monte-Carlo simulations could be used to develop methods to accommodate this non-linear response when it is caused by the transmitted data rather than by ambient light. This situation will most often arise when orthogonal frequency division multiplexing (OFDM) is used as a modulation scheme. In OFDM, data is transmitted by modulating several orthogonal carriers. This increases the amount of data that can be transmitted in the system's bandwidth. However, adding subcarriers means that OFDM has a high peak transmitted power. Furthermore, the process of separating the subcarriers relies upon the assumption that the system has a linear response. At the moment, the state-of-the-art method of dealing with the SiPM non-linearity when OFDM is employed is to use a Volterra series non-linear equalizer [13,14]. However, this standard adaptive method relies upon a large number of parameters. In the future, the understanding of the origins of the SiPMs non-linearity arising from the Monte Carlo simulations might lead to the development of a specific method to deal with the SiPM non-linearity when OFDM is being used. This would hopefully be simpler to implement and/or improve the systems performance when compared to the existing state-of-the-art system.

If SiPMs become the photodetectors of choice in receivers, then manufacturers will need to determine the relative importance of SiPM parameters such as PDE, number of microcells, recovery time and output pulse width. The impact of these parameters could be investigated experimentally using those SiPMs that are already commercially available. However, experiments are difficult to perform reliably, and the available SiPMs represent a small range of possible parameter values and other parts of the system, in particular the transmitter, which can have an impact on the experimental results. These considerations mean that the best way to compare the performance of SiPMs with different parameter combinations is using a detailed numerical simulation, which has been shown to generate results which agree with experimental results.

Author Contributions: Conceptualization, W.M. and S.C.; methodology, W.M.; software, W.M.; validation, W.M. and S.C.; investigation, S.C.; data curation, W.M.; writing—original draft preparation, S.C.; writing—review and editing, W.M. and S.C.; supervision, S.C.; project administration, S.C.; funding acquisition, S.C. All authors have read and agreed to the published version of the manuscript.

Funding: This research has been supported by the UK Engineering and Physical Sciences Research Council (EPSRC) under Grant EP/R00689X/1.

Data Availability Statement: The data is available from steve.collins@eng.ox.ac.uk.

Conflicts of Interest: The authors declare no conflict of interest.

References

1. Haas, H.; Elmirghani, J.; White, I. Optical Wireless Communication. *Philos. Trans. R. Soc. A* **2020**, *378*, 20200051. [CrossRef] [PubMed]
2. Khalighi, M.-A.; Hamza, T.; Bourennane, S.; Leon, P.; Opderbecke, J. Underwater Wireless Optical Communications Using Silicon Photo-Multipliers. *IEEE Photon. J.* **2017**, *9*, 1–10. [CrossRef]
3. Leon, P.; Roland, F.; Brignone, L.; Opderbecke, J.; Greer, J.; Khalighi, M.A.; Hamza, T.; Bourennane, S.; Bigand, M. A new underwater optical modem based on highly sensitive Silicon Photomultipliers. In Proceedings of the OCEANS 2017, Aberdeen, UK, 19–22 June 2017. [CrossRef]

4. Zhang, L.; Chitnis, D.; Chun, H.; Rajbhandari, S.; Faulkner, G.; O'Brien, D.; Collins, S. A Comparison of APD- and SPAD-Based Receivers for Visible Light Communications. *J. Light. Technol.* **2018**, *36*, 2435–2442. [CrossRef]
5. Ahmed, Z.; Zhang, L.; Faulkner, G.; O'Brien, D.; Collins, S. A Shot-Noise Limited 420 Mbps Visible Light Communication System using Commercial Off-the-Shelf Silicon Photomultiplier (SiPM). In Proceedings of the 2019 IEEE International Conference on Communications Workshops (ICC Workshops), Shanghai, China, 20–24 May 2019.
6. Ahmed, Z.; Singh, R.; Ali, W.; Faulkner, G.; O'Brien, D.; Collins, S. A SiPM-Based VLC Receiver for Gigabit Communication Using OOK Modulation. *IEEE Photonics Technol. Lett.* **2020**, *32*, 317–320. [CrossRef]
7. Zhang, L.; Tang, X.; Sun, C.; Chen, Z.; Li, Z.; Wang, H.; Jiang, R.; Shi, W.; Zhang, A. Over 10 attenuation length gigabits per second underwater wireless optical communication using a silicon photomultiplier (SiPM) based receiver. *Opt. Express* **2020**, *28*, 24968. [CrossRef] [PubMed]
8. Khalighi, M.A.; Akhouayri, H.; Hranilovic, S. Silicon-Photomultiplier-Based Underwater Wireless Optical Communication Using Pulse-Amplitude Modulation. *IEEE J. Ocean. Eng.* **2019**, *45*, 1611–1621. [CrossRef]
9. Tang, X.; Zhang, L.; Sun, C.; Chen, Z.; Wang, H.; Jiang, R.; Li, Z.; Shi, W.; Zhang, A. Underwater Wireless Optical Communication Based on DPSK Modulation and Silicon Photomultiplier. *IEEE Access* **2020**, *8*, 204676–204683. [CrossRef]
10. Matthews, W.; Ahmed, Z.; Ali, W.; Collins, S. A 3.45 Gigabits/s SiPM-Based OOK VLC Receiver. *IEEE Photonics Technol. Lett.* **2021**, *33*, 487–490. [CrossRef]
11. Ali, W.; Faulkner, G.; Ahmed, Z.; Matthews, W.; Collins, S. Giga-Bit Transmission Between an Eye-Safe Transmitter and Wide Field-of-View SiPM Receiver. *IEEE Access* **2021**, *9*, 154225–154236. [CrossRef]
12. Li, Y.; Hua, Y.; Henderson, R.K.; Chitnis, D. A Photon Limited SiPM Based Receiver for Internet of Things. In Proceedings of the 2021 Asia Communications and Photonics Conference (ACP), Shanghai, China, 24–27 October 2021. [CrossRef]
13. Zhang, L.; Jiang, R.; Tang, X.; Chen, Z.; Chen, J.; Wang, H. A Simplified Post Equalizer for Mitigating the Nonlinear Distortion in SiPM Based OFDM-VLC System. *IEEE Photon. J.* **2021**, *14*, 1–7. [CrossRef]
14. Huang, S.; Chen, C.; Bian, R.; Haas, H.; Safari, M. 5 Gbps Optical Wireless Communication using Commercial SPAD Array Receivers. *Opt. Lett.* **2022**, *47*, 2294–2297. [CrossRef] [PubMed]
15. Li, Y.; Chitnis, D. A real-time SiPM based receiver for FSO communication. In Proceedings of the Next-Generation Optical Communication: Components, Sub-Systems, and Systems XI, San Francisco, CA, USA, 3 March 2022.
16. Matthews, W.; Collins, S. The negative impact of anode resistance on SiPMs as VLC receivers. In Proceedings of the 2022 17th Conference on Ph. D Research in Microelectronics and Electronics (PRIME), Sardinia, Italy, 12–15 June 2022. [CrossRef]
17. Acerbi, F.; Gundacker, S. Understanding and simulating SiPMs. In *Nuclear Instruments and Methods in Physics Research Section A: Accelerators, Spectrometers, Detectors and Associated Equipment*; Elsevier: Amsterdam, The Netherlands, 2019; Volume 926, pp. 16–35.
18. Gnecchi, S.; Dutton, N.A.W.; Parmesan, L.; Rae, B.R.; McLeod, S.J.; Pellegrini, S.; Grant, L.A.; Henderson, R.K. A Simulation Model for Digital Silicon Photomultipliers. *IEEE Trans. Nucl. Sci.* **2016**, *63*, 1343–1350. [CrossRef]
19. He, C.; Ahmed, Z.; Collins, S. Signal Pre-Equalization in a Silicon Photomultiplier-Based Optical OFDM System. *IEEE Access* **2021**, *9*, 23344–23356. [CrossRef]
20. Huang, S.; Safari, M. Hybrid SPAD/PD Receiver for Reliable Free-Space Optical Communication. *IEEE Open J. Commun. Soc.* **2020**, *1*, 1364–1373. [CrossRef]
21. Huang, S.; Safari, S. SPAD-Based Optical Wireless Communication With Signal Pre-Distortion and Noise Normalization. *IEEE Trans. Commun.* **2022**, *70*, 2593–2605. [CrossRef]
22. Zhang, L.; Jiang, R.; Tang, X.; Chen, Z.; Li, Z.; Chen, J. Performance Estimation and Selection Guideline of SiPM Chip within SiPM-Based OFDM-OWC System. *Photonics* **2022**, *9*, 637. [CrossRef]
23. Hinrichs, M.; Berenguer, P.W.; Hilt, J.; Hellwig, P.; Schulz, D.; Paraskevopoulos, A.; Bober, K.L.; Freund, R.; Jungnickel, V. A Physical Layer for Low Power Optical Wireless Communications. *IEEE Trans. Green Commun. Netw.* **2020**, *5*, 4–17. [CrossRef]
24. Otte, A.N.; Garcia, D.; Nguyen, T.; Purushotham, D. Characterization of three high efficiency and blue sensitive silicon photomultipliers. *Nucl. Instruments Methods Phys. Res. Sect. A Accel. Spectrometers Detect. Assoc. Equip.* **2017**, *846*, 106–125. [CrossRef]
25. Onsemi.com 2022. Introduction to the Silicon Photomultiplier (SiPM) AND977O/D. Available online: https://www.onsemi.com/pub/Collateral/AND9770-D.PDF (accessed on 8 November 2022).
26. Onsemi.com. 2020. J-Series SiPM Sensors Datasheet. Available online: https://www.onsemi.com/pub/Collateral/MICROJ-SERIES-D.PDF (accessed on 20 June 2022).
27. He, C.; Lim, Y. Silicon Photomultiplier (SiPM) Selection and Parameter Analysis in Visible Light Communications. In Proceedings of the 31st Wireless and Optical Communications Conference (WOCC), Shenzhen, China, 11–12 August 2022. [CrossRef]

 *photonics*

Article

Performance Estimation and Selection Guideline of SiPM Chip within SiPM-Based OFDM-OWC System

Long Zhang, Rui Jiang *, Xinke Tang, Zhen Chen, Zhongyi Li and Juan Chen

Peng Cheng Laboratory, Shenzhen 518055, China
* Correspondence: jiangr01@pcl.ac.cn

Abstract: The orthogonal frequency division multiplexing (OFDM), which has high spectral efficiency, is an attractive solution for silicon photomultiplier (SiPM)-based optical wireless communication (OWC) systems to boost data rates. However, the currently available SiPMs are not optimized for implementing the OFDM receiver. Incorporating different types of SiPM at the OFDM receiver results in different data rates at the same condition. Therefore, the receiver designer requires a method for predicting the performance of SiPMs and then selects the best one to build the optimum receiver. In this paper, we first investigate the origin of SiPM's power-dependent frequency response. The investigation outcome is then used to create a method for predicting the subcarrier SNR. Combining the estimated subcarrier SNR with the bit-loading scheme, we finally propose a general approach for estimating the fundamental OFDM data rate an SiPM chip can support at a given received power. Results are then presented that can be used by the future receiver designer as a guideline to find the best type of SiPM to build the optimum OFDM receiver.

Keywords: silicon photomultiplier; optical wireless communication; OFDM

Citation: Zhang, L.; Jiang, R.; Tang, X.; Chen, Z.; Li, Z.; Chen, J. Performance Estimation and Selection Guideline of SiPM Chip within SiPM-Based OFDM-OWC System. *Photonics* **2022**, *9*, 637. https://doi.org/10.3390/photonics9090637

Received: 5 August 2022
Accepted: 30 August 2022
Published: 5 September 2022

Publisher's Note: MDPI stays neutral with regard to jurisdictional claims in published maps and institutional affiliations.

1. Introduction

Optical wireless communication (OWC), which uses the unlicensed spectrum to provide high-speed wireless communication, is expected to play an important role in the 6G network [1]. The OWC system requires a highly sensitive photodetector [2]. The P-i-N photodiode (PIN PD) and avalanche photodiode (APD) are the commonly used photodetectors in OWC. Unfortunately, the PIN has no internal gain, and the APD generates excess shot noise, which limits their sensitivity. SiPM is an array of single-photon avalanche diodes (SPAD) with an internal gain larger than 10^6 and without excess noise [3,4]. Results show that the sensitivity of SiPM can approach the Poisson limit, which is significantly more sensitive than the two commonly used PDs in OWC [4,5]. With this advantage, incorporating SiPM in the OWC receiver can significantly improve the OWC system's achievable SNR at the same received power. However, SiPM suffers from dead time, during which the fired SPAD microcell is unable to detect another incident photon [3,4]. The dead time generates a nonlinear output response and a bandwidth limit, both of which limit the data rate that the SiPM-based receiver can achieve [5].

OFDM, which modulates the quadrature amplitude modulation (QAM) symbols on the orthogonal subcarriers, has a high spectral efficiency [6]. Using OFDM in an SiPM-based OWC system can maximize its achievable data rate. It has been demonstrated that the data rate that the SiPM-based OFDM-OWC system can support varies from 500 Mbps to 5 Gbps [7]. Although different signal optimization algorithms result in different data rate enhancement, the fundamental performance of an SiPM-based OFDM-OWC system is primarily determined by the SiPM chip used to build the receiver. A receiver designer has to be able to select the best type of SiPM from those that are available. Moreover, if SiPMs become an integral part of OWC receivers, their manufacturers will be interested in optimizing their performance for this sizeable potential market. To help future SiPM and receiver designers, this paper proposes a general approach to estimate the performance of OFDM receivers that incorporate an SiPM and points out a guideline to select and design the best type of SiPM for implementing the optimum OFDM receiver.

In Section 2, the origin of SiPM's power-dependent frequency response is investigated. A method for estimating the subcarrier SNR is then described in Section 3. In Section 4, an approach developed to estimate the data rate of the SiPM-based OFDM system is presented. The results obtained using this approach to estimate the performance of SiPMs with various numbers of SPADs, recovery time constant, and photon detection efficiency (PDE) are discussed in Section 5. Finally, conclusions are drawn in Section 6.

2. Power-Dependent Frequency Response of SiPM

SiPM has a power-dependent frequency response [8], and hence the frequency resource of an SiPM-based OFDM system can use changes with the received power. To build an accurate approach for estimating SiPM's performance within an OFDM system, the origin of SiPM's power-dependent frequency response has to be understood.

SiPM is a solid-state photon counter which consists of a number of parallel-connected photon-counting microcells. Each microcell includes a SPAD and a quenching resistor. All the microcells within SiPM share a common output. Once a photon is detected, a current pulse generated by the fired microcell is added to the common output. Hence, the total output signal of an SiPM is a sum of the current pulses associated with different fired microcells. At an incident light intensity L, the average number of fired SPAD microcells N_f within observation time T is given by [9]:

$$N_f = \frac{N \times T \times \lambda_{in}}{1 + \lambda_{in} \times \tau} \tag{1}$$

$$\lambda_{in} = \frac{A_{SPAD} \times \eta \times L}{E_{ph}} \tag{2}$$

where N is the total number of SPAD microcells within SiPM, A_{SPAD} is the total area of a single SPAD, E_{ph} refers to the single photon energy, λ_{in} is the arrived photon rate, η refers to the PDE, and τ is the dead time which is 2.2 times the recovery time constant of a single SPAD microcell, τ_d [9].

The most commonly used equivalent circuit model of an SiPM is illustrated in Figure 1 [10,11]. According to the status of each microcell, the SiPM is divided into active and passive components. In the active part, each fired SPAD is modeled as a resistor R_d in parallel with a capacitor C_d. The resistor represents the internal resistance of the diode space-charge and quasi-neutral region, and the capacitor represents the junction capacitance of the inner depletion layer [10,11]. In the passive part, the inactive SPAD is modeled as a capacitor only. The integrated quenching resistor is modeled by a resistor R_q associated with its parallel stray capacitance C_q. The recovery time constant τ_d is given by $\tau_d = R_q(C_d + C_q)$. A further parasitic capacitance C_g across the two-pixel terminals is also introduced in the equivalent model, accounting for the presence of the metal grid paths spanning over the entire surface of the semiconductor device [10]. When N_f photons are detected by an SiPM with N microcells, the resistor and capacitor values in Figure 1 are [10]:

$$C_{d,N_f} = C_d \times N_f, \qquad R_{q,N_f} = \frac{R_q}{N_f}, \qquad C_{q,N_f} = C_q \times N_f$$
$$C_{d,N_p} = C_d \times (N - N_f), \quad R_{q,N_p} = \frac{R_q}{(N - N_f)}, \quad C_{q,N_p} = C_q \times (N - N_f) \tag{3}$$
$$R_{d,N_f} = \frac{R_d}{N_f}$$

where N_f in the subscript represents the number of fired SPADs corresponding to the active components and N_p in the subscript refers to number of inactive SPADs corresponding to the passive components.

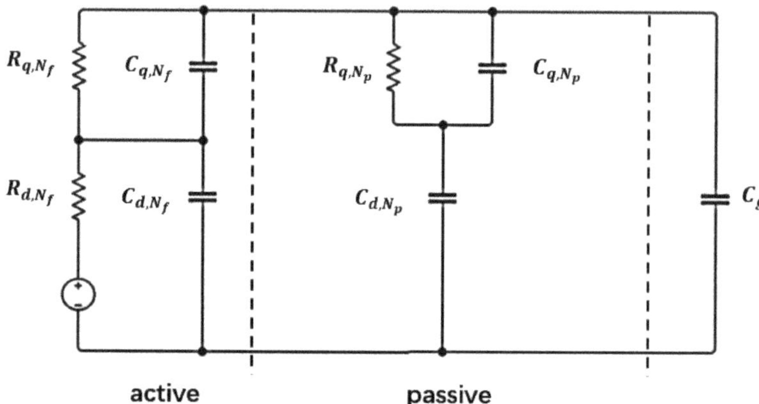

Figure 1. The equivalent circuit of an SiPM chip [10,11].

The transfer function of the SiPM can be modeled by an exponential decay [12]:

$$h(t) = \frac{1}{\tau_s} e^{-\frac{t}{\tau_s}} \tag{4}$$

The frequency response of the SiPM, $1/(j\omega\tau_s + 1)$, is therefore determined by the RC time constant τ_s. For OFDM transmission, the incident light power over the SiPM has to make SiPM's output pulses overlap to form a signal envelope for accurate signal detection. The envelope is proportional to the incident light level which makes the SiPM behave like a highly sensitive APD [3]. Since the output signal is contributed by the fired microcells, no current passing through R_{q,N_p} and hence the passive microcells can be regarded as a parasitic capacitor C_{eq}. The RC constant τ_s can therefore be expressed as:

$$\tau_s = \frac{R_{q,N_f} \times R_{d,N_f}}{R_{q,N_f} + R_{d,N_f}} \times (C_{q,N_f} + C_{d,N_f} + C_{eq} + C_g) \tag{5}$$

where C_{eq} is given by $C_{eq} = \left(C_{d,N_p} \times C_{q,N_p}\right) / \left(C_{d,N_p} + C_{q,N_p}\right)$.

The accuracy of Equation (5) has been validated using an off-the-shelf SiPM (On semiconductor 30035). The RC time constant of the tested SiPM was measured using LD LP520-SF15 and an oscilloscope (MSO7104B). The bandwidth of the tested LD is larger than 1 GHz. In the measurement, the LD is modulated using a rectangular waveform with a frequency of 100 kHz. The purple spots are the measured RC time constant of the tested SiPM at the standard output against different received powers, and the blue curve is the estimated RC time constant based on Equation (5). The detailed parameters of the tested SiPM are listed in Table 1, obtained from [13,14]. Results in Figure 2 suggest that Equation (5) can accurately estimate the power-dependent RC constant. In addition, since τ_s in Equation (4) is a function of the number of fired SPAD microcells, the frequency response of SiPM is power-dependent. With the number of fired SPADs increasing, more diode resistors become parallel and the number of diode parasitic capacitors is reduced, which makes the RC constant become reduced with the increased optical power.

Table 1. Key parameters in simulation [13,14].

Parameters	Value	Parameters	Value
C_d	160 fF	C_g	10 fF
C_q	11.6 fF	N	5676
R_q	200 kΩ	τ_d	45 ns
R_d	50 Ω	η	22%@520 nm
C_{fast}	2.5 fF	A_{SPAD}	3.07 $\times$ 3.07 mm^2

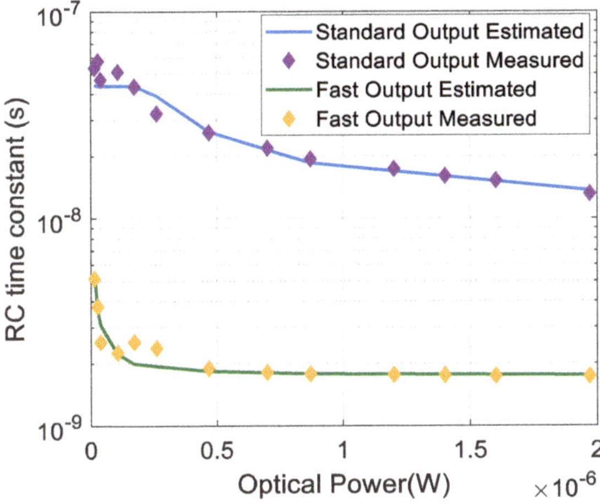

Figure 2. The measured RC time constant of the tested SiPM at different incident optical power using the standard output and the fast output.

According to the relationship between the 3 dB bandwidth and the RC time constant, $B_{3dB} = 1/(2\pi\tau_s)$ [12], the maximum 3 dB bandwidth of the tested SiPM is limited to ~10 MHz. To improve the time resolution, the manufacturer added a series-connected capacitor C_{fast} at the output, as shown in Figure 3. Because the equivalent diode capacitance is reduced by an order of magnitude, the SiPM using the fast output has a much smaller RC constant than using the standard output. The equivalent diode capacitance is:

$$C_{eq,d} = \frac{C_d \times C_{fast}}{C_d + C_{fast}} \tag{6}$$

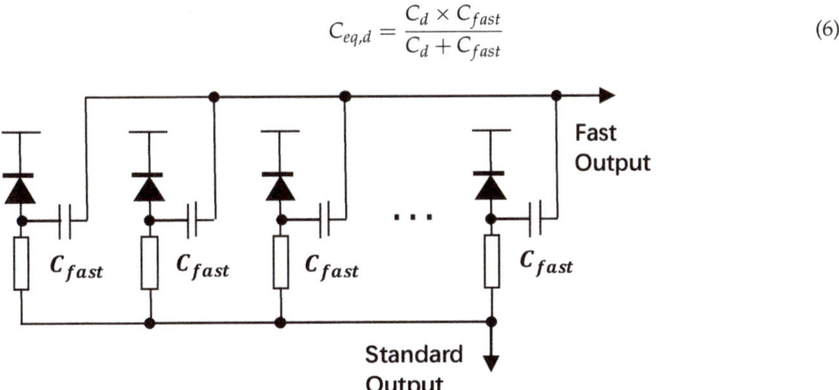

Figure 3. Schematic of SiPM with fast output [13].

The yellow spots in Figure 2 are the measured RC time constant of the tested SiPM using the fast output and the green curve shows the estimated RC time constant calculated by Equation (5) after substituting $C_{eq,d}$ into Equation (3). The agreement between the estimated results and the measured data indicate that Equations (4)–(6) can be used to estimate the frequency response of the tested SiPM when fast output has been used. Additionally, compared with the standard output, the 3 dB bandwidth was improved by a factor of 10 when fast output was used.

3. Method of Predicting Subcarrier SNR

For estimating the OFDM data rate, the SNR at each subcarrier is required. The estimated frequency response, which shows the signal power at different frequencies, can be used as prior knowledge for the SNR estimation.

The OFDM signal is a superposition of the modulated subcarriers. Without performing power loading, the transmitted power is equal across subcarriers. Regarding each subcarrier as an independent single carrier transmission, the number of photons detected by SiPM at an individual subcarrier within a sample duration can be estimated by Equation (1). The dominant noise source in an SiPM-based receiver is shot noise [3]. Assuming the channel response is flat, the SNR at each subcarrier can be expressed as [15]:

$$SNR = \frac{N_s}{\sqrt{N_s + N_b}} \tag{7}$$

where N_s is the number of detected signal photons and N_b is the number of detected background photons. However, in practice, the SNRs at different subcarriers degrade with the increase in frequency due to the channel response. Therefore, Equation (7) can only be valid for estimating the maximum subcarrier SNR.

The frequency response shows the signal power at different frequencies, which is related to the signal counts N_s at the SiPM-based receiver. When the background counts (N_b) are negligible, the SNR at the SiPM-based receiver is close to a square root of N_s according to Equation (7). Therefore, the normalized SNR at the SiPM-based receiver is a square root of the normalized channel gain in frequency response. For generating the optical OFDM signal, the clipping process is required to prevent the negative intensity and the nonlinear distortion at the cost of introducing the clipping noise and distortion, which also affects the subcarrier SNR [16]. Combining the maximum subcarrier SNR, the normalized SNR, and the correction factor caused by clipping, the SNRs at all the subcarriers $SNR(f)$ can be estimated, which is given by:

$$SNR(f) = 10 \times \log_{10} SNR_{\max} + 10 \times \log_{10} \sqrt{F(f)} + \alpha(f) \tag{8}$$

where $SNR_{\max}$ refers to the maximum subcarrier SNR. $F(f)$ is the normalized frequency response of the whole system which takes the frequency response of all the devices within the system into account. The frequency response of the SiPM at different received power can be estimated using Equations (4) and (5), and the frequency response of other devices can be obtained from the datasheet provided by the manufacturer; $\alpha(f)$ is the correction factor which is related to the impact of clipping. The fundamental performance of an SiPM-based OWC system is determined by the SiPM chip incorporated in the receiver. Optimizing the clipping level will provide additional SNR gain for a given SiPM [7]. Since this work concentrates on figuring out the SiPM chip design and selection guideline instead of investigating the signal optimization method, for simplicity, the clipping level that makes the signal distortion and clipping noise negligible is used in this study. Therefore, $\alpha(f)$ is set to 0 in this work.

In order to verify the accuracy of using Equation (8) for predicting the subcarrier SNR, an experiment was performed. The experimental setup is shown in Figure 4. For simplicity, the DC-biased optical OFDM (DCO-OFDM) is used in this work. Compared with the asymmetrically clipped optical OFDM (ACO-OFDM), DCO-OFDM has higher spectral efficiency. The DCO-OFDM signal with 512 subcarriers and a cyclic prefix (CP) length of 16 was generated in MATLAB. This signal was conditioned for transmission by clipping any values outside the allowed operating range. The clipping level was set to $\pm 2.5\sigma$ to minimize the signal distortion and noise from clipping [16,17]. The conditioned digital signal was then converted to the analog signal through AWG (AWG5202). The maximum amplitude of the output signal from the AWG was limited to 500 mV. A Mini-circuit ZHL6A+ amplifier with a 3 dB bandwidth of 500 MHz was used to maximize the transmitted signal. Combining with a DC bias (50 mA) at the LD mount, the amplified signal drove a 520 nm LD LP520-SF15 to generate the optical signal. An adjustable optical attenuator was placed after the LD to change the transmitted optical power. To reduce the impact of ambient light, this experiment was conducted in the dark, and a 1 nm optical filter with a 3 mm × 3 mm aperture was placed over the SiPM. The ambient light measured after the optical filter was less than 2 nW. The output signal from the SiPM's fast output was captured by an oscilloscope (MSO7104B). The captured signal was recovered back to the binary bits after synchronization, fast Fourier transform (FFT), equalization, and M-QAM demodulation. SNR estimation was implemented using the reference pilot symbols modulated in binary phase-shift keying (BPSK). The SNRs at different frequency subcarriers were obtained based on error vector magnitude estimation (EVM) [18].

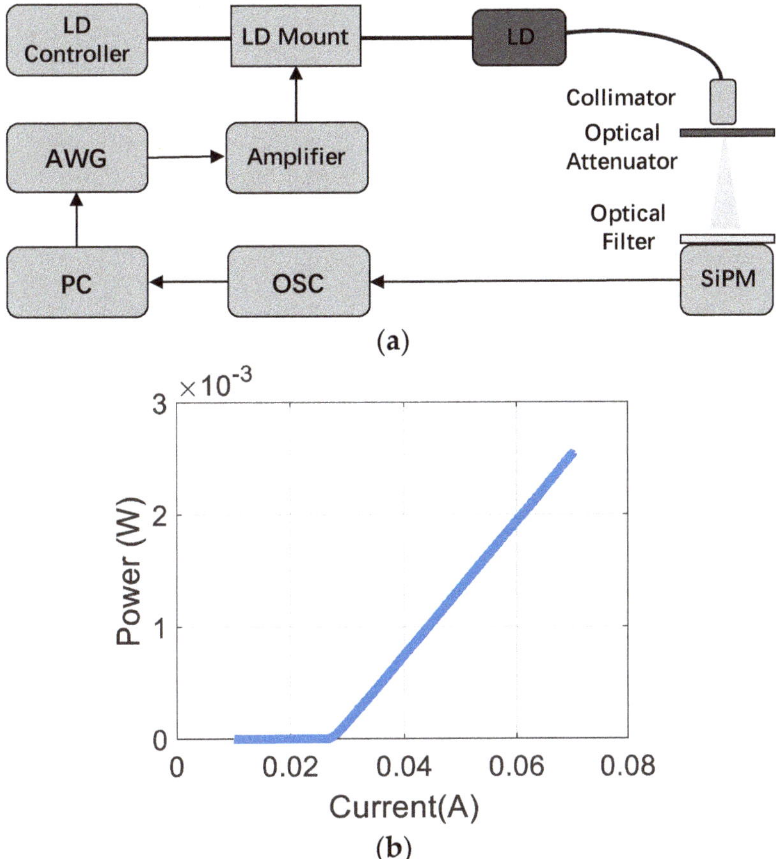

Figure 4. (a) Experimental setup (AWG: Arbitrary Waveform Generator, LD: laser diode, OSC: oscilloscope). (b) Measured optical power from the tested LD at different bias currents.

The discrete spots in Figure 5 are the measured maximum subcarrier SNRs achieved at various received optical powers. The solid curve in Figure 5 is calculated based on Equation (7). The good agreement between the estimated results and the measured data suggests that Equation (7) can be used to estimate the maximum subcarrier SNR when the clipping levels make the clipping noise and clipping distortion negligible.

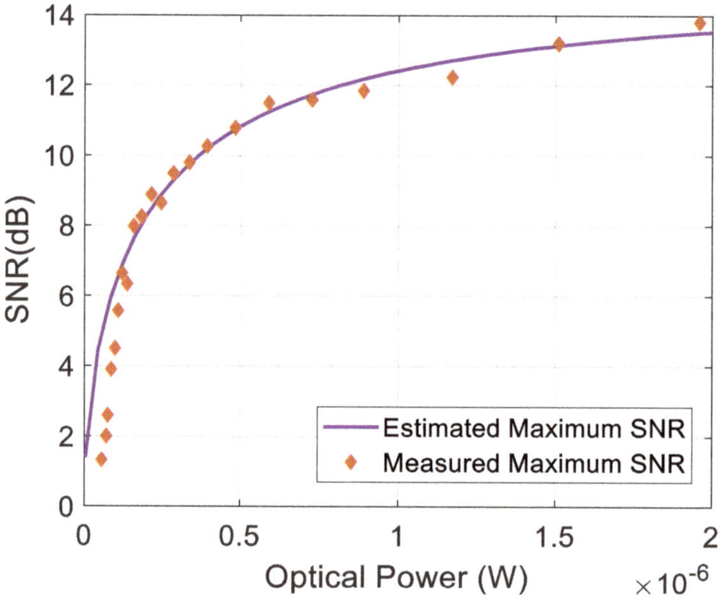

Figure 5. The maximum subcarrier SNR against various received optical power using a clipping level of ±2.5σ.

The dashed curve in Figure 6 shows the estimated SNRs of the tested system at different frequency subcarriers, which are calculated based on Equation (8). In order to obtain an accurate estimation of the tested system's channel response, the frequency response of the amplifier $G_a(f)$ and other electronics $G_o(f)$ used in Figure 4 was also taken into account. The fitting function for the tested amplifier and other electronics is given by:

$$G_a(f) = a_0 + a_1 \times \cos(wf) + b_1 \times \sin(wf)$$
$$+a_2 \times \cos(2wf) + b_2 \times \sin(2wf) \qquad (9)$$
$$+a_3 \times \cos(3wf) + b_3 \times \sin(3wf)$$

$$G_o(f) = a \times f^b + c \qquad (10)$$

where f is the frequency and the best fitting values to the measured $G_a(f)$ and $G_o(f)$ in dB are listed in Table 2. The correction factor α is set to 0 since the signal distortion and noise introduced by clipping can be ignored at the clipping level of ±2.5σ [16]. Results in Figure 6 show that the estimated subcarrier SNR approximately matches the measured results, which suggests that Equation (8) can be used to predict the subcarrier SNR.

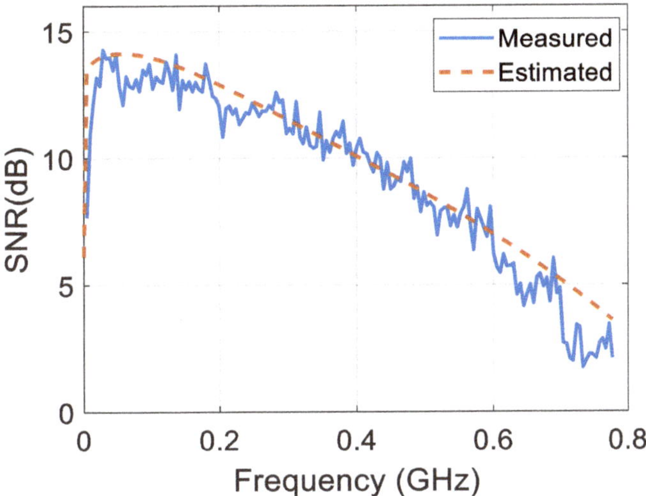

Figure 6. Subcarrier SNR measured at a received optical power of 2 μW using a clipping level of ±2.5σ.

Table 2. Fitting parameters.

Parameters	Value	Parameters	Value
a_0	−30.76	b_3	−3.961
a_1	22.15	w	1.618×10^{-9}
a_2	18.11	a	−61.55
a_3	10.08	b	−0.1706
b_1	−0.5856	c	3.275
b_2	−2.362	-	-

4. Data Rate Estimation and Verification

After applying the bit-loading algorithm to the estimated subcarrier SNR, the achievable OFDM data rate using an SiPM-based receiver can be predicted. The bit-loading process allocates different numbers of bits to the subcarriers according to the subcarrier SNR [18]. The SNR required to obtain a target BER at different QAM sizes is calculated based on Gaussian distribution. After the DFT process, the signal-dependent noise within the SiPM-based receiver becomes signal independent over the frequency subcarrier, making the QAM constellation diagram at each subcarrier also follow the Gaussian distribution [19]. Therefore, the typical bit-loading scheme can be directly used in the SiPM-based OFDM system. In this paper, the Levin–Campello (LC) algorithm is used for bit loading [18]. Table 3 lists the SNRs required by M-QAM to obtain a BER of 3.8×10^{-3}, considering an FEC overhead of 7% [20].

Table 3. SNR required by M-QAM to obtain a BER of 3.8×10^{-3}.

M-QAM	Required SNR (dB)
2-PSK	6.8
4-QAM	9.8
8-QAM	14.4
16-QAM	16.5

The spectral efficiency of the DCO-OFDM system after performing bit loading is given by [17]:

$$E = \frac{\sum_{k=0}^{N_{FFT}/2-1} \mathrm{sgn}(M_k)\log_2 M_k}{N_{FFT} + N_{CP}} \tag{11}$$

where N_{FFT} is the FFT size, M_k is the constellation size on the kth subcarrier after bit loading, N_{cp} is the size of the CP in the time domain, and $sgn(x)$ is the sign function. Then, the system's data rate is:

$$D = 2 \times B \times E \tag{12}$$

where B is the single-sided bandwidth of the system which can be calculated as $B = \frac{1}{2 \times T_s}$ and T_s is the sampling period.

A flow chart representing the approach used to estimate the data rate that an SiPM can support at a given received power with OFDM is shown in Figure 7. This approach starts with inputting the incident light intensity L, sample time T, the specification of a given SiPM of interest, and the frequency response of other devices within the system. Then, the number of SPADs fired within SiPM at a given received power is calculated based on Equations (1) and (2). The frequency response of the SiPM-based receiver achieved at a given received power can then be estimated using Equations (4) and (5). Combined with the maximum subcarrier SNR, the subcarrier SNR is calculated by Equation (8). After applying the bit-loading algorithm, the spectral efficiency and hence the data rate can be predicted based on Equation (12).

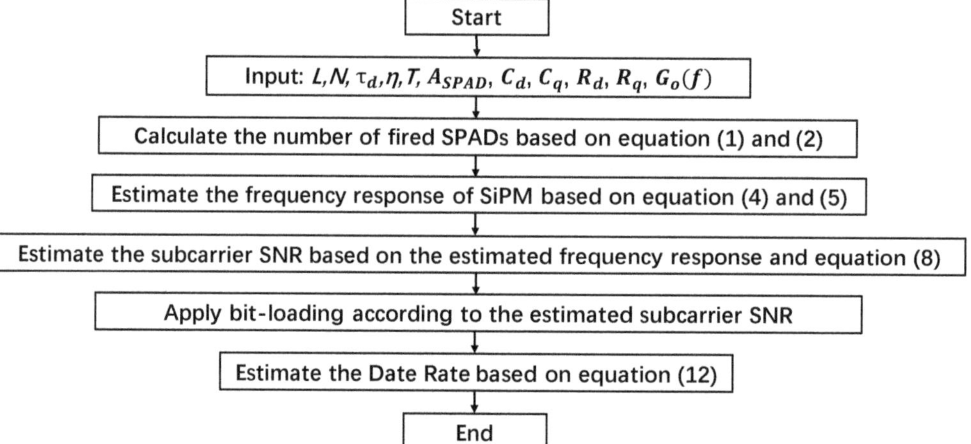

Figure 7. A general approach to estimate the achievable OFDM data rate using a SiPM-based OFDM receiver.

The dashed line in Figure 8 shows the estimated data rates at different received optical power calculated based on the data rate estimation approach shown in Figure 7. Due to that the number of allocated bits at a given subcarrier do not increase linearly with SNR, the dashed curve is not linear. The red spots are the measured results achieved by the tested system shown in Figure 4. The good agreement between the estimated and the measured results suggests that this approach can estimate the OFDM data rate achieved by an SiPM-based receiver at different received optical power.

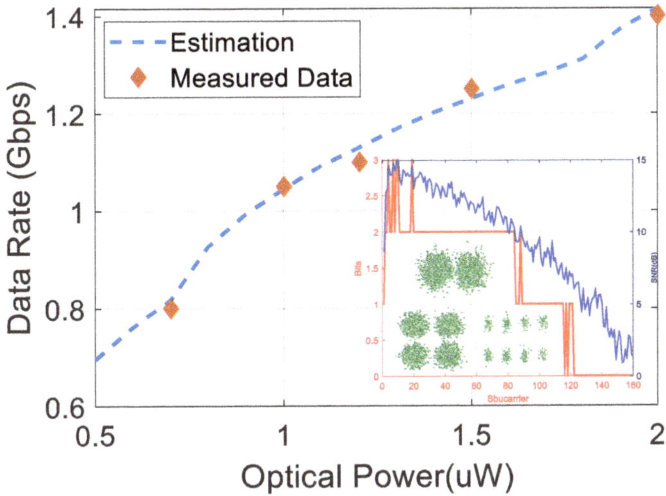

Figure 8. Data rate achieved at different received optical power with bit loading. The inset shows the bits assigned to each subcarrier achieving a data rate of 1.4 Gbps at the received optical power of 2 μW.

5. Discussion

Since SiPMs are a relatively new technology, the newly released product has significantly better characteristics than their predecessors, which offers the prospect of achieving even higher data rates at the same received power. The approach validated in Section 4 is used in this section to investigate the possible performance of SiPMs with a higher number of SPADs, shorter recovery time constant, and higher PDE. In the simulation, the same parameters and system configuration in Figure 4 are used.

Due to that each SPAD microcell is inactive during its dead time, SiPM has a nonlinear response. This nonlinear response limits the subcarrier SNR of the SiPM-based receiver according to Equations (1) and (7). An SiPM with a larger number of SPADs will have more available SPADs to detect the incident photons and hence will be less impacted by the dead time. Increasing the number of SPADs within SiPM can therefore increase the SNR at each subcarrier. However, this approach also reduces the bandwidth of the SiPM due to the increased number of passive microcells and their associated passive capacitance. Figure 9 shows the estimated data rate at a PDE of 22% with various recovery time constants against different numbers of SPADs. Results suggest that the bandwidth reduction is negligible compared with the SNR increment. For example, the SNR is improved by 3 dB at the reduction in the bandwidth by 1% when the number of SPADs increases from 5000 to 20,000. Consequently, increasing the number of SPADs within SiPM eventually boosts the data rate, as shown in Figure 9.

The recovery time constant, which determines the length of dead time, is a function of R_q, C_d, and C_q. Reducing the recovery time constant can therefore reduce SiPM's nonlinearity and increase SiPM's bandwidth simultaneously, according to Equations (1) and (5). At present, the typical recovery time constant of a commercially available SiPM varies from 15 ns to 50 ns [13]. Results in Figure 9 show the possible data rates can be achieved with typical recovery time constant values [13], which confirms that the data rate increases with the reduction in the recovery time constant.

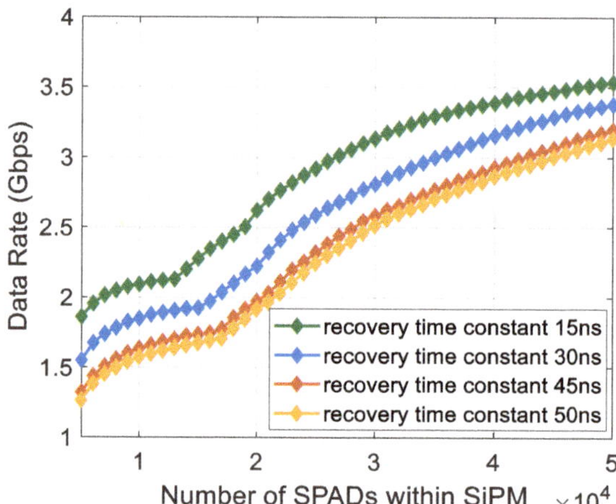

Figure 9. The data rate can be achieved with a larger number of SPADs and a shorter recovery time constant at a PDE of 22% with a received optical power of 2 μW.

Figure 10 shows the estimated data rate against different numbers of SPADs at a recovery time constant of 15 ns with different PDEs. Equation (7) suggests that increasing the number of received signal photon counts can improve the SNR. As a result, increasing SiPM's PDE improves the SNR and hence the achievable data rate at the same received power, as shown in Figure 10. In [7], a commercially available SiPM, which has a peak PDE of 48%, recovery time constant of 15 ns, and contains 1.5×10^4 SPADs, achieved a data rate of 3 Gbps at a received power of 2 μW operated at a PDE of 36% using DCO-OFDM. This experimental result is quite close to our prediction shown in Figure 10. The validated performance improvement confirms that the SiPM with a larger number of SPADs, shorter recovery time, and higher PDE can achieve a higher data rate. When the number of SPADs is higher than 4×10^4, at a recovery time constant of 15 ns and a PDE of 48%, the data rate of the SiPM-based OFDM-OWC system can be higher than 4 Gbps with the same received power. Under the same condition, the data rate that can be supported by conventional PIN is below 500 Mbps, and the typical APD is 3 Gbps [21]. This result suggests that the increased sensitivity made available by the use of SiPM can be exploited to boost the OWC system's data rate.

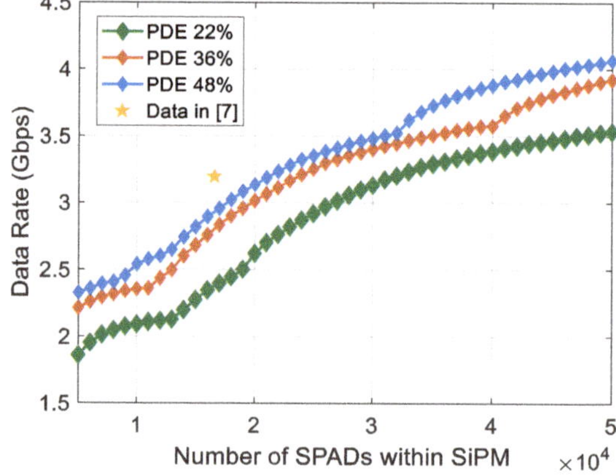

Figure 10. Data rate can be achieved by the SiPM with different PDEs at the received optical power of 2 μW [7].

6. Conclusions

In an SiPM-based OFDM-OWC system, the SiPM chip used at the receiver dominates the achievable data rate. The electrical property of SiPM determines the frequency resource used for transmitting the OFDM symbol, and the nonlinearity of the SiPM determines the subcarrier SNR.

In this paper, a model for estimating the power-dependent RC constant and hence the frequency response of the SiPM was derived according to SiPM's equivalent circuit model and showed to agree with results obtained from a commercially available SiPM. Since the normalized SNR at the SiPM-based receiver is a square root of the normalized channel gain in frequency response, a method for predicting the subcarrier SNR was then presented based on the estimated frequency response and the maximum subcarrier SNR. The subcarrier SNR estimation method validated in the paper was then used to create a general approach for predicting the achievable OFDM data rate using a given SiPM. Results presented in this paper show that this approach accurately predicts the OFDM data rate using an SiPM-based receiver.

The validated estimation approach was then used to predict the performance of SiPMs with a higher number of SPADs, shorter recovery time constant, and higher PDE. The results show that the best type of SiPM to use in OFDM receivers is one with a large number of SPADs, short recovery time constant, and large PDE. An SiPM with 4×10^4 SPADs, 15 ns recovery time, and 48% PDE should be possible to support a data rate of 4 Gbps. Further work is needed to confirm these predictions and to explore the possibility of achieving even higher data rates.

Author Contributions: Methodology, L.Z.; validation, Z.C. and Z.L.; data curation, J.C.; formal analysis, L.Z. and R.J.; writing—original draft preparation, L.Z.; writing—review and editing, L.Z., X.T., R.J., and J.C.; visualization, L.Z.; supervision, L.Z. and R.J.; funding acquisition, R.J. All authors have read and agreed to the published version of the manuscript.

Funding: This work was supported by the National Key Basic Research Program of China (62101291) and the PCL Key Project (PCL2021A05).

Institutional Review Board Statement: Not applicable.

Informed Consent Statement: Not applicable.

Data Availability Statement: Not applicable.

Conflicts of Interest: The authors declare no conflict of interest.

References

1. You, X.; Wang, C.X.; Huang, J.; Gao, X.; Zhang, Z.; Wang, M.; Huang, Y.; Zhang, C.; Jiang, Y.; Wang, J.; et al. Towards 6G Wireless Communication Networks: Vision, Enabling Technologies, and New Paradigm Shifts. *Sci. China Info. Sci.* **2021**, *64*, 1–74. [CrossRef]
2. Chi, N.; Zhou, Y.; Wei, Y.; Hu, F. Visible Light Communication in 6G: Advances, Challenges, and Prospects. *IEEE Veh. Technol. Mag.* **2020**, *15*, 93–102. [CrossRef]
3. Zhang, L.; Chitnis, D.; Chun, H.; Rajbhandari, S.; Faulkner, G.; O'Brien, D.; Collins, S. A Comparison of APD- and SPAD-Based Receivers for Visible Light Communications. *J. Light. Technol.* **2018**, *36*, 2435–2442. [CrossRef]
4. Huang, S.; Patanwala, S.M.; Kosman, J.; Henderson, R.K.; Safari, M. Optimal Photon Counting Receiver for Sub-Dead-Time Signal Transmission. *J. Light. Technol.* **2020**, *38*, 5225–5235. [CrossRef]
5. Li, Y.; Safari, M.; Henderson, R.; Haas, H. Nonlinear Distortion in SPAD-Based Optical OFDM Systems. In Proceedings of the 2015 IEEE Globecom Workshops (GC Wkshps), San Diego, CA, USA, 6–10 December 2015; pp. 1–6. [CrossRef]
6. He, C.; Ahmed, Z.; Collins, S. Signal Pre-Equalization in a Silicon Photomultiplier-Based Optical OFDM System. *IEEE Access* **2021**, *9*, 23344–23356. [CrossRef]
7. Huang, S.; Chen, C.; Bian, R.; Haas, H.; Safari, M. 5 Gbps Optical Wireless Communication using Commercial SPAD Array Receivers. *Opt. Lett.* **2022**, *47*, 2294–2297. [CrossRef] [PubMed]
8. Chen, Z.; Tang, X.; Sun, C.; Li, Z.; Shi, W.; Wang, H.; Zhang, L.; Zhang, A. Experimental Demonstration of Over 14 AL Underwater Wireless Optical Communication. *IEEE Photon- Technol. Lett.* **2021**, *33*, 173–176. [CrossRef]
9. Matthews, W.; Ahmed, Z.; Ali, W.; Collins, S. A 3.45 Gigabits/s SiPM-Based OOK VLC Receiver. *IEEE Photonics Technol. Lett.* **2021**, *33*, 487–490. [CrossRef]

10. Acerbi, F.; Gundacker, S. Understanding and simulating SiPMs. *Nucl. Instrum. Methods Phys. Res. A, Accel. Spectrom. Detect. Assoc. Equip.* **2019**, *926*, 16–35. [CrossRef]

11. Seifert, S.; van Dam, H.T.; Huizenga, J.; Vinke, R.; Dendooven, P.; Lohner, H.; Schaart, D.R. Simulation of Silicon Photomultiplier Signals. *IEEE Trans. Nucl. Sci.* **2009**, *56*, 3726–3733. [CrossRef]

12. Ali, W.; Manousiadis, P.; O'Brien, D.C.; Turnbull, G.; Samuel, I.; Collins, S. A Gigabit VLC receiver that incorporates a flu-orescent antenna and a SiPM. *J. Lightwave Technol.* **2022**, *40*, 5369–5375. [CrossRef]

13. J-Series SiPM Sensors Datasheet, Mar. 2020. Available online: https://www.onsemi.com/pub/Collateral/MICROJ-SERIES-D.PDF (accessed on 10 December 2021).

14. Otte, A.N.; Garcia, D.; Nguyen, T.; Purushotham, D. Characterization of three high efficiency and blue sensitive silicon pho-tomultipliers. *Nucl. Instruments Methods Phys. Res. Sect. A Accel. Spectrometers Detect. Assoc. Equip.* **2017**, *846*, 106–125. [CrossRef]

15. Zappa, F.; Tisa, S.; Tosi, A.; Cova, S. Principles and features of single-photon avalanche diode arrays. *Sens. Actuators A: Phys.* **2007**, *140*, 103–112. [CrossRef]

16. Dimitrov, S.; Sinanovic, S.; Haas, H. Clipping Noise in OFDM-Based Optical Wireless Communication Systems. *IEEE Trans. Commun.* **2012**, *60*, 1072–1081. [CrossRef]

17. Dimitrov, S.; Haas, H. *Principles of LED Light Communications: Towards Networked Li-Fi*; Cambridge University Press: Cambridge, UK, 2015.

18. Tsonev, D.; Chun, H.; Rajbhandari, S.; McKendry, J.J.D.; Videv, S.; Gu, E.; Haji, M.; Watson, S.; Kelly, A.E.; Faulkner, G.; et al. A 3-Gb/s single-LED OFDM-based wireless VLC link using a gallium nitride microLED. *IEEE Photonics Technol. Lett.* **2014**, *26*, 637–640. [CrossRef]

19. Almer, O.; Tsonev, D.; Dutton, N.A.; Al Abbas, T.; Videv, S.; Gnecchi, S.; Haas, H.; Henderson, R.K. A SPAD-Based Visible Light Communications Receiver Employing Higher Order Modulation. In Proceedings of the IEEE Global Communications Conference (GLOBECOM), San Diego, CA, USA, 6–10 December 2015.

20. Forward Error Correction for High Bit-Rate DWDM Submarine Systems. ITU-T Recommendation G.975.1. 2004. Available online: https://www.itu.int/rec/T-REC-G.975.1-200402-I/en (accessed on 10 December 2021).

21. Jukić, T.; Steindl, B.; Zimmermann, H. 400um Diameter APD OEIC in 0.35um BiCMOS. *IEEE Photonics Technol. Lett.* **2016**, *28*, 2004–2007. [CrossRef]

photonics

MDPI

Article

Synchronous Clock Recovery of Photon-Counting Underwater Optical Wireless Communication Based on Deep Learning

Haodong Yang [1], Qiurong Yan [1,*], Ming Wang [1], Yuhao Wang [1], Peng Li [2] and Wei Wang [2]

[1]   College of Information Engineering, Nanchang University, Nanchang 330031, China
[2]   The State Key Laboratory of Transient Optics and Photonics, Xi'an Institute of Optics and Precision Mechanics, Chinese Academy of Sciences, Xi'an 710119, China
*   Correspondence: yanqiurong@ncu.edu.cn

Abstract: In photon-counting underwater optical wireless communication (UOWC), the recovery of the time slot synchronous clock is extremely important, and it is the basis of symbol synchronization and frame synchronization. We have previously proposed a time slot synchronous clock extraction method based on single photon pulse counting, but the accuracy needs to be further improved. Deep learning is very effective for feature extraction; synchronous information is already implicit in the discrete single photon pulse signal output by single photon avalanche diode (SPAD), which is used as a communication receiver. Aiming at this characteristic, a method of time slot synchronous clock recovery for photon-counting UOWC based on deep learning is proposed in this paper. Based on the establishment of the underwater channel model and SPAD receiver model, the Monte Carlo method is used to generate discrete single photon pulse sequences carrying synchronous information, which are used as training data. Two neural network models based on regression problem and classification problem are designed to predict the phase value of the time slot synchronous clock. Experimental results show that when the average number of photons per time slot is eight, photon-counting UOWC with a data rate of 1Mbps and a bit error rate (BER) of 5.35×10^{-4} can be achieved.

Keywords: underwater optical wireless communication (UOWC); photon-counting; deep learning; time slot synchronous clock

Citation: Yang, H.; Yan, Q.; Wang, M.; Wang, Y.; Li, P.; Wang, W. Synchronous Clock Recovery of Photon-Counting Underwater Optical Wireless Communication Based on Deep Learning. *Photonics* **2022**, *9*, 884. https://doi.org/10.3390/photonics9110884

Received: 27 September 2022
Accepted: 18 November 2022
Published: 21 November 2022

Publisher's Note: MDPI stays neutral with regard to jurisdictional claims in published maps and institutional affiliations.

1. Introduction

Nearly two-thirds of the earth's surface is covered by the ocean, but these ocean resources remain largely unexploited. With the growing demand for resources, people's activities in the ocean are increasing rapidly, so the demand for related underwater information transmission technology is also more urgent [1–4]. There are currently three main underwater wireless communication technologies; namely, underwater ultrasonic communication, underwater radio frequency (RF) communication and underwater optical communication. Although underwater ultrasonic communication can achieve a transmission distance of several kilometers or more, the communication rate is very low (Kbps) and cannot meet the needs of large-capacity data interaction [5,6]. As another option, underwater RF communication can provide a data transmission rate up to tens of Mbps. Its transmission distance is very limited due to severe attenuation underwater; usually, only a few meters can be transmitted [7,8]. Underwater optical wireless communication (UOWC) can provide an ultra-high data rate (Gbps), and has the characteristics of low time delay and high energy efficiency. Therefore, UOWC has become a hot research topic [9–12].

Due to the complexity of the underwater channel, the optical signal will be affected by scattering and absorption when transmitted in an underwater channel, so the optical signal is very weak while arriving at the receiving end. At present, many researchers have already realized high-speed UOWC. A 21 m UOWC with a data rate of 5.5 Gbps has been achieved in [13]. In [14], the authors realized a UOWC distance of 34.5 m with a data rate of 2.70 Gbps. Shen et al. realized a UOWC distance of over 20 m at a high data rate of up to 1.5 Gbps by using a 450 nm LD [15]. In [16], the authors proposed a UOWC system based on a convolutional neural network (CNN) demodulator and analyzed the BER performance of the system. Cui et al. experimentally demonstrated the performance of CNN-based signal decoders in UOWC systems [17]. In [18,19], the authors used a CNN combiner for feature analysis and combination, and a CNN demodulator to recover the transmitted information. Although a high sensitivity avalanche photon diode (APD) has been used as the receiver detector in the above articles and the communication rate is up to Gbps, the transmission distance is limited to tens of meters. Improving the transmission distance of UOWC from tens to hundreds of meters is a challenging task. In recent years, in order to realize long-distance UOWC, some researchers have used a single photon avalanche diode (SPAD) to detect an optical signal at the receiving end. Compared with APD, SPAD is more sensitive and can detect a single photon as the photon-counting receiver. In 2016, through simulation analysis, Wang et al. found that SPAD-based UOWC is feasible when it transmits more than 100 m in clean ocean and 300 m in pure sea water [20]. In [21], a synchronization method based on photon-counting UOWC was proposed, which achieved a communication distance of more than 100 m with a symbol error rate (SER) of 10^{-4}. In [22], UOWC with a communication distance of 46 m based on a multi-pixel photon counter (MPPC) detector was realized. UOWC with a communication distance of 21 m and a data rate of 312.03Mbps based on MPPC was achieved in [23]. In [24], Huang et al. realized UOWC with a communication distance of 1000 m and a data rate of 10 bps based on photon-counting detectors. In [25], the author proposed a multiple light emitting diode (LED) chips parallel transmission scheme for UOWC based on SPAD, and proved that the system can significantly improve the BER performance. In [26], a real-time, high-speed UPCC system was designed and experimentally validated based on SPAD. A novel deep learning-aided signal detection scheme for an SPAD-based UOWC system was proposed in [27]. In [28], Hema et al. proposed a deep learning based signal detection system for channel estimation and increases in transmission distance. In [29], a novel fully connected deep neural network (FC-DNN)-based receiver was proposed and experimentally demonstrated in a UOWC system. All the above literatures use SPAD as the receiver detector, and all have realized relatively long-distance UOWC, but none of them have specified how to realize synchronization and recover data from the received discrete single photon pulse. Clock recovery is the basis of data recovery, including three-level synchronous clock recovery for time slots, symbols and frames. How to recover the time slot synchronous clock from the discrete random single photon pulse is the key to the demodulation of the photon-counting communication signal. In 2019, Dr. Yan et al. proposed a time slot synchronous clock recovery method based on pulse counting [30]. The time slot synchronous clock extracted by this method has low precision, especially in long-distance UOWC. When the average number of photons per bit is very small, the phase error of the extracted time slot synchronous clock is very large, resulting in a lot of errors in the recovered baseband signal.

Deep neural networks have recently achieved exciting successes in computer vision, speech recognition and natural language processing because of their powerful data learning capabilities. Deep learning models can be used not only for classification and regression, but also for feature extraction. Fast online recovery and powerful nonlinear mapping capabilities are the main advantages of deep learning methods.

In photon-counting UOWC, the synchronous clock information is implied in the discrete single photon pulse sequence output by the SPAD receiver. In view of this characteristic, we proposed a novel time slot synchronous clock recovery method based on deep learning for photon-counting UOWC. The time slot synchronous clock can be directly extracted from the received discrete single photon pulse by a trained deep neural network, so the baseband communication data can be quickly recovered. In order to verify our proposed method, a photon-counting UOWC experimental system was built.

2. System Design and Principle

2.1. System Design

The photon-counting UOWC system based on deep learning is shown in Figure 1. The sending end performs On-Off Keying (OOK) modulation on the data to be sent and then loads it on the drive circuit, which will drive the LED to turn on and off. The light signal emitted by the LED passes through a collimating lens, which converts the originally divergent light signal into a parallel light signal, and then passes through the underwater channel. Due to the influence of absorption and scattering, the light signal is very weak while arriving at the receiving end. The receiving end uses a focusing lens to focus the optical signal into the SPAD. The discrete single photon pulse signal output by SPAD passes through a starting position determination module, and then a suitable starting sampling point is found. At the same time, the sampling module starts sampling the discrete single photon pulse output by SPAD, and the sequence after sampling is converted into binary sequence; then, the binary sequence is input into the previously trained neural network. The trained neural network extracts the feature of the binary sequence and recovers the time slot synchronous clock. Finally, the baseband signal is recovered according to the time slot synchronous clock.

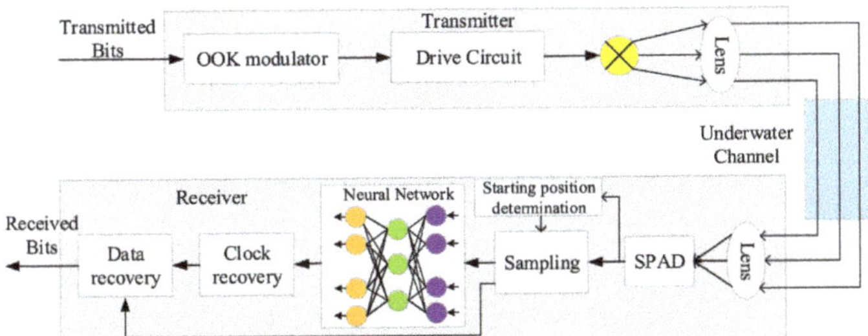

Figure 1. Photon-counting UOWC system based on deep learning.

2.2. Principle

2.2.1. Principle of Recovering Time Slot Synchronous Clock Based on Deep Learning

After detecting the optical signal carrying information, SPAD outputs a discrete single photon pulse signal. The phase of time slot synchronous clock is hidden in the discrete single photon pulse signal; we designed a neural network and let it learn to extract the hidden phase. As shown in Figure 2, the initial phases of the time slot synchronous clocks are set between 0 and 2π, and the initial phases of these time slot synchronous clocks are evenly divided into M types. The underwater photon-counting wireless optical communication is simulated by the Monte Carlo method, and single photon pulse signals with different initial phases carrying the synchronization head information are obtained. These single photon pulse signals are sampled and converted into binary sequences as the data part of the training set. When single photon pulse signals with different initial phases are obtained, the phases of the corresponding time slot synchronous clocks are also different. In this paper, the phase with the smallest phase error from the ideal time slot synchronous clock phase among the M kinds of time slot synchronous clock phases uniformly classified is used as the class label of the current single photon pulse signal, and the class label is processed by one-hot encoding. After the training data are obtained through the above method, the training data are fed into the built neural network, and the weights and biases in the network are continuously trained and optimized. After the network converges, the trained network can be used to extract features from discrete single photon pulses, and then to identify and reconstruct the phase of the time slot synchronous clock from the extracted features, as shown in Figure 3.

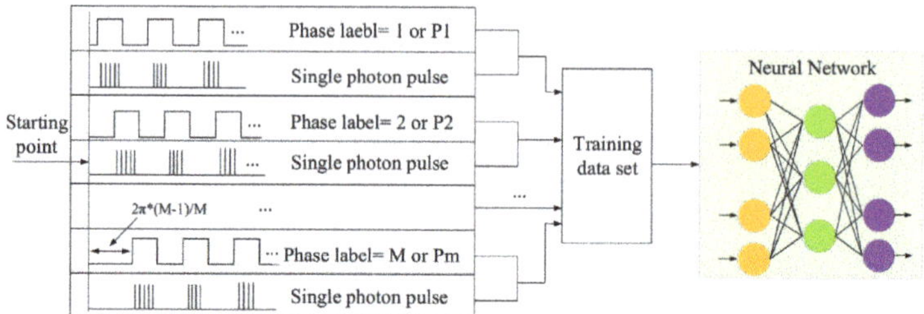

Figure 2. Principle of neural network trained to extract the hidden phase.

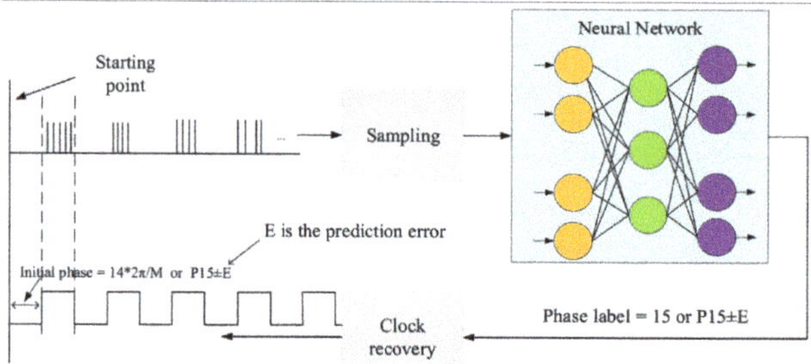

Figure 3. Principle of recovering time slot synchronous clock based on deep learning.

2.2.2. Frame Structure Design

In the actual photon-counting UOWC system, how to determine the starting position of a segment of data is a difficult problem. Here we design a data frame format, which can quickly find the appropriate starting position and provide the premise for data recovery. The format of the data frame is shown in Figure 4. A complete data frame consists of a series of starting position estimation sequence (SPES), silent time 1, synchronization header, frame header, valid data and silent time 2. The SPES and silent time are used to find a suitable starting position. The synchronization header sequence is a square wave signal carrying synchronization information. The function of the frame header is to mark the starting position of the valid data; the valid data are placed behind the frame header. At the end of the data frame, there is a silent time to distinguish the two different data frames.

SPES	Silent time 1	Syn head	Frame head	Data	Silent time 2

Figure 4. Frame structure.

2.2.3. Principle of Starting Position Determination

According to the data frame structure designed above, the timing diagram of how to find the starting position is shown in Figure 5. The SPAD outputs discrete random electrical pulses after detecting the light signal. We designed a counter to count the single photon pulse output by SPAD, and a timer counts the silent time of the signal output by SPAD. When the counter count value reaches a certain value and the silent time count value also reaches a certain value, we consider this moment as the starting position. At this time, pulling up the sampling enable signal starts the sampling module. The sampling module starts sampling the single photon pulse signal output by SPAD and converts the sampled sequence into binary sequence. Then, the Nf_s/b_r sampling points in front of the binary sequence are fed into the trained neural network, so that the time slot synchronous clock phase value can be restored, and the time slot synchronous clock can be restored according to the phase value, where f_s represents the sampling clock frequency, and b_r represents the baud rate of the modulated signal.

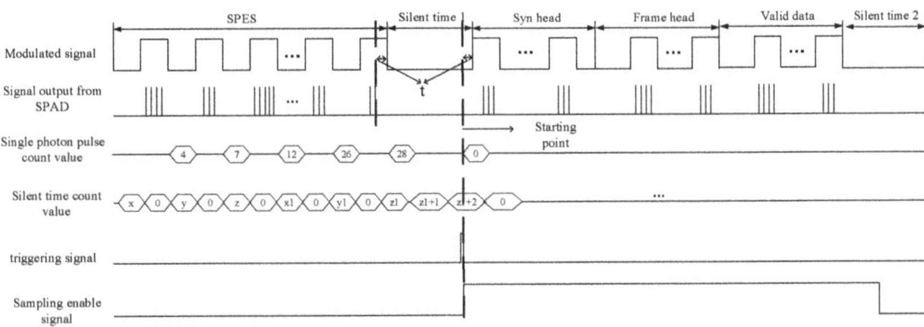

Figure 5. Timing diagram of starting position determination.

3. Design of Neural Network and Production of Training Data

3.1. Design of Classification and Regression Neural Network

For the deep learning method to recover the time slot synchronous clock phase value, we propose two neural network models to predict the phase value of a time slot synchronous clock. The clock phase determination network (CPD-Net) belongs to the classification network, and the clock phase reconstruction network (CPR-Net) belongs to the regression network. In both types of networks, we use a certain number of hidden layers. The benefit is to improve the accuracy of phase identification and extraction, and ultimately reduce the BER of the system. When the number of photons in a unit time slot is relatively large, a four-layer network can achieve good results. However, when the number of photons is low, there are fewer features that can be identified. In order to accurately identify features and recover the synchronous clock when the number of photons is small, we add more fully connected layers. Experiments show that when the number of photons in a unit time slot is small, the use of several layers of fully connected layers has a good effect on feature recognition, and can even reduce the overhead caused by the length of the synchronization head.

3.1.1. Structure of CPD-Net

The structure of CPD-Net is shown in Figure 6. In order to extract the phase value of the time slot synchronous clock more quickly and accurately, we designed a shallow time slot synchronous clock recognition network. This network has a small amount of calculation and high recognition accuracy. The network consists of two convolutional layers and eight fully connected layers, and each layer is followed by a ReLU activation function. The first two convolutional layers are used for phase feature extraction, the last eight fully connected layers are used for phase classification, and finally, the phase value with the highest probability is output in the output layer. To sum up, the model of CPD-Net can be expressed as:

$$\hat{Q}_l = f_t(f_{t-1}(\ldots f_1(I_n))) \tag{1}$$

where I_n represents the input data, and $f_t(t = 1,2, \ldots ,10)$ represents the mathematical operation of each layer of the network.

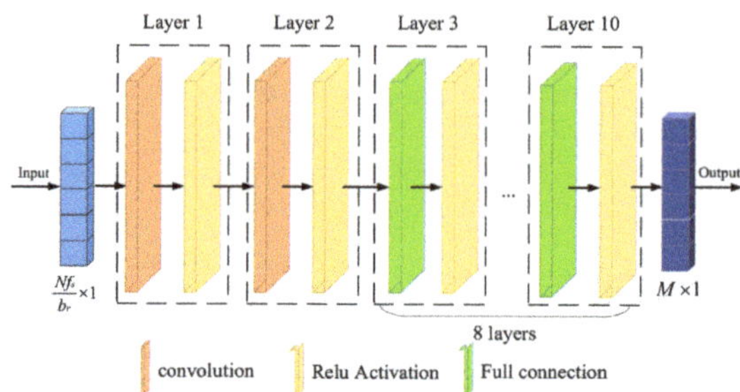

Figure 6. The structure of CPD-Net.

In general, the purpose of training a neural network is to reduce the loss function by optimizing the parameters of the network. Since CPD-Net is a typed network, the cross-entropy error is used here as the loss function, which can be expressed as:

$$l_{cpd} = -\sum_{i=1}^{M} Q_i * \log(P_i) \tag{2}$$

where Q_i is the expected output value, and P_i is the predicted probability of network output.

3.1.2. Structure of CPR-Net

Although the aforementioned CPD-Net can provide high-accuracy clock phase recognition, its phase is always a fixed category, and the precision of phase extraction is not high enough. In response to this problem, we have proposed a regression-based CPR-Net; the network structure is shown in Figure 7. The network is mainly composed of four convolutional layers and eight fully connected layers, and the ReLU activation function is used behind each layer. Four convolutional layers constitute an encoder and decoder. The first two convolutional layers are used as encoder to extract features from the discrete single photon pulse, the latter two convolutional layers are used as decoder to reconstruct the time slot synchronous clock phase value from the extracted features. Eight fully connected layers gradually reduce the dimension of the data output by the decoder. Finally, the phase prediction value is output in the output layer. In summary, the model of CPR-Net can be expressed as:

$$\hat{S}_l = f_n(f_{n-1}(\ldots f_1(U_n))) \tag{3}$$

where U_n represents the input data and $f_n(n = 1,2\ldots,12)$ represents the mathematical operation of each layer of the network. Since CPR-Net is a regression network, the cross entropy is no longer used as the loss function. The mean square error is used as the loss function, so its loss function can be expressed as:

$$l_{cpr} = \frac{1}{n}\sum_{i=1}^{n} (s_i - g_i)^2 \tag{4}$$

where g_i is the expected output value and s_i is the actual output value of the network.

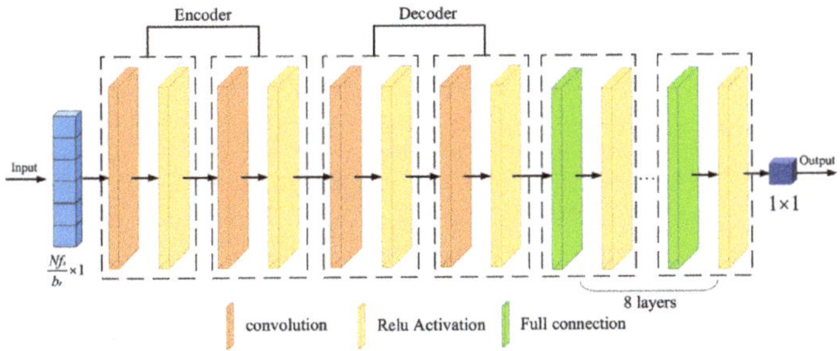

Figure 7. The structure of CPR-Net.

3.2. Underwater Channel Model and SPAD Receiver

The relationship between the LED emission power and the average number of emitted photons can be expressed as:

$$P_t = \frac{N_t h v}{T},$$ (5)

where P_t is the LED emission power, N_t is the average number of photons emitted by LED, h is the Planck constant, v is the frequency of light and T is the time slot interval.

Optical signal transmission in an underwater channel will be affected by absorption and scattering. Since the absorption factor $a(\lambda)$ and the scattering factor $b(\lambda)$ are the main causes of channel attenuation, the accumulated total attenuation coefficient $c(\lambda) = a(\lambda) + b(\lambda)$ in the underwater channel. According to the accumulated total attenuation coefficient, we use the Beer model to model the attenuation of underwater optical signal transmission; then, the number of photons Nr at the receiving end can be expressed as:

$$N_r = N_t e^{-c(\lambda)L},$$ (6)

where L represents the underwater communication distance.

The number of photons after SPAD optical detection can be expressed as:

$$\lambda_s = \mu_p(N_r + N_a T) + N_b T,$$ (7)

where μ_p is the photoelectric conversion efficiency of SPAD, N_a is the number of background light noise photons per unit time and N_b is the dark count of SPAD per unit time.

Generally speaking, the number of photons output by SPAD can be modeled as a Poisson statistical distribution under ideal circumstances, and the probability of detecting y photons can be expressed as:

$$Pideal(y) = \frac{\lambda_s{}^y e^{-\lambda_s}}{y!},$$ (8)

When considering the dead time of SPAD, the number of photons detected by SPAD no longer obey the standard Poisson statistical distribution. At this time, the probability of detecting y photons can be given by [19]:

$$P(y) = \begin{cases} \sum_{i=0}^{y} \frac{\lambda_s{}^i(1-(i+1)\delta)^i}{i!} e^{-\lambda_s(1-(i+1)\delta)} \\ \quad - \sum_{i=0}^{y-1} \frac{\lambda_s{}^i(1-i\delta)^i}{i!} e^{-\lambda_s(1-i\delta)}, & \text{if } y < y_{\max} \\ 0, & \text{if } y \geq y_{\max} \end{cases}$$ (9)

where $\delta = \frac{\tau}{T}$, τ is the dead time, and the maximum number of photons detected by SPAD $y_{\max} = \lfloor T/\tau \rfloor + 1$, $\lfloor k \rfloor$ represents the largest integer less than k.

3.3. The Production Process of Training Data

As shown in Figure 8, according to the above underwater channel model and SPAD receiver model, we can use the Monte Carlo method to simulate the process of photon-counting UOWC to create the training data set. To prevent overfitting, we use Dropout in the network. Our training set is large enough to contain rich 01 sequences. During the test phase, we generate a large number of test single photon pulses, which also work well. This means that our network does not have the problem of overfitting.

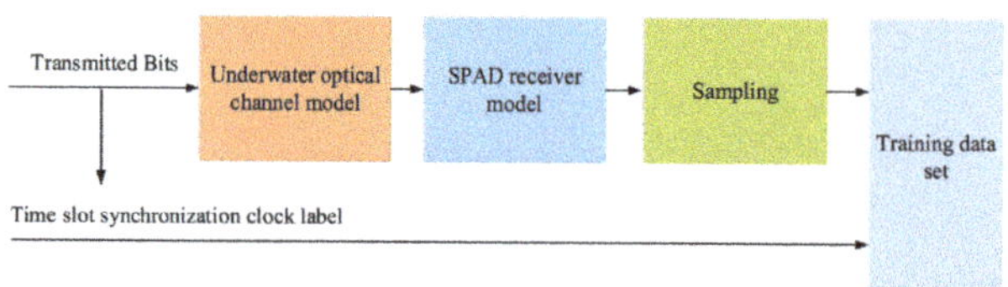

Figure 8. Production process of training data.

3.3.1. Training Data Production of CPD-Net

In the training data production process of CPD-Net, we stipulated that the initial phase value of the time slot synchronous clock was in the range of *0-2π*, and the initial phase of the time slot synchronous clock was divided into *M* types evenly. The initial phase difference of two adjacent time slot synchronous clocks is *2π/M*, and the *M* time slot synchronous clocks with different initial phases are numbered. We send synchronization header data of length *N* bits on the sender side, and orderly increase the initial phase of sending synchronization header data. The value of each increase is $\pi b_s/f_s$; the number of the corresponding time slot synchronous clock can be expressed as:

$$l_d = f([|0 - P_h|, |\frac{2\pi}{M} - P_h|, \ldots, |\frac{2\pi(M-1)}{M} - P_h|]),\tag{10}$$

The function of *f(x)* is to find the subscript with the smallest value in the array, and P_h represents the initial phase value of the transmitted signal. For example, we divide the initial phase of the time slot synchronous clock into 20 kinds; then, every two adjacent time slot synchronous clocks are separated by $\pi/10$. At this time, the initial phase of the transmitted signal is $\pi/40$, and according to Equation (8), the time slot synchronous clock phase number corresponding to the transmitted signal is one.

When the optical signal passes through the underwater channel model and the SPAD receiver model, the single photon pulse signal output by the SPAD is sampled and binarized; the Nf_s/b_r numbers in front of the binary sequence are taken as the data part of the training data set. At the same time, the number of the corresponding time slot synchronous clock is used as the label part of the training data set. It is very important that the label data is one-hot encoded.

3.3.2. Training Data Production of CPR-Net

The training data production process of CPR-Net and CPD-Net are basically the same. Because CPR-Net is a network that deals with regression problems, the label in the training data of CPR-Net is no longer the phase category of the time slot synchronous clock, and the label of CPR-Net training data can be expressed as:

$$l_c = P_h,\tag{11}$$

4. Simulation and Water Tank Experiment Results

4.1. Simulation Results

In order to verify the feasibility of the method we proposed, we conducted simulation experiments on the photon-counting UOWC system based on deep learning and gave some simulation results. In the simulation experiment, we selected two typical water qualities,

Jerlov IB and Jerlov II; their attenuation coefficients are given in Table 1. In order to simulate real deep-sea communication, we did not consider the background light in the simulation process, but only considered the non-standard Poisson noise and dark counting noise in the process of SPAD detecting photons. The detailed parameters of the simulation are given in Table 2.

Table 1. Jerlov IB, II water quality attenuation parameters.

	Jerlov IB	**Jerlov II**
$a(\lambda)(m^{-1})$	0.064	0.087
$b(\lambda)(m^{-1})$	0.08	0.216
$c(\lambda)(m^{-1})$	0.144	0.303

Table 2. Simulation parameters of photon-counting UOWC system based on deep learning.

Parameters	**Values**
Modulation	On-Off Keying (OOK)
Baud rate (b_r)	10 Mbps
Length of synchronization header (N)	10
Phase type of synchronization clock (M)	20
Sampling frequency (f_s)	200 MHz
Wavelength	450 nm
Power of LED	1 W
Max efficiency	35%
Dark count rate	25 Hz
Pulse width	5 ns
Dead-time	8 ns

The results of the performance evaluation of two network models are shown in Figure 9. The training data sets used for training and testing are different. It can be seen from the figure that, whether it is in Jerlov IB or Jerlov II water quality, as the transmission distance increases, the probability of CPD-Net recognizing the correct time slot synchronous clock phase value decreases. For CPR-Net, with the increase of the transmission distance, the error between the predicted phase value output by the network and the ideal phase will increase.

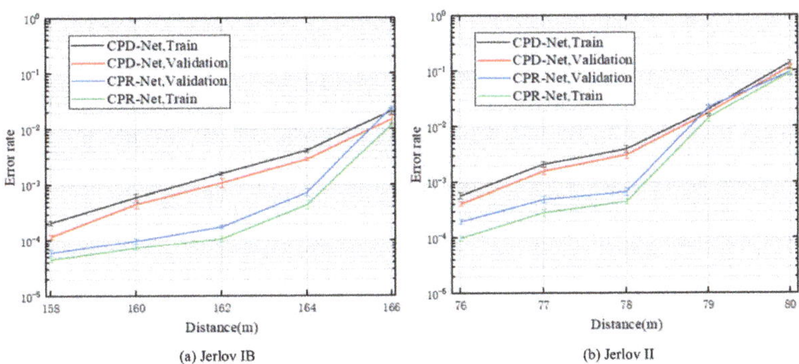

Figure 9. Performance evaluation of CPD-Net and CPR-Net under different water quality.

As shown in Figure 10, we compared the pulse counting method and the deep learning method. It can be seen from the figure that under the same conditions, the system communication BER using the deep learning method will be lower. We can also see from the figure that the BER is lower when using regression-based CPR-Net to reconstruct the time slot synchronous clock phase value. The reason for this result is that the time slot synchronous clock phase predicted by CPR-Net is more accurate, so the system communication BER using CPR-Net recovery time slot synchronous clock is lower.

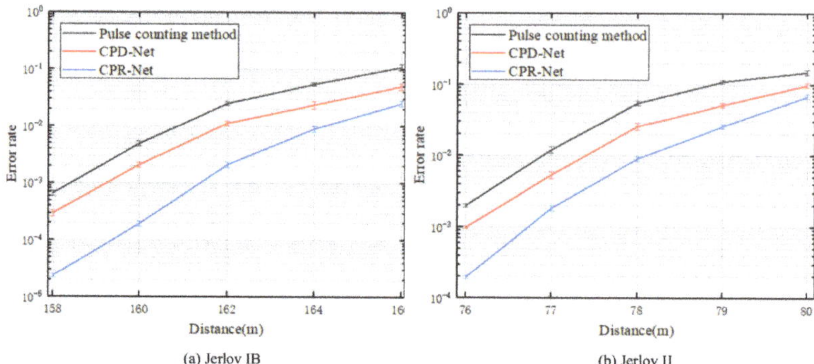

Figure 10. The influence of different time slot synchronous clock recovery methods on system communication BER.

4.2. Results of Water Tank Experiment

The water tank experiment system is shown in Figure 11. At the sending end, we use a Field Programmable Gate Array (FPGA) (ALINX, AX516) to perform OOK modulation on the received data, and then drive the LED to turn on and off. A diaphragm is placed behind the LED, which is mainly used to adjust the optical power of the LED. We use a tank full of water to simulate the underwater channel; the length of the tank is about 1.5 m. Because the length of the water tank is not sufficient, in order to simulate deep-sea communication, an attenuator is placed behind the diaphragm to attenuate the light signal emitted by the LED and the tank is opaque. At the receiving end, we use another FPGA (ALTERA, DE2-115) to sample the single photon pulse signal output by the SPAD (THORLABS, SPCM 20A), and send the sampled data to the computer via Ethernet for time slot synchronous clock and data recovery.

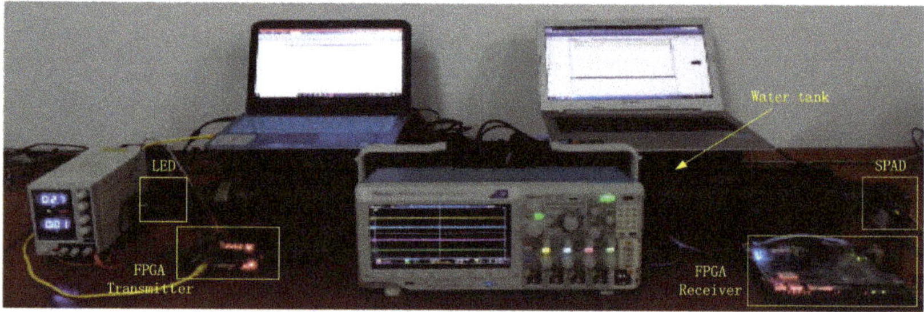

Figure 11. Diagram of the water tank experimental system.

4.2.1. Parameter Settings for Training Data Production

In order to use the trained network in our water tank experiment system, we produced a large amount of training data through the method in Chapter 3, and then used these training data to continuously train CPD-Net and CPR-Net. It is especially important that the parameters in the process of making training data are completely consistent with the parameters of the actual water tank experimental system. In order to count the background light and dark count noise in the experimental system, this section makes a statistical analysis of the noise photon pulse output by the SPAD when the LED at the sending end is off. Figure 12 gives the statistical distribution of the number of noise photons within 1000 time slots, where the time slot frequency is 1 Hz. It can be seen from the figure that the average number of noise photons in the experimental system in 1 s is only four. When the communication baud rate is 1Mbps, the number of noise photons within each bit is basically negligible, so the background light noise and dark count noise are not considered when making the training data set in this section. The detailed parameters of the water tank experimental system are given in Table 3.

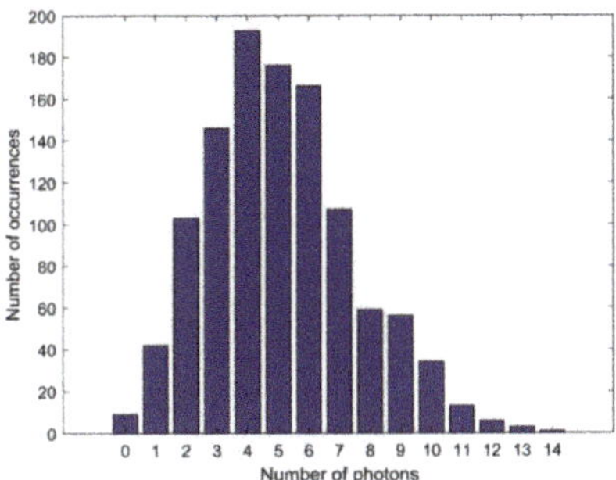

Figure 12. Statistical distribution of the number of noise photons in 1000 time slots.

Table 3. Detailed parameters of the water tank experimental system.

Parameters	Values
SPAD number	SPCM 20A
Modulation	On-Off Keying (OOK)
Baud rate (b_r)	1 Mbps
Length of synchronization header (N)	40
Phase type of synchronization clock (M)	50
Sampling frequency (f_s)	50 MHz
Wavelength	450 nm
Max efficiency	35%
Dark count rate	25 Hz
Pulse width	20 ns
Dead-time	40 ns

4.2.2. Analysis of Network Performance

As shown in Figure 13, we presented the training results of CPD-Net and CPR-Net when the average number of photons per time slot was different. It can be seen from the figure that as the average number of photons in each time slot increases, the probability of CPD-Net identifying the wrong time slot synchronous clock phase value gradually decreases; the error between the phase value predicted by CPR-Net and the phase value of the ideal time slot synchronous clock becomes smaller.

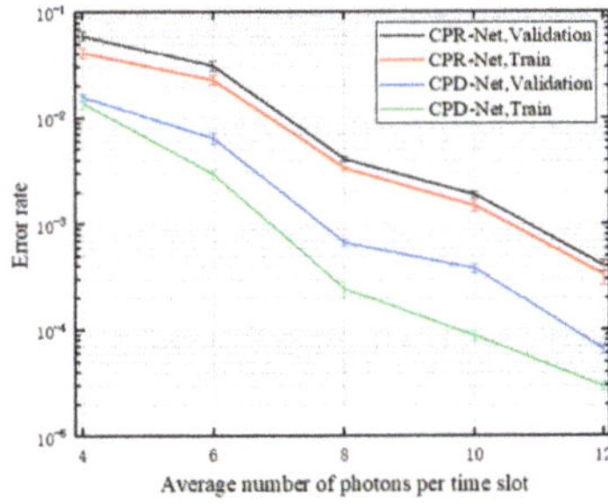

Figure 13. The impact of the average number of photons per time slot on the performance evaluation of CPR-Net and CPD-Net.

4.2.3. Analysis of Experimental Results

In order to compare the traditional pulse counting method with the deep learning method proposed by us, we adopted the deep learning method and the traditional pulse counting method to realize time slot synchronous clock and data recovery, respectively, in the water tank experimental system; the corresponding system communication BER is shown in Figure 14. It can be seen from the figure that the system communication BER using the deep learning method is lower. This phenomenon also proves that the time slot synchronous clock recovered by the deep learning method is more accurate. It is worth noting that the BER of the system using CPR-Net is lower than the BER of the system using CPD-Net, which directly shows that the method of using CPR-Net to predict the time slot synchronous clock phase value is better.

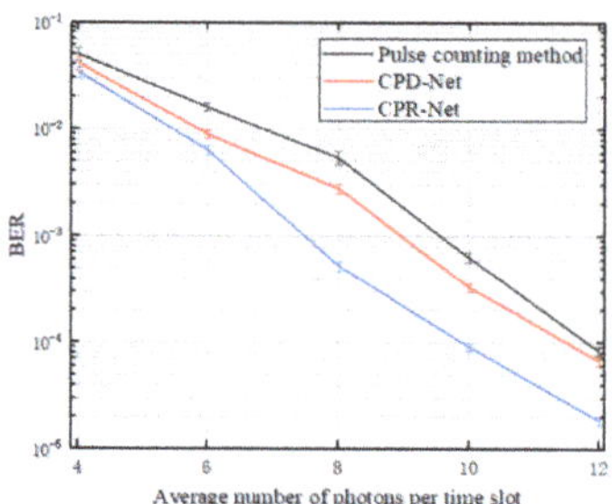

Figure 14. The influence of different time slot synchronous clock recovery methods on the communication BER of the water tank experimental system.

5. Conclusions

Aiming at the shortcomings of existing photon-counting UOWC time slot synchronous clock recovery methods, a time slot synchronous clock recovery scheme for photon-counting UOWC based on deep learning is proposed in this paper. By establishing an underwater channel model and SPAD receiver model, a large amount of training data is produced by using the Monte Carlo method based on these two models, and two neural network models based on the classification problem and regression problem are designed to predict the phase value of the time slot synchronous clock. When the network training is complete, we use the trained network to recover the time slot synchronous clock in the photon-counting UOWC system, and then recover the data. Both simulation and water tank experiment results show that the deep learning method we proposed is better than the existing pulse counting method and can effectively reduce the BER of the photon-counting UOWC system. Another point is that the method of using regression-based CPR-Net to reconstruct the time slot synchronous clock phase is better.

Author Contributions: Conceptualization, H.Y. and Q.Y.; methodology, M.W.; validation, H.Y., Q.Y. and M.W.; formal analysis, H.Y.; investigation, Y.W.; resources, P.L. and W.W.; data curation, H.Y. and M.W.; writing—original draft preparation, H.Y.; writing—review and editing, Q.Y.; visualization, Y.W.; supervision, P.L. and W.W.; project administration, Y.W., P.L. and W.W.; funding acquisition, Q.Y. All authors have read and agreed to the published version of the manuscript.

Funding: This research was funded by National Natural Science Foundation of China (No. 61565012), China Postdoctoral Science Foundation (No. 2015T80691), the Science and Technology Plan Project of Jiangxi Province (No. 20151BBE50092), the Funding Scheme to Outstanding Young Talents of Jiangxi Province (No. 20171BCB23007).

Data Availability Statement: The data presented in this study are available on request from the corresponding author.

Conflicts of Interest: The authors declare no conflict of interest.

References

1. Kaushal, H.; Kaddoum, G. Underwater optical wireless communication. *IEEE Access* **2016**, *4*, 1518–1547. [CrossRef]
2. Khalighi, M.A.; Uysal, M. Survey on free space optical communication: A communication theory perspective. *IEEE Commun. Surv. Tutor.* **2014**, *16*, 2231–2258. [CrossRef]
3. Arnon, S. Underwater optical wireless communication network. *Opt. Eng.* **2010**, *49*, 015001. [CrossRef]
4. Akyildiz, F.; Pompili, D.; Melodia, T. Underwater acoustic sensor networks: Research challenges. *Ad Hoc Netw.* **2005**, *3*, 257–279. [CrossRef]
5. Sozer, E.M.; Stojanovic, M.; Proakis, J.G. Underwater acoustic networks. *IEEE J. Ocean. Eng.* **2000**, *25*, 72–83. [CrossRef]
6. Wang, J.-M.; Lu, C.-H.; Li, S.-B.; Xu, Z.-Y. 100 m/500 Mbps underwater optical wireless communication using an NRZ-OOK modulated 520 nm laser diode. *Opt. Express* **2019**, *27*, 12171–12181. [CrossRef]
7. Cochenour, B.; Mullen, L.; Muth, J. Temporal response of the underwater optical channel for high-bandwidth wireless laser communications. *IEEE J. Ocean. Eng.* **2013**, *38*, 730–742. [CrossRef]
8. Xu, J.; Kong, M.W.; Lin, A.B.; Song, Y.H.; Yu, X.Y.; Qu, F.Z.; Han, J.; Deng, N. OFDM-based broadband underwater wireless optical communication system using a compact blue LED. *Opt. Commun.* **2016**, *369*, 100–105. [CrossRef]
9. Nakamura, K.; Mizukoshi, I.; Hanawa, M. Optical wireless transmission of 405 nm, 1.45 Gbit/s optical IM/DD-OFDM signals through a 4.8 m underwater channel. *Opt. Express* **2015**, *23*, 1558–1566. [CrossRef]
10. Oubei, H.M.; Duráan, J.R.; Janjua, B.; Wang, H.-Y.; Tsai, C.-T.; Chi, Y.-C.; Ng, T.K.; Kuo, H.-C.; He, J.-H.; Alouini, M.-S.; et al. Wireless Optical Transmission of 450 nm, 3.2 Gbit/s 16-QAM-OFDM Signals over 6.6 m Underwater Channel. In Proceedings of the Conference on Lasers and Electro-Optics, San Jose, CA, USA, 5–10 June 2016; p. SW1F.1.
11. Xu, J.; Song, Y.; Yu, X.; Lin, A.; Kong, M.; Han, J.; Deng, N. Underwater wireless transmission of high-speed QAM-OFDM signals using a compact red-light laser. *Opt. Express* **2016**, *24*, 8097–8109. [CrossRef]
12. Xu, J.; Lin, A.; Yu, X.; Kong, M.; Song, Y.; Qu, F.; Han, J.; Jia, W.; Deng, N. High-speed underwater wireless optical communication using a compact OFDM-modulated green laser diode. *IEEE Photonics Technol. Lett.* **2016**, *28*, 2133–2136. [CrossRef]
13. Chen, Y.-F.; Kong, M.-W.; Ali, T.; Wang, J.-L.; Sarwar, R.; Han, J.; Guo, C.-Y.; Sun, B.; Deng, N.; Xu, J. 26 m/5.5 Gbps air-water optical wireless communication based on an OFDM-modulated 520-nm laser diode. *Opt. Express* **2017**, *25*, 14760–14765. [CrossRef]
14. Liu, X.-Y.; Yi, S.-Y.; Zhou, X.-L.; Fang, Z.-L.; Qiu, Z.-J.; Hu, L.-G.; Cong, C.-X.; Zheng, L.-R.; Liu, R.; Tian, P.-F. 34.5 m underwater optical wireless communication with 2.70 Gbps data rate based on a green laser diode with NRZ-OOK modulation. *Opt. Express* **2017**, *25*, 27937–27947. [CrossRef]
15. Shen, C.; Guo, Y.-J.; Oubei, H.M.; Ng, T.K.; Liu, G.-Y.; Park, K.-H.; Ho, K.-T.; Alouini, M.-S.; Ooi, B.S. 20-meter underwater wireless optical communication link with 1.5 Gbps data rate. *Opt. Express* **2016**, *24*, 25502–25509. [CrossRef]
16. Ma, W.; Lu, H.; Chen, D.; Jin, J.; Wang, J. Orbital angular momentum underwater wireless optical communication system based on convolutional neural network. *J. Opt.* **2022**, *24*, 065701. [CrossRef]
17. Cui, X.; Yin, X.; Chang, H.; Liao, H.; Chen, X.; Xin, X.; Wang, Y. Experimental study of machine-learning-based orbital angular momentum shift keying decoders in optical underwater channels. *Opt. Commun.* **2019**, *452*, 116–123. [CrossRef]
18. Lu, H.; Jiang, M.; Cheng, J. Deep learning aided robust joint channel classification, channel estimation, and signal detection for underwater optical communication. *IEEE Trans. Commun.* **2020**, *69*, 2290–2303. [CrossRef]
19. Lu, H.; Chen, W.; Jiang, M. Deep Learning Aided Misalignment-Robust Blind Receiver for Underwater Optical Communication. *IEEE Wirel. Commun. Lett.* **2021**, *10*, 1984–1988. [CrossRef]
20. Wang, C.; Yu, H.-Y.; Zhu, Y.-J. A long distance underwater visible light communication system with single photon avalanche diode. *IEEE Photonics J.* **2017**, *8*, 7906311. [CrossRef]
21. Hu, S.-Q.; Mi, L.; Zhou, T.-H.; Chen, W.-B. 35.88 attenuation lengths and 3.32 bits/photon underwater optical wireless communication based on photon-counting receiver with 256-PPM. *Opt. Express* **2018**, *26*, 21685–21699. [CrossRef]
22. Shen, J.-N.; Wang, J.-L.; Yu, C.-Y.; Chen, X.; Wu, J.-Y.; Zhao, M.-M.; Qu, F.-Z.; Xu, Z.-W.; Han, J.; Xu, J. Single LED-based 46-m underwater wireless optical communication enabled by a multi-pixel photon counter with digital output. *Opt. Commun.* **2019**, *438*, 78–82. [CrossRef]
23. Wang, J.; Yang, X.; Lv, W.; Yu, C.; Wu, J.; Zhao, M.; Qu, F.; Xu, Z.; Han, J.; Xu, J. Underwater wireless optical communication based on multi-pixel photon counter and OFDM modulation. *Opt. Commun.* **2019**, *451*, 181–185. [CrossRef]
24. Huang, J.; Wen, G.; Dai, J.; Zhang, L.; Wang, J. Channel model and performance analysis of long-range deep sea wireless photon-counting communication. *Opt. Commun.* **2020**, *473*, 125989. [CrossRef]
25. Wang, C.; Yu, H.Y.; Zhu, Y.J.; Wang, T.; Ji, Y.W. Multi-LED parallel transmission for long distance underwater VLC system with one SPAD receiver. *Opt. Commun.* **2018**, *410*, 889–895. [CrossRef]
26. Huang, J.; Li, C.; Dai, J.; Shu, R.; Zhang, L.; Wang, J. Real-Time and High-Speed Underwater Photon-Counting Communication Based on SPAD and PPM Symbol Synchronization. *IEEE Photonics J.* **2021**, *13*, 7300209. [CrossRef]
27. Jiang, R.; Sun, C.-M.; Zhang, L.; Tang, X.-K.; Wang, H.-J.; Zhang, A.-D. Deep Learning Aided Signal Detection for SPAD-Based Underwater Optical Wireless Communications. *IEEE Access* **2020**, *8*, 20363–20374. [CrossRef]
28. Hema, R.; Sudha, S.; Ananthi, A.; Harinie, G.; Gayathri, K. Deep Learning Based Signal Detection in Underwater Wireless Optical Communication. *J. Crit. Rev.* **2020**, *7*, 1423–1436.

29. Du, Z.; Deng, H.; Dai, Y.; Hua, Y.; Jia, B.; Qian, Z.; Xiong, J.; Lyu, W.; Zhang, Z.; Ma, D.; et al. Experimental demonstration of an OFDM-UWOC system using a direct decoding FC-DNN-based receiver. *Opt. Commun.* **2022**, *508*, 127785. [CrossRef]
30. Yan, Q.-R.; Li, Z.-H.; Hong, Z.; Zhan, T.; Wang, Y.-H. Photon-Counting Underwater Wireless Optical Communication by Recovering Clock and Data From Discrete Single Photon Pulses. *IEEE Photonics J.* **2019**, *11*, 7905815. [CrossRef]

Communication

CGA-VLP: High Accuracy Visible Light Positioning Algorithm Using Single Square LED with Geomagnetic Angle Correction

Chen Yang [1], Shangsheng Wen [2,*], Danlan Yuan [3], Junye Chen [1], Junlin Huang [4] and Weipeng Guan [2]

[1]   School of Automation Science and Engineering, South China University of Technology, Guangzhou 510641, China
[2]   School of Materials Science and Engineering, South China University of Technology, Guangzhou 510641, China
[3]   School of Information Engineering, South China University of Technology, Guangzhou 510641, China
[4]   School of Mathematics, South China University of Technology, Guangzhou 510641, China
*   Correspondence: shshwen@scut.edu.cn

Abstract: Visible light positioning (VLP), benefiting from its high accuracy and low cost, is a promising technology for indoor location-based services. In this article, the theoretical limits and error sources of traditional camera-based VLP systems are analyzed. To solve the problem that multiple LEDs are required and auxiliary sensors are imperfect, a VLP system with a single square LED which can correct the geomagnetic angle obtained from a geomagnetic sensor is proposed. In addition, we conducted a static positioning experiment and a dynamic positioning experiment integrated with pedestrian dead reckoning on an Android platform to evaluate the effectiveness of the proposed method. According to the experimental results, when the horizontal distance between the camera and the center of the LED is less than 120 cm, the average positioning error can be retained within 10 cm and the average positioning time on the mobile phone is 39.64 ms.

Keywords: visible light positioning; geomagnetic sensors; single square LED

Citation: Yang, C.; Wen, S.; Yuan, D.; Chen, J.; Huang, J.; Guan, W. CGA-VLP: High Accuracy Visible Light Positioning Algorithm Using Single Square LED with Geomagnetic Angle Correction. *Photonics* **2022**, *9*, 653. https://doi.org/10.3390/photonics9090653

Received: 15 August 2022
Accepted: 12 September 2022
Published: 14 September 2022

Publisher's Note: MDPI stays neutral with regard to jurisdictional claims in published maps and institutional affiliations.

1. Introduction

In recent years, indoor location-based services and applications, including personal localization and navigation, object searching, and robotics, have grown rapidly [1]. Moreover, indoor positioning is still a challenging problem since the performance of a Global Positioning System (GPS) decreases remarkably in an indoor environment due to the obstruction of the walls during signal transmission. A few techniques and devices have been proposed for indoor positioning systems (IPS) to improve the performance of indoor positioning. Firstly, wireless signals (such as Wi-Fi, Bluetooth, radio frequency identification, and ZigBee) are focused on and researched extensively, then the potential of light emitting diodes (LEDs) in indoor positioning is explored. In the past few years, many algorithms for LED-based positioning have been proposed and verified by experiments, presenting better positioning results or lower costs compared to those based on wireless signals [2].

Unlike traditional radio-based technology, visible light positioning (VLP) is a type of indoor positioning technology based on visible light communication (VLC) [3–5]. LEDs can transmit data over the air by modulating at a high frequency that is invisible to the human eye but perceivable by an image sensor (IS) or photodiode (PD). Using PD, or photo detector, which provides a converted current from the illumination, methodologies can be classified according to their received optical signals, namely, received signal strength (RSS) [6,7], time of arrival (TOA) [8]/time difference of arrival (TDOA) [9], angle of arrival (AOA), and fingerprinting [10]. Although PD is a common receiver of optical signals, it is not an ideal VLP device. Firstly, it is sensitive to the light intensity and diffuse reflection of the light signal, which is detrimental to high accuracy localization [11] (p. 1). Secondly, its detection area is too small [12] (p. 1) and will thus increase the LEDs needed. By contrast, camera-based VLP is favored by both industry and commerce due to its high positioning accuracy and good compatibility with user devices such as mobile robots and smartphones. Some state-of-the-art (SOTA), camera-based VLP systems have achieved centimeter-level accuracy on commodity smartphones [13] or mobile robots [14].

However, there are still some practical limitations. One of the most urgent issues is that VLP normally requires multiple LEDs in the camera's field of view (FOV) [15–17], which means that lamps need to be densely distributed, and the effective positioning area becomes small. In order to reduce the number of LEDs required in the process, additional Micro-Electro-Mechanical System (MEMS) sensors are generally chosen to provide orientation information. However, as shown in previous research [18,19], another positioning error source owing to the inaccurate azimuth angle is introduced. In [11], with the employment of the inertial measurement unit (IMU) as the variable, a pair of comparative experiments was conducted. The error tripled when using the IMU due to the algorithm compensation and measurement error.

In this article, we put forward a single-LED localization system based on IS and geomagnetic sensor (GS). The LED used in this system is square in shape, which is common in daily life. Unlike the circular LEDs widely used before in VLP, which have numerous symmetry axes and offer less usable point features, being one feature of a circular LED [20], the square LED can provide not only displacement information but also rotation information which can effectively correct the geomagnetic angle obtained from the GS. In [21] (p. 12), the experiment illustrated that raw measurement of the heading can vastly deviate from the true value, with angle errors up to 60 degrees, which shows that the correction is not redundant. The innovative contributions are highlighted as follows:

1. We propose a VLP scheme based on the corrected geomagnetic angle (CGA-VLP) in which we relax the assumption on the minimum number of observable LEDs efficiently to one and improve the robustness in the harsh environment.
2. The proposed methodology can correct the geomagnetic angles obtained from GS, which could be further applied to other algorithms.
3. The scheme is evaluated in static and real-time environments through a tailor-made Android application and modulation drive, with pedestrian dead reckoning (PDR) functioning when LED is out of the camera's FOV. The accuracy and real-time performance are both excellent for real applications.

The rest of this article is organized as follows: the second section illustrates the proposed CGA-VLP system. The verification results are then presented in the third section. Finally, we render our conclusions.

2. Methodology

2.1. Overall Structure

The architecture of the proposed CGA-VLP system is shown in Figure 1. The modulated LED lamps with VLC functions are used as transmitters. The images are caught by Complementary Metal-Oxide-Semiconductor (CMOS) IS vertically and decoded to obtain their unique identities which are related to their global coordinates. The geomagnetic angle can be obtained from the GS, then corrected by using the geometric relation of the square LED in the images, which will be illustrated in detail in the next subsection. For a comprehensive understanding of VLC, we refer readers to our previous work [3]. PDR is another solution for IPS, which will be explained and fused with CGA-VLP in the experiment section.

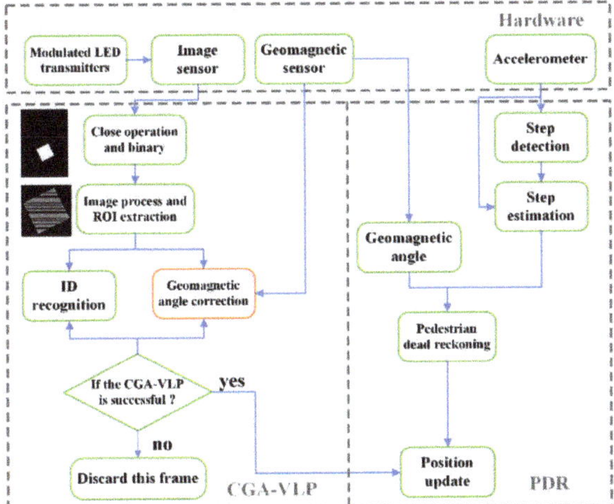

Figure 1. The block diagram of the proposed positioning system.

2.2. The Principle of Imaging Positioning

The model of the proposed VLP system is shown in Figure 2. The world coordinate (denoted as {W}), image coordinate (denoted as {I}), and pixel coordinate (denoted as {P}) are defined as follows, respectively. The origin point of the image coordinate system is the intersection point between the optical axis of the camera and the imaging plane of the image sensor. The relationship between the pixel coordinate and the image coordinate system can be denoted by the following formula:

$$\begin{cases} m = (i - i_0)dm \\ n = (j - j_0)dn \end{cases}, \qquad (1)$$

where (i_0, j_0) are the coordinates of the image sensor in the pixel coordinate system, located in the center of the image. The unit transformation of two coordinate systems is 1 pixel = dm mm and 1 pixel = dn mm.

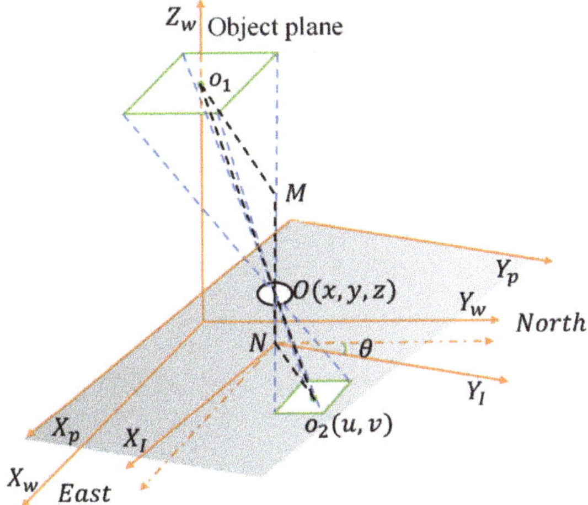

Figure 2. The positioning system model.

The original point of the world coordinate system is the vertical projection of the lamp center to the ground, and X_w, Y_w are on the plane of the ground. The direction of the Y_w axis can be arbitrary. For the sake of simplicity, it is set parallel with the direction of north in our scenario. The dashed coordinate system is to assist explanation and has no physical meaning. It shares the same origin point with the image coordinate system and is parallel with the world coordinate. The unit is pixel.

$$\begin{bmatrix} u' \\ v' \end{bmatrix} = \begin{bmatrix} \cos\theta & -\sin\theta \\ \sin\theta & \cos\theta \end{bmatrix} \begin{bmatrix} u \\ v \end{bmatrix}. \tag{2}$$

Through digital image processing, the centroid coordinates (u, v) of LED are easy to obtain, which can then be transferred to the coordinates (u', v') in the virtual coordinate system through Equation (2). In addition, θ denotes the included angle of the two coordinate systems. Ignoring the tiny deviation of the x and y axes of the photosensitive device, dm and dn are approximately equal to k. According to triangular similarity, the following equations can be obtained:

$$\frac{NO_2}{MO_1} = \frac{NO}{MO} = \frac{k \cdot u'}{x} = \frac{k \cdot v'}{y}, \tag{3}$$

$$\mu = \frac{u'}{x} = \frac{v'}{y}, \tag{4}$$

where (x, y, z) are the coordinates of the camera lens in the world coordinate system, while μ is the conversion ratio of the pixel coordinate system and world coordinate system which can be calculated by employing the actual size and the image size of the LED. Through the above process, the 2D position of the mobile phone can be determined.

According to the imaging principle:

$$\frac{1}{f} = \frac{1}{MO} + \frac{1}{NO}, \tag{5}$$

where f is the focal length. z is accessible if the focal length is known. However, the camera of a smartphone usually has the function of automatically adjusting the focal length in order to obtain clearer images, thus making it only valid at one time. Therefore, in this article, we do not measure the focal length and the height of the camera is not considered.

2.3. Geomagnetic Angle Correction

To obtain the rotation angle about the z axis in single-LED-based VLP algorithms, several methods can be used, such as utilizing a mark [22,23], or adopting sensors as assistance [24]. However, marks on the lamp will affect the illumination as well as the aesthetics. Currently, mobile phones always embed GS, which means that no additional equipment is needed if the geomagnetic angle is used. However, the indoor magnetic field is the superposition of the geomagnetic field and the interference field caused by the steel structure, elevators, cables, doors, and windows, so the geomagnetic angle detected indoors is always inaccurate [25] (p. 2). To correct the geomagnetic angle, the angle information of the square lamp is utilized. For the sake of simplicity, the LED is placed in a specific posture with one side of the square parallel to the direction pointing north. The initial posture of the phone is set with the geomagnetic angle equating to zero, where the photo of the LED captured vertically by the camera will resemble the square denoted as ABCD in Figure 3. When the phone spins, the LED revolves in the opposite direction on the picture. In this way, the clockwise rotation angle of the phone is exactly the geomagnetic angle; represented in the image is the rotation angle of the square denoted as θ in Figure 3. According to the similar triangle principle, it is easy to compute:

$$\gamma_1 = \alpha = 90° - \theta. \tag{6}$$

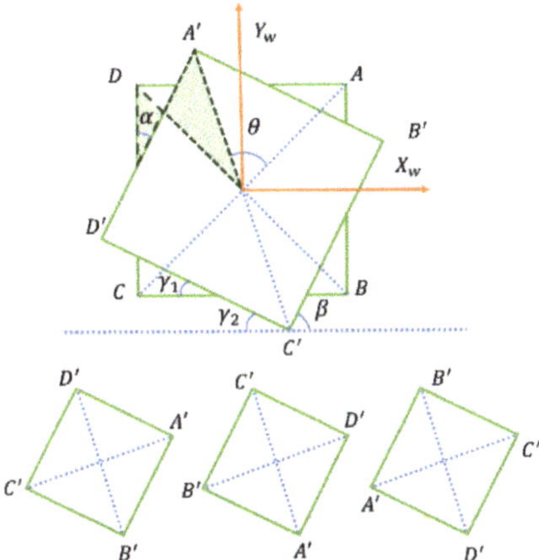

Figure 3. Four possible situations.

Since γ_1 and γ_2 are corresponding angles, they are certainly equal in value and γ_2 is a reliable angle that we can measure from the image.

Corresponding to the four cases shown in Figure 3, there are four possible situations and the real projection cannot be distinguished from the image. The four possible values for real rotation angle γ are illustrated in Equation (7):

$$\gamma = \gamma_2, \gamma = \gamma_2 + 90^\circ, \gamma = \gamma_2 + 180^\circ, \gamma = \gamma_2 + 270^\circ, \tag{7}$$

which will be compared to the value obtained from the geomagnetic sensor, then the angle with the smallest difference will be selected as the corrected geomagnetic angle. In the practical application, there may be difficulty in the installation of lamps according to the above settings. However, it does not matter as long as the γ_0, which indicates the included angle between one side of the lamp and due north, is noted. It can be calculated through the shot taken when the phone is in the initial posture, namely, β shown in Figure 3. In addition, γ_2 should subtract γ_0 before being used in Equation (7). After that, the algorithm is the same.

3. Experiments and Analysis

3.1. Receiver

The system is made up of a receiver and transmitters. The receiver is a mobile device, namely, Huawei P10. The IS used in this experiment is the embedded front camera. The exposure time of the camera was set as 0.05 ms to ensure the stripes and the edges of the LED are clear. This parameter may vary depending on the aperture size of different cameras. The resolution is optional but was selected as 1920 × 1080 in our experiment, which is what we recommend so that the picture of this resolution can meet the requirements of clarity while not being too large. With the rolling shutter effect, the exposure of the camera is conducted in a row-by-row manner instead of exposing the whole image at a single moment, so the flash of the LED will form stripes in the image. The image is processed successively by close operation, gray processing, binarization, and region of interest (ROI) extraction. To eliminate the interference of other lamps, the ROI with a shape close to a square and a size within a certain range is selected and decoded. In addition, the contours in the ROI will be detected using Canny operators, then the Hough transformation is employed to extract the lines, with which the geomagnetic angle will be corrected, as shown in Figure 3. Thus, the precise position of the camera can be obtained. Due to the stripes, there will be many horizontal lines, so the angle close to zero calculated from the picture needs to be discarded. If the sides of the square are also parallel with the sides of the picture, the contour cannot be distinguished from the stripes. Therefore, under that circumstance, the correction will be abolished and the raw geomagnetic angle will be used.

In our experiments, all data capturing and processing are performed on the mobile device, through a tailor-made application, as shown in Figure 4a. The application can display direction, positioning results, and positioning mode in real time. In addition, the data can be exported in a table for further analysis. When there is no LED in the camera's FOV, the application will execute PDR which will be introduced in a later section.

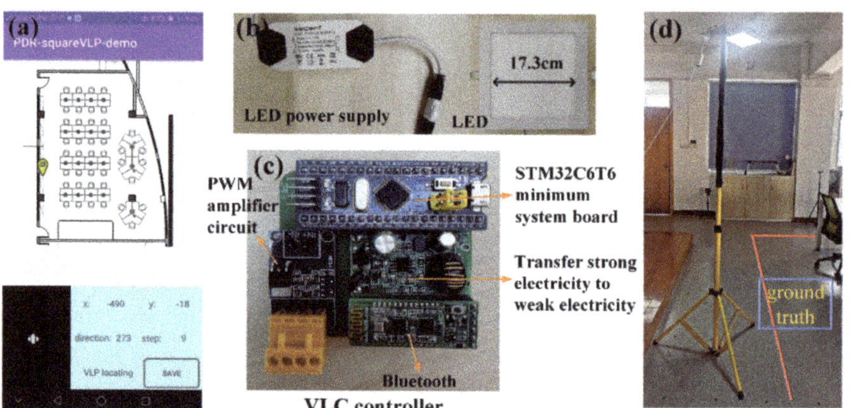

Figure 4. System setup. (**a**) The interface of the software; (**b**) The LED and the power supply; (**c**) The VLC controller; (**d**) The scenario of the experiment.

3.2. Transmitters

The transmitters are modulated LEDs mounted to the light pole, as shown in Figure 4d. To modulate LED more conveniently, a tailor-made VLC controller module that integrates the Bluetooth modulation function was designed.

The principle of the VLC controller is revealed in Figure 5. The alternating current is converted to direct current by the LED power supply, as shown in Figure 4b. The buck module is responsible for the power supply of the Bluetooth and the MCU, namely, STM32C6T6. In addition, the Pulse Width Modulation (PWM) amplifier circuit amplifies the PWM signal outputted by the MCU to the rated voltage of the LED. The current is then modulated by the VLC controller to illume the LED and transmit the signal simultaneously. The Bluetooth can communicate with mobile phones and then transfer the instructions to the MCU which controls the on–off state of the LED. For convenience, the off-the-shelf modules are organized on a printed circuit board (PCB), at the corner of which the power interface and the interface for LED are gathered, as shown in Figure 4c.

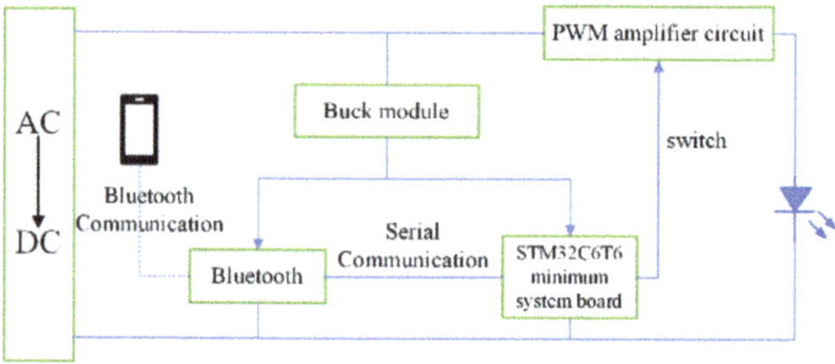

Figure 5. The block diagram of the VLC controller.

The time-varying switch state of the lamp represents binary data sequences which consist of the header and unique identification (ID). After modulation, the LED will send the specified data circularly at the same time of illumination.

3.3. CGA-VLP System Positioning Accuracy

To evaluate the position accuracy of the proposed positioning system, two series of experiments were performed. The first series was to test the stationary positioning performance of CGA-VLP. The square lamp was installed horizontally 260 cm above the ground, with the mobile phone placed flat on the ground.

We chose 108 evenly spaced points around the LED center and calculated the positioning results employing CGA-VLP and GA-VLP, respectively. The limited horizontal range of positioning was 150 cm from the center of the LED.

Figure 6 shows the positioning results and the corresponding errors. The results of CGA-VLP are displayed in red. CGA-VLP's positioning effect is significantly better than GA-VLP's, namely, the blue ones, especially when the mobile phone is farther away from the center of the lamp.

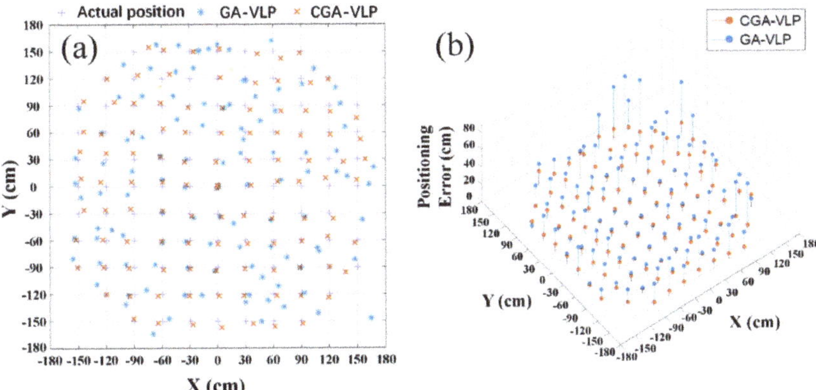

Figure 6. Positioning results and corresponding errors. (**a**) Positioning results; (**b**) Positioning errors.

We calculated the average error of the positioning results under the same horizontal distance, as shown in Figure 7. When the horizontal distance is under 120 cm, the maximum average positioning error of CGA-VLP is 8.5 cm. Even if the horizontal distance reaches 175 cm, the average positioning errors of CGA-VLP can still be maintained below 20 cm. In addition to the error of installing lamps, the experimental error comes from the combined action of the error of the ROI position and the error of the rotation angle. When the mobile phone is farther away from the center of the positioning area, the error of the ROI position increases. After multiplying by the rotation angle with error, the positioning error will sharply increase. By contrast, with the corrected geomagnetic angle, the positioning error will not increase as much.

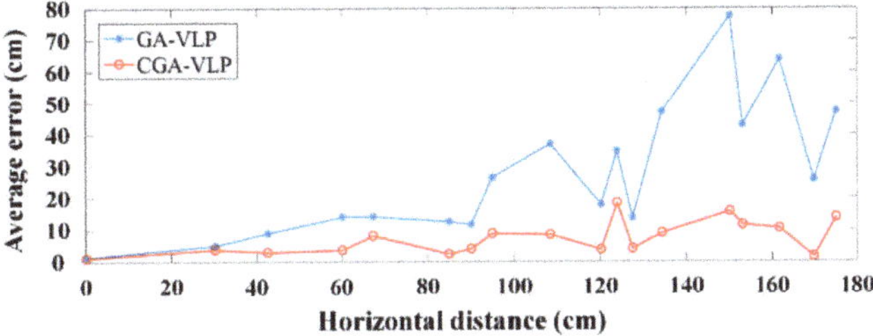

Figure 7. Average error when comparing CGA-VLP and GA-VLP.

3.4. Dynamic Positioning

As mentioned above, the effective positioning area of VLP is confined by the camera's FOV. Once the LED cannot be captured, the positioning cannot be executed. The demand for density of LEDs deems that VLP is not suitable for realistic application independently. Luckily, PDR is another solution for IPS. Generally, a PDR algorithm consists of three phases: step detection (SD), step length estimation (SLE), and position–solution update (PSU). Benefiting from the popularity of smartphones, the methodology is adored due to its simplicity and low cost [26]. Equation (8) illustrates the mechanism of PDR:

$$\begin{cases} x_n = x_{n-1} + l \cdot cos\gamma \\ y_n = y_{n-1} + l \cdot sin\gamma \end{cases}' \tag{8}$$

where (x_n, y_n) is the current position coordinates, (x_{n-1}, y_{n-1}) is the position coordinates of the previous moment, and l is the size of each step. Unlike the schemes requiring signal generators installed in the environment before experiments, PDR uses sensors attached to the users to estimate relative positions to previous or known position, so it is more susceptible to cumulative error.

In this section, we fused VLP and PDR to adapt to real application scenarios. PDR was employed when VLP could not work, and VLP can correct the cumulative error of PDR. The flow chart of the scheme is shown in Figure 1. Peak detection [27] was adopted for SD, while the Weinberg model [28] was adopted for SLE. The heading angle obtained through the GS and the estimated stride length were then combined for PSU.

During the test, the smartphone maintains a horizontal state, with the top of the smartphone pointing to the moving direction. Limited by the size of the experimental site, the route was set as a 12×6 m rectangle. Three rounds around the rectangular path were completed in each experiment, so the route would be 108 m in total. For test operations, we equipped our laboratory with four LEDs, with one on each side of the rectangle. One of them is shown in Figure 4d, with a corner of the ground truth marked using red lines. More detailed parameters can be found in Table 1. We performed our experiment with two positioning methods simultaneously with the only difference between the methods being whether CGA-VLP was used when the LED was in the camera's FOV. In order to show our experiment device and scene more clearly, we also made a simple demonstration video (see Video S1).

Table 1. Parameters of the Experiments.

LED Specifications	
Coordinates of LED1 (cm)	$(-490, -28)$
Coordinates of LED2 (cm)	$(-1225, -300)$
Coordinates of LED3 (cm)	$(-1002, -620)$
Coordinates of LED4 (cm)	$(22, -148)$
Rated voltage of the LED	72 V
Power of the LED	18 W
Mobile Phone Specifications	
Frame rate	5 fps
Sampling rate of the accelerometer	250 Hz
Resolution	1920×1080
Camera exposure time	0.05 ms

3.4.1. Accuracy of the Dynamic Positioning

The positioning results of different methods are represented by different colors, with the points connected by straight lines to show the trajectories. To prevent overlap, the results of the three laps are plotted separately. The positioning track corrected by CGA-VLP is roughly close to the actual route, while the track for pure-PDR increasingly deviates from the ground truth as the route lengthens, with the final error reaching 3 m. Missing detection, false detection, and wrong step estimation will all affect the distance, and inaccurate direction will cause the trajectory to drift. It is clear that the two trajectories are the same where there is no LED, but in the fusion positioning, the trajectory was pulled back to the actual route by the VLP before the PDR error is further expanded. As shown in the purple box in Figure 8a, the error of drift is corrected by VLP in time. In the purple box of Figure 8b, the positioning distance exceeds the actual walking distance, which is corrected through VLP, thus preventing the accumulation of the error. In addition, in the similar position of Figure 8c, the positioning distance is shorter than the real distance, which is also corrected by VLP.

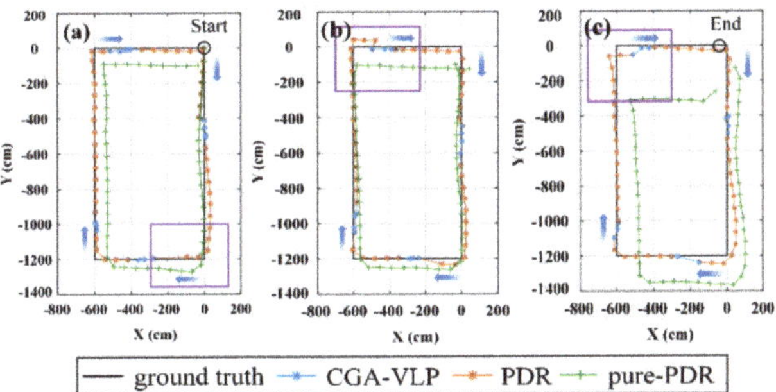

Figure 8. The positioning results. (**a**) The results of the first lap; (**b**) The results of the second lap; (**c**) The results of the third lap.

3.4.2. Real-Time Performance of the Dynamic Positioning

In this subsection, we focus on real-time performance. To reduce the burden of the phone, the frame rate was set as 5 fps, which was sufficient for calculating CGA-VLP several times when passing the lamp at normal speed. The data of seven experiments were recorded, of which 220 frames were with LED. The program execution time of the key steps of CGA-VLP was separately recorded and is shown in Figure 9. The mean time for correcting the rotation angle and decoding and extracting the ROI is 4.1415 ms, 16.5 ms, and 15.4679 ms, respectively. In addition, the total delay is 39.64 ms, on average, which also includes the time to create a new picture and convert it from bitmap format to RGB format, the time to make logical judgments in the main function, and the time to modify the UI. With variation due to the status of the phone and the quality of the photo, the calculation time fluctuated greatly on both sides of the average value. Despite this, the application of positioning when walking at a normal speed can be satisfied with the real-time performance of the algorithm. As shown in Figure 8, when people walk past the LED, the program can stably position through CGA-VLP several times, which precisely reflects the position of the pedestrian and correct errors.

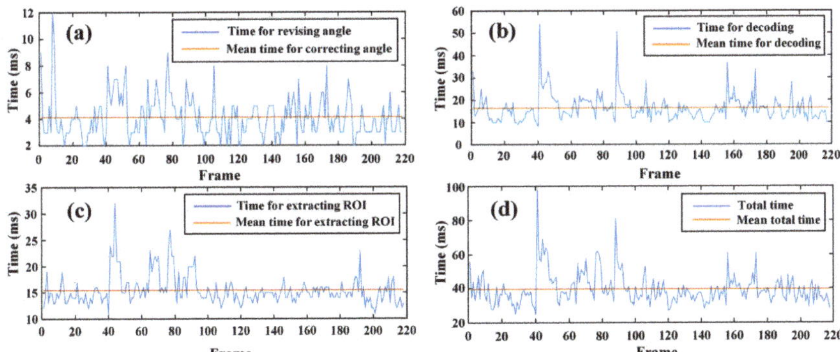

Figure 9. Program execution time. (**a**) The time for revising the geomagnetic angle; (**b**) The time for extracting the ROI; (**c**) The time for decoding; (**d**) The total time for program execution.

3.4.3. Accuracy of the CGA

In this subsection, the accuracy and error source of the proposed CGA are discussed. The errors of the corrected heading angle were calculated and presented in Figure 10, with the mean error 3.3°. The heading angle error shows randomness with several values over 10°, but according to the cumulative distribution function, more than 95% of errors are less than 8°. When the error of the geomagnetic sensor is within the range of 45° above and below the true value, the operation of correcting the geomagnetic angle is effective and does not contain systematic errors, to some extent ensuring the robustness of the algorithm. However, admittedly, there are still several limitations causing calculation error. Firstly, the proposed CGA-VLP requires the imaging plane and the square lamp to be parallel, which is almost impossible to achieve in the dynamic scenario, where the tester holds the phone while walking. Secondly, there may be errors in the process of extracting LED edges and calculating angles. Thirdly, the hand-held mobile phone will shake during walking. However, we think the practicality of the proposed method is acceptable since the blue lines shown in Figure 8 are close to the actual path.

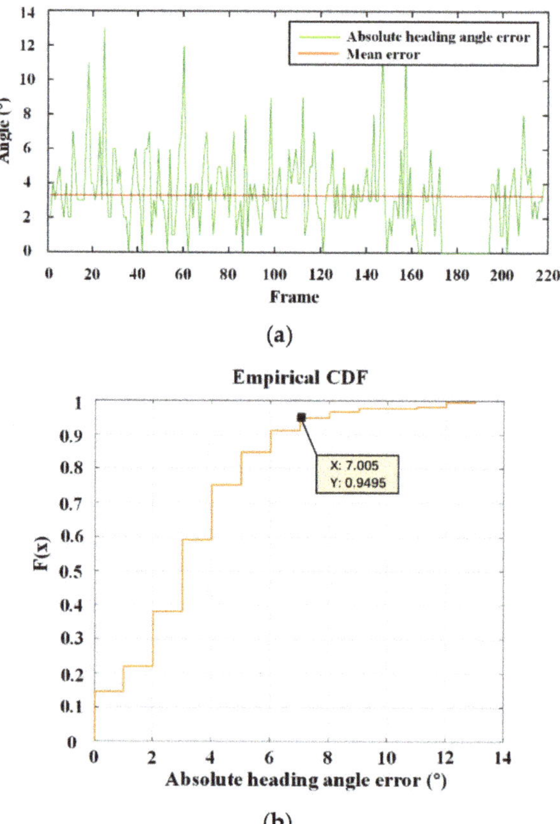

Figure 10. The absolute heading angle error. (**a**) The absolute heading angle error; (**b**) The cumulative distribution function of the absolute heading angle error.

3.5. Discussion

In this study, we propose a novel CGA-VLP algorithm which utilizes the GS to relax the number of observable LEDs required for positioning to one as well as correct the geomagnetic angle, thus ensuring accuracy. Consistent with previous research [25,29,30], the effect of correction of the geomagnetic angle when used for indoor positioning is remarkable. However, unlike these methods using filtering to correct the geomagnetic angle, the rotation information contained in the picture is utilized in our algorithm. In [25], after correction by the proposed 1D CNN-Kalman, 95% heading angle errors are less than 9°. The authors of [30] show that the mean error of the heading angle corrected by the Adaptive Cubature Kalman Filter is approximately 6°. By contrast, our CGA algorithm shows advantages both in terms of average error and cumulative error. In [23], the average positioning error is 2.3 cm, but the positioning area is 0.8 m × 0.8 m, with the calculation time 60 ms in a low-end embedded platform. In [11], the 2D error of 95% of points reaches 9 cm. In [31], the positioning error is up to 8.7 cm considering 90% of points. In general, the CGA-VLP has advantages in accuracy and delay. It is worth mentioning that if the CGA-VLP is used in a robot where the camera could be fixed and stable, the error caused by unstrict parallelism between LED and the imaging plane may be avoidable; we plan to further explore this in our future research. In addition, the correction is limited when the stripes caused by VLC are parallel with the side of the square, which is also an aspect we want to improve.

4. Conclusions

In this article, we proposed a VLP system with a single square LED which can correct the geomagnetic angle obtained from the GS. The static experiment showed that although the positioning error would increase as the phone moved farther away from the center of the LED, it could still be maintained reliably within 10 cm when the horizontal distance was less than 120 cm, while the positioning error for GA-VLP reached 40 cm. The algorithm was also tested and verified in a dynamic scenario fusing PDR. Positioning ability and real-time performance were both sufficiently excellent for live applications. The total delay was 39.64 ms, on average. In the future, we expect to explore the practical application of the proposed CGA-VLP in robots, improve its performance in terms of the effective positioning area and practicality, and implement tight fusion of PDR and VLP with the Kalman filter to improve its accuracy.

Supplementary Materials: The following supporting information can be downloaded at https://www.mdpi.com/article/10.3390/photonics9090653/s1, Video S1: Demo for the fusion of CGA-VLP and DPR.

Author Contributions: Methodology: C.Y. and W.G.; Experiment and Analysis, C.Y. and D.Y.; Writing, C.Y.; Visualization, J.H. and C.Y.; Hardware, C.Y.; Proofreading, W.G., J.C. and S.W; Funding, S.W. All authors have read and agreed to the published version of the manuscript.

Funding: This work was funded in part by the Guangdong Science and Technology Project under Grant 2017B010114001, in part by the National Undergraduate Innovative and Entrepreneurial Training Program under Grants 202010561158, 202010561155, 202110561162, 202110561165, and 202110561163, and in part by the Guangdong Provincial Training Program of Innovation and Entrepreneurship for Undergraduates under Grant S202010561272.

Institutional Review Board Statement: Not applicable.

Informed Consent Statement: Not applicable.

Data Availability Statement: Not applicable.

Conflicts of Interest: The authors declare no conflict of interest.

References

1. Alletto, S.; Cucchiara, R.; Del Fiore, G.; Mainetti, L.; Mighali, V.; Patrono, L.; Serra, G. An indoor location-aware system for an IoT-based smart museum. *IEEE Internet Things J.* **2016**, *3*, 244–253. [CrossRef]
2. Zhuang, Y.; Hua, L.C.; Qi, L.N.; Yang, J.; Cao, P.; Cao, Y.; Wu, Y.P.; Thompson, J.; Haas, H. A survey of positioning systems using visible LED lights. *IEEE Commun. Surv. Tutor.* **2018**, *20*, 1963–1988. [CrossRef]
3. Song, H.Z.; Wen, S.S.; Yang, C.; Yuan, D.L.; Guan, W.P. Universal and effective decoding scheme for visible light positioning based on optical camera communication. *Electronics* **2021**, *10*, 1925. [CrossRef]

4. Chow, C.W.; Chen, C.Y.; Chen, S.H. Enhancement of signal performance in LED visible light communications using mobile phone camera. *IEEE Photonics J.* **2015**, *7*, 7903607. [CrossRef]

5. Chen, Y.; Ren, Z.M.; Han, Z.Z.; Liu, H.L.; Shen, Q.X.; Wu, Z.Q. LED based high accuracy indoor visible light positioning algorithm. *Optik* **2021**, *243*, 166853. [CrossRef]

6. Sun, X.; Zhuang, Y.; Huai, J.; Hua, L.; Chen, D.; Li, Y.; Cao, Y.; Chen, R. RSS-based visible light positioning using non-linear optimization. *IEEE Internet Things J.* **2022**, *9*, 14134. [CrossRef]

7. Chen, Y.; Zheng, H.; Liu, H.; Han, Z.Z.; Ren, Z.M. Indoor High Precision Three-Dimensional Positioning System Based on Visible Light Communication Using Improved Hybrid Bat Algorithm. *IEEE Photonics J.* **2020**, *12*, 6802513. [CrossRef]

8. Wang, T.Q.; Sekercioglu, Y.A.; Neild, A.; Armstrong, J. Position accuracy of Time-of-Arrival based ranging using visible light with application in indoor localization systems. *J. Lightwave Technol.* **2013**, *31*, 3302–3308. [CrossRef]

9. Jung, S.-Y.; Hann, S.; Park, C.-S. TDOA-based optical wireless indoor localization using LED ceiling lamps. *IEEE Trans. Consum. Electron.* **2011**, *57*, 1592–1597. [CrossRef]

10. Shi, C.; Niu, X.; Li, T.; Li, S.; Huang, C.; Niu, Q. Exploring Fast Fingerprint Construction Algorithm for Unmodulated Visible Light Indoor Localization. *Sensors* **2020**, *20*, 7245. [CrossRef]

11. Huang, H.Q.; Lin, B.; Feng, L.H.; Lv, H.C. Hybrid indoor localization scheme with image sensor-based visible light positioning and pedestrian dead reckoning. *Appl. Opt.* **2019**, *58*, 3214–3221. [CrossRef]

12. Wang, Y.; Hussain, B.; Yue, C.P. Arbitrarily tilted receiver camera correction and partially blocked LED image compensation for indoor visible light positioning. *IEEE Sens. J.* **2022**, *22*, 4800–4807. [CrossRef]

13. Fang, J.B.; Yang, Z.; Long, S.; Wu, Z.Q.; Zhao, X.M.; Liang, F.N.; Jiang, Z.L.; Chen, Z. High-speed indoor navigation system based on visible light and mobile phone. *IEEE Photonics J.* **2017**, *9*, 8200711. [CrossRef]

14. Guan, W.; Huang, L.; Wen, S.; Yan, Z.; Liang, W.; Yang, C.; Liu, Z. Robot localization and navigation using visible light positioning and SLAM fusion. *J. Lightwave Technol.* **2021**, *39*, 7040–7051. [CrossRef]

15. Xu, J.J.; Gong, C.; Xu, Z.Y. Experimental indoor visible light positioning systems with centimeter accuracy based on a Commercial smartphone camera. *IEEE Photonics J.* **2018**, *10*, 7908717. [CrossRef]

16. Guan, W.; Zhang, X.; Wu, Y.; Xie, Z.; Li, J.; Zheng, J. High precision indoor visible light positioning algorithm based on double LEDs using CMOS image sensor. *Applied Sciences.* **2019**, *9*, 1238. [CrossRef]

17. Liang, Q.; Lin, J.H.; Liu, M. Towards robust visible light positioning under LED shortage by visual-inertial fusion. In Proceedings of the 10th International Conference on Indoor Positioning and Indoor Navigation (IPIN), Pisa, Italy, 30 September–3 October 2019.

18. Li, F.; Zhao, C.S.; Ding, G.Z.; Gong, J.; Liu, C.X.; Zhao, F.; Assoc Comp, M. A reliable and accurate indoor localization method using phone inertial sensors. In Proceedings of the 14th ACM International Conference on Ubiquitous Computing (UbiComp), Carnegie Mellon University, Pittsburgh, PA, USA, 5–8 September 2012.

19. Li, M.Y.; Mourikis, A.I. Online temporal calibration for camera-IMU systems: Theory and algorithms. *Int. J. Rob. Res.* **2014**, *33*, 947–964. [CrossRef]

20. Liang, Q.; Liu, M. A tightly coupled VLC-inertial localization system by EKF. *IEEE Robot. Autom. Lett.* **2020**, *5*, 3129–3136. [CrossRef]

21. Xie, B.; Chen, K.; Tan, G.; Lu, M.; Liu, Y.; Wu, J.; He, T. Lips: A light intensity-based positioning system for indoor environments. *ACM Trans. Sens. Netw.* **2016**, *12*, 1–27. [CrossRef]

22. Zhang, R.; Zhong, W.D.; Qian, K.M.; Zhang, S. A single LED positioning system based on circle projection. *IEEE Photonics J.* **2017**, *9*, 7905209. [CrossRef]

23. Li, H.P.; Huang, H.B.; Xu, Y.Z.; Wei, Z.H.; Yuan, S.C.; Lin, P.X.; Wu, H.; Lei, W.; Fang, J.B.; Chen, Z. A fast and high-accuracy real-time visible light positioning system based on single LED lamp with a beacon. *IEEE Photonics J.* **2020**, *12*, 7906512. [CrossRef]

24. Ji, Y.; Xiao, C.; Gao, J.; Ni, J.; Cheng, H.; Zhang, P.; Sun, G. A single LED lamp positioning system based on CMOS camera and visible light communication. *Opt. Commun.* **2019**, *443*, 48–54. [CrossRef]

25. Hu, G.H.; Wan, H.; Li, X.X. A High-precision magnetic-assisted heading angle calculation method based on a 1D convolution neural network (CNN) in a complicated magnetic environment. Micromachines. *Micromachines* **2020**, *11*, 642. [CrossRef]

26. Harle, R. A survey of indoor inertial positioning systems for pedestrians. *IEEE Commun. Surv. Tutor.* **2013**, *15*, 1281–1293. [CrossRef]

27. Fang, S.H.; Wang, C.H.; Huang, T.Y.; Yang, C.H.; Chen, Y.S. An enhanced ZigBee indoor positioning system with an ensemble approach. *IEEE Commun. Lett.* **2012**, *16*, 564–567. [CrossRef]

28. Weinberg, H. *An-602 Using the adxl202 in Pedometer and Personal Navigation Applications*; Analog Devices Inc.: Norwood, MA, USA, 2002.

29. Wang, Y.; Zhao, H.D. Improved Smartphone-Based Indoor Pedestrian Dead reckoning Assisted by Visible Light Positioning. *IEEE Sen. J.* **2019**, *19*, 2902–2908. [CrossRef]

30. Geng, J.J.; Xia, L.Y.; Wu, D.J. Attitude and heading estimation for indoor positioning based on the adaptive cubature Kalman filter. *Micromachines* **2021**, *12*, 79. [CrossRef]

31. Yan, Z.H.; Guan, W.P.; Wen, S.S.; Huang, L.Y.; Song, H.Z. Multirobot cooperative localization based on visible light positioning and odometer. *IEEE Trans. Instrum. Meas.* **2021**, *70*, 7004808. [CrossRef]

Article

Indoor Visible-Light 3D Positioning System Based on GRU Neural Network

Wuju Yang, Ling Qin *, Xiaoli Hu and Desheng Zhao

College of Information Engineering, Inner Mongolia University of Science and Technology, Baotou 014010, China; yangwuju@stu.imust.edu.cn (W.Y.); huxiaoli@imust.edu.cn (X.H.); 2018978@imust.edu.cn (D.Z.)
* Correspondence: qinling1979@imust.edu.cn

Abstract: With the continuous development of artificial intelligence technology, visible-light positioning (VLP) based on machine learning and deep learning algorithms has become a research hotspot for indoor positioning technology. To improve the accuracy of robot positioning, we established a three-dimensional (3D) positioning system of visible-light consisting of two LED lights and three photodetectors. In this system, three photodetectors are located on the robot's head. We considered the impact of line-of-sight (LOS) and non-line-of-sight (NLOS) links on the received signals and used gated recurrent unit (GRU) neural networks to deal with nonlinearity in the system. To address the problem of poor stability during GRU network training, we used a learning rate attenuation strategy to improve the performance of the GRU network. The simulation results showed that the average positioning error of the system was 2.69 cm in a space of 4 m $\times$ 4 m $\times$ 3 m when only LOS links were considered and 2.66 cm when both LOS and NLOS links were considered with 95% of the positioning errors within 7.88 cm. For two-dimensional (2D) positioning with a fixed positioning height, 80% of the positioning error was within 9.87 cm. This showed that the system had a high anti-interference ability, could achieve centimeter-level positioning accuracy, and met the requirements of robot indoor positioning.

Keywords: robot; visible-light positioning (VLP); three-dimensional (3D); line-of-sight (LOS) and non-line-of-sight (NLOS) links; gated recurrent units (GRU) neural networks; learning rate decay strategy

Citation: Yang, W.; Qin, L.; Hu, X.; Zhao, D. Indoor Visible-Light 3D Positioning System Based on GRU Neural Network. *Photonics* **2023**, *10*, 633. https://doi.org/10.3390/photonics10060633

Received: 17 April 2023
Revised: 26 May 2023
Accepted: 28 May 2023
Published: 31 May 2023

1. Introduction

With the progress of human beings and the development of technology, the application scenarios of robots have become more complex and diversified, and robots need to complete more difficult and intelligent work. In order to improve the efficiency and performance of robots, the positioning and navigation of autonomous robots are essential. At present, wireless positioning technologies such as wireless local area networks (WLANs), Bluetooth, radio frequency identification (RFID), ZigBee, and ultra-wideband (UWB) are commonly used for indoor positioning [1–5], but these wireless technologies generally have disadvantages such as high electromagnetic radiation, high deployment costs, and low positioning accuracy [6]. Compared with these wireless technologies, visible-light has the advantages of abundant bandwidth resources, no electromagnetic pollution, and low equipment costs, and it can achieve lighting and positioning at the same time. As a new type of wireless positioning technology, visible-light positioning based on LED has become a research hotspot in the field of wireless positioning [7].

In recent years, with the development of artificial intelligence, machine learning and deep learning algorithms, with their strong self-learning and generalization abilities, have become able to provide accurate positioning results in the context of VLP, and increasing numbers of people have applied them to indoor visible-light positioning. Abu Bakar et al. [8] use a weighted k-nearest neighbor (WKNN) algorithm for localization in a fingerprint recognition technique based on received signal strength (RSS). The results show that the positioning accuracy of the WKNN algorithm is better than that of the multi-layer perceptron (MLP)-based regressor. In addition to using a single ML algorithm, multiple ML algorithms can also be used for fusion localization. Huy Q. Tran et al. [9] use a dual-functional ML algorithm leveraging machine learning classification (MLC) and machine learning regression (MLR) functions to improve localization accuracy under the negative effects of multipath reflection. They use ML classification functions to divide the floor of a room into two separate zones. Then, the regression function of the ML algorithm is used to predict the position of the optical receiver. ML algorithms can also analyze and optimize other parameters. Sheng Zhang et al. [10] use neural networks to reduce position offset errors caused by uneven initial delay patterns of off-the-shelf LEDs. However, applying ML algorithms in VLP also has limitations, as they often require propagation of near-ideal behavior of the model and its parameters to perform well, and ML algorithms are too data-dependent and require a lot of time to be measured offline. We can obtain the data set by linear fitting, which can effectively reduce the offline measurement time. In addition, the parameter settings in the ML algorithm have a great influence on the positioning results, and the best model is not obtained frequently. Therefore, we need to call parameters or process them using optimization algorithms.

At present, most work on indoor visible-light localization has focused on two-dimensional positioning, assuming a fixed receiver height and ignoring positional errors due to height variation [11–13]. In the future, robots will need to complete a variety of difficult actions, so their positioning height cannot be limited, and they will require accurate and reliable three-dimensional positioning covering indoor areas. However, some 3D visible-light positioning systems use hybrid algorithms [14–16], which greatly increase the complexity of the system. In order to improve the accuracy of robot positioning and reduce system complexity, we proposed a three-dimensional indoor visible-light localization system based on a GRU neural network. We use three PDs as receivers and two LED light sources as transmitters, each LED sends signals of different frequencies. The signals collected by PD are filtered to obtain two signals of different frequencies. When data is processed, it is usually processed sequentially, so the collected data can be considered a kind of sequential data. Recurrent neural networks are very efficient for data with sequential properties, and can mine time series information from the data. The GRU network is a variant of the recurrent network, which can automatically extract effective features from experimental data, so as to obtain high positioning performance, and the localization model structure is simple and converges easily. In this study, considering the influence of LOS and NLOS links on the received signal strength, the GRU algorithm was applied to a three-dimensional indoor visible-light positioning system; a fingerprint database was established using the optical power value and position data received by the PD and then substituted into the GRU neural network to train the model; and, finally, the position information was predicted by the trained model, and the feasibility of the proposed algorithm was proved by simulations. As far as we know, the traditional 3D VLP positioning method requires the use of three or more LEDs to accurately position and does not consider wall reflections [17,18]. In addition, they usually use multiple localization algorithms in the selection of positioning algorithms, which makes the VLP system model complex. Compared to these articles, our proposed VLP system uses only two LEDs and a new receiver model, which is lower cost and easier to implement. We only use one positioning algorithm to achieve accurate three-dimensional positioning, and the system complexity is low. We analyze the influence of positioning height on the VLP system.

The rest of this paper is organized as follows: Section 2 describes the composition of the visible-light localization system model. Section 3 describes the principles of the GRU neural network. Then, in Section 4, the application of the GRU neural network for visible-light localization is described. Finally, the positioning results are discussed in Section 5, and the performance of the visible-light positioning system is analyzed.

2. Visible-Light Positioning Model

2.1. System Model

The indoor visible-light localization model designed in this study is shown in Figure 1. The room size was set to 4 m × 4 m × 3 m, and the corner of the room was used as the origin to establish the Cartesian coordinate system of the space. We used two LEDs as transmitters, placed on the ceiling, and each LED sent signals of different frequencies.

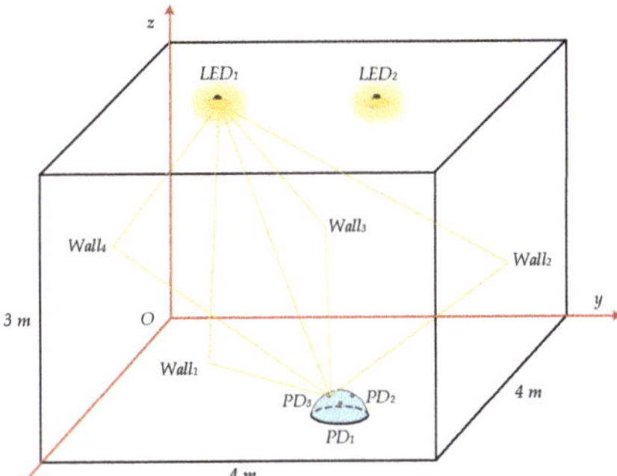

Figure 1. Indoor visible-light positioning model.

To fully receive the signal sent by the transmitter, we used three PDs as receivers, which were located in front of the robot's head, on the left at the rear, and on the right at the rear. The model structure of the robot head receiver is shown in Figure 2, which represents the robot head as a hemispherical model, and the three PDs on the head and the top center point are equidistant. In this robot head receiver model, the top center point O was used as the test point; r is the radius of the hemisphere; l is the length of the arc between point O and PD_i; $\alpha_i (i = 1, 2, 3)$ is the azimuth angle of PD_i; θ is the central angle of the arc between point O and PD_i; and $\beta (0 < \beta < 90°)$ is the elevation angle of PD_i, which can be expressed as the following.

$$\beta = \theta = l/r \tag{1}$$

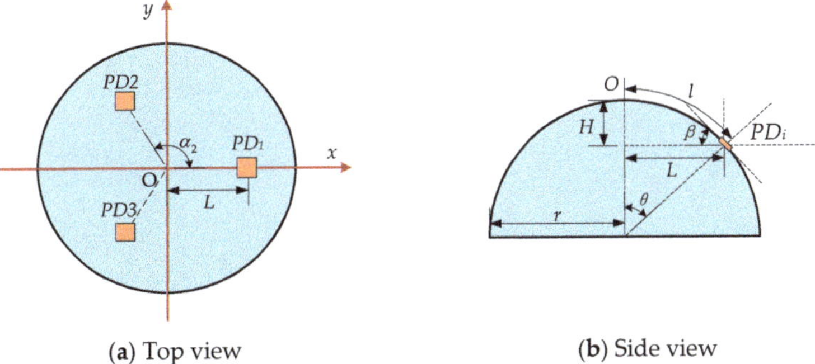

(a) Top view　　　　　　　　　　(b) Side view

Figure 2. Robot head receiver model structure.

Therefore, the relationship between the position (x_i, y_i, z_i) of PD_i and the position (x_0, y_0, z_0) of the top center point O is

$$\begin{cases} x_i = x_0 + L\cos(\alpha_i) \\ y_i = y_0 + L\sin(\alpha_i) \\ z_i = z_0 - H \end{cases} , \tag{2}$$

where L is the horizontal distance between point O and PD_i, and H is the vertical distance between point O and PD_i. L and H can be expressed as the following.

$$L = r\sin(\beta),\tag{3}$$

$$H = r(1 - \cos(\beta)).\tag{4}$$

2.2. Channel Model

The indoor visible-light channel model is shown in Figure 3 for the direct link model and the reflected link model, respectively. For an LOS link model, the indoor optical signal transmission link is short, so the attenuation of the optical signal caused by absorption and scattering is small. However, for an NLOS link model, because the indoor walls, floors, and other objects with reflection characteristics cause the diffuse reflection of the optical signal, the optical signal transmission link becomes longer, increasing the attenuation of the optical signal. Therefore, we considered the transmission of optical signals through LOS and NLOS links. This not only conformed to the real-world environment but also allowed further study of the adverse effects of reflection on system performance, making the positioning system more reliable and practical.

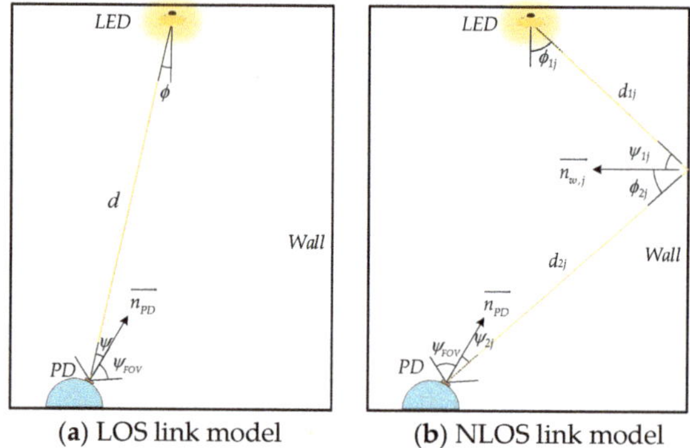

(a) LOS link model   (b) NLOS link model

Figure 3. Indoor visible-light channel model.

In the LOS link model, the relationship between the received power P_{LOS} of the PD and the LED transmitted power P_t can be expressed as [19].

$$P_{LOS} = P_t H_{LOS}(0),\tag{5}$$

where $H_{LOS}(0)$ is the DC gain of the LOS link. Assuming that the LEDs obey the Lambert radiation model, $H_{LOS}(0)$ can be expressed as [20]

$$H_{LOS}(0) = \begin{cases} \frac{(m+1)A_{PD}}{2\pi d^2} \cos^m(\phi) T_s(\psi) g(\psi) \cos(\psi), & 0 \leq \psi \leq \psi_{FOV}, \\ 0, & \psi > \psi_{FOV} \end{cases}\tag{6}$$

where A_{PD} is the effective receiving area of the PD; d is the distance from the PD to the LED; m is the Lambertian emission order; ϕ is the emission angle of the LED; $T_s(\psi)$ is the optical filter gain; $g(\psi)$ is the gain of the optical concentrator; and ψ and ψ_{FOV} are the incidence and field-of-view (FOV) angles of the PD, respectively. m and $g(\psi)$ can be expressed as [21]

$$m = -\frac{\ln(2)}{\ln(\cos(\phi_{1/2}))},\tag{7}$$

$$g(\psi) = \begin{cases} \dfrac{n^2}{\sin^2(\psi_{FOV})}, & 0 \leq \psi \leq \psi_{FOV} \\ 0, & \psi > \psi_{FOV} \end{cases}, \tag{8}$$

where $\phi_{1/2}$ is the semi-angle at half-power of the LED emitters, and n is the internal refractive index of the optical concentrator. In this paper, two LEDs placed on the ceiling were used as light sources, and three PDs placed on the hemispherical surface were used as receivers. Each PD had a certain inclination angle, and the radiating angle cosine of the LED and the incidence angle cosine of the inclined PD could be expressed as [22]

$$\cos(\phi) = \frac{h}{d}, \tag{9}$$

$$\cos(\psi) = \frac{\overrightarrow{v_{PD\_LED}} \cdot \overrightarrow{n_{PD}}}{\left\|\overrightarrow{v_{PD\_LED}}\right\|\left\|\overrightarrow{n_{PD}}\right\|}, \tag{10}$$

where h is the vertical height of the LED in relation to the PD; $\overrightarrow{v_{PD\_LED}}$ is the direction vector from the PD to the LED; and $\overrightarrow{n_{PD}}$ is the normal vector of the PD receiving surface, which can be expressed as

$$\overrightarrow{n_{PD}} = (\cos(\alpha_r)\sin(\beta_r), \sin(\alpha_r)\sin(\beta_r), \cos(\beta_r)), \tag{11}$$

where α_r and β_r are the azimuth and tilt angles of the PD, respectively. If the LED position coordinates were (x_t, y_t, z_t), and the PD position coordinates were (x_r, y_r, z_r), then from Equations (10) and (11) we could obtain the incidence angle cosine of the inclined PD to receive LED light as follows:

$$\cos(\psi) = \frac{(x_t - x_r)\cos(\alpha_r)\sin(\beta_r) + (y_t - y_r)\sin(\alpha_r)\sin(\beta_r) + (z_t - z_r)\cos(\beta_r)}{\sqrt{(x_t - x_r)^2 + (y_t - y_r)^2 + (z_t - z_r)^2}}. \tag{12}$$

In a primary reflective NLOS link, the relationship between the received power P_{NLOS} of the PD and the LED transmitted power P_t can be expressed as

$$P_{NLOS} = P_t H_{NLOS}(0), \tag{13}$$

where $H_{NLOS}(0)$ is the DC gain of the primary reflected NLOS link, which can be expressed as [23]

$$H_{NLOS}(0) = \begin{cases} \sum_{j}^{N} \dfrac{A_{PD}(m+1)\rho\Delta A}{2\pi^2 d_{1j}^2 d_{2j}^2} \cos^m(\phi_{1j})\cos(\psi_{1j})\cos(\phi_{2j})\cos(\psi_{2j})T_s(\psi_{2j})g(\psi_{2j}), & 0 \leq \psi_{2j} \leq \psi_{FOV} \\ 0, & \psi_{2j} > \psi_{FOV} \end{cases}, \tag{14}$$

where N indicates the number of all reflective walls divided by ΔA as the area element; ρ is the reflectivity of the wall; d_{1j} is the distance between the LED and the wall reflective element; d_{2j} is the distance between the wall reflective element and the PD; ϕ_{1j} is the LED emission angle; ψ_{1j} and ϕ_{2j} are the incidence and emission angles of the wall reflective element, respectively; and ψ_{2j} is the incidence angle of the PD. If the normal vector $\overrightarrow{n_{w,j}}$ of the wall reflecting element is

$$\overrightarrow{n_{w,j}} = (\cos(\alpha_{w,j})\sin(\beta_{w,j}), \sin(\alpha_{w,j})\sin(\beta_{w,j}), \cos(\beta_{w,j})), \tag{15}$$

where $\alpha_{w,j}$ and $\beta_{w,j}$ are the azimuth and tilt angles of the wall reflector element, respectively, then the cosine corresponding to $\phi_{1j}, \psi_{1j}, \phi_{2j}$, and ψ_{2j} can be expressed as

$$\cos(\phi_{1j}) = \frac{h_{1j}}{d_{1j}}, \tag{16}$$

$$\cos(\psi_{1j}) = \frac{(x_t - x_{w,j})\cos(\alpha_{w,j})\sin(\beta_{w,j}) + (y_t - y_{w,j})\sin(\alpha_{w,j})\sin(\beta_{w,j}) + (z_t - z_{w,j})\cos(\beta_{w,j})}{d_{1j}}, \tag{17}$$

$$\cos(\phi_{2j}) = \frac{(x_r - x_{w,j})\cos(\alpha_{w,j})\sin(\beta_{w,j}) + (y_r - y_{w,j})\sin(\alpha_{w,j})\sin(\beta_{w,j}) + (z_r - z_{w,j})\cos(\beta_{w,j})}{d_{2j}}, \tag{18}$$

$$\cos(\psi_{2j}) = \frac{(x_{w,j} - x_r)\cos(\alpha_r)\sin(\beta_r) + (y_{w,j} - y_r)\sin(\alpha_r)\sin(\beta_r) + (z_{w,j} - z_r)\cos(\beta_r)}{d_{2j}}, \tag{19}$$

where h_{1j} is the vertical height of the LED in relation to the wall reflector element, and $(x_{w,j}, y_{w,j}, z_{w,j})$ are the position coordinates of the wall reflector element.

In the VLP system, each LED is installed in a vertical ceiling downward fashion, with its half-power half-angle set to $30°$, which means the amount of light that the ceiling receives directly from the LED bulb is limited. We design the robot's shell with a low-reflectivity material, so we do not take into account the reflection of the robot itself. The receiver is mounted on the robot's head, and the reflection from the floor is blocked by the robot. In addition, because the optical power reflected more than twice will be less than the noise power, it can be ignored [24]. In this study, only the primary reflection of the four walls of the room was considered, which can reduce the complexity of the light propagation path. This is simpler for VLP system design and implementation. Compared with multiple reflections, the transmission path stability of NLOS transmission is higher, and the signal quality and stability are relatively better. The received power P_r of the PD during the transmission of the indoor LED light signal in the LOS link and NLOS link model could be expressed as [25] the following:

$$P_r = P_{LOS} + P_{NLOS}. \tag{20}$$

3. GRU Neural Network Model

As general recursion neural networks (RNNs) present the problems of long-term dependence and gradient explosion [26], Hochreiter and Schmidhuber proposed the long short-term memory (LSTM) neural network in 1997. This network contains input, forget, and output gates that control input, memory, and output values, respectively [27]. Therefore, the LSTM network can effectively solve the problem of gradient vanishing and gradient explosion and is highly effective for large-scale problem processing; thus, it is widely used. The GRU network was proposed by Kyunghyun Cho et al. in 2014. This is a highly effective variant of the LSTM network [28], and the basic GRU unit structure is shown in Figure 4.

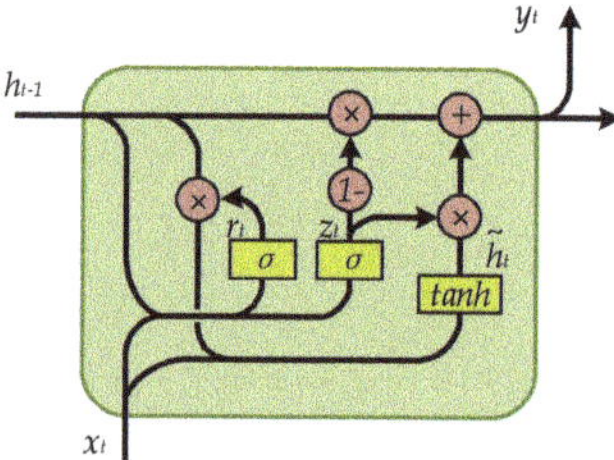

Figure 4. The basic unit structure of the GRU.

In a classical GRU network, the forward propagation equation at moment t is as follows:

$$r_t = \sigma(x_t \cdot W_{rx} + h_{t-1} \cdot W_{rh} + b_r), \tag{21}$$

$$z_t = \sigma(x_t \cdot W_{zx} + h_{t-1} \cdot W_{zh} + b_z), \tag{22}$$

$$\widetilde{h}_t = \tanh(x_t \cdot W_{hx} + (r_t * h_{t-1}) \cdot W_{hh} + b_h), \tag{23}$$

$$h_t = (1 - z_t) * h_{t-1} + z_t * \widetilde{h}_t, \tag{24}$$

$$y_t = \sigma(W_o \cdot h_t + b_o), \tag{25}$$

where $\cdot$ and $*$ denote matrix multiplication and matrix dot product, respectively; W_{rx}, W_{rh}, W_{zx}, W_{zh}, W_{hx}, W_{hh}, and W_o are the hidden layer weights; b_r, b_z, b_h, and b_o are the hidden layer biases; x_t is the input at moment t; h_{t-1} is the hidden layer output state at moment $t-1$; r_t and z_t are the reset gate and update gate, respectively; $\widetilde{h}_t$ is the candidate set state at moment t; h_t is the hidden layer output state at moment t; y_t is the output at moment t; and σ and tanh are activation functions. In general, σ is a sigmoid function, which can be expressed as

$$\sigma(x) = \frac{1}{1 + e^{-x}}, \tag{26}$$

and tanh is a tangent function, which can be expressed as

$$\tanh(x) = \frac{e^x - e^{-x}}{e^x + e^{-x}}. \tag{27}$$

As with LSTM networks, GRU networks can also overcome the long-term dependency problem of traditional RNNs; however, the GRU network integrates the input and forget gates of the LSTM network into a single update gate, so the only two gates in the GRU network are the reset and update gates. In Equation (21), the reset gate r_t controls the extent to which the hidden layer output state h_{t-1} at moment $t-1$ is passed to the candidate set $\widetilde{h}_t$ at moment t. In Equation (22), the update gate z_t determines the extent to which the output state h_{t-1} at moment $t-1$ is carried to moment t. In Equation (23), the candidate set state $\widetilde{h}_t$ uses the reset gate r_t to store past information. This is because the output of the reset gate will proceed through the sigmoid function, and each element in its output matrix is between 0 and 1, so the reset gate will control the size of the gate opening; a value closer to 1 indicates that more information is memorized. In Equation (24), the update gate z_t determines how much of the candidate set state information $\widetilde{h}_t$ at moment t and h_{t-1} at moment $t-1$ will be retained, and the retained information is used as the output state information h_t of the hidden layer at moment t. For Equation (25), using the hidden layer output state h_t at moment t as the output y_t at moment t is generally straightforward, i.e.,

$$y_t = h_t. \tag{28}$$

The output at time t is passed to time $t+1$ to continue forward propagation as the input at time $t+1$.

We compared the commonly used recurrent neural networks, employing identical parameter settings. As shown in Table 1, ensuring prediction accuracy, the model complexity of the GRU network is lower than that of the LSTM model, which not only reduces the training parameters, but also accelerates the network training time.

Table 1. Comparison of positioning algorithms.

Positioning Algorithm	Mean Squared Error	Average Error (m)	Maximum Error (m)	Training Parameters	Training Time (s)
SimpleRNN	0.08891	1.02182	1.99923	5475	147.86
GRU	0.00038	0.02666	0.75596	16,923	172.91
LSTM	0.00045	0.03554	0.46776	21,675	234.57

4. Positing Process

4.1. Construction of Fingerprint Database

The robot moves in an indoor space area, and the maximum height during its activities is uncertain. In this study, we took the average height of a person, 1.7 m, as the maximum height during robot activity. Therefore, a volume of 4 m × 4 m × 1.7 m in the room was used as the positioning space, divided into sections of 0.18 m × 0.18 m × 0.18 m. The four vertices of each small square area after division were used as reference points, the robot head receiver model was placed at each reference point, and the top center point coincided with the reference point. We used three PDs to acquire optical signals and then filtered them. Thus, we obtained two signals of different frequencies and calculated their optical power values. Finally, we recorded the optical power value and position coordinates obtained at the reference point in the fingerprint database. The fingerprint data at the k-th reference point can be expressed as:

$$F_k = \begin{bmatrix} P_{k11} & P_{k12} & P_{k21} & P_{k22} & P_{k31} & P_{k32} & x_k & y_k & z_k \end{bmatrix}, \tag{29}$$

where $P_{kij}(i = 1, 2, 3; j = 1, 2)$ is the optical power value of the j-th LED light source received by the i-th PD at the k-th reference point, and (x_k, y_k, z_k) are the position coordinates at the k-th reference point. Therefore, the VLP fingerprint database F_{db} could be constructed as

$$F_{db} = \begin{bmatrix} F_1 & F_2 & \cdots & F_N \end{bmatrix}^T, \tag{30}$$

where N is the number of reference points.

After dividing the positioning space into 0.18 m × 0.18 m × 0.18 m sections, the data obtained at the reference point were used as the training set. In addition, the positioning space was divided into 0.24 m × 0.24 m × 0.24 m sections, and the data obtained at this reference point were used as the test set. The training set was used to train the network model and provide it with a predictive ability, and the test set was used to evaluate the performance of the trained network model.

4.2. Data Preprocessing

GRU neural networks are very sensitive to input data, so we needed to normalize the input data. This process involved mapping the input data onto the same dimension, so that data of different dimensions had equal importance in the network. This not only improved the speed of network convergence, but also eliminated the influence of dimensions on the final result. We normalized the input data using

$$x_{norm} = \frac{x - x_{\min}}{x_{\max} - x_{\min}}, \tag{31}$$

where x is the input data for the training set, $x_{\min}$ is the minimum value of all input data in the training set, $x_{\max}$ is the maximum value of all input data in the training set, and x_{norm} is the normalized input data.

In addition, the GRU network required three-dimensional tensor inputs, so the input data needed to be converted into three-dimensional tensors before they were fed into the network. The input of the network was the optical power data, so the power data needed to be converted into three-dimensional tensors. The converted k-th power data could be represented as

$$I_k = \left[\begin{bmatrix} P_{k11} & P_{k12} \end{bmatrix} \begin{bmatrix} P_{k21} & P_{k22} \end{bmatrix} \begin{bmatrix} P_{k31} & P_{k32} \end{bmatrix}\right]. \tag{32}$$

Then, the input data could be expressed as

$$I = \begin{bmatrix} I_1 & I_2 & \cdots & I_n \end{bmatrix}^T, \tag{33}$$

where n is the number of input data, and the shape of input data is $(n, 3, 2)$.

4.3. Selection of Performance Indicators

We used the mean squared error (MSE) and root mean squared error (RMSE) to evaluate the performance of the GRU network and VLP models.

The loss and evaluation functions of the GRU network model used MSE, which could effectively represent the error between the predicted and actual output of the network. In the process of neural network training, the gradient obtained by the loss function was input into the optimizer for gradient descent, and then the network weight was updated by backpropagation. We repeatedly trained the network to continuously improve its predictive capabilities. Finally, the test set was substituted into the trained network model for evaluation, and the network performance was evaluated by MSE. The MSE was calculated as follows:

$$e_{MSE} = \frac{1}{N} \sum_{i=1}^{N} \left[(\hat{x}_i - x_i)^2 + (\hat{y}_i - y_i)^2 + (\hat{z}_i - z_i)^2 \right], \tag{34}$$

where N is the number of sample sets, (x_i, y_i, z_i) are the true values of the i-th sample point of the sample set, and $(\hat{x}_i, \hat{y}_i, \hat{z}_i)$ are the predicted values of the i-th sample point of the sample set.

In the positioning process, the RMSE could better reflect the relationship between the predicted and true positions, so the RMSE was used to calculate the VLP error. The RMSE between the true and predicted coordinates of the k-th reference point could be expressed as

$$e_k = \sqrt{(\hat{x}_k - x_k)^2 + (\hat{y}_k - y_k)^2 + (\hat{z}_k - z_k)^2}, \tag{35}$$

where (x_k, y_k, z_k) are the true coordinates of the k-th reference point in the test set, and $(\hat{x}_k, \hat{y}_k, \hat{z}_k)$ are the predicted coordinates of the k-th reference point in the test set. Therefore, the average positioning error was

$$\bar{e} = \frac{1}{N} \sum_{k=1}^{N} e_k. \tag{36}$$

4.4. Building the GRU Network Model

We used the Python 3.9 compiler for the experiments and Tensorflow 2.6 and the Keras 2.6 deep learning framework to build the GRU network models. When building a network model, its initial weights are random, and so the predictions of the trained model differ each time. Therefore, in order to achieve reproducible experimental results, we had to fix the random seed before building the network model. In addition, in the process of network model construction, one must manually configure the number of GRU network layers and the number of neurons in the network layer. Furthermore, before training the network, one must also set the hyperparameters, such as the learning rate, number of iterations, and batch size. These parameters affect the complexity and performance of a model, so they need to be set appropriately. Below, we present the comparison and analysis of different hyperparameter values.

To explore the influence of the number of neurons on the accuracy of the model, we compared the values at intervals of eight.

As shown in Figure 5, the average positioning error was lower when the number of neurons in the GRU network layer was 24. However, the complexity of the model also increased when the number of neurons exceeded 24, and the average positioning error did not change significantly with an increase in the number of neurons. Therefore, the number of neurons in the GRU layer of the network model was set to 24.

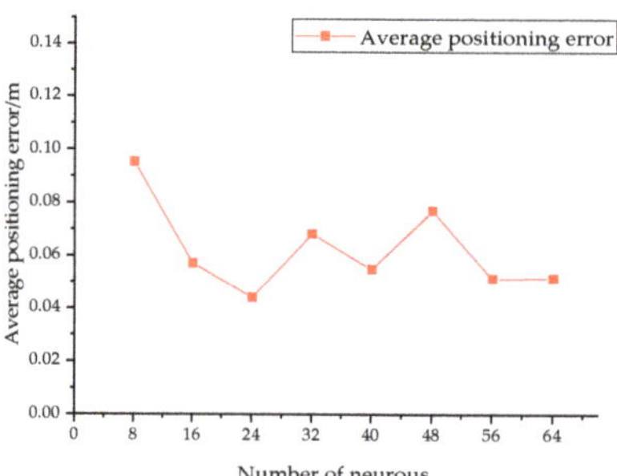

Figure 5. Average localization error for different numbers of neurons.

After settling on 24 network neurons, we analyzed the influence of the number of GRU network layers on the model performance.

From Table 2, one can see that the mean squared error and average localization error of the GRU network were smaller when the number of layers was two, and the model performance was improved. Furthermore, as the number of network layers increased, the error increased. When the number of layers is greater than two, increasing the number of layers of the network requires assigning more weights and training time to the network, which will lead to increased complexity of the network model and overfitting of the model, reducing the accuracy of the model. Therefore, we set the number of layers in the GRU network to two.

Table 2. The influence of the number of GRU network layers on the accuracy of the model.

Number of Network Layers	Mean Squared Error	Average Error (m)
1	0.00483	0.11334
2	0.00082	0.04432
3	0.00203	0.08636
4	0.00231	0.07098
5	0.00467	0.14691

The batch size is the number of samples selected for training at one time, and backpropagation is performed by calculating the gradient of these samples, so it affects the degree of optimization and speed of a model.

In this study, the compared batch sizes were 16, 32, 64, 128, and 256. From Table 3, one can see that when the batch size was too small, the gradient of calculation was unstable due to the paucity of samples, and the network did not easily converge, causing the model accuracy to decrease. However, the network generalization ability was reduced when the batch size was too large, though the network model error did not change significantly. Table 3 also shows that the training time decreased as the batch size increased. According to our comparative analysis, the model was more effective when the batch size was set to 128.

Table 3. The influence of batch size on the accuracy of the model.

Batch Size	Mean Squared Error	Average Error (m)	Training Time (s)
16	0.00714	0.15795	2015.49
32	0.00143	0.08539	1036.73
64	0.00176	0.07599	617.59
128	0.00082	0.04432	388.47
256	0.00106	0.06665	247.98

Table 4 shows the effect of the learning rate on the model performance. The model performance was more favorable when the learning rate was set to 0.01, and the decreasing curve of the network loss function is shown in Figure 6.

Table 4. The influence of the learning rate on the accuracy of the model.

Learning Rate	Mean Squared Error	Average Error (m)
0.005	0.00091	0.04544
0.010	0.00082	0.04432
0.015	0.00151	0.07569
0.020	0.00193	0.08529
0.025	0.00724	0.18912

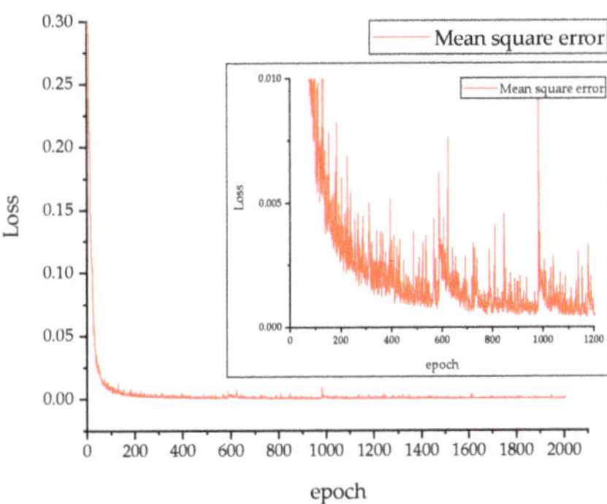

Figure 6. Loss function decline curve.

Figure 6 shows that when the number of iterations was around 950, the downward curve of the loss function was relatively flat, and there was no downward trend in subsequent iterations. To prevent overfitting and reduce training time, the maximum number of iterations of the network set to 950.

During network training, the gradient descent was slow when the learning rate was too small; thus, the training time needed to be increased to bring the model closer to the local optimum. However, the gradient decreased quickly when the learning rate was too large. Oscillation is easy in the later stage of training, but stabilization to local optimality is not straightforward, and gradient explosion may occur. In order to ensure that the network converged quickly at the beginning of training and more effectively at the end of training, we proposed a strategy to adjust the learning rate dynamically. Thus, the learning rate decay curve could be expressed as:

$$lr(epoch) = \frac{a}{1 + \exp(c(epoch - b))}, \tag{37}$$

where *epoch* is the iteration number of network training, and *a*, *b*, and *c* are set values, satisfying $a > 0$, $b > 0$, and $c > 0$. Here, *a* is the upper convergence boundary of the learning rate decay curve, and the value of $lr(0)$ is $a/(1 + \exp(-bc))$ when *epoch* = 0. If

$\exp(-bc) << 1$, $lr(0)$ is closer to a. Therefore, a can be regarded as the initial learning rate. In this study, $a = 0.01$ was adopted. The value denoted as b is the inflection point of the curve; lr is larger in the interval of $epoch \in [0, b)$, so the gradient descent is faster and the network converges rapidly. Additionally, lr decreases continuously after $epoch = b$, so the gradient descent slows down, which effectively suppresses the gradient oscillation it the late training period, and the network is more easily stabilized to the local optimum. The component c is related to the decrease in the curve at the inflection point; the higher the value of c, the faster the curve falls at the inflection point. Based on continuous testing, the average positioning error was small when $a = 0.01$, $b = 700$, and $c = 0.02$, and the corresponding learning rate decay curve is shown in Figure 7.

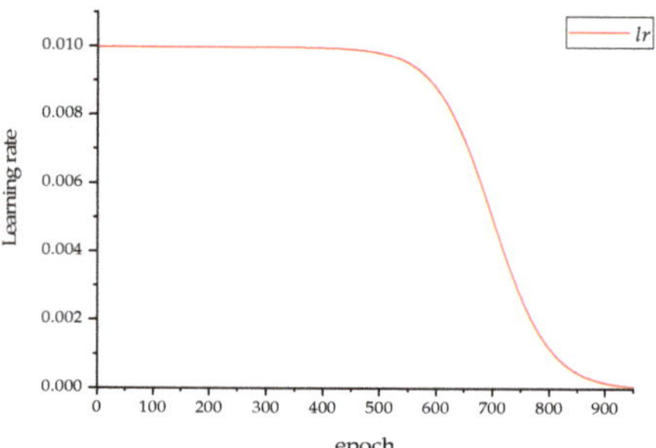

Figure 7. Learning rate decay curve.

As shown in Table 5, the learning rate decay strategy proposed in this paper corresponded to a higher VLP system accuracy, indicating that the method was effective.

Table 5. The effect of the proposed learning rate decay strategy and the learning rate setting of 0.1 on the accuracy of the model.

Learning Rate	Mean Squared Error	Average Error (m)	Training Time (s)
0.01	0.00075	0.04131	169.09
lr	0.00038	0.02660	172.91

Therefore, the GRU network model was constructed according to the parameters established above, and its structure is shown in Figure 8.

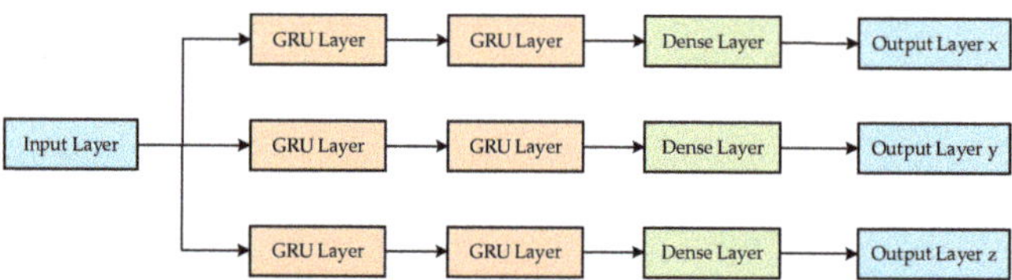

Figure 8. Structure of the GRU network model.

The model contained an input layer and three output layers, that is, the power data were input into the network, and the output comprised three coordinates. The hidden layer used three identical network structures, each containing two GRU network layers. In order to transform the data format of the GRU layer output into the final output data format, a dense layer was added before the output layer, and the network model parameters are shown in Table 6.

Table 6. GRU network model parameters.

Parameter	Value
Number of neurons in the GRU layer	24
Number of neurons in the dense layer	1
Batch size	128
Number of iterations	950
Learning rate	Equation (37)
Optimizer	Adam

5. Simulation Results and Analysis

To verify the localization performance of the proposed algorithm, a simulation environment was built according to the indoor visible-light localization model in Figure 1. We placed the hemispherical surface receiver model at each reference point in the positioning space and used three PDs on the hemispherical surface to acquire the signals sent by the two LEDs. The simulation parameters are shown in Table 7.

Table 7. Main parameters of simulation experiment.

Parameter	Value
Room size (length $\times$ width $\times$ height)	$4\,\text{m} \times 4\,\text{m} \times 3\,\text{m}$
Height of positioning space	$0\text{–}1.7\,\text{m}$
(Training, testing) partition	$(0.18, 0.24)\,\text{m}$
LED position (x, y, z)	$(1, 2, 3); (3, 2, 3)$
LED semi $-$ angle at half $-$ power ($\phi_{1/2}$)	$30°$
Amplitude of LED signal	$10\,\text{V}$
Frequency of LED signal	$4\,\text{KHz and }5\,\text{KHz}$
Effective area of PD (A_{PD})	$10^{-4}\,\text{m}^2$
Azimuth angle of PDs ($\alpha_1, \alpha_2, \alpha_3$)	$0°, 135°, 225°$
Radius of the robot receiver model (r)	$0.15\,\text{m}$
Arc length from PD to the top center point (l)	$0.05\,\text{m}$
Gain of optical filter $T_s(\psi)$	1
Refractive index of optical concentrator (n)	1.5
FOV of PD (ψ_{FOV})	$90°$
Refractive index (ρ)	0.8
Reflection surface element area (ΔA)	$0.0225\,\text{m}^2$
Filter sampling frequency	$15\,\text{KHz}$
Type of filter	Butterworth bandpass filter

In the simulation, the LED emitted a cosine AC signal, and to ensure that the LED communicated while achieving normal lighting, we added a DC bias to the LED signal. At the receiving end, the phase of the AC signal received by the PD was related to the transmission path of the signal, and the phase of the received signal differed each iteration. To be realistic, a phase shift of kT was implemented for the LED emission signal in the simulation, where $k \in [0, 1)$ is a randomly generated value and T is the LED emission signal period.

We obtained the simulated fingerprint data from the VLP model, and the sizes of the training and testing sets were 5290 and 2312, respectively. The training set was substituted into the GRU neural network to train the model, and after the training was completed, the testing set was substituted into the trained model to predict the position. The three-dimensional positioning predicted using the GRU network model for the LOS link and LOS + NLOS link scenarios is shown in Figure 9.

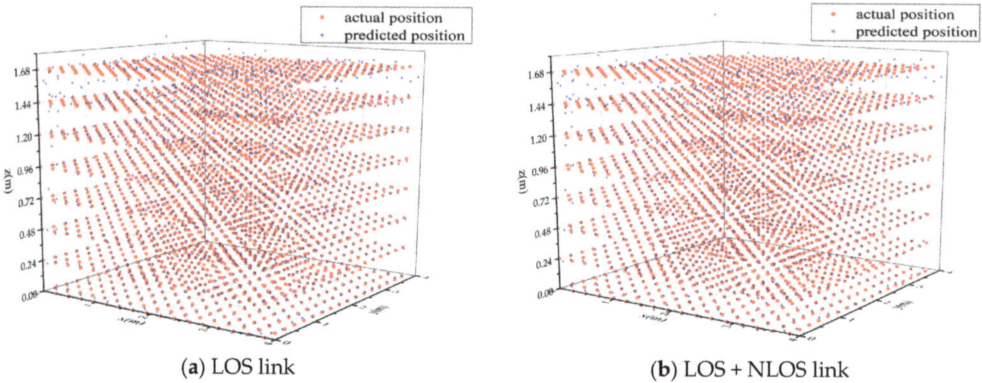

(a) LOS link   (b) LOS + NLOS link

Figure 9. Three-dimensional positioning predicted by GRU model.

Figure 9a,b show that as the positioning height increases, the deviation of the predicted location point from the actual location point increases. The positioning results in the corners are relatively poor. In addition, by comparing the positioning results in the z-axis direction of the LOS link and the LOS + NLOS link at a positioning height of 1.68 m, we find that the positioning results in the LOS + NLOS link are better.

Table 8 shows that the average localization error of the VLP model was 2.69 cm when only the LOS link case was considered, while the average localization error was 2.66 cm when both the LOS and NLOS link cases were considered. Figure 10 indicates that 95% of the positioning error was within 7.88 cm, showing that the model achieved centimeter-level positioning accuracy and met the needs of indoor positioning for robots.

Table 8. Performance comparison of 3D indoor visible-light localization models under different links.

Link	Mean Squared Error	Average Error (m)
LOS	0.00045	0.02687
LOS + NLOS	0.00038	0.02660

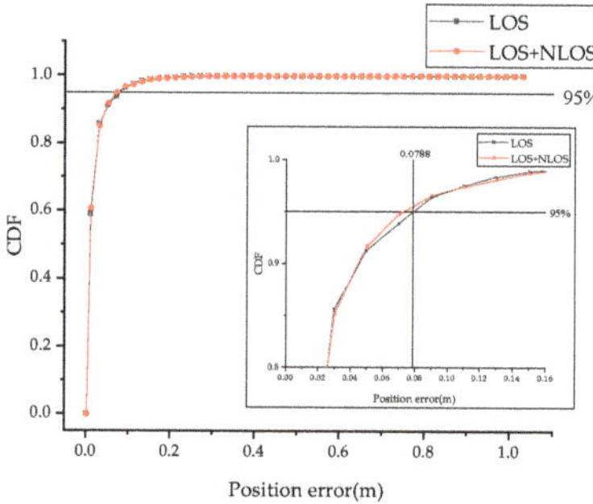

Figure 10. Cumulative distribution of positioning errors for LOS and LOS + NLOS links in 3D visible-light positioning system.

In the study, we used the same GRU network structure to make separate predictions for x, y, and z coordinates. To study the GRU network's prediction of x, y, and z coordinates, we analyze each coordinate error distribution separately. As can be seen from Figure 11, 90% of the errors in LOS + NLOS links are within 0.0265 m. Among them, the error in predicting the x-coordinate is the largest. As can be seen from Figure 1, the arrangement of

LEDs in the x-axis direction has a greater influence on the optical signal received by the receiver.

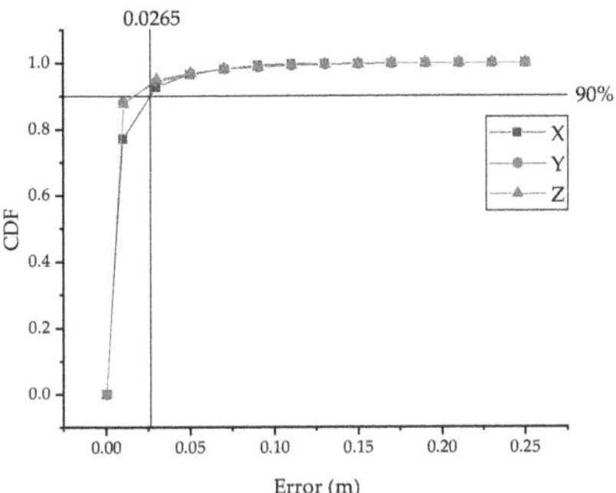

Figure 11. Cumulative distribution of errors for predicting three coordinates in a LOS + NLOS link.

To analyze the influence of the height on the accuracy of the model, we compared the two-dimensional positioning errors of the planes corresponding to different positioning heights. Table 9 shows the average and maximum positioning errors corresponding to the two-dimensional planes with the receiver placed at different heights under the LOS and LOS + NLOS link scenarios. When the positioning height was 0.24 m, the average positioning error of the model was the smallest for both LOS and LOS + NLOS links: the minimum values were 1.32 cm and 1.34 cm, respectively, and the maximum errors were 8.72 cm and 6.9 cm, respectively. However, when the positioning height was 1.68 m, the average positioning error of the model was the highest for both LOS and LOS + NLOS links, with maximum values of 7.75 cm and 7.84 cm, respectively, and maximum errors of 101.65 cm and 75.6 cm, respectively.

Table 9. Comparison of 2D positioning errors at different positioning heights for LOS and LOS + NLOS links.

Height (m)	LOS		LOS + NLOS	
	Average Error (m)	Maximum Error (m)	Average Error (m)	Maximum Error (m)
0	0.01672	0.08093	0.01771	0.08095
0.24	0.01324	0.08719	0.01347	0.06899
0.48	0.01420	0.08867	0.01384	0.09652
0.72	0.01752	0.13633	0.01614	0.14298
0.96	0.01976	0.23178	0.01946	0.22707
1.20	0.02436	0.22531	0.02308	0.24067
1.44	0.03169	0.18135	0.03071	0.18333
1.68	0.07747	1.01654	0.07839	0.75597

Figure 12 shows that 80% of the positioning errors were within 9.87 cm for different positioning heights under the LOS link and LOS + NLOS link scenarios, and 80% of the positioning errors were within 3.44 cm for positioning heights below 1.44 m. Moreover, the CDF curve of the positioning error produced by the proposed algorithm for the LOS and LOS + NLOS link scenarios was small, which indicated that the algorithm had a good generalization ability and robustness for locating different links. Therefore, we will only discuss the positioning results for LOS + NLOS links.

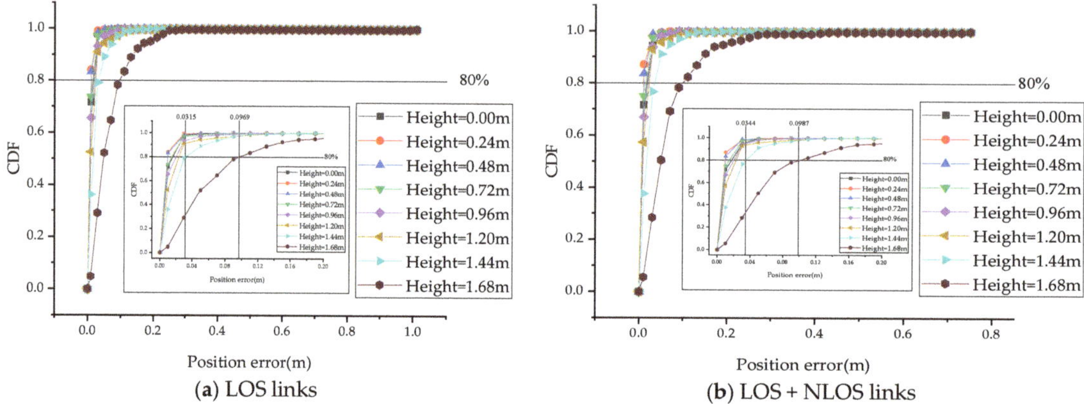

Figure 12. Cumulative distribution of two-dimensional positioning errors at different heights.

Figure 13 shows that when the positioning height was low, the errors were basically the same. When the positioning plane increased to a certain height, the positioning error also increased, and when the positioning height increased from 1.44 m to 1.68 m, this trend was more obvious. An analysis of Equations (15) and (27) reveals that the positioning error was mainly due to measurement errors related to the dc gain $H_{LOS}(0)$ and $H_{NLOS}(0)$ of the channel. When the positioning height increased, the emission angle of the LED light source also increased, and, according to Equations (4) and (12), this led to the higher attenuation of the optical signal, thereby increasing the error of the optical signal received by the PD and reducing the positioning accuracy.

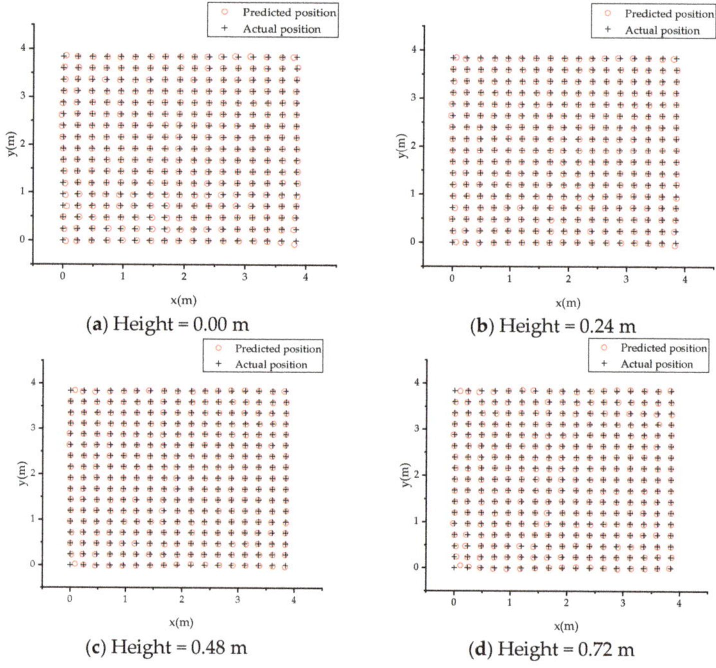

Figure 13. *Cont.*

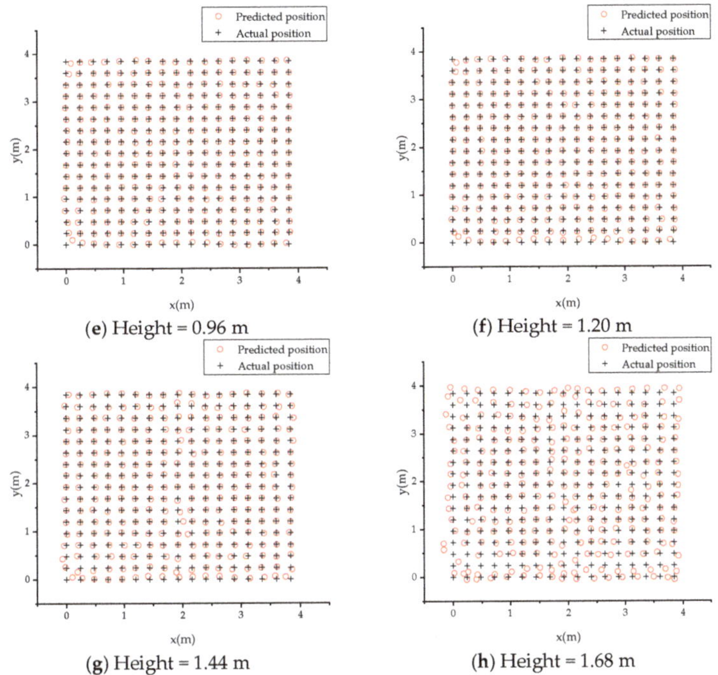

Figure 13. Comparison chart of 2D positioning results on different positioning heights under LOS + NLOS link.

6. Conclusions

We proposed an indoor visible-light three-dimensional positioning system based on a GRU neural network that solved the problem of the low positioning accuracy of existing robots. After the GRU network model was established, a learning rate attenuation strategy was proposed to improve the performance of the GRU network. A receiver placed on the robot's head was used to collect optical power data and then predict position coordinates from the trained GRU neural network. The experimental results showed that the average 3D positioning error was 2.69 cm when considering only LOS links, while the average error was 2.66 cm when considering LOS and NLOS links at the same time, and 95% of the positioning error was within 7.88 cm. For two-dimensional positioning with a fixed positioning height, 80% of the positioning error was within 9.87 cm. When the positioning height was 0.24 m, the average positioning error of the model under LOS and LOS + NLOS link scenarios was 1.32 cm and 1.34 cm, respectively. Therefore, the proposed method could achieve centimeter-level positioning accuracy to meet the needs of indoor robot positioning.

Author Contributions: L.Q.: conceptualization, investigation, supervision, resources, and writing (review). W.Y.: conceptualization, methodology, investigation, data curation, formal analysis, software, writing (original draft and editing), and validation. X.H. and D.Z.: resources and visualization. All authors have read and agreed to the published version of the manuscript.

Funding: This research was funded by the National Natural Science Fund Projects of China (62161041) and the Applied Technology Research and Development Fund Project of Inner Mongolia Autonomous Region (2021GG0104).

Institutional Review Board Statement: Not applicable.

Informed Consent Statement: Not applicable.

Data Availability Statement: Not applicable.

Conflicts of Interest: The authors declare no conflict of interest.

References

1. Liu, H.; Darabi, H.; Banerjee, P.; Liu, J. Survey of wireless indoor positioning techniques and systems. *IEEE Trans. Syst. Man Cybern. Part C (Appl. Rev.)* **2007**, *37*, 1067–1080. [CrossRef]
2. Zhuang, Y.; Yang, J.; Li, Y.; Qi, L.; El-Sheimy, N. Smartphone-based indoor localization with bluetooth low energy beacons. *Sensors* **2016**, *16*, 596. [CrossRef] [PubMed]
3. Ruiz, A.R.; Granja, F.S.; Honorato, J.C.; Rosas, J.I. Accurate pedestrian indoor navigation by tightly coupling foot-mounted IMU and RFID measurements. *IEEE Trans. Instrum. Meas.* **2011**, *61*, 178–189. [CrossRef]
4. Yan, D.; Kang, B.; Zhong, H.; Wang, R. Research on positioning system based on Zigbee communication. In Proceedings of the 2018 IEEE 3rd Advanced Information Technology, Electronic and Automation Control Conference (IAEAC), Chongqing, China, 12–14 October 2018; pp. 1027–1030.
5. Monica, S.; Ferrari, G. UWB-based localization in large indoor scenarios: Optimized placement of anchor nodes. *IEEE Trans. Aerosp. Electron. Syst.* **2015**, *51*, 987–999. [CrossRef]
6. Lin, P.; Hu, X.; Ruan, Y.; Li, H.; Fang, J.; Zhong, Y.; Zheng, H.; Fang, J.; Jiang, Z.L.; Chen, Z. Real-time visible light positioning supporting fast moving speed. *Opt. Express* **2020**, *28*, 14503–14510. [CrossRef] [PubMed]
7. Chizari, A.; Jamali, M.V.; Abdollahramezani, S.; Salehi, J.A.; Dargahi, A. Visible light for communication, indoor positioning, and dimmable illumination: A system design based on overlapping pulse position modulation. *Optik* **2017**, *151*, 110–122. [CrossRef]
8. Bakar, A.H.; Glass, T.; Tee, H.Y.; Alam, F.; Legg, M. Accurate visible light positioning using multiple-photodiode receiver and machine learning. *IEEE Trans. Instrum. Meas.* **2020**, *70*, 7500812. [CrossRef]
9. Tran, H.Q.; Ha, C. Improved visible light-based indoor positioning system using machine learning classification and regression. *Appl. Sci.* **2019**, *9*, 1048. [CrossRef]
10. Zhang, S.; Zhong, W.D.; Du, P.; Chen, C. Experimental demonstration of indoor sub-decimeter accuracy VLP system using differential PDOA. *IEEE Photonics Technol. Lett.* **2018**, *30*, 1703–1706. [CrossRef]
11. Liu, R.; Liang, Z.; Yang, K.; Li, W. Machine learning based visible light indoor positioning with single-LED and single rotatable photo detector. *IEEE Photonics J.* **2022**, *14*, 7322511. [CrossRef]
12. Hao, X.; Sun, W.; Chen, J.; Yu, C. Vertical measurable displacement approach for altitude accuracy improvement in 3D visible light positioning. *Opt. Commun.* **2021**, *490*, 126914.
13. Alonso-González, I.; Sánchez-Rodríguez, D.; Ley-Bosch, C.; Quintana-Suárez, M.A. Discrete indoor three-dimensional localization system based on neural networks using visible light communication. *Sensors* **2018**, *18*, 1040. [PubMed]
14. Jia, C.; Yang, T.; Wang, C.; Sun, M. High-Accuracy 3D Indoor Visible Light Positioning Method Based on the Improved Adaptive Cuckoo Search Algorithm. *Arab. J. Sci. Eng.* **2022**, *47*, 2479–2498.
15. Hsu, L.S.; Chow, C.W.; Liu, Y.; Yeh, C.H. 3D Visible Light-Based Indoor Positioning System Using Two-Stage Neural Network (TSNN) and Received Intensity Selective Enhancement (RISE) to Alleviate Light Non-Overlap Zones. *Sensors* **2022**, *22*, 8817. [CrossRef]
16. Zhang, Z.; Chen, H.; Zeng, W.; Cao, X.; Hong, X.; Chen, J. Demonstration of Three-Dimensional Indoor Visible Light Positioning with Multiple Photodiodes and Reinforcement Learning. *Sensors* **2020**, *20*, 6470. [CrossRef] [PubMed]
17. Zhao, H.-X.; Wang, J.-T. A novel three-dimensional algorithm based on practical indoor visible light positioning. *IEEE Photonics J.* **2019**, *11*, 6101308. [CrossRef]
18. Chen, Y.; Guan, W.; Li, J.; Song, H. Indoor real-time 3-D visible light positioning system using fingerprinting and extreme learning machine. *IEEE Access* **2019**, *8*, 13875–13886. [CrossRef]
19. Nguyen, N.T.; Suebsomran, A.; Sripimanwat, K.; Nguyen, N.H. Design and simulation of a novel indoor mobile robot localization method using a light-emitting diode positioning system. *Trans. Inst. Meas. Control* **2016**, *38*, 305–314. [CrossRef]
20. Wang, Z.; Liang, Z.; Li, X.; Li, H. Indoor Visible Light Positioning Based on Improved Particle Swarm Optimization Method with Min-Max Algorithm. *IEEE Access.* **2022**, *10*, 130068–130077. [CrossRef]
21. Li, H.; Wang, J.; Zhang, X.; Wu, R. Indoor visible light positioning combined with ellipse-based ACO-OFDM. *IET Commun.* **2018**, *12*, 2181–2187. [CrossRef]
22. Seguel, F.; Krommenacker, N.; Charpentier, P.; Soto, I. A novel range free visible light positioning algorithm for imaging receivers. *Optik* **2019**, *195*, 163028.
23. Ghassemlooy, Z.; Popoola, W.; Rajbhandari, S. *Optical Wireless Communications: System and Channel Modelling with Matlab®*; CRC Press: Boca Raton, FL, USA, 2019.
24. Fan, K.; Komine, T.; Tanaka, Y.; Nakagawa, M. The effect of reflection on indoor visible-light communication system utilizing white LEDs. In Proceedings of the 5th International Symposium on Wireless Personal Multimedia Communications, Honolulu, HI, USA, 27–30 October 2002; Volume 2, pp. 611–615.
25. Tran, H.Q.; Ha, C. Fingerprint-based indoor positioning system using visible light communication—A novel method for multipath reflections. *Electronics* **2019**, *8*, 63. [CrossRef]
26. Bengio, Y.; Simard, P.; Frasconi, P. Learning long-term dependencies with gradient descent is difficult. *IEEE Trans. Neural Netw.* **1994**, *5*, 157–166. [PubMed]
27. Hochreiter, S.; Schmidhuber, J. Long short-term memory. *Neural Comput.* **1997**, *9*, 1735–1780. [CrossRef]

28. Cho, K.; Van Merriënboer, B.; Gulcehre, C.; Bahdanau, D.; Bougares, F.; Schwenk, H.; Bengio, Y. Learning phrase representations using RNN encoder-decoder for statistical machine translation. *arXiv* **2014**, arXiv:1406.1078.

Article

A Visible Light 3D Positioning System for Underground Mines Based on Convolutional Neural Network Combining Inception Module and Attention Mechanism

Bo Deng [1], Fengying Wang [2,*], Ling Qin [1] and Xiaoli Hu [1]

[1]   College of Information Engineering, Inner Mongolia University of Science and Technology, Baotou 014010, China; dengbo@stu.imust.edu.cn (B.D.); qinling1979@imust.edu.cn (L.Q.); huxiaoli@imust.edu.cn (X.H.)
[2]   Engineering Training Center (Innovation and Entrepreneurship Education College), Inner Mongolia University of Science and Technology, Baotou 014010, China
*   Correspondence: wangfengying@imust.edu.cn

Abstract: To improve the accuracy of personnel positioning in underground coal mines, in this paper, we propose a convolutional neural network (CNN) three-dimensional (3D) visible light positioning (VLP) system based on the Inception-v2 module and efficient channel attention mechanism. The system consists of two LEDs and four photodetectors (PDs), with the four PDs on the miner's helmet. Considering the height fluctuation of PD and the impact of wall reflection on the received light power, we adopt the Inception module to perform a multi-scale extraction of the features of the received light power, thus solving the limitation of the single-scale convolution kernel on the positioning accuracy. In order to focus on the information that is more critical to positioning among the numerous input features, giving different features of the optical power data corresponding weights, we use an efficient channel attention mechanism to make the positioning model more accurate. The simulation results show that the average positioning error of the system was 1.63 cm in the space of 6 m × 3 m × 3.6 m when both the line-of-sight (LOS) and non-line-of-sight (NLOS) links were considered, with 90% of the localization errors within 4.55 cm. During the experimental stage, the average positioning error was 11.12 cm, with 90% of the positioning errors within 28.75 cm. These show that the system could achieve centimeter-level positioning accuracy and meet the requirements for underground personnel positioning in coal mines.

Keywords: visible light positioning (VLP); coal mines; three-dimensional (3D); Inception; efficient channel attention; convolutional neural network (CNN)

Citation: Deng, B.; Wang, F.; Qin, L.; Hu, X. A Visible Light 3D Positioning System for Underground Mines Based on Convolutional Neural Network Combining Inception Module and Attention Mechanism. *Photonics* **2023**, *10*, 918. https://doi.org/10.3390/photonics10080918

Received: 6 July 2023
Revised: 31 July 2023
Accepted: 7 August 2023
Published: 9 August 2023

1. Introduction

With the continuous complexity of the coal mine working environment and the improvement of safety requirements, the research on underground positioning technology in coal mines has become an important field for coal mine safety management and production efficiency improvement. During underground operations, inaccurate personnel positioning can lead to the mislocation or misjudgment of a miner's position, thereby increasing the risk of accidents. For example, suppose the positioning system misjudges a miner's location. In that case, it may cause the worker to mistakenly enter a hazardous area or approach dangerous equipment, increasing the likelihood of experiencing accidents and incidents. Moreover, accurate personnel localization is critical for the emergency rescue of a coal fire, landslides, or other accidents. Inaccurate positioning can impede rescuers from quickly and accurately locating trapped personnel, resulting in a delayed emergency response time and intensifying the difficulty and risk of rescue efforts. Accurate personnel location can help to monitor and manage miners' working status and duration. Ensuring the precise positioning of underground coal mine personnel can improve the management of their entry and exit, thereby reducing the safety risk and management difficulty in coal mines. Therefore, realizing the accurate positioning of underground personnel in coal mines is essential to ensure the safe production and efficient operation of coal mines. Various positioning methods have been proposed to address the underground personnel positioning challenge, including Wi-Fi positioning, Bluetooth positioning, radio frequency identification (RFID), ultra-wideband (UWB) positioning, and others [1–4]. Wi-Fi positioning technology has disadvantages such as complex hotspot acquisition and high power consumption; Bluetooth positioning technology usually relies on Bluetooth hotspots deployed in space, which requires precise arrangement and adjustment and increases the complexity of system deployment and maintenance; RFID technology was first applied to personnel positioning under the mines, but it has disadvantages such as a small transmission range and low positioning accuracy; and UWB technology has a higher positioning accuracy, but due to the broadband characteristics of UWB, it may produce interference with other wireless signals, affecting the positioning accuracy and reliability. Moreover, realizing high-precision UWB positioning requires specialized hardware equipment, which increases the cost and deployment difficulties. Compared with these wireless technologies, visible light communication (VLC) utilizes the visible light spectrum for data transmission and communication and has advantages, including unrestricted operation within the wireless spectrum, a high bandwidth capacity, strong anti-interference capabilities, and enhanced security. Moreover, the prevalence of lighting devices within underground coal mine environments facilitates the deployment of visible light positioning (VLP). By leveraging existing lighting infrastructure, VLP presents a forward-looking solution to the challenge of locating personnel in underground coal mines.

According to the different receivers, VLP is usually divided into an imaging type [5] and a non-imaging type [6]. Imaging-based VLP employs a camera or image sensor to capture visible light signals, utilizing image processing and computer vision technology to determine the device's position. The device's location is determined by analyzing the captured image's features, textures, or markers. However, this approach necessitates complex hardware, thus increasing the system's overall complexity and cost. On the other hand, non-imaging VLP does not rely on image data directly but utilizes parameters that are extracted from the received visible light signal for localization. This method primarily relies on signal measurement and processing techniques, such as Time of Arrival (TOA), Time Difference of Arrival (TDOA), Angle of Arrival (AOA), and received signal strength (RSS) [7–10]. Among these techniques, the fingerprint localization method based on received signal strength has garnered extensive research attention due to its utilization of simple hardware equipment and its high localization accuracy.

Machine learning and deep learning technologies have been widely used in the mining industry, bringing many advantages to coal mine production and management. Jo et al. [11] proposed an IoT technology prediction system for air quality pollutants in underground mines. The system collects real-time air quality data using various sensors deployed in underground mines and employs machine learning algorithms to analyze and predict the data. Wang et al. [12] summarized the advantages and challenges of applying machine learning and deep learning to classify microseismic events in mines, which provides reliable technical support for mine safety and geologic disaster prevention. Li et al. [13] proposed a hierarchical deep learning framework based on images used for coal and gangue detection. This framework employs deep learning algorithms and utilizes a hierarchical structure to solve the problem of coal and gangue differentiation in coal mines. These studies indicate that introducing machine learning and deep learning technology provides more intelligent and automated coal mine production and management solutions, thus effectively improving efficiency, safety, and sustainability. Therefore, combining deep learning and visible light positioning technology is feasible to accurately position people who are underground in coal mines. More and more researchers are also applying deep learning to visible light localization. By selecting suitable deep learning models and optimizing them for specific positioning tasks, researchers can improve the models' learning and generalization abilities, thereby enhancing the positioning accuracy and opening up new possibilities.

Chen et al. [14] proposed a long short-term memory fully connected network (LSTM-FCN)-based localization algorithm for implementing a VLP system with a single LED and multiple photodetectors (PDs). Lin et al. [15] proposed a model replication technique utilizing a position cell model to generate additional position samples and augment the diversity of the training data. Wei et al. [16] developed a method employing a metaheuristic algorithm to optimize the initial weights and thresholds of the extreme learning machine (ELM), thereby improving localization accuracy. However, the use of an optimization-seeking algorithm adds complexity to the model. Zhang et al. [17] presented a 3D indoor visible light positioning system based on an artificial neural network with a hybrid phase difference of arrival (PDOA) and RSS approach, enhancing the system stability in light signal intensity variations and reducing the impact of modeling inaccuracies. However, the effect of reflection was not considered. Presently, most visible light positioning studies focus solely on 2D localization [18–20]. However, a reliable 3D localization method is crucial for locating people underground in mines. This is because the heights of the miners vary according to the job's requirements, and height fluctuations can impact the positioning accuracy. Conventional 3D positioning methods typically require at least three LEDs for positioning [21,22]. These LEDs emit signals and communicate with a receiver to determine the target's location. However, this method has several limitations. First, multiple LEDs need to be installed, increasing the complexity and cost of the system. Second, since the signals emitted by the LEDs are reflected on surfaces such as the walls in the mine, the traditional method ignores the effect of such reflections on the localization results. This can lead to an increase in localization errors, especially in complex underground mine environments. In addition, the PD's tilt and the PD height's fluctuation can also impact the positioning accuracy, which are factors that are often not adequately considered in conventional methods. Some existing 3D visible light positioning systems employ hybrid algorithms [23–25], increasing the system complexity. To address these challenges and enhance the accuracy and simplicity of underground mine localization, this paper proposes a convolutional neural network (CNN) 3D visible light positioning system based on the Inception-v2 module [26] and efficient channel attention (ECA) module [27]. In this study, two LEDs were utilized as emitters and four PDs were used as receivers, and the effects of the wall reflections and PDs' tilts on localization were considered. Conventional convolutional neural networks often rely on stacking deeper convolutional layers to improve performance, which increases the model's parameter count and the risk of overfitting. This paper employs the Inception module, enabling parallel operations of multiple convolutional and pooling layers with varying sizes. This approach yields multiple feature representations of the input and reduces the computational complexity. Additionally, the ECA module assigns weights to different channel features, extracting the most critical features and ultimately enhancing the localization accuracy.

Its simplicity and ease of implementation characterize the proposed algorithmic model in this paper. Simulation experiments have validated its efficacy in localizing personnel in underground mines. The rest of this paper is organized as follows: Section 2 elucidates the components of the visible light positioning model. Section 3 expounds the structure and principles of Inception-ECANet. Section 4 explores the network parameters that influence localization. Section 5 presents the simulation and experimental results. Lastly, Section 6 provides a conclusion to the study.

2. Visible Light Positioning Model

2.1. System Model

The visible light positioning system and receiver model designed in this study are shown in Figure 1. In a space of 6 m × 3 m × 3.6 m, two LEDs are placed at the tunnel's ceiling. These LEDs serve as both a source of illumination and a means to transmit positioning signals. Within the positioning space, the LEDs emit signals of identical frequency. The PDs on the miner's helmet acts as a receiver to receive positioning signals. The receiver is designed as a symmetric multi-PD model to adapt to various positioning scenarios effectively. The central PD_0 is positioned at the receiver's midpoint, while the three inclined $PD_i (i = 1, 2, 3)$ are symmetrically arranged around PD_0. The positional relationship between the horizontal $PD_0(x_P, y_P, z_P)$ and the tilted $PD_i(x_{Pi}, y_{Pi}, z_{Pi})$ is [28]

$$\begin{cases} x_{Pi} = x_P + l\cos\theta\cos\alpha_i \\ y_{Pi} = y_P + l\cos\theta\cos\alpha_i \\ z_{Pi} = z_P + l\sin\theta \end{cases}, \tag{1}$$

where l is the length of the line segment from PD_0 to PD_i, which is parallel to the inclined plane; θ is the elevation angle of PD_i; and α_i is the angle between the projection of the line connecting PD_0 and PD_i in the *xoy* plane and the positive direction of the *x*-axis.

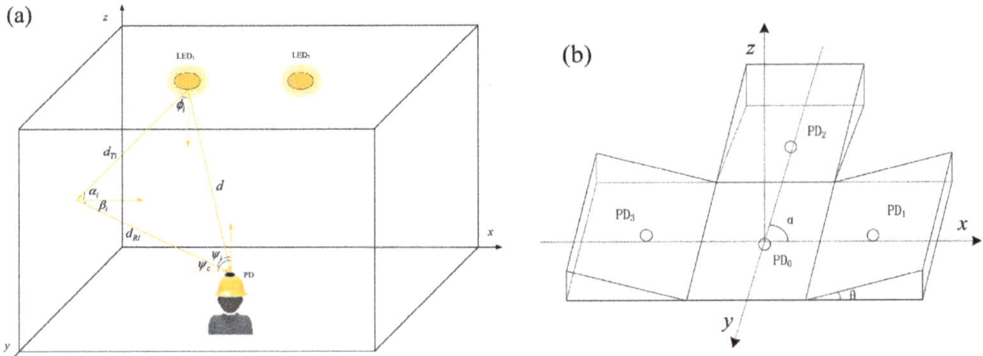

Figure 1. (**a**) Visible light positioning system model and (**b**) receiver model.

2.2. Channel Model

Indoor visible light communication systems can be categorized into line-of-sight (LOS) propagation and non-line-of-sight (NLOS) propagation. For the LOS link model, the signal propagates directly from the source to the receiver without interference from obstacles. Assuming that the LED light source radiation adheres to the Lambert distribution, the channel gain of the LOS link model is

$$H_{LOS}(0) = \begin{cases} \frac{A_r(m+1)}{2\pi d^2}\cos^m(\phi)T_s(\psi)g(\psi)\cos(\psi), 0 \le \psi \le \psi_c \\ 0 \qquad\qquad , \text{else} \end{cases}, \tag{2}$$

where A_r is the light detection area of the PD receiver; d is the linear distance between the LED lamp and the PD receiver; ϕ is the LED lamp emission angle; $T_s(\psi)$ is the transmittance of the light filter; $g(\psi)$ is the optical concentrator gain; ψ_c is the field of view of the receiver; and m is the number of Lambert emission levels, which correlates with the LED's half power angle $\phi_{1/2}$, and the relationship is

$$m = \frac{-\ln 2}{\ln(\cos\phi_{1/2})}, \tag{3}$$

The gain of the optical concentrator can be expressed as

$$g(\psi) = \begin{cases} \frac{n^2}{\sin^2\psi_c}, 0 \le \psi \le \psi_c \\ 0 \quad, \psi > \psi_c \end{cases}, \tag{4}$$

where n is the refractive index of the optical concentrator. The received power of the receiver can be expressed as

$$P_r = P_t \times H_{LOS}(0), \tag{5}$$

where P_t is the emitted power of the LED. Most of the investigated positioning methods assume that the PD is positioned horizontally and that the LED lamp's emission and incidence angles are equal. However, during the actual positioning process, the receiver may experience tilting due to the miner's body movement. Consequently, the emission angle ϕ of the LED lamp and the incidence angle ψ of the tilted PD undergo changes and can be expressed as follows:

$$\phi = \cos^{-1}\left(\frac{h}{d}\right), \tag{6}$$

$$\psi = \cos^{-1}\left(\frac{\vec{v}_{tilt} \cdot \vec{n}_{tilt}}{\left\|\vec{v}_{tilt}\right\| \cdot \left\|\vec{n}_{tilt}\right\|}\right), \tag{7}$$

where h is the vertical distance from the LED to the plane where the PD above the miner's head is located, $\vec{v}_{tilt}$ is the vector from the LED to the PD, and $\vec{n}_{tilt}$ is the normal vector of the inclined plane. Let the coordinates of the LED be $(x_{txd}, y_{txd}, z_{txd})$, and let the coordinates of the PD be $(x_{rxd}, y_{rxd}, z_{rxd})$; then, the direction vector is $\vec{v}_{tilt} = (x_{txd} - x_{rxd}, y_{txd} - y_{rxd}, z_{txd} - z_{rxd})$. If the normal vector of the horizontal PD when it is vertically up is $\vec{n} = (0, 0, 1)$, then according to the geometric relationship, the normal vector of the tilted PD is $\vec{n}_{tilt} = (\cos(\alpha_t)\sin(\theta_t), \sin(\alpha_t)\sin(\theta_t), \cos(\theta_t))$, where α_t is the azimuth of the PD and θ_t is the tilt angle of the PD.

In indoor localization scenarios, it is crucial to consider both the LOS links and the influence of the wall reflections. However, for the NLOS links, reflections beyond the first order have a minimal impact on the visible light positioning. As a result, this paper focuses solely on evaluating the impact of the primary reflection. To accomplish this, we divide the surface of each wall into q microelements, each with an area denoted as ΔA. The channel gain of the NLOS link can be expressed as [29]

$$H_{NLOS}^{(1)} = \begin{cases} \frac{m+1}{2\pi^2} \sum_{i=1}^{q} \frac{A_r \rho \Delta A \cos^m(\phi_i)\cos(\alpha_i)\cos(\beta_i)\cos(\psi_i)}{d_{Ti}^2 d_{Ri}^2}, & 0 \leq \psi_i \leq \psi_c \\ 0 & , \text{else} \end{cases} \tag{8}$$

where q is the total number of reflective elements; p is the reflection coefficient; ΔA is the area of reflective elements; d_{Ti} is the distance from the LED to the i-th reflective element; d_{Ri} is the distance from the i-th reflective element to the receiver; ϕ_i is the emission angle of the i-th reflection; α_i and β_i are the horizontal angle between the i-th reflective point and the LED line and the horizontal angle between the i-th reflective point and the receiver line, respectively; and ψ_i is the angle of incidence of the i-th reflection. In indoor visible light positioning, the received power P_r of the PD can be expressed as follows when considering the light transmission through the LOS link and NLOS link:

$$P_r = P_t(H_{LOS}(0) + H_{NLOS}(0)). \tag{9}$$

3. Inception-ECANet Model

3.1. Convolutional Neural Network

Inspired by biological vision systems, convolutional neural networks combine multi-layer convolution and pooling operations with a full connection layer to extract the features and classify the input data. The convolutional layer filters the input through convolutional operations and extracts the local features of the input data. The pooling layers are used to downsample the data, reducing the parameter count while maintaining spatial invariance. The fully connected layer maps the high-level features to different output classes. In this study, the optical power data under investigation is one-dimensional. When applied to one-dimensional data, CNNs extract local and global features from the input sequence, capturing the pattern and association information. However, one-dimensional CNNs also possess limitations. The fixed perceptual field sizes of 1D convolutional operations prevent dynamic adjustment according to the sequence length, resulting in constraints when handling long-term dependencies and contextual information. Longer sequence inputs may necessitate larger convolutional kernels and deeper networks to capture more comprehensive feature representations. This parameter-sharing property of 1D convolutional layers increases the model's parameter count. To address these challenges, this paper introduces the Inception structure and combines it with the ECA mechanism, thereby enhancing the representation capability of the improved model.

3.2. Inception Structure

The structure of Inception-v2 is shown in Figure 2. Unlike the traditional sequential connection of convolutional and pooling layers, the Inception-v2 module employs a distinct approach [26]. It simultaneously conducts convolution and pooling operations of varying sizes, such as 1×1, 3×3, and 5×5, enabling the network model to capture both global information (through 3×3 convolution) and local information (through 1×1 convolution). By utilizing parallel convolutional layers, the Inception-v2 module performs feature extraction on the input data, operating on the convolutional kernels of different scales and combining their outputs. This approach facilitates the extraction of information regarding the received optical power at multiple scales in the time domain, addressing the limitation of localization accuracy imposed by single-scale convolution kernels.

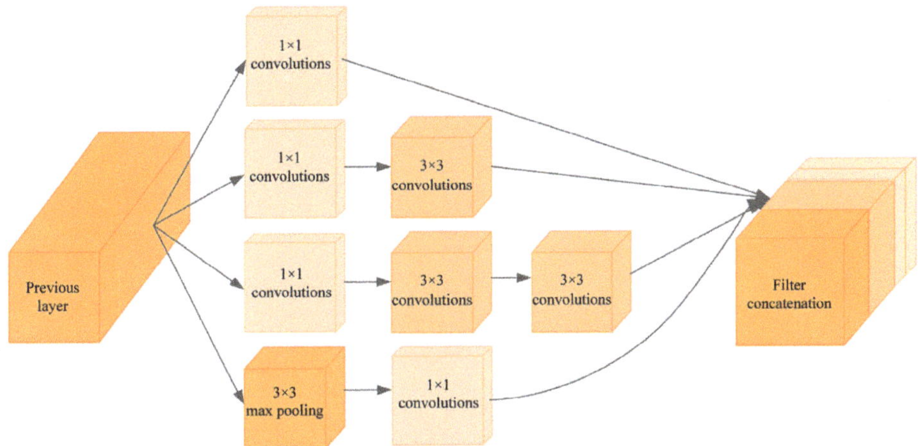

Figure 2. Inception-v2 architecture.

3.3. ECA Mechanism

After the Inception module processed the input data, the positioning model obtained some optical power information with different characteristic dimensions. In order to further obtain more and higher-dimensional feature information and give more weight to the more important features, attention mechanisms need to be used. The attention mechanism is a common technique in deep learning that enhances the model's focus on the input and extracts crucial feature information. This mechanism emulates the attention mechanism that is observed in the human visual system, enabling the model to automatically select and weigh the relevant parts of the input. This study utilizes the ECA mechanism, shown in Figure 3, to extract the important weights for each channel in the input feature map by adaptively weighting the channel dimensions [27]. Incorporating this mechanism aids in reinforcing the representation of essential features and improving the model's attention toward key features, thereby enhancing its overall performance. With the ECA mechanism, it is possible to comprehensively capture the optical power features from the input and utilize them more effectively, facilitating more accurate learning and inference by the model.

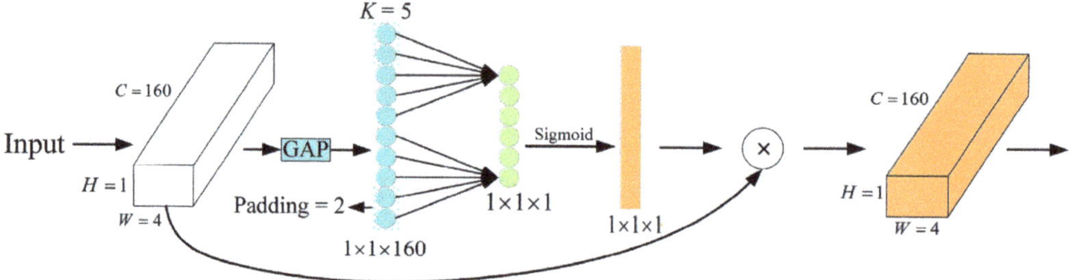

Figure 3. Efficient channel attention module.

The ECA module, an ultra-lightweight attention module, significantly enhances the performance of deep neural networks without increasing the model complexity. One of its key advantages over the traditional SENet [30] module lies in its improved local cross-channel interaction strategy. The ECA module achieves moderate cross-channel interaction by directly establishing connections between the channels and weights, reducing the model complexity while preserving performance. Despite introducing only a small number of parameters, the ECA module yields substantial performance improvements. Additionally, the ECA module employs an adaptive method to determine the size of the one-dimensional convolutional kernel. Specifically, it utilizes a fast 1D convolution of size K to facilitate local cross-channel interactions, with K representing the coverage of such interactions. To avoid a manual adjustment of K, the ECA module utilizes an adaptive approach to set its size proportionally to the channel dimension, generating attention weights as outlined in Algorithm 1.

Algorithm 1. The ECA module generates attention-weighting processes.

Input: feature map x of dimension $H \times W \times C$,
1 Define $t = \text{int}(\text{abs}((\log(C,2) + b)/gamma))$, $(b = 1, gamma = 2)$
2 Set the size of the adaptive convolution kernel k,
  $k = t$ if $t \,\%\, 2$ else $t + 1$
3 Global average pooling of the input feature map x,
  y = tf.keras.layers.GlobalAveragePooling1D(x)
4 A 1-dimensional convolution operation with a convolution kernel of size k is performed on the output y,
  Conv = tf.keras.layers.Conv1D(1,kernel size = k, padding= 'same')
5 Sigmoid activation function is used to map the weights between (0,1),
  y = tf.sigmoid(y)
6 Weighting the attention weights to the original input to obtain the final output,
  y = Multiply()([x, y])
Output: $1 \times W \times C$ channel weighted feature y.

3.4. Inception-ECANet Network Framework

We propose a novel combined model called the Inception-ECANet for visible light 3D positioning. The overall architecture of the model is shown in Figure 4. The model takes one-dimensional optical power data as the input and processes them through a convolutional layer with a large convolutional kernel. This layer effectively extracts valuable information from the original optical power data. After the initial convolutional block processing, the model obtains information across different feature dimensions. To further capture the multi-scale and comprehensive features, the Inception structure is incorporated, combined with the ECA mechanism, which allows for the appropriate weighting of the different channel features. Subsequently, a maximum pooling layer is added to reduce the computational burden and parameter count, and to eliminate redundant information, thereby enhancing the model's computational efficiency and generalization capability. A flattening layer is introduced after the pooling layer to establish connectivity with the neurons in the fully connected layer. Finally, the output layer produces three coordinate values as the model's output.

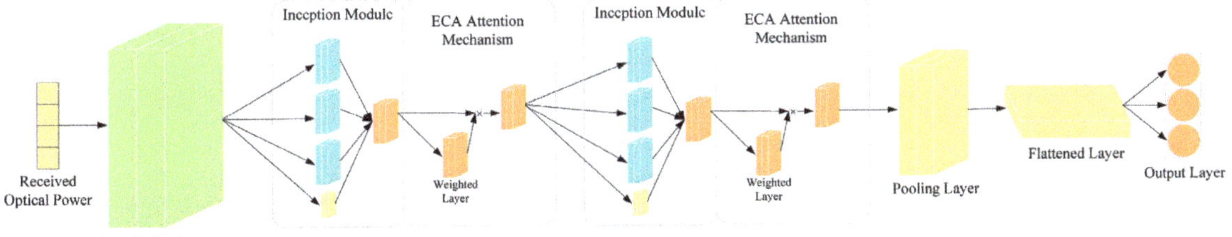

Figure 4. Inception-ECANet network architecture.

4. Positioning Process

4.1. Building a Fingerprint Database

The miner will move randomly during work, and the height of the PD above the miner's head varies. In this study, we set the maximum height of the PD to 1.8 m. Two LEDs are chosen as the radiation light sources, positioned at different locations above the miner's head. The coordinates of the LEDs are denoted as L1 (2, 1.5, and 3.6) and L2 (4, 1.5, and 3.6), respectively. To achieve the accurate positioning of the miner, the positioning process is divided into an offline phase and an online phase. In the offline phase, the positioning space of 6 m × 3 m × 1.8 m is divided into smaller spaces of 0.2 m × 0.2 m × 0.2 m. For each small space, the center point of the top square area is selected as the reference point, and four PDs are used at each reference point to receive the signal emitted by the LED light source. A fingerprint database is constructed by recording each reference point's optical power values and their corresponding location coordinates. The fingerprint data of the i-th sampling point, denoted as R_i, can be expressed as

$$R_i = \left(P_{ij}, P_{ij}, P_{ij}, P_{ij}, x_i, y_i, z_i \right) \tag{10}$$

where P_{ij} is the optical power received by the j-th PD at the i-th reference point, and (x_i, y_i, z_i) are the 3D location coordinates of the i-th reference point. Thus, the complete fingerprint database can be expressed as $R_{db} = (R_1, R_2, R_3, \cdots, R_N)^{\mathrm{T}}$, and N is the number of reference points.

During the online localization phase, the received optical power values from the PD are utilized to predict the real-time position coordinates of the miners. To evaluate the effectiveness of the localization system, the localization space was further partitioned into smaller units measuring 0.25 m × 0.25 m × 0.25 m. The data collected from these reference points served as the testing set. Through the testing set evaluation, we could objectively assess the performance and accuracy of the positioning system.

4.2. Inception-ECANet Parameter Selection

During the design of the Inception-ECANet model, numerous key parameters require optimization, such as the number of convolutional layers, the size and quantity of convolutional kernels, the learning rate, the choice of the optimizer, the number of iterations, the batch size, and the selection of activation functions. These parameters significantly affect the overall accuracy and computational efficiency of the network. As a result, selecting the appropriate parameter configuration meticulously is vital for constructing the localization model. The following sections will compare and analyze various hyperparameter values to identify the optimal parameter combination.

The choice of the batch size significantly affects the accuracy of the localization model. In this study, the compared batch sizes were 16, 32, 64, 128, and 256, and the results are shown in Table 1. From Table 1, one can see that using a larger batch size enables a better utilization of parallel computing, resulting in faster training. However, this may lead to instability in the parameter updates and increase the likelihood of the model converging to local optima. On the other hand, selecting a smaller batch size introduces more noise during training, as each parameter update is based on a smaller number of samples. This leads to slower training and requires more iterations to achieve the same performance level. Considering the available computational resources and training effects, we chose a batch size of 128 to train the model in this study.

Table 1. The influence of batch size on the accuracy of the model.

Batch Size	Average Error/m	Maximum Error/m	Training Time/s
16	0.01849	0.26080	1548.75
32	0.01745	0.22399	920.64
64	0.01716	0.23451	474.68
128	0.01634	0.14717	271.76
256	0.01763	0.19478	187.77

The number of convolutional kernels in the Inception module significantly influences the model's complexity and representational power. Choosing a smaller number of convolutional kernels results in a more simple and abstract feature representation. On the other hand, a larger number of convolutional kernels enhances the model's ability to transform features, leading to a richer and more complex representation. However, excessive convolutional kernels can lead to model overfitting and slower training. Therefore, when designing the Inception module, a trade-off must be made between the localization accuracy and the model complexity when choosing the number of convolutional kernels. Based on this consideration and after repeated experimental comparisons, the model parameters used in this paper were chosen as shown in Table 2.

Table 2. Model parameters.

Layer Name		(Convolution Kernel) Size	Number of Convolution Kernels	(Convolution Kernel) Step Size
Convolutional layer		3×1	128	1×1
Inception module	Convolutional layer	1×1	16	1×1
	Convolutional layer	$1 \times 1/3 \times 1$	16/64	1×1
	Convolutional layer	$1 \times 1/3 \times 1/3 \times 1$	16/32/64	1×1
	Pooling layer	$3 \times 1/1 \times 1$	16	1×1
Attention module		5×1	1	1×1
Inception module	Convolutional layer	1×1	32	1×1
	Convolutional layer	$1 \times 1/3 \times 1$	16/48	1×1
	Convolutional layer	$1 \times 1/3 \times 1/3 \times 1$	16/32/48	1×1
	Pooling layer	$3 \times 1/1 \times 1$	32	1×1
Attention module		5×1	1	1×1
Pooling layer		2×1	—	2×1
Fully connected layer		3	—	—

During model training, the input data consist of optical power data from various heights. To ensure training stability and improve the convergence speed, it is essential to normalize the original input data. The mean–variance normalization expression is as follows:

$$R_{\mathrm{r}}' = \frac{R_{\mathrm{r}} - \mu}{\sigma}, \tag{11}$$

where $R_r{'}$ is the normalized data, R_r is the received optical power, and μ and σ are the mean and standard deviation of the sample data.

Several optimization algorithms, including stochastic gradient descent (SGD), Adagrad, RMSprop, Adam, and Adadelta, were compared by adjusting the learning rate to determine the most suitable one for the localization model. The impact of the different optimization algorithms on the root mean square error of the localization model at various learning rates is shown in Figure 5. Notably, the Adam optimization algorithm achieved the smallest root mean square error at a learning rate of 0.001. Therefore, we selected the Adam algorithm with a learning rate of 0.001 to train Inception-ECANet, which can improve the convergence speed and performance of the model, making it better suited for localization tasks.

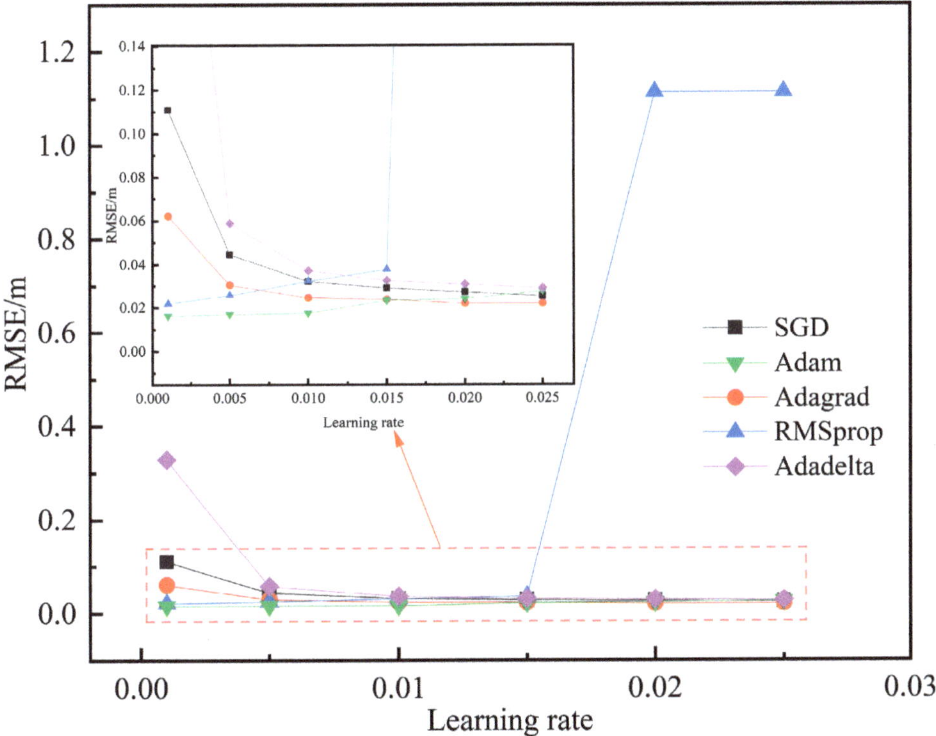

Figure 5. The influence of various optimization algorithms on positioning performance at different learning rates.

To assess the performance of the Inception-ECANet model and improve the prediction accuracy, we incorporated a loss function as a learning criterion to guide the training process. The loss function plays a vital role in training and evaluating the localization model. For prediction problems, the mean square error (MSE) is a widely employed loss function. Reducing the mean square error facilitates the performance optimization of the positioning model, leading to an enhanced prediction accuracy. Its mathematical expression is

$$E_{\mathrm{MSE}} = \frac{1}{N}\sum_{i=1}^{N}\left[(\hat{x}_i - x_i)^2 + (\hat{y}_i - y_i)^2 + (\hat{z}_i - z_i)^2\right], \tag{12}$$

where N is the number of reference points, $(\hat{x}_i, \hat{y}_i, \hat{z}_i)$ are the predicted position coordinates of the i-th reference point of the positioning model, and (x_i, y_i, z_i) are the real position coordinates of the first reference point.

Once the localization model is trained, it is essential to assess whether the model meets the localization accuracy requirements. To achieve this, the validation set is used to perform localization predictions on the trained model. The magnitude of the localization error is then analyzed using the root mean square error (RMSE), which offers a comprehensive measure of the prediction error by calculating the square root of the average error between the predicted and actual values. The mathematical expression of the RMSE is

$$E_{\text{RMSE}} = \sqrt{\frac{1}{N}\sum_{i=1}^{N}\left[(\hat{x}_i - x_i)^2 + (\hat{y}_i - y_i)^2 + (\hat{z}_i - z_i)^2\right]}. \tag{13}$$

5. Simulation and Experimental Analysis

5.1. Simulation Analysis

To assess the performance of the proposed Inception-ECANet localization method, we conducted modeling and simulation using the Python3.9 compiler. The Inception-ECANet was implemented in TensorFlow 2.10 and trained on an NVIDIA RTX 4090. During the training process, we utilized the mean square error as the loss function and employed the Adam optimization algorithm with an initial learning rate of 0.001. The model was trained for 1400 epochs, using a batch size of 128. For the localization space, measuring 6 m $\times$ 3 m $\times$ 1.8 m, we uniformly divided it into smaller spaces with side lengths of 0.2 m. Each small space's center point in the top square area was selected as a reference point. The received optical power value and the position coordinates of these reference points were used as the training set data to train the Inception-ECANet model, thus establishing a prediction model for the visible light positioning method in mines. Subsequently, the localization space was further divided into small spaces with a side length of 0.25 m. The received optical power values and the coordinates from these points were utilized as the testing set data to evaluate the performance of the trained localization model. The simulation parameters are shown in Table 3.

Table 3. Simulation parameters.

Parameter	Value
Room size/m $\times$ m $\times$ m	$6 \times 3 \times 3.6$
Height of positioning space/m	0–1.8
Position of LED/m	(2, 1.5, 3.6); (4, 1.5, 3.6)
Power of each LED bulb P_t/W	15
Field of view $\psi_c/\left(^\circ\right)$	90
Half power angles of LED $\phi_{1/2}/\left(^\circ\right)$	70
Tilt angle of PD $\theta/\left(^\circ\right)$	30
Azimuth of PDs $(\alpha_1, \alpha_2, \alpha_3)/\left(^\circ\right)$	0,90,180
Effective area of PD A_r/m^2	0.0001
Gain of optical filter $T_s(\psi)$	1
Reflection coefficient ρ	0.7
Reflection surface element area $\Delta A/\text{m}^2$	0.01
Distance from PD_0 to PD_i l/m	0.05
Refractive index of optical concentrator n	1.5

The training and validation sets are selected to train and test the localization model, and the predicted 3D localization distribution of the model obtained is shown in Figure 6. In order to visually represent the localization error, the localization error distribution of the PD located at different heights is shown in Figure 7.

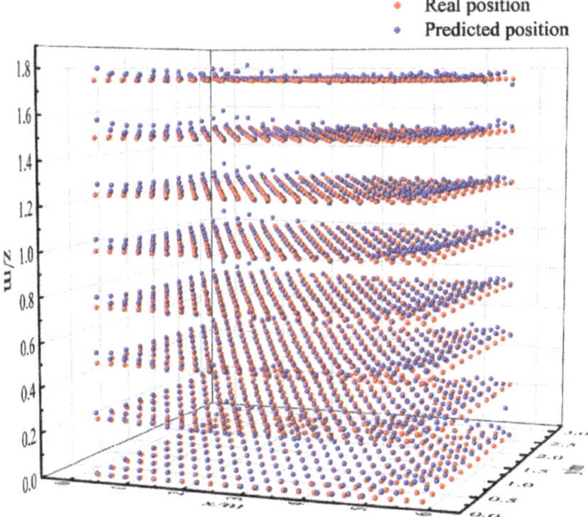

Figure 6. The model's predictions of the 3D positioning distribution.

From Figures 6 and 7, it is evident that the proposed positioning model exhibits exceptional performance in 3D space. With an average positioning error of only 1.63 cm and a maximum positioning error of 14.71 cm, the model achieves centimeter-level accuracy, meeting the precise requirements of mine positioning. Additionally, it was observed that the positioning model exhibits larger errors in the edge and corner regions. These errors can be attributed to the longer path that light must travel to reach these areas and the greater angular deviation from the photodetector. When the light enters the photodetector at a steeper angle, it fails to be fully captured, resulting in the attenuation of the light intensity and an increase in the localization error.

To investigate the impact of different submodules in Inception-ECANet on the localization accuracy, we conducted experiments, and the comparison results of the localization errors after incorporating various submodules are presented in Table 4.

Table 4. Comparison result of positioning error after adding different submodules.

Positioning Algorithm	Average Error/m	Maximum Error/m	Training Time/s
CNN	0.02249	0.25562	218.01
CNN + ECA	0.01825	0.20091	228.40
CNN + Inception-v2	0.01724	0.18514	275.36
CNN + Inceptionv2 + ECA	0.01634	0.14717	295.42

As seen in Table 4, adding two submodules significantly enhances the localization accuracy of the positioning model. Regarding the individual submodules, the CNN + Inception-v2 module demonstrates a higher accuracy than the CNN + ECA module, indicating the superior effectiveness of the Inception-v2 module in improving the localization accuracy. The Inception-v2 module's advantage lies in its utilization of a multi-scale convolutional kernel, which enables the extraction of more detailed and informative features. In contrast, the attention mechanism employs a single convolutional kernel, resulting in limited improvements in the localization accuracy. Furthermore, the combination of these two submodules shows more significant improvements in the localization accuracy compared to each submodule alone. Upon incorporating the Inception-v2 module and the ECA module, the average localization error is reduced by 27.35%, and the maximum localization error is reduced by 42.43%. These outcomes signify that the fusion of the Inception-v2 module and the ECA module enhances the network's feature extraction capability, thereby improving the localization accuracy of the model.

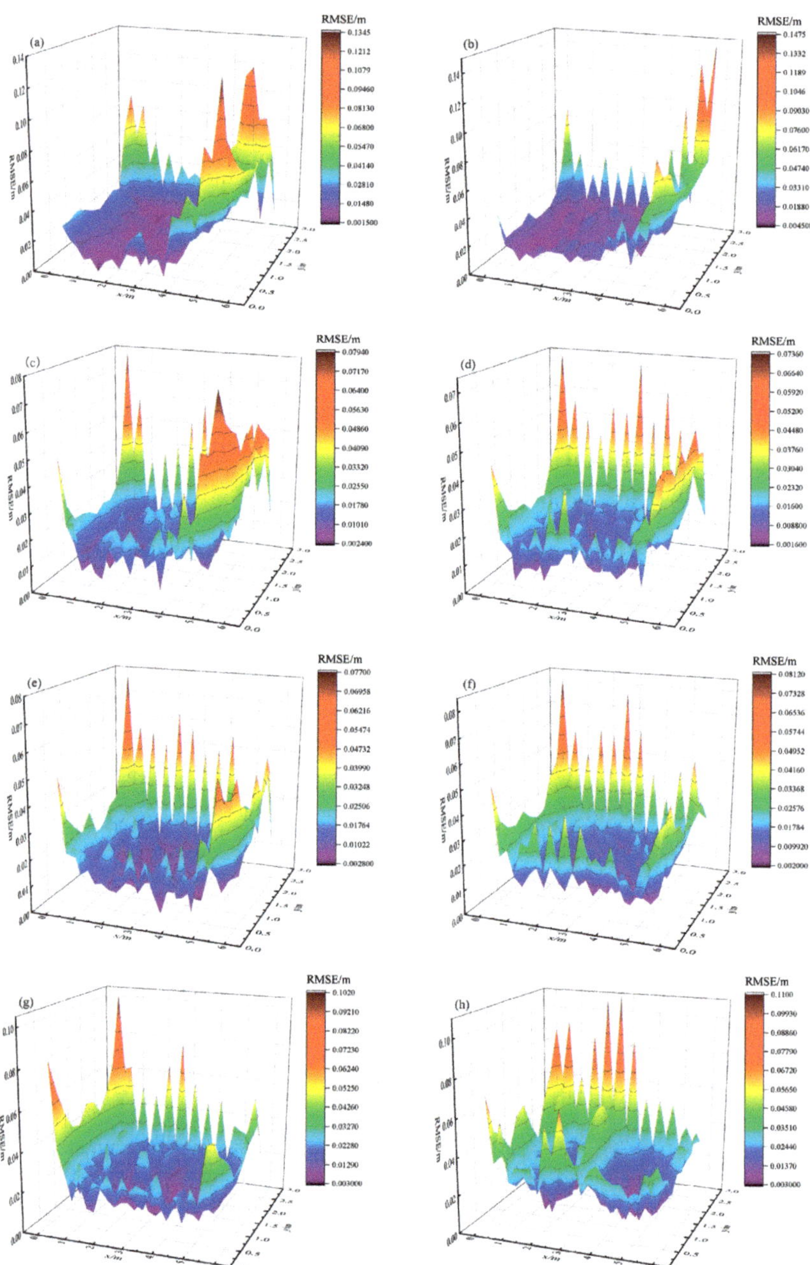

Figure 7. Positioning error distribution of receiving plane at different heights. (**a**) Height = 0 m; (**b**) height = 0.25 m; (**c**) height = 0.5 m; (**d**) height = 0.75 m; (**e**) height = 1.0 m; (**f**) height = 1.25 m; (**g**) height = 1.5 m; (**h**) height = 1.75 m.

5.2. Experimental Analysis

The proposed positioning model in this paper demonstrates favorable performance in the personnel positioning within underground coal mines under simulated conditions. However, it is essential to acknowledge the disparities between the actual application environment and the simulation conditions. To further validate the effectiveness of the proposed positioning model, we constructed a simulated experimental scenario with dimensions of 6 m × 3 m × 3.6 m, as shown in Figure 8. During the experiment, two LEDs with a 15 W emitting power served as the emitters. LED1 was positioned at coordinates (2, 1.5, and 3.6), while LED2 was located at (4, 1.5, and 3.6). The experimental space was enclosed with a black cloth to simulate real-world conditions, and four S1133 silicon photodiodes were utilized as the receiving terminals.

Figure 8. Simulation experiment scene.

We used a stand to position the PD at various height positions to acquire data, simulating the receiver's height variation during the miner's work. During experiments, we uniformly divided the length and width of the positioning space with 0.2 m spacing and selected four typical heights (0 m, 0.6 m, 1.2 m, and 1.8 m) to collect data at the divided reference points. The collected data were then used as the training set to train the model. To validate the accuracy of the localization model, we further divided the localization space at a spacing of 0.25 m and used the collected data as the validation set. To reduce the impact of LED light fluctuations on the results, we performed ten acquisitions of optical power data at each reference point and used the average value as the input for the localization model.

After testing, the positioning model exhibited an average positioning error of 11.12 cm in 3D space, with a maximum positioning error of 59.54 cm. Furthermore, 90% of the positioning error fell within 28.75 cm. The cumulative distribution of the positioning error is shown in Figure 9.

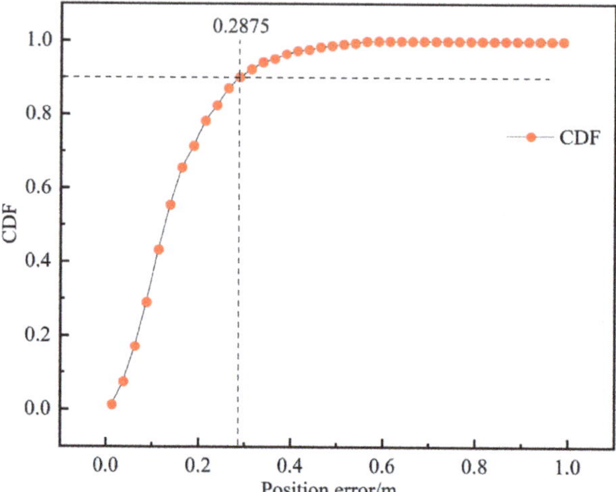

Figure 9. Cumulative distribution of positioning errors.

To investigate the impact of height on the positioning accuracy, we compared the positioning errors at various heights, as detailed in Table 5. The results in Table 5 show that the receiver height significantly influences the positioning errors. This effect can be attributed to the increased light deficit area between the two light sources as the height increases, resulting in a more significant variability of optical power values across the receiving plane. This variability has implications for the regularity and similarity of received data, subsequently affecting the data fitting during network training and the accuracy of the predicted results on the validation set, thus leading to an increase in the localization error.

Table 5. Three-dimensional positioning errors at different heights.

Height/m	Average Error/m	Maximum Error/m	Minimum Error/m
0	0.08729	0.47930	0.01303
0.6	0.11902	0.57127	0.01223
1.2	0.12369	0.56821	0.00649
1.8	0.11129	0.59536	0.01730

The proposed algorithm in this study was compared with several other localization methods, namely the Backpropagation Neural Network (BPNN), Recurrent Neural Network (RNN), long short-term memory network (LSTM), and CNN. The localization errors of these localization methods are shown in Table 6. The results clearly demonstrate that the algorithm proposed in this paper significantly enhances the localization accuracy. In comparison to the BPNN, the proposed algorithm reduced the average localization error by 33.35% and reduced the maximum localization error by 32.55%. Similarly, when compared with the RNN, the average localization error was reduced by 48.19%, and the maximum localization error decreased by 58.13%. In contrast, in comparison to the LSTM, the average localization error was reduced by 49.56%, and the maximum localization error decreased by 56.56%. Moreover, compared with the CNN, the proposed algorithm achieved a reduction of 13.96% in the average localization error and a reduction of 27.70% in the maximum localization error.

Table 6. Positioning errors of different neural network localization methods.

Positioning Algorithm	Average Error/m	Maximum Error/m
BPNN	0.16684	0.88266
RNN	0.21462	1.42188
LSTM	0.22046	1.37061
CNN	0.12924	0.82350
Inception-ECANet	0.11120	0.59536

To provide a more intuitive demonstration of the localization effect, a comparison of the cumulative distribution of localization errors among the five algorithms is shown in Figure 10. It can be observed that 90% of the localization errors of the proposed localization method are less than 28.75 cm. In contrast, for the other four localization methods (BPNN, RNN, LSTM, and CNN), 90% of their localization errors are below 44.28 cm, 58.53 cm, 61.13 cm, and 34.28 cm, respectively. This comparison highlights that the proposed Inception-ECANet localization method exhibits significantly lower localization errors overall.

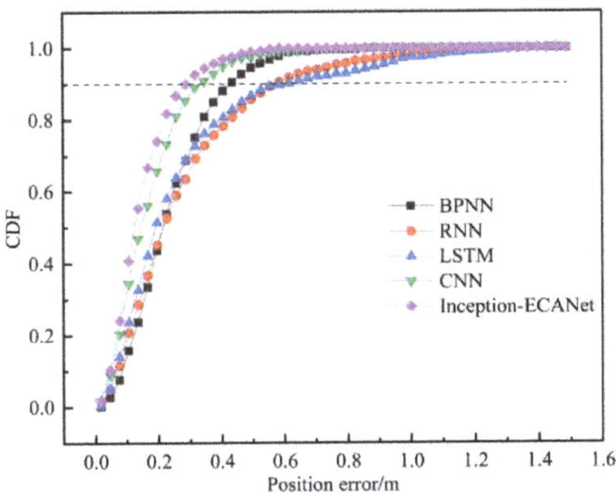

Figure 10. Cumulative distribution of positioning errors for different neural network localization methods.

6. Conclusions

We proposed a convolutional neural network visible light 3D localization system for localizing underground coal mine personnel by combining the Inception-v2 and ECA modules. The system employed two LEDs as transmitting base stations and four PDs mounted on miners' helmets as receivers. The optical power data acquired from the receivers are used to train the Inception-ECANet model, enabling a precise prediction of the position coordinates. The simulation results demonstrate that within a 6 m $\times$ 3 m $\times$ 3.6 m space, the Inception-ECANet localization method achieves an average error of 1.63cm and a maximum error of 14.71 cm, with 90% of the localization errors below 4.55 cm. An experimental validation further confirmed the effectiveness of the proposed method, achieving an average error of 11.12 cm and a maximum error of 59.54 cm within the same-sized localization space. It was worth noting that compared to four other positioning methods (BPNN, RNN, LSTM, and CNN), the proposed positioning method in this paper demonstrates outstanding performance. The research results show that when using this method, 90% of the positioning errors are within 28.75 cm, which is far superior to the other four positioning methods. Compared to the BPNN, the algorithm reduced the average positioning error by 33.35%. Similarly, compared to the RNN, the average positioning error was reduced by 48.19%. Compared to the LSTM, the average positioning error was reduced by 49.56%. Furthermore, the proposed algorithm reduced the average positioning error by 13.96% compared to the CNN. Through a comprehensive comparative analysis, it can be seen that the positioning method proposed in this paper exhibits lower positioning errors, which further validates the superiority and practicality of the proposed positioning algorithm in underground personnel positioning in coal mines.

Author Contributions: F.W., conceptualization, resources, supervision, and writing—review and editing. B.D., formal analysis, methodology, software, validation, data curation, and writing—original draft preparation. L.Q. and X.H., resources and project administration. All authors have read and agreed to the published version of the manuscript.

Funding: This research was funded by the National Natural Science Foundation of China (62161041), the Natural Science Foundation of Inner Mongolia (2022MS06012), the Inner Mongolia Key Technology Tackling Project (2021GG0104), and the Basic Research Funds for Universities directly under the Inner Mongolia Autonomous Region (2023XKJX010).

Institutional Review Board Statement: Not applicable.

Informed Consent Statement: Not applicable.

Data Availability Statement: Not applicable.

Conflicts of Interest: The authors declare no conflict of interest.

References

1. Xue, J.; Zhang, J.; Gao, Z.; Xiao, W. Enhanced WiFi CSI Fingerprints for Device-Free Localization With Deep Learning Representations. *IEEE Sens. J.* **2023**, *23*, 2750–2759. [CrossRef]
2. Kumari, S.; Siwach, V.; Singh, Y.; Barak, D.; Jain, R.; Rani, S. A Machine Learning Centered Approach for Uncovering Excavators' Last Known Location Using Bluetooth and Underground WSN. *Wirel. Commun. Mob. Comput.* **2022**, *2022*, 9160031. [CrossRef]
3. Cavur, M.; Demir, E. RSSI-based hybrid algorithm for real-time tracking in underground mining by using RFID technology. *Phys. Commun.* **2022**, *55*, 101863. [CrossRef]
4. Yuan, X.; Bi, Y.; Hao, M.; Ji, Q.; Liu, Z.; Bao, J. Research on Location Estimation for Coal Tunnel Vehicle Based on Ultra-Wide Band Equipment. *Energies* **2022**, *15*, 8524. [CrossRef]
5. Huynh, P.; Yoo, M. VLC-Based Positioning System for an Indoor Environment Using an Image Sensor and an Accelerometer Sensor. *Sensors* **2016**, *16*, 783. [CrossRef]
6. Steendam, H.; Wang, T.Q.; Armstrong, J. Theoretical Lower Bound for Indoor Visible Light Positioning Using Received Signal Strength Measurements and an Aperture-Based Receiver. *J. Light. Technol.* **2017**, *35*, 309–319. [CrossRef]
7. Zhao, H.X.; Wang, J.T. A Novel Three-Dimensional Algorithm Based on Practical Indoor Visible Light Positioning. *IEEE Photonics J.* **2019**, *11*, 6101308. [CrossRef]
8. Du, P.; Zhang, S.; Chen, C.; Alphones, A.; Zhong, W.-D. Demonstration of a Low-Complexity Indoor Visible Light Positioning System Using an Enhanced TDOA Scheme. *IEEE Photonics J.* **2018**, *10*, 7905110. [CrossRef]
9. De-La-Llana-Calvo, Á.; Lázaro-Galilea, J.; Gardel-Vicente, A.; Rodríguez-Navarro, D.; Bravo-Muñoz, I. Indoor positioning system based on LED lighting and PSD sensor. In Proceedings of the 2019 International Conference on Indoor Positioning and Indoor Navigation (IPIN), Pisa, Italy, 30 September–3 October 2019; pp. 1–8.
10. Li, H.; Wang, J.; Zhang, X.; Wu, R. Indoor visible light positioning combined with ellipse-based ACO-OFDM. *IET Commun.* **2018**, *12*, 2181–2187. [CrossRef]
11. Jo, B.; Khan, R.M.A. An Internet of Things System for Underground Mine Air Quality Pollutant Prediction Based on Azure Machine Learning. *Sensors* **2018**, *18*, 930. [CrossRef] [PubMed]
12. Jinqiang, W.; Basnet, P.; Mahtab, S. Review of machine learning and deep learning application in mine microseismic event classification. *Min. Miner. Depos.* **2021**, *15*, 19–26. [CrossRef]
13. Li, D.; Zhang, Z.; Xu, Z.; Xu, L.; Meng, G.; Li, Z.; Chen, S. An Image-Based Hierarchical Deep Learning Framework for Coal and Gangue Detection. *IEEE Access* **2019**, *7*, 184686–184699. [CrossRef]
14. Chen, H.; Han, W.; Wang, J.; Lu, H.; Chen, D.; Jin, J.; Feng, L. High accuracy indoor visible light positioning using a long short term memory-fully connected network based algorithm. *Opt. Express* **2021**, *29*, 41109–41120. [CrossRef]
15. Lin, D.-C.; Chow, C.-W.; Peng, C.-W.; Hung, T.-Y.; Chang, Y.-H.; Song, S.-H.; Lin, Y.-S.; Liu, Y.; Lin, K.-H. Positioning Unit Cell Model Duplication With Residual Concatenation Neural Network (RCNN) and Transfer Learning for Visible Light Positioning (VLP). *J. Light. Technol.* **2021**, *39*, 6366–6372. [CrossRef]
16. Wei, F.; Wu, Y.; Xu, S.; Wang, X. Accurate visible light positioning technique using extreme learning machine and meta-heuristic algorithm. *Opt. Commun.* **2023**, *532*, 129245. [CrossRef]
17. Zhang, S.; Du, P.; Chen, C.; Zhong, W.-D.; Alphones, A. Robust 3D Indoor VLP System Based on ANN Using Hybrid RSS/PDOA. *IEEE Access* **2019**, *7*, 47769–47780. [CrossRef]

18. Song, S.H.; Lin, D.C.; Liu, Y.; Chow, C.W.; Chang, Y.H.; Lin, K.H.; Wang, Y.C.; Chen, Y.Y. Employing DIALux to relieve machine-learning training data collection when designing indoor positioning systems. *Opt. Express* **2021**, *29*, 16887–16892. [CrossRef]
19. Liu, R.; Liang, Z.; Yang, K.; Li, W. Machine Learning Based Visible Light Indoor Positioning With Single-LED and Single Rotatable Photo Detector. *IEEE Photonics J.* **2022**, *14*, 7322511. [CrossRef]
20. Bakar, A.H.A.; Glass, T.; Tee, H.Y.; Alam, F.; Legg, M. Accurate Visible Light Positioning Using Multiple-Photodiode Receiver and Machine Learning. *IEEE Trans. Instrum. Meas.* **2021**, *70*, 7500812. [CrossRef]
21. Shao, S.; Khreishah, A.; Khalil, I. Enabling Real-Time Indoor Tracking of IoT Devices Through Visible Light Retroreflection. *IEEE Trans. Mob. Comput.* **2020**, *19*, 836–851. [CrossRef]
22. Sato, T.; Shimada, S.; Murakami, H.; Watanabe, H.; Hashizume, H.; Sugimoto, M. ALiSA: A Visible-Light Positioning System Using the Ambient Light Sensor Assembly in a Smartphone. *IEEE Sens. J.* **2022**, *22*, 4989–5000. [CrossRef]
23. Guan, W.; Huang, L.; Hussain, B.; Yue, C.P. Robust Robotic Localization Using Visible Light Positioning and Inertial Fusion. *IEEE Sens. J.* **2022**, *22*, 4882–4892. [CrossRef]
24. Chaochuan, J.; Ting, Y.; Chuanjiang, W.; Mengli, S. High-Accuracy 3D Indoor Visible Light Positioning Method Based on the Improved Adaptive Cuckoo Search Algorithm. *Arab. J. Sci. Eng.* **2021**, *47*, 2479–2498. [CrossRef]
25. Hsu, L.S.; Chow, C.W.; Liu, Y.; Yeh, C.H. 3D Visible Light-Based Indoor Positioning System Using Two-Stage Neural Network (TSNN) and Received Intensity Selective Enhancement (RISE) to Alleviate Light Non-Overlap Zones. *Sensors* **2022**, *22*, 8817. [CrossRef] [PubMed]
26. Szegedy, C.; Vanhoucke, V.; Ioffe, S.; Shlens, J.; Wojna, Z. Rethinking the Inception Architecture for Computer Vision. In Proceedings of the 2016 IEEE Conference on Computer Vision and Pattern Recognition (CVPR), Las Vegas, NV, USA, 27–30 June 2016; IEEE: Piscataway, NJ, USA, 2016; pp. 2818–2826. [CrossRef]
27. Wang, Q.; Wu, B.; Zhu, P.; Li, P.; Zuo, W.; Hu, Q. ECA-Net: Efficient Channel Attention for Deep Convolutional Neural Networks. In Proceedings of the 2020 IEEE/CVF Conference on Computer Vision and Pattern Recognition (CVPR), Seattle, WA, USA, 13–19 June 2020; pp. 11531–11539.
28. Wang, K.; Liu, Y.; Hong, Z. RSS-based visible light positioning based on channel state information. *Opt. Express* **2022**, *30*, 5683–5699. [CrossRef] [PubMed]
29. Yang, W.; Qin, L.; Hu, X.; Zhao, D. Indoor Visible-Light 3D Positioning System Based on GRU Neural Network. *Photonics* **2023**, *10*, 633. [CrossRef]
30. Hu, J.; Shen, L.; Sun, G. Squeeze-and-Excitation Networks. In Proceedings of the 2018 IEEE/CVF Conference on Computer Vision and Pattern Recognition, Salt Lake City, UT, USA, 18–23 June 2018; pp. 7132–7141.

Article

On the Optimisation of Illumination LEDs for VLP Systems

José Miguel Menéndez [1,2,*] **and Heidi Steendam** [1]

[1] TELIN/IMEC, Ghent University, 9000 Ghent, Belgium
[2] Department of Electrical and Computer Engineering, ESPOL University, Campus Gustavo Galindo, Guayaquil P.O. Box 09-01-5863, Ecuador
[*] Correspondence: josemiguel.menendezsanchez@ugent.be or jmmnend@espol.edu.ec

Abstract: Recent studies have explored the synergy of illumination and positioning using indoor lighting infrastructure. While these studies mainly focused on the analysis of the performance of visible light positioning, these works did not consider the illumination aspects of such combined systems. In this paper, we analyse the illumination aspects based on the main illumination characteristics defined in the European Standard EN 12464-1, i.e., the horizontal illuminance and the uniformity of illuminance. As in the standard, we distinguish between a task area, where visual activities are performed that demand higher illuminance and uniformity, and a surrounding area that borders the former. In our analysis, we derive simple rules of thumb to determine the number and placement of LEDs to satisfy the constraints on the horizontal illuminance and uniformity for a given area.

Keywords: VLP; indoor navigation; horizontal illuminance; illuminance uniformity

Citation: Menéndez, J.M.; Steendam, H. On the Optimisation of Illumination LEDs for VLP Systems. *Photonics* **2022**, *9*, 750. https://doi.org/10.3390/photonics9100750

Received: 31 August 2022
Accepted: 3 October 2022
Published: 10 October 2022

Publisher's Note: MDPI stays neutral with regard to jurisdictional claims in published maps and institutional affiliations.

1. Introduction

Visible light LEDs are progressively replacing traditional incandescent and fluorescent light sources, due to their energy efficiency and longer life time [1]. In contrast to the traditional light sources, LEDs can easily be modulated up to GHz, making them suitable for communication purposes. Furthermore, LEDs are typically mounted on the ceiling, implying most visible light communication (VLC) links contain a line-of-sight (LOS) component. Therefore, visible light LEDs are considered for indoor positioning. Several works have already investigated the accuracy of visible light positioning (VLP) systems [2–4], and reported excellent performance. They considered different approaches to estimating the position, i.e., based on the received signal strength (RSS) [5,6], angle-of-arrival (AOA) [7,8], time-of-arrival (TOA) [9,10] or based on the receiver type such as charge-coupled device (CCD) cameras or photo diode (PD) [11,12].

In the literature, it is stated that to have a low-cost solution, visible light communication and positioning can be combined with illumination after some minor adaptations in the infrastructure needed to modulate the LEDs. In those works, the synergy between positioning and illumination is seen from the positioning viewpoint, e.g., system parameters are optimised to achieve best positioning performance. In other words, the VLP system is prioritised over what is supposed to be the primary function of the illumination system, i.e., to provide adequate illumination to allow visual activities to be carried out safely. However, these studies make assumptions that could affect the level of visual comfort within the area where the system is evaluated, such as the placement of the LEDs on the ceiling, the optical power of the transmitters or the use of stand alone LEDs.

To the author's best knowledge, only a few works deal with the optimisation of illumination, and these works mainly explore the process of designing and manufacturing light sources [13–15]. So, in order to offer design engineers simple guidelines for joint optimisation of illumination and positioning, we focus in this paper on the illumination aspects of a combined illumination and communication/positioning system. Taking as reference the EN 12464-1 standard, which stipulates and regulates all aspects to be considered for adequate lighting in indoor work areas, we analyse the horizontal illuminance and uniformity of illuminance in a given area. We scrutinise how parameters such as the number of LEDs (or LED arrays) and the spacing between them affect the illuminance of the room. In contrast to [16] where a metaheuristic planning algorithm is used, in which the placement of the LEDs is obtained through simulations, we provide in this paper simple analytical expressions for the optimal range of the number of LEDs and spacing between LEDs (or LED arrays) to comply with the requirements of the standard in terms of horizontal illuminance and uniformity. The resulting guidelines can be used in a broad range of situations, i.e., for rooms with different sizes, number of LEDs and Lambertian order of the LEDs.

The rest of the paper is structured as follows. Section 2 gives the system description and discusses the criteria for the evaluation of the system. A thorough analysis of the horizontal illuminance and uniformity is given in Section 3. Finally, the conclusions are given in Section 4.

2. System Description and Evaluation Criteria

In this paper, we assume L white LEDs are attached to the ceiling. We consider two scenarios: in the first scenario, we assume LEDs are grouped in $P_A = P_{A,x} \times P_{A,y}$ arrays of LEDs, where $P_{A,x}$ and $P_{A,y}$ are the number of arrays in the x- and y-direction, respectively, and L_A is the number of LEDs per array, so that $L = L_A P_{A,x} P_{A,y}$, while in the second scenario, the L LEDs are stand alone, which corresponds to the special case where $L_A = 1$. Define $\mathcal{S}_\ell$ as the set of LEDs within array $\ell = 1, \ldots, P_A$, with $\mathcal{S}_\ell \cap \mathcal{S}_{\ell'} = \varnothing, \ell \neq \ell'$. Further, define the set $\mathcal{S}$ of all L LEDs as $\mathcal{S} = \cup_{\ell=1}^{P_A} \mathcal{S}_\ell$. We assume that the centre of array ℓ has coordinates $v_{A,\ell} = (x_{A,\ell}, y_{A,\ell}, z_{A,\ell})^T$, Figure 1. The coordinates of LED $i \in \mathcal{S}_\ell$ within array ℓ are $v_{\ell,i}^{LED} = v_{A,\ell} + v_{\ell,i}$, where $v_{\ell,i} = (x_{\ell,i}, y_{\ell,i}, z_{\ell,i})^T$ is the relative displacement of LED $i \in \mathcal{S}_\ell$ with respect to the centre of the array. We assume LED i within array ℓ has Lambertian order $m_{\ell,i}$, which is connected to the semi-angle $\Phi_{1/2,\ell,i}$ of the LED i at which half optical power is reached through $m_{\ell,i} = -\ln 2 / \ln \left(\cos \left(\Phi_{1/2,\ell,i} \right) \right)$. We assume the coordinates of all LEDs are known, and all LEDs point straight downwards, i.e., their normal is $N_{\ell,i} = (0, 0, -1)^T, \forall i \in \mathcal{S}_\ell, \forall \ell = 1, \ldots, P_A$.

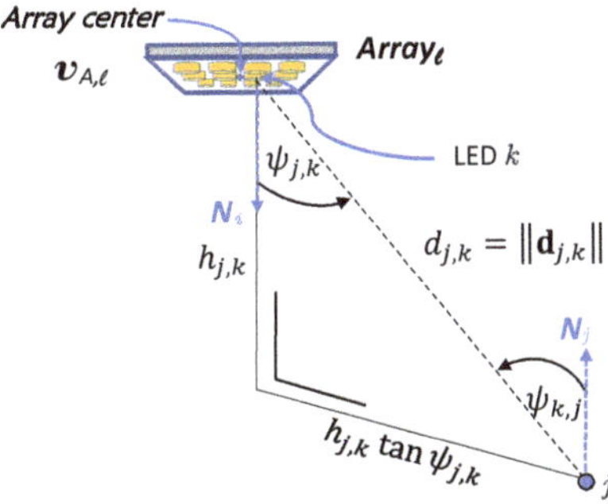

Figure 1. Geometrical definitions in the illumination system, k corresponds to the pair (ℓ, i).

Visual comfort, an important aspect to be considered especially for indoor environments, allows people to perform visual tasks efficiently and accurately. A lack or excess of light directly impacts this comfort. This comfort is measured through the illuminance level and its distribution in the area, where the illuminance, E_h, is a function of the luminous intensity that measures the light intensity I emitted by a light source in a particular direction per unit solid angle [17], and is expressed by:

$$I = \frac{\partial}{\partial \Omega} \Phi_v,$$

(1)

where Ω is the spatial angle over which the luminous flux, Φ_v, is distributed. For LEDs with Lambertian pattern, the radiation intensity, $I(\psi)$ in [cd], of an LED in the direction ψ is given by

$$I(\psi) = I_0 \cos^m(\psi), \tag{2}$$

where the centre luminous intensity I_0 equals $I_0 = \frac{m+1}{2\pi}\Phi_v$, and m is the Lambertian order of the light source. We analyse the amount of light from LED $k \in \mathcal{S}$ with Lambertian order m_k; i.e., k corresponds to a pair (ℓ, i), falling on a point j in a horizontal plane $\mathcal{P}$, with the vector between the LED k and point j equal to $\boldsymbol{d}_{k,j}$. The corresponding horizontal illuminance E_h, in [lx], depends on the orientation of the light source with respect to the lit point j, i.e., the angle $\psi_{j,k}$ between the normal vector $\boldsymbol{N}_k$ of the LED and $\boldsymbol{d}_{k,j}$, as well as on the orientation of the lit point j in the horizontal plane $\mathcal{P}$ with respect to the LED k, i.e., the angle $\psi_{k,j}$ between the normal vector $\boldsymbol{N}_j$ of the plane $\mathcal{P}$ at point j and $\boldsymbol{d}_{k,j}$ [18]. The horizontal illuminance $E_h(\psi_{j,k}, \psi_{k,j})$ is expressed by

$$E_h(\psi_{j,k}, \psi_{k,j}) = \sum_{i=k}^{L} \frac{I\left(\psi_{j,k}\right)}{d_{j,k}^2} \cos(\psi_{k,j}), \tag{3}$$

with $d_{j,k} = ||\boldsymbol{d}_{j,k}||$. In this paper, we assumed that the LED points straight downwards and the plane $\mathcal{P}$ is parallel to the ceiling, implying $\psi_{j,k} = \psi_{k,j}$.

In this paper, we evaluate the illumination requirements for indoor work areas. To this end, we consider the EN 12464-1 standard [19], which specifies the average illuminance $\overline{E}_h$ as well as the uniformity $\mathcal{U}$ of illumination, with the uniformity given by $\mathcal{U} = \min\{E_h\}/\overline{E}_h$. Depending on the intended use of the areas, e.g., office, production and warehouse, the standard defines the lighting levels and distinguishes between task area and surrounding area. Table 1 shows the required illumination levels in the surrounding area for the given average illuminance in the task area. The uniformity of illumination also depends on the area type, and is constrained by $\mathcal{U}_t \geq 0.7$ for the task area, and $\mathcal{U}_s \geq 0.5$ for the surrounding area. For example, in an office area environment, the average illuminance must range between 300 and 500 lux.

Table 1. Values of the illuminance and uniformity in the task and surrounding areas according to the standard EN 12464-1:2007 [19].

$\overline{E}_h$ in Task Areas (in lux)	$\overline{E}_h$ in Surrounding Areas (in lux)
$\geq$750	500
500	300
300	200
$\leq$200	$\overline{E}_{h,\text{task}}$
$\mathcal{U}_t \geq 0.7$	$\mathcal{U}_s \geq 0.5$

3. Analysis of Horizontal Illuminance and Uniformity

In this section, we analyse how parameters such as the number and position of LEDs affect the horizontal illumination and uniformity in the lighting system. We then outline some rules of thumb according to which LED arrays can be optimally installed to meet the requirements of adequate illumination. We consider that both the lit area that is evaluated and the ceiling area have dimensions $X_{\max} \times Y_{\max}$ with the latter located parallel and at a distance $Z_{\max}$ above the first. Without loss of generality, we set $Z_{\max} = 2$ m, luminous flux $\Phi_{v,k} = \Phi_v$ and Lambertian order $m_k = m_S$, $\forall k \in \mathcal{S}$, unless otherwise specified.

As the uniformity consists of the ratio of the minimum value $\min\{E_h\}$ and the average $\overline{E}_h$ of the illuminance, we will first concentrate on the effect of the number and placement of the LEDs on $\overline{E}_h$ and $\min\{E_h\}$ separately, and then extend the analysis to the uniformity. We first take a closer look at the average illuminance $\overline{E}_h$.

3.1. Average Horizontal Illuminance

Let us first consider the dependency of the average illuminance on the placement of the LEDs. To this end, we consider a square area where $L = P_L^2$ LEDs with $\Phi_v^{typ} = 270$ lm are attached to the ceiling in a square grid with spacing δ_L between the LEDs. In Figure 2, we show the average illuminance $\overline{E}_h$, assuming $X_{max} = Y_{max} = \{10, 13, 15\}$ m, $P_L = \{12, 15\}$ and $\delta_L \in [0, \delta_{L,max}]$ where $\delta_{L,max} = \frac{X_{max}}{P_L - 1}$ is the maximum spacing for which all LEDs are within the area of interest. We only consider contributions from light radiated towards positions inside the considered area, and neglect reflection against walls and other objects. As can be observed, the average illuminance is relatively independent of the placement of the LEDs. For larger spacings δ_L, the average illuminance slowly decays, because LEDs will be placed closer to the edges of the considered area, implying more light is lost as it is radiated towards a position outside the area of interest. The reduction will be larger when more LEDs are close to the edges of the area of interest, i.e., when the number of LEDs is large and the area of interest is relatively small. While the effect of the placement of the LEDs on the average illuminance in most cases is limited, the figure reveals that the impact of the number of LEDs is much more important. Hence, in the following, we analyse the dependency of the average illuminance on the number of LEDs.

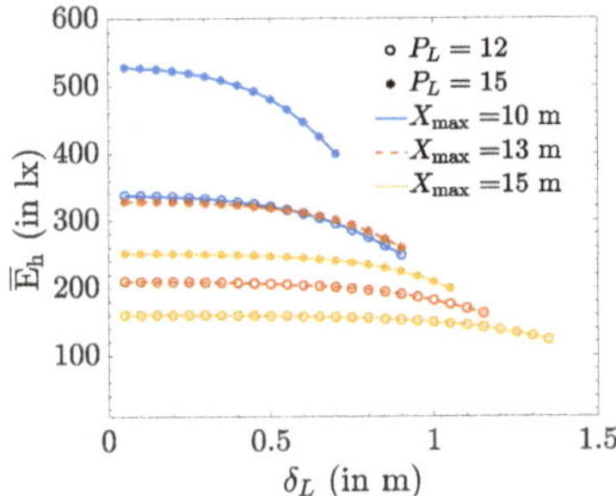

Figure 2. Horizontal illuminance average $\overline{E}_h$ for $X_{max} = Y_{max} = \{10, 13, 15\}$ m, $P_L = \{12, 15\}$ and $\Phi_v^{typ} = 270$ lm.

Assume we measure the illuminance at a distance Z_{max} below the ceiling, where the LEDs are attached. The horizontal illuminance E_h corresponding to a single LED, at a point at vertical distance Z_{max} below the LED and seeing the LED from an incident angle ψ and azimuth angle α is given by (3):

$$E_h(\psi, \alpha) = \frac{(m_S + 1)\Phi_v}{2\pi Z_{max}^2} \cos^{m_S+3} \psi, \tag{4}$$

with Φ_v being the luminous flux output of the LED. Using this expression, we now will formulate an upper and lower bound on the average illuminance in a rectangular area with size $X_{max} \times Y_{max}$. We assume the LED is placed in the centre of this area. To find the upper bound, we compute the total illuminance in a circular area with radius R_{max}. A straightforward choice for R_{max} is $R_{max,1} = \frac{1}{2}\sqrt{X_{max}^2 + Y_{max}^2}$, corresponding to the smallest circle that encloses the rectangular area with size $X_{max} \times Y_{max}$. The resulting total illuminance within the circular area equals $E_{h,tot}(\psi_{max})$, where $\psi_{max} = \arctan \frac{R_{max}}{Z_{max}}$ and

$$\begin{aligned}
E_{h,tot}(\psi) &= \int_0^{2\pi} d\alpha \int_0^{\psi} E_h(\tilde{\psi}, \alpha) Z_{max}^2 \tan \tilde{\psi} \sec^2 \tilde{\psi} d\tilde{\psi} \\
&= \Phi_v \left(1 - \cos^{(m_S+1)} \psi\right)
\end{aligned} \tag{5}$$

is the total illuminance in a circular area with radius R corresponding to a maximum radiation angle $\psi = \arctan \frac{R}{Z_{max}}$. As the circular area is larger than the rectangular area, this total illuminance is the upper bound to the total illuminance in the rectangular area. Dividing the total illuminance $E_{h,tot}(\psi_{max})$ by the area $\mathcal{A}_{max} = X_{max}Y_{max}$ of the rectangle therefore results in an upper bound on the average illuminance:

$$\overline{E}_h \leq \frac{\Phi_v}{\mathcal{A}_{max}}\left(1 - \cos^{(m_s+1)}\psi_{max}\right) \triangleq \overline{E}_{h,up}. \tag{6}$$

However, when the rectangular area is not close to a square area, this upper bound is far from tight. Therefore, we propose a tighter bound when the area is not close to a square. Assume $X_{max} \neq Y_{max}$. Let us consider the circle with radius $R_{max,2} = \frac{1}{2}\max(X_{max}, Y_{max})$. Although not all parts of the rectangular area are enclosed in this circle, the parts of the rectangle not contained in the circle lie at distance $> R_{max,2}$ from the LED. At the same time, some parts of the circle will not be enclosed in the rectangle. In this latter case, the distance between a position in those parts and the LED is in the interval $[R_{min}, R_{max,2}]$, with $R_{min} = \frac{1}{2}\min(X_{max}, Y_{max})$. Taking into account that the illuminance (4) reduces with the distance to the LED, it follows that the total illuminance in the circular area with radius $R_{max,2}$ is larger than that of the rectangular area, provided that the circular area is larger than the rectangular area, i.e., when $\frac{\min(X_{max}, Y_{max})}{\max(X_{max}, Y_{max})} < \frac{\pi}{4}$. In this case, using $R_{max,2}$ to compute (6) yields a tighter upper bound as $R_{max,2} < R_{max,1}$. In conclusion, the upper bound is given by (6), with $\psi_{max} = \arctan\frac{R_{max}}{Z_{max}}$ where

$$R_{max} = \begin{cases} R_{max,1}, & \text{for } \frac{\min(X_{max}, Y_{max})}{\max(X_{max}, Y_{max})} \geq \frac{\pi}{4} \\ R_{max,2}, & \text{otherwise} \end{cases}. \tag{7}$$

In the derivation of (5), we assumed that the LED was positioned above the centre of the receiver area. This LED position maximises the amount of light inside the rectangular area $\mathcal{A}_{max}$. However, in practice, LEDs will be distributed over the area to achieve good uniformity of lighting. When an LED is not positioned in the centre of the rectangular area, the amount of light that falls outside the area will increase, implying (6) with ψ_{max} computed using (7) can also serve as an upper bound for the total illumination within the rectangular area for other LED positions.

To find a lower bound on the average illumination, we consider the circular area with radius R_{min}. Assuming the LED is placed in the centre of the rectangular area, this radius corresponds to the largest circle that is enclosed in the rectangular area. Hence, assuming the LED is placed in the centre, the resulting total illuminance within the circular area, i.e., $\overline{E}_h(\psi_{min})$ (5) with $\psi_{min} = \arctan\frac{R_{min}}{Z_{max}}$, will be smaller than the total illuminance in the rectangular area as the parts of the rectangle that fall outside this circular area are neglected in the computation of the total illuminance. Further, to take into account that the total illuminance reduces when the LED is placed closer to the boundaries, we consider the worst case position for the LED. This worst case position is a corner of the rectangular area, resulting in a total illuminance $\frac{1}{4}\overline{E}_h(\psi_{min})$, as 75% of the light is radiated to directions outside the rectangular area. While this illuminance is a strict lower bound on the total illuminance, it also strongly underestimates the true average illuminance in practical scenarios where LEDs are distributed over the area. To obtain a tighter bound, we therefore consider the situation where the LED is positioned at the boundary of the area, more specifically in the middle of the smallest side of the rectangle. In that case, only 50% of the radiated light is lost. Dividing the resulting total illuminance by the area $\mathcal{A}_{max}$, we obtain the following (approximate) lower bound on the average illuminance:

$$\overline{E}_h \geq \frac{\Phi_v}{2\mathcal{A}_{max}}\left(1 - \cos^{(m_s+1)}\psi_{min}\right) \triangleq \overline{E}_{h,low}. \tag{8}$$

When L LEDs are placed in the area, the average illuminance will be bounded by:

$$L\overline{E}_{h,low} \leq \overline{E}_h \leq L\overline{E}_{h,up}. \tag{9}$$

Let us assume we want an average illuminance equal to $\overline{E}_h = \mathcal{E}$ lx, then we obtain the following bounds on the required number L of LEDs:

$$L \geq \frac{\mathcal{A}_{\max}\mathcal{E}}{\Phi_v(1 - \cos^{m_S+1}\psi_{\max})} \triangleq L_{\min} \tag{10}$$

and

$$L \leq \frac{2\mathcal{A}_{\max}\mathcal{E}}{\Phi_v(1 - \cos^{m_S+1}\psi_{\min})} \triangleq L_{\max}. \tag{11}$$

To illustrate, we consider a rectangular area with $X_{\max} = 10$ m, $Y_{\max} = \frac{X_{\max}}{1.5}$ m and $Z_{\max} = 2$ m, and LEDs with $m_S = 1$ and $\Phi_v = 270$ lm. For an average illuminance $\mathcal{E} = 300$ lx, we obtain $L \in [L_{\min,300}, L_{\max,300}] = [86, 202]$ and for $\mathcal{E} = 500$ lx, $L \in [L_{\min,500}, L_{\max,500}] = [144, 336]$.

Let us compare the resulting bounds on L with simulation results for the average illuminance. In this simulation, we placed $L = P_L^2$ LEDs in the rectangular area mentioned above. The L LEDs are attached separately to the ceiling in a rectangular grid of size $2\Omega_{c,x} \times 2\Omega_{c,y}$, with $\frac{\Omega_{c,x}}{\Omega_{c,y}} = 1.5$, $2\Omega_{c,x} \leq X_{\max}$ m and $2\Omega_{c,y} \leq Y_{\max}$ m, where the centre of the grid is the centre of the ceiling. The spacing between the LEDs equals $\delta_{I,x} = \frac{2}{P_L-1}\Omega_{c,x}$ and $\delta_{I,y} = \frac{2}{P_L-1}\Omega_{c,y}$. The spatial average $\overline{E}_h$ of the horizontal illuminance is shown in Figure 3 for different values of $\Omega_{c,x}$ and P_L, assuming $\Phi_v^{typ} = 270$ lm. We observe that the required number of LEDs to achieve an average horizontal illuminance between 300 and 500 lx varies with $(\Omega_{c,x}, \Omega_{c,y})$, i.e., with the distribution of the LEDs over the ceiling. This is explained as LEDs near the boundaries will leak more light outside the receiver area than LEDs in the centre of the receiver area. Therefore, less LEDs will be needed to reach a given average illuminance when the LEDs are co-located in the centre, although it is clear that this will result in a worse uniformity than distributed LEDs. An average horizontal illuminance between 300 and 500 lx is obtained for $P_L \in [10, 14]$, or $L \in [100, 196]$. This falls within the interval predicted by the bounds on L, i.e., $[L_{\min,300}, L_{\max,500}] = [86, 336]$. Hence, the bounds (10) and (11) are suitable for obtaining an approximation for the number of required LEDs, where the lower bound is appropriate when the LEDs are all centered in the room, while the upper bound is more suitable when the LEDs are distributed over the whole receiver area.

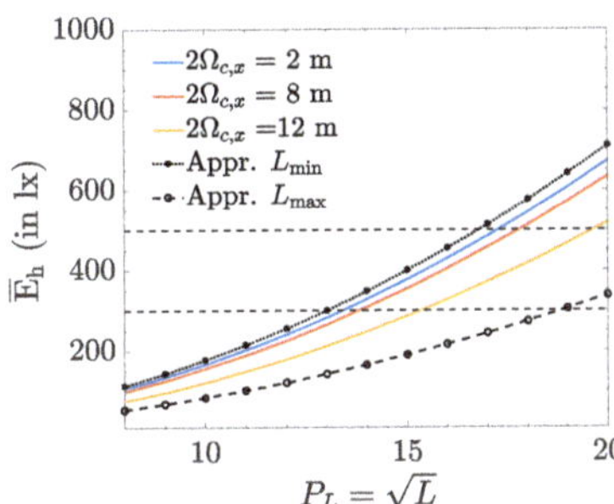

Figure 3. Number of LEDs required to achieve $\overline{E}_h$, $\Phi_v^{typ} = 270$ lm.

In our example, the number L of LEDs required to achieve an average illuminance in the interval $[300, 500]$ lx was of the order 100–300. It is obvious that in practice, to reduce installation costs, these LEDs will not be attached individually to the ceiling. Instead, they will be grouped in luminaries containing several LEDs. In the following, we therefore assume that each luminary consists of an array of L_A LEDs, resulting in $P_{A,x} \times P_{A,y}$ arrays that need to be distributed over the ceiling, with $P_{A,x}P_{A,y}L_A = L$. We assume that the arrays are placed in a rectangular grid with spacing $\delta_{A,x}$ and $\delta_{A,y}$ between the centers of the arrays. Within each array, we further assume that the spacing between the LEDs is small, so each luminary can be considered as a virtual LED with luminous flux $\Phi_v = L_A \Phi_v^{typ}$.

3.2. Minimum Horizontal Illuminance

In the above analysis, we showed that the average illuminance imposes conditions on the required number of LEDs. In the following, we show that the optimal placement of the luminaries is determined by the uniformity of illumination, and more specifically by the minimum horizontal illuminance $E_{h,min}$. Taking into account that the required uniformity differs for the task area and surrounding area, we first define the task area $\mathcal{A}_t$ as the central area delimited by X_{max}^T and Y_{max}^T, while the surrounding area $\mathcal{A}_s$ is the strip at the edges of the receiver area, as illustrated in Figure 4.

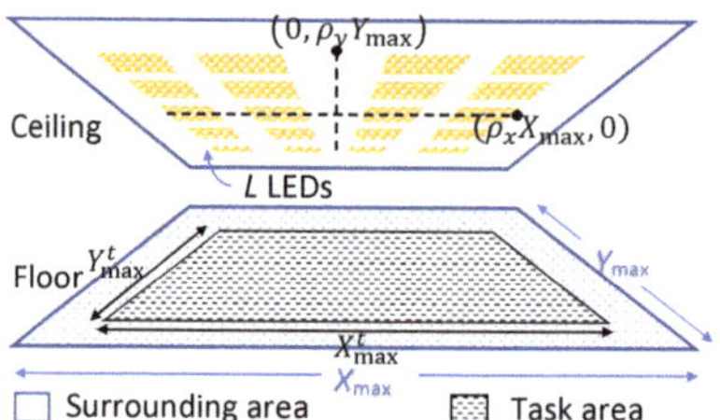

Figure 4. Receiver area lit by L LEDs uniformly grouped in $P_{A,x} \times P_{A,y}$ arrays on the ceiling within an area of $\rho_x X_{max} \times \rho_y Y_{max}$.

First, we want to demonstrate that $E_{h,min}$ is essentially independent of the size of the area, and that it is mainly determined by the four parameters defined in Figure 5a, i.e., the distance $\delta_{A,x}$ and $\delta_{A,y}$ between the luminaries, and the distance Δ_x and Δ_y between the luminaries and the border of the area. To illustrate that, for given $(\delta_{A,x}, \delta_{A,y}, \Delta_x, \Delta_y)$, the minimum is highly independent of the size of the area, we determine $E_{h,min}$ for different values of $P_{A,x}$ and $P_{A,y}$ with

$$
\begin{aligned}
X_{max} &= (P_{A,x} - 1)\delta_{A,x} + 2\Delta_x \\
Y_{max} &= (P_{A,y} - 1)\delta_{A,y} + 2\Delta_y.
\end{aligned}
\tag{12}
$$

The resulting $E_{h,min}$ is shown in Figure 5b. As can be observed, the minimum of the horizontal illuminance is indeed essentially independent of $P_{A,x}$ and $P_{A,y}$ and thus of the size of the considered area, whereas it strongly depends on $(\delta_{A,x}, \delta_{A,y}, \Delta_x, \Delta_y)$ and , hence, the placement of the luminaries. Therefore, we will take a closer look at the impact of these four parameters on $E_{h,min}$.

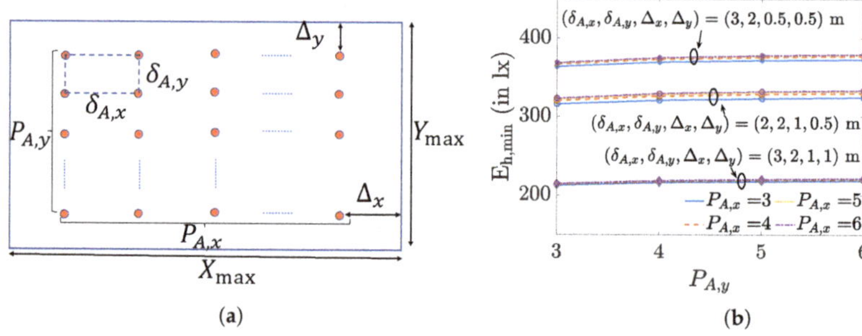

Figure 5. (**a**) Receiver area layout of size $X_{max} \times Y_{max}$ with $P_{A,x} \times P_{A,y}$ arrays and (**b**) $E_{h,min}$ for different $P_{A,x}$, $P_{A,y}$, $\delta_{A,x}$, $\delta_{A,y}$, Δ_x and Δ_y.

Because of the complexity of the expressions for the horizontal illuminance, i.e., the horizontal illuminance depends on a large number of parameters, a simple analytical expression for this minimum is not available, implying the minimum must be obtained through a two-dimensional search over the considered area. As such, a two-dimensional search comes with high complexity and gives no insight into the optimisation problem, in the following, we derive approximate expressions for the position of the minimum, with which we are able to obtain (an approximation of) the value of the minimum in an analytical way. Taking into account the dependency of the horizontal illuminance on the distance, it is clear that the minimum illuminance in an area is found in the part of the area surrounded by the least number of LEDs, so close to the corners of the considered (task or surrounding) area. Therefore, we restrict our attention to the evaluation area $\mathcal{A}_e$ shown in Figure 6, which includes the four arrays closest to the boundary (indicated by the red dots). In the case of the surrounding area, the minimum is determined by the parameters $(\delta_{A,x}, \delta_{A,y}, \Delta_x^s, \Delta_y^s)$. As all arrays are assumed to be inside the boundaries of the considered area, Δ_x^s and Δ_y^s are non-negative, and the relationship between $(\delta_{A,x}, \delta_{A,y}, \Delta_x^s, \Delta_y^s)$ and $(X_{\max}, Y_{\max})$ is given by (12), i.e., $\Delta_x^s = \Delta_x$ and $\Delta_y^s = \Delta_y$. On the other hand, in the case of the task area, $E_{h,\min}$ is determined by the parameters $(\delta_{A,x}, \delta_{A,y}, \Delta_x^t, \Delta_y^t)$, where Δ_x^t and Δ_y^t can be negative if some of the arrays closest to the corner of the task area are located in the surrounding area. Assuming the task area has size $X_{\max}^t \times Y_{\max}^t$ with $X_{\max}^t = \zeta_x^t X_{\max}$ and $Y_{\max}^t = \zeta_y^t Y_{\max}$, with $\zeta_x^t, \zeta_y^t \in [0, 1]$, and $P_{A,x}^t$ and $P_{A,y}^t$ arrays are located inside the task area, it follows that

$$
X_{\max}^t =
\begin{cases}
(P_{A,x}^t - 1)\delta_{A,x} + \Delta_x^t & \text{if } P_{A,x}^t = P_{A,x} \\
P_{A,x}^t \delta_{A,x} + \Delta_x^t & \text{if } P_{A,x}^t < P_{A,x}
\end{cases}
$$

$$
Y_{\max}^t =
\begin{cases}
(P_{A,y}^t - 1)\delta_{A,y} + \Delta_y^t & \text{if } P_{A,y}^t = P_{A,y} \\
P_{A,y}^t \delta_{A,y} + \Delta_y^t & \text{if } P_{A,y}^t < P_{A,y}
\end{cases}
, \tag{13}
$$

where the first lines in (13) correspond to the case that all LEDs in the x ($P_{A,x}^t = P_{A,x}$) and y directions ($P_{A,y}^t = P_{A,y}$), respectively, are located in the task area, while in the second case, some of the LEDs are in the surrounding area. When $P_{A,x}^t < P_{A,x}$ ($P_{A,y}^t < P_{A,y}$), the resulting Δ_x^t (Δ_y^t) will be negative. Depending on the values of $(\delta_{A,x}, \delta_{A,y}, \Delta_x^{s/t}, \Delta_y^{s/t})$, the minimum illuminance will be in the corner m_c of the considered area (see Figure 7a), close to the centre m_m of the four arrays (see Figure 7b) or on the boundary of the considered area, close to the middle m_b of the two nearest arrays (see Figure 7c), where the minimum can be on the boundary in the x direction ($m_{b,x}$) or in the y direction ($m_{b,y}$), as defined in Figure 6. In our analysis, we will restrict our attention to these four positions, i.e., m_c, m_m, $m_{b,x}$ and $m_{b,y}$. In the remainder of this section, we drop the superscript s/t in $\Delta_x^{s/t}$ and $\Delta_y^{s/t}$ for notational simplicity.

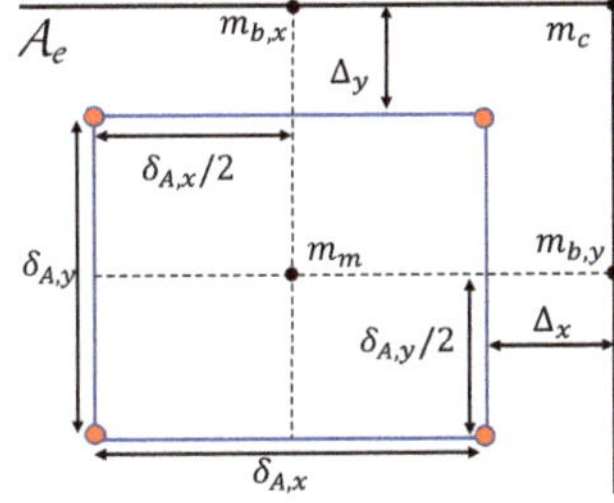

Figure 6. Parameter definition for the evaluation area $\mathcal{A}_e$.

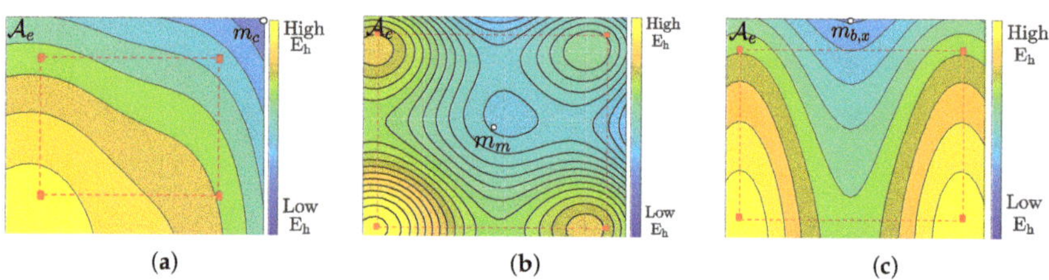

(a)   **(b)**   **(c)**

Figure 7. Illumination patterns having the minimum (**a**) in the corner m_c of the considered area, (**b**) close to the centre m_m of the four arrays, (**c**) on the boundary of the considered area, approximately halfway between the two nearest arrays (here $m_{b,x}$).

To determine which of the four reference positions corresponds to the minimum for a given 4-tuple $(\delta_{A,x}, \delta_{A,y}, \Delta_x, \Delta_y)$, we evaluate the horizontal illuminance at these points. The horizontal illuminance at a distance $Z_{\max}$ below an array and at a horizontal distance d from the array is equal to

$$E_h(d, Z_{\max}) = C\left(Z_{\max}^2 + d^2\right)^{-\frac{m_s+3}{2}} \triangleq C\mathcal{L}(d), \tag{14}$$

where C is a factor that is independent of the distance d. As we want to compare illuminance levels at different positions in a horizontal plane at vertical distance $Z_{\max}$ from the array, we ignore in the following analysis the common factor C and use the function $\mathcal{L}(d)$ in the comparison. Combining the contributions from the four nearest arrays, we obtain the illuminance

$$\mathcal{L}_4 = \mathcal{L}(d_1) + \mathcal{L}(d_2) + \mathcal{L}(d_3) + \mathcal{L}(d_4) \tag{15}$$

where d_i, $i = 1, \ldots, 4$ are the distances between the reference point and the four considered arrays. These distances are given in Table 2 for the different reference points. To further simplify (15), we only consider the contributions from the nearest arrays, i.e., with the smallest d_i:

$$\mathcal{L}_4 \approx \beta_{\min}\mathcal{L}(d_{\min}) \tag{16}$$

where $d_{\min} = \min(d_1, d_2, d_3, d_4)$ and $\beta_{\min} \in \{1, 2, 4\}$ is the number of terms in (15) corresponding to arrays at distance $d_{\min}$ to the reference point. Using the resulting approximations (16), we will determine the reference point at which the horizontal illuminance is the smallest:

$$\hat{m}_o = \arg \min_{m_o \in \mathcal{I}_4} \left(\mathcal{L}_{4,c}, \mathcal{L}_{4,m}, \mathcal{L}_{4,b,x}, \mathcal{L}_{4,b,y}\right) \tag{17}$$

where $\hat{m}_o \in \mathcal{I}_4 = \{m_m, m_c, m_{b,x}, m_{b,y}\}$. In the following, the subscript 'o' is used to indicate the parameters corresponding to this optimal reference point. Note that when $\Delta_x < -\frac{\delta_{A,x}}{2}$ or $\Delta_y < -\frac{\delta_{A,y}}{2}$, m_m, $m_{b,x}$ and/or $m_{b,y}$ will fall outside the considered area, in which case we will omit the corresponding reference positions in our comparison. Taking into account (14), this comparison will result in inequalities of the form

$$\gamma_o(Z_{\max}^2 + d_{\min,o}^2) \geq \gamma_j(Z_{\max}^2 + d_{\min,j}^2) \tag{18}$$

with $\gamma_j = (\beta_{\min,j})^{-\frac{2}{m_s+3}}$, and $j \in \mathcal{I}_4 \setminus \{\hat{m}_o\}$. From Table 2 and Equations (12) and (13), it follows that these inequalities give constraints on the spacing $\delta_{A,x}$ and $\delta_{A,y}$ between the arrays. More specifically, we can identify regions for each reference point in the $(\delta_{A,x}, \delta_{A,y})$ plane, bounded by the conic sections following from the inequalities (18). This is illustrated in Figure 8a, in which we show the reference position that results in the minimum horizontal illuminance as a function of $\delta_{A,x}$ and $\delta_{A,y}$, as well as the ellipses $\mathcal{E}_{c-b,x}$,

$\mathcal{E}_{c-b,y}$, $\mathcal{E}_{m-b,x}$ and $\mathcal{E}_{m-b,y}$, that form the decision regions between m_m and m_c with $m_{b,x}$ and $m_{b,y}$, and the hyperbola $\mathcal{H}_{b,x-b,y}$ that bounds the decision region between $m_{b,x}$ and $m_{b,y}$. In Figure 8b,c, we show the true $E_{h,min}$ for the surrounding area and task area, respectively, and for $X_{max} = 11$ m, $Y_{max} = 10$ m, $Z_{max} = 2$ m, $P_{A,x} = 3$, $P_{A,y} = 3$ and $m_S = 1$, along with their delimiting conic sections. To find the true $E_{h,min}$ in the simulation, we used the true illuminance (14) and the contributions of all $P_{A,x}P_{A,y}$ luminaries, where the minimum $E_{h,min}$ is found through an exhaustive search. As can be expected, the dependency of the minimum illuminance to the spacing between the luminaries is different for the two areas. For small $(\delta_{A,x}, \delta_{A,y})$, i.e., when the luminaries are tightly clustered in the centre of the ceiling, the corners of the area are poorly illuminated, resulting in a low $E_{h,min}$ in the surrounding area. At the same time, the centre of the area is better lit, implying $E_{h,min}$ in the task area is much higher. We observe in the figures that $E_{h,min}$ reaches a maximum value at some $(\delta_{A,x}, \delta_{A,y})$, but the position of the maximum is different for the task and the surrounding area. However, for both the task area and the surrounding area, the maximum $E_{h,min}$ is located on their respective hyperbolas, $\mathcal{H}_{b,x-b,y}^t$ and $\mathcal{H}_{b,x-b,y}^s$, in the segment determined by the intersections between the ellipses $\mathcal{E}_{m-b,x}$ and $\mathcal{E}_{m-b,y}$, and the ellipses $\mathcal{E}_{c-b,x}$ and $\mathcal{E}_{c-b,y}$.

Table 2. Distances between the arrays and the reference points.

	m_c	m_m	$m_{b,x}$	$m_{b,y}$
d_1^2	$\Delta_x^2 + \Delta_y^2$	$\left(\frac{\delta_{A,x}}{2}\right)^2 + \left(\frac{\delta_{A,y}}{2}\right)^2$	$\left(\frac{\delta_{A,x}}{2}\right)^2 + \Delta_y^2$	$\Delta_x^2 + \left(\frac{\delta_{A,y}}{2}\right)^2$
d_2^2	$(\delta_{A,x} + \Delta_x)^2 + \Delta_y^2$	$\left(\frac{\delta_{A,x}}{2}\right)^2 + \left(\frac{\delta_{A,y}}{2}\right)^2$	$\left(\frac{\delta_{A,x}}{2}\right)^2 + \Delta_y^2$	$\Delta_x^2 + \left(\frac{\delta_{A,y}}{2}\right)^2$
d_3^2	$\Delta_x^2 + (\delta_{A,y} + \Delta_y)^2$	$\left(\frac{\delta_{A,x}}{2}\right)^2 + \left(\frac{\delta_{A,y}}{2}\right)^2$	$\left(\frac{\delta_{A,x}}{2}\right)^2 + (\delta_{A,y} + \Delta_y)^2$	$(\delta_{A,x} + \Delta_x)^2 + \left(\frac{\delta_{A,y}}{2}\right)^2$
d_4^2	$(\delta_{A,x} + \Delta_x)^2 + (\delta_{A,y} + \Delta_y)^2$	$\left(\frac{\delta_{A,x}}{2}\right)^2 + \left(\frac{\delta_{A,y}}{2}\right)^2$	$\left(\frac{\delta_{A,x}}{2}\right)^2 + (\delta_{A,y} + \Delta_y)^2$	$(\delta_{A,x} + \Delta_x)^2 + \left(\frac{\delta_{A,y}}{2}\right)^2$

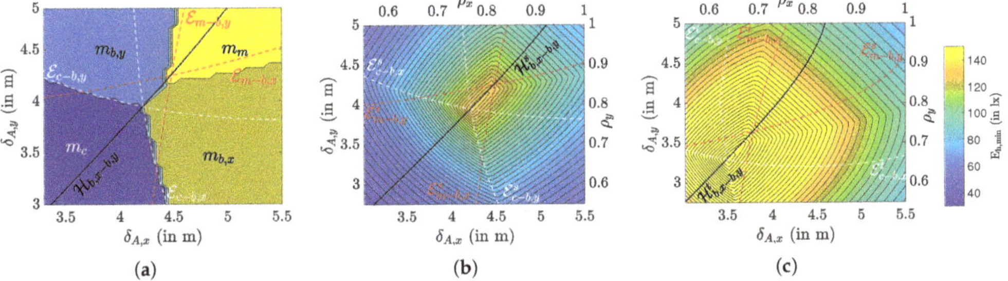

Figure 8. For $X_{max} = 11$ m, $Y_{max} = 10$ m, $Z_{max} = 2$ m, $P_{A,x} = 3$, $P_{A,y} = 3$ and $m_S = 1$ (**a**) shows the decision regions for $(m_m, m_c, m_{b,x}, m_{b,y})$ in the $(\delta_{A,x}, \delta_{A,y})$ plane to obtain the minimum horizontal illuminance $E_{h,min}$ in both (**b**) the surrounding area and (**c**) the task area.

3.3. Uniformity

Using the analysis of Section 3.1, we are able to determine the range for the total number L of LEDs required to satisfy the constraints on the average horizontal illuminance. The next step is to determine how these LEDs must be grouped in luminaries, i.e., arrays of LEDs, to satisfy the uniformity constraints. It is obvious that when the number of luminaries is large, the uniformity constraints will be satisfied, although this will come with a large installation cost. Hence, we are interested in finding (1) the minimum number of luminaries for which in both areas the uniformity is larger than the threshold $\mathcal{U}_a^{(th)}$, $a \in \{s, t\}$, and (2) the range over which the spacing $(\delta_{A,x}, \delta_{A,y})$ may vary for a given number of luminaries. In the following, we define $K_S = \frac{X_{max}}{Y_{max}}$ as the ratio between the dimensions of the area, and $K_A = \frac{P_{A,x}}{P_{A,y}}$ as the ratio of the number of arrays in each dimension.

In Figure 9, we show the uniformity as a function of the spacing $(\delta_{A,x}, \delta_{A,y})$ for both the task and surrounding area, for $X_{\max} = 17$ m, $P_{A,x} = 6$, $K_S = 1.3$, $K_A = 6/4$, $m_S = 1$ and $Z_{\max} = 2$ m. The uniformity is computed using (14) and takes into account the contributions from all luminaries, where to obtain $E_{h,\min}$, we use the estimated position (17) of the minimum. We also show in both figures the region in which the uniformity exceeds the respective thresholds, i.e., $\{(\delta_{A,x}, \delta_{A,y}) \mid \mathcal{U}_t \geq \mathcal{U}_t^{(th)}\}$ (bounded by the red curve) for the task area, and $\{(\delta_{A,x}, \delta_{A,y}) \mid \mathcal{U}_s \geq \mathcal{U}_s^{(th)}\}$ (bounded by the green curve) for the surrounding area, as well as the compliance region $\mathcal{CR}$ (bounded by the dotted curve), which is the region for $(\delta_{A,x}, \delta_{A,y})$ where the uniformity constraints in both the task and surrounding area are satisfied: $\mathcal{CR} = \{(\delta_{A,x}, \delta_{A,y}) \mid \mathcal{U}_t \geq \mathcal{U}_t^{(th)}\} \cap \{(\delta_{A,x}, \delta_{A,y}) \mid \mathcal{U}_s \geq \mathcal{U}_s^{(th)}\}$. This compliance region changes when we alter one or more of the following parameters: the area size $X_{\max} \times Y_{\max}$, the number of arrays $(P_{A,x}, P_{A,y})$, the Lambertian order m_S or the vertical distance $Z_{\max}$. In some cases, the compliance region will be empty, e.g., if the number of arrays is not sufficiently large to meet the uniformity constraints irrespective of the spacing between the arrays. In general, if the number of arrays is sufficiently large, the compliance region will result in a relatively large range of potential spacings $(\delta_{A,x}, \delta_{A,y})$. Unfortunately, due to the complexity of the problem, no analytical expression for the boundaries of this compliance region can be derived.

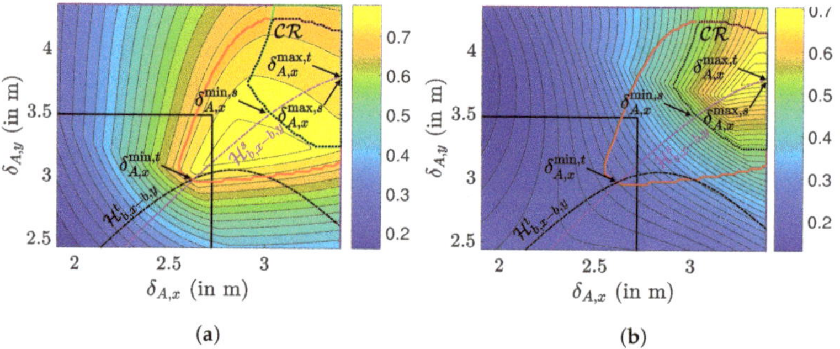

Figure 9. Uniformity distribution for $X_{\max} = 17$ m, $K_S = 1.3$, $K_A = 6/4$, $Z_{\max} = 2$ m and $m_S = 1$. (a) $\mathcal{U}_t$ and (b) $\mathcal{U}_s$.

We will therefore approach the problem from a slightly different viewpoint. Let us consider the maximum uniformity in the task and the surrounding area. Firstly, this maximum will give an indication if the compliance region is empty. Indeed, if the maximum value of the uniformity is below the threshold in the task or surrounding area, the compliance region will be empty. However, even if for both the task area and surrounding area the maximum is above the threshold, it is not guaranteed that the compliance region is non-empty, i.e., if the regions for the task area and surrounding area do not overlap. Secondly, for a given number of luminaries, the maximum uniformity will result in the highest comfort to the user. Therefore, we look for the values of $(\delta_{A,x}, \delta_{A,y})$ where for given $\delta_{A,x}$, the corresponding value for $\delta_{A,y}$ results in the largest uniformity, i.e., the ridge in the uniformity distribution. As we can observe in the figure, the ridge of maximum values of the uniformity for a given $\delta_{A,x}$ in the surrounding area approximately corresponds to the hyperbola $\mathcal{H}_{b,x-b,y}^s$. This is obvious, as in the previous section, the maximum value of $E_{h,\min}^s$ for given $\delta_{A,x}$ was shown to be on this hyperbola, and the average horizontal illuminance is largely independent of the array spacing. For the task area, the ridge of maxima of $E_{h,\min}^t$ will fall on the hyperbola $\mathcal{H}_{b,x-b,y}^t$ provided that no luminaries are located outside the task area. However, when for this maximum $E_{h,\min}^t$ some luminaries are outside the task area, the hyperbola $\mathcal{H}_{b,x-b,y}^t$ no longer follows the ridge of maxima. Regardless of the placement of the luminaries, we observed in our simulations that the hyperbola

$\mathcal{H}^s_{b,x-b,y}$ of the surrounding area roughly follows the ridge of maxima in the task area. Therefore, we will consider for both areas the hyperbola $\mathcal{H}^s_{b,x-b,y}$ to determine the value of $\delta_{A,y}$ corresponding to the maximum uniformity for a given $\delta_{A,x}$.

To obtain the range of $\delta_{A,x}$ and corresponding $\delta_{A,y}$ for which the uniformity constraints are satisfied, we determine the intersections between the hyperbola and the regions where the uniformity is above the threshold in the task and surrounding area, respectively, i.e., for the task area, $\delta_{A,x} \in [\delta_{A,x}^{\min,t}, \delta_{A,x}^{\max,t}]$ when the point $(\delta_{A,x}, \delta_{A,y})$ on the hyperbola satisfies $\mathcal{U}_t \geq \mathcal{U}_t^{(th)}$ and for the surrounding area $\delta_{A,x} \in [\delta_{A,x}^{\min,s}, \delta_{A,x}^{\max,s}]$ when the point $(\delta_{A,x}, \delta_{A,y})$ on the hyperbola satisfies $\mathcal{U}_s \geq \mathcal{U}_s^{(th)}$, as indicated in Figure 9. The range for $\delta_{A,x}$ where the uniformity constraints are satisfied for both regions, i.e., where the spacing $(\delta_{A,x}, \delta_{A,y})$ belongs to the compliance region, is then given by the intersection of the two intervals: $\delta_{A,x}^{CR} \in [\delta_{A,x}^{\min}, \delta_{A,x}^{\max}]$, where $\delta_{A,x}^{\min} = \max\{\delta_{A,x}^{\min,s}, \delta_{A,x}^{\min,t}\}$ and $\delta_{A,x}^{\max} = \min\{\delta_{A,x}^{\max,s}, \delta_{A,x}^{\max,t}\}$. As can be observed in the figure, the upper bound $\delta_{A,x}^{\max}$ may coincide with the maximum possible spacing $\overline{\delta}_{A,x} = \frac{X_{\max}}{P_{A,x}-1}$, or in some situations with $\overline{\delta}_{A,y} = \frac{Y_{\max}}{P_{A,y}-1}$. Let us therefore first compute the points $(\overline{\delta}_{A,x}, \delta_{A,y}(\overline{\delta}_{A,x}))$ and $(\delta_{A,x}(\overline{\delta}_{A,y}), \overline{\delta}_{A,y})$ on the hyperbola $\mathcal{H}^s_{b,x-b,y}$ corresponding to the maximum spacing $\overline{\delta}_{A,x}$ and $\overline{\delta}_{A,y}$. After some straightforward derivations, it follows that $\delta_{A,y}(\overline{\delta}_{A,x}) > \overline{\delta}_{A,y}$ if $K_S = \frac{X_{\max}}{Y_{\max}} > \frac{P_{A,x}-1}{P_{A,y}-1}$, and that $\delta_{A,x}(\overline{\delta}_{A,y}) > \overline{\delta}_{A,x}$ if $K_S = \frac{X_{\max}}{Y_{\max}} < \frac{P_{A,x}-1}{P_{A,y}-1}$. Hence, when $K_S > \frac{P_{A,x}-1}{P_{A,y}-1}$, the hyperbola will first cross the maximum $\overline{\delta}_{A,x}$, implying this is the tightest maximum spacing, while when $K_S = \frac{X_{\max}}{Y_{\max}} > \frac{P_{A,x}-1}{P_{A,y}-1}$, then the hyperbola first crosses the maximum $\overline{\delta}_{A,y}$, which then is the tightest maximum spacing. Hence, the following upper bound on $\delta_{A,x}$ can be formulated:

$$\delta_{A,x}^{area} = \begin{cases} \overline{\delta}_{A,x} & \text{if } K_S < \frac{P_{A,x}-1}{P_{A,y}-1} \\ \delta_{A,x}(\overline{\delta}_{A,y}) & \text{if } K_S > \frac{P_{A,x}-1}{P_{A,y}-1} \end{cases} \tag{19}$$

with

$$\delta_{A,x}(\overline{\delta}_{A,y}) = \frac{(P_{A,x}-1)X_{\max}}{(P_{A,x}-1)^2-1} - \sqrt{\frac{X_{\max}^2}{((P_{A,x}-1)^2-1)^2} - \frac{Y_{\max}^2}{(P_{A,y}-1)^2((P_{A,x}-1)^2-1)}}. \tag{20}$$

Taking into account that $Y_{\max} = \frac{X_{\max}}{K_S}$, it follows that both bounds in (19) are linear in $X_{\max}$.

Next, we determine the other bounds for $\delta_{A,x}$ for given parameter settings, i.e., for different K_A, K_S, $P_{A,x}$, $P_{A,y}$, $X_{\max}$, $Y_{\max}$, $Z_{\max}$ and m_S. To this end, let us compute the true uniformity $\mathcal{U}_t$ and $\mathcal{U}_s$, i.e., without use of any approximations, and determine the bounds using an exhaustive search. In Figure 10, we show the resulting range of $\delta_{A,x}$, i.e., the green areas, as a function of the size $X_{\max}$ of the area for different values of $P_{A,x}$, $K_A = \frac{P_{A,x}}{P_{A,y}}$ and $K_S = \frac{X_{\max}}{Y_{\max}}$, for $Z_{\max} = 2$ m and $m_S = 1$. As can be observed in the figure, the bounds $\delta_{A,x}^{\min}$ and $\delta_{A,x}^{\max}$ depend in a piece-wise linear way on the size $X_{\max} \times \frac{X_{\max}}{K_S}$ of the area. This piece-wise linear behaviour could be expected, as to keep an as large as possible uniformity, scaling the area while keeping the number of luminaries equal will result in a scaling of the spacing between the luminaries, resulting in a linear behaviour. The piece-wise character of the bounds is because at some point the lower (upper) bound corresponding to the task area becomes tighter than that of the surrounding area, or vice versa. Therefore, we approximate the range for $\delta_{A,x}$ as a function of $X_{\max}$ as a piece-wise linear upper and lower bound, as shown in Figure 11. The first bound is (19), due to the maximum allowed spacing to keep all luminaries within the considered area. The spacing $\delta_{A,x}^{area}$ serves as an upper bound on the spacing for the task area as well as for the surrounding area in case the uniformity constraints still hold for this maximum possible spacing, which occurs when the considered area is sufficiently small. When the area size is larger, the uniformity constraints no longer will be met for the maximum possible spacing $\overline{\delta}_{A,x}$. In that case, the upper bound on $\delta_{A,x}$ will be determined by $\min\{\delta_{A,x}^{\max,s}, \delta_{A,x}^{\max,t}\}$. In our simulations, we always found that $\delta_{A,x}^{\max,s} > \delta_{A,x}^{\max,t}$, implying we can ignore the bound $\delta_{A,x}^{\max,s}$. To find the expressions for the linear equations for $\delta_{A,x}^{\max,t}$, $\delta_{A,x}^{\min,t}$ and $\delta_{A,x}^{\min,s}$, we carried out a large number of simulations for various values for the parameters $P_{A,x}$, $X_{\max}$, $K_A = \frac{P_{A,x}}{P_{A,y}}$, $K_S = \frac{X_{\max}}{Y_{\max}}$, $Z_{\max}$ and $m_S = -\ln 2 / \ln(\cos(\Phi_{1/2}))$, and matched the coefficients of the linear equations to the simulated bounds. This resulted in the empirical equations

$$
\begin{aligned}
\tilde{\delta}_{A,x}^{\min,t} &= \left(\frac{1.406 P_{A,x} - 1.275}{(P_{A,x} - 0.5)(P_{A,x} - 1)(K_A + 14)} X_{\max} + \frac{9.059 P_{A,x} - 18.487}{(P_{A,x} + 12.78)(P_{A,x} - 1)(K_A + 9)} \right)(\zeta_x^t + 0.2)(K_S + 7) \\
\tilde{\delta}_{A,x}^{\min,s} &= \left(\frac{(1.054 - 0.011 P_{A,x})(K_A + 14)}{15(P_{A,x} - 0.721)} X_{\max} + \frac{0.243 P_{A,x} - 2.201}{P_{A,x} - 0.964} \right) - \frac{K_S - 1}{10} - \frac{10\zeta_x^t - 8}{P_{A,x}^2} \\
\tilde{\delta}_{A,x}^{\max,t} &= \frac{P_{A,x} - 0.6}{P_{A,x} - 1} \mathcal{K}
\end{aligned}
\tag{21}
$$

where $\mathcal{K} = \frac{(K_S + 4(K_A - 1)K_S + 7)}{2(K_A^2 + 2)} \frac{Z_{\max} \sin \Phi_{1/2}}{\mathcal{U}_t^{(th)}}$, and $\zeta_x^t = \frac{X_{\max}^t}{X_{\max}}$ follows from the definition of the size of the task area. The resulting bounds are shown in Figure 10. As can be observed, the bounds match well the true bounds on $\delta_{A,x}$ for given area size. The bounds were tested for the parameter ranges given in Table 3, and were found to be accurate. Hence, the empirical bounds (21) can be used for at least these parameter settings. Taking into account that it is required that $\tilde{\delta}_{A,x}^{\max,t} \leq \delta_{A,x}^{area} \leq \overline{\delta}_{A,x} = \frac{X_{\max}}{P_{A,x} - 1}$, it follows that $\mathcal{K}(P_{A,x} - 0.6) \leq X_{\max}$. This yields the following lower bound on the number of luminaries:

$$
P_{A,x} \geq \left\lceil \frac{X_{\max}}{\mathcal{K}} + 0.6 \right\rceil \stackrel{\Delta}{=} P_{A,x}^{\min},
\tag{22}
$$

and consequently

$$
P_{A,y} \geq \frac{P_{A,x}^{\min}}{K_A} \stackrel{\Delta}{=} P_{A,y}^{\min}.
\tag{23}
$$

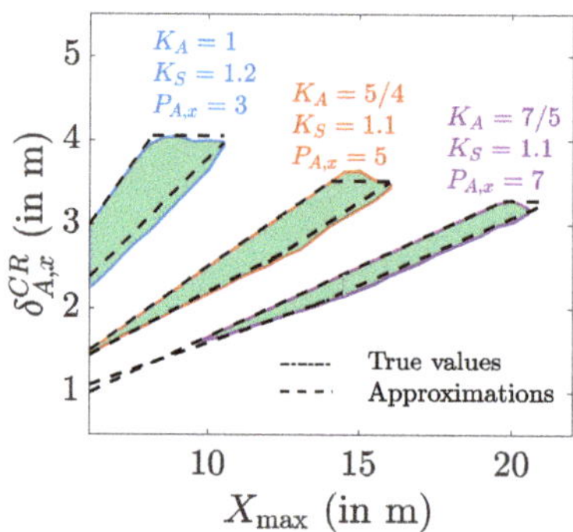

Figure 10. Boundary intervals for the spacing $\delta_{A,x}^{CR}$ in the compliance region for different K_A, K_S and $P_{A,x}$ with $\zeta_x^t = \zeta_y^t = 0.9$, $Z_{\max} = 2$ m and $m_S = 1$.

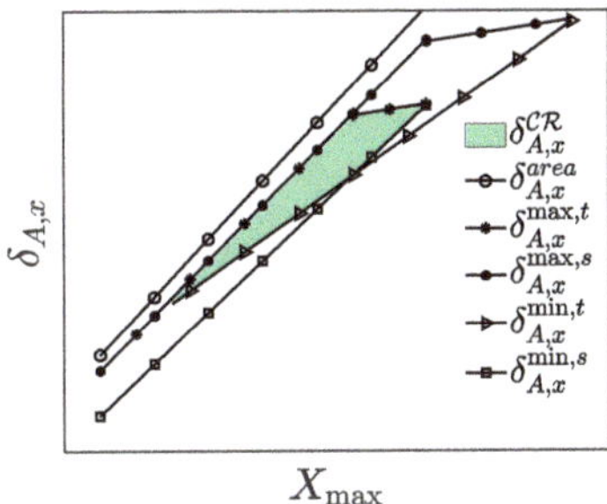

Figure 11. Linearised bounds on $\delta_{A,x}$.

Table 3. Range of parameters for which the approximations (21) and (22) are valid.

Parameter		Range
$X_{\max}$	$\in$	$[5,25]$ m
K_S	$\in$	$[1,1.7]$
$Y_{\max}$	$=$	$\frac{X_{\max}}{K_S}$
$Z_{\max}$	$\in$	$[2,4]$ m
$P_{A,x}$	$=$	$\{3,\cdots,7\}$
$P_{A,y}$	$=$	$\{3,\cdots,7\}$
K_A	$=$	$\frac{P_{A,x}}{P_{A,y}}$
$\mathcal{U}_t^{(th)}$	$=$	0.7
$\mathcal{U}_s^{(th)}$	$=$	0.5
ζ_x^t	$\in$	$[0.75,0.9]$
ζ_y^t	$=$	ζ_x^t
$\Phi_{1/2}$	$\in$	$[40,70]^\circ$

To verify the resulting bounds on $\delta_{A,x}$, let us consider the same example we used to determine the range for the number of LEDs to achieve an average illuminance in the range $[300,500]$ lx, i.e., with $X_{\max} = 10$ m, $Z_{\max} = 2$ m, $K_S = \frac{\Omega_{c,x}}{\Omega_{c,y}} = 1.5$, $\Phi_v^{typ} = 270$ lm and $\Phi_{1/2} = \frac{\pi}{3}$ (i.e., $m_S = 1$). By using (10) and (11), we determined that the number L of LEDs needed to have an average horizontal illumination between 300 and 500 lux ranges in $L \in [86,336]$. In the following, we assume $L = 180$. These LEDs must be grouped in luminaries, in a grid of $P_{A,x} \times P_{A,y}$ luminaries. In our example, we assume $P_{A,x} = P_{A,y}$, i.e., $K_A = \frac{P_{A,x}}{P_{A,y}} = 1$. Further, the dimensions of the task area are selected as $\zeta_x^t = \zeta_y^t = 0.8$ with $\zeta_x^t = \frac{X_{\max}^t}{X_{\max}}$ and $\zeta_y^t = \frac{Y_{\max}^t}{Y_{\max}}$. For the task area, we assume the threshold for uniformity is $\mathcal{U}_t^{(th)} = 0.7$ and for the surrounding area $\mathcal{U}_s^{(th)} = 0.5$. This yields $\mathcal{K} = 3.51$. Applying the bounds (22) and (23), it follows that the number of luminaries is lower bounded by $P_{A,x}^{\min} = 3$ and $P_{A,y}^{\min} = 3$. This gives $P_L = \frac{L}{P_{A,x}P_{A,y}} = 20$ LEDs per luminary. As $K_S = 1.5 > \frac{P_{A,x}-1}{P_{A,y}-1} = 1$, it follows that the bound $\delta_{A,x}^{area}$ (19) is determined by the maximum spacing $\bar{\delta}_{A,y}$. This yields $\delta_{A,x}^{area} = \delta_{A,x}(\bar{\delta}_{A,y}) = 3.94$ m. On the other hand, $\tilde{\delta}_{A,x}^{\max,t} = \frac{P_{A,x}-0.6}{P_{A,x}-1}\mathcal{K} = 4.20$ m, implying $\delta_{A,x}^{\max} = \min(\delta_{A,x}^{area}, \tilde{\delta}_{A,x}^{\max,t}) = 3.94$ m. Similarly, we compute the lower bounds using Equation (21) to obtain $\tilde{\delta}_{A,x}^{\min,t} = 3.57$ m and $\tilde{\delta}_{A,x}^{\min,s} = 3.70$ m, and thus $\delta_{A,x}^{\min} = \max(\tilde{\delta}_{A,x}^{\min,t}, \tilde{\delta}_{A,x}^{\min,s}) = 3.70$ m. This results in the interval $\delta_{A,x}^{CR} \in [3.70,3.94]$ m.

To check if this interval corresponds to the true compliance region for the given parameter values, we computed the true uniformity and average illuminance. The uniformity for the task and surrounding area are shown in Figure 12a, and the average horizontal illuminance for the task area $\bar{E}_{h,t}$, surrounding area $\bar{E}_{h,s}$ and the whole area of interest $\bar{E}_h$ in Figure 12b. In the figure, we show the true compliance interval $\delta_{A,x}^{CR} = [3.50,3.94]$ in which the uniformity in both the task and surrounding area are above the threshold. As can be observed, the predicted compliance region $\tilde{\delta}_{A,x}^{CR}$ is contained in the true compliance region $\delta_{A,x}^{CR}$, $\tilde{\delta}_{A,x}^{CR} \subset \delta_{A,x}^{CR}$, although the predicted interval slightly underestimates the true interval. We noticed in our simulations that when the difference $K_A - K_S > 0.4 \times K_S$, the predicted interval tends to underestimate the interval, although the predicted interval still results in a system setup that satisfies the constraints on the illuminance level and uniformity. In Figure 13, the compliance intervals are shown for different values for K_A, for $K_S = 1.5$. As can be observed, the predicted and the true compliance interval match very well for $K_A \approx K_S$. As in practice, the number of luminaries is scaled with the dimensions of the area, i.e., $K_A \approx K_S$, this implies the proposed approximations predict the range of potential spacings between the luminaries well.

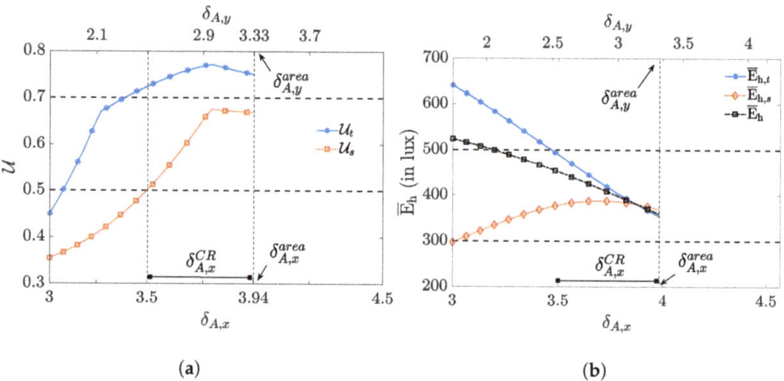

Figure 12. True values of (**a**) uniformity: $\mathcal{U}_t$ and $\mathcal{U}_s$, and (**b**) the horizontal illuminance average: $\overline{E}_{h,t}$, $\overline{E}_{h,s}$ and $\overline{E}_h$, for $X_{\max} = 10$ m, $K_S = 1.5$, $Z_{\max} = 2$ m, $K_A = 1$, $\Phi_v^{typ} = 270$ lm, $\Phi_{1/2} = \frac{\pi}{3}$, $\zeta_x^t = \zeta_y^t = 0.8$, $P_L = 20$ LEDs, $P_{A,x} = 3$ and $P_{A,y} = 3$.

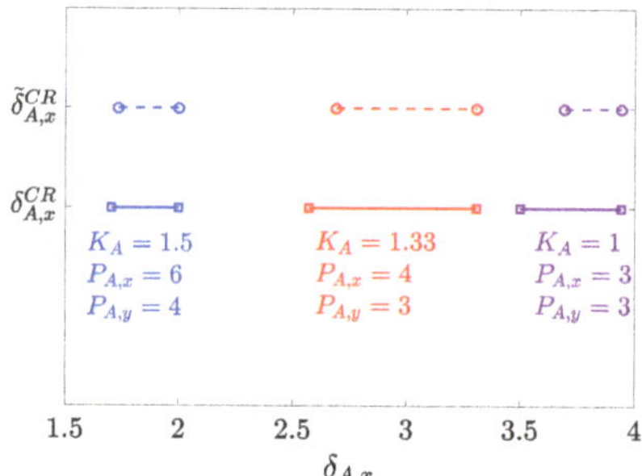

Figure 13. True range of $\delta_{A,x}^{CR}$ versus predicted range $\tilde{\delta}_{A,x}^{CR}$, for $X_{\max} = 10$ m, $K_S = 1.5$, $Z_{\max} = 2$ m, $\Phi_v^{typ} = 270$ lm, $\Phi_{1/2} = \frac{\pi}{3}$, $\zeta_x^t = \zeta_y^t = 0.8$, $P_L = 20$ LEDs and for different K_A.

4. Conclusions

In this paper, we considered the illumination for indoor areas, in the context of a combined illumination and positioning or communication system. To this end, we analysed the effect of the number and placement of LEDs on the main illumination characteristics, i.e., the average horizontal illuminance and uniformity, as defined in the DIN EN 12464-1 standard. We summarise our main contributions:

- The average horizontal illuminance is largely independent of the placement of the LEDs, but depends strongly on the number of LEDs. On the other hand, the uniformity mainly depends on the placement of the LEDs.

- The number of LEDs needed to reach a given average illuminance is bounded by Equations (10) and (11).
- The minimum of the horizontal illuminance is crucial to determine the uniformity. To avoid a two-dimensional search for the minimum value, we analysed the position of the minimum. We found that the minimum value can be found at one out of four positions, as illustrated in Figure 7. To determine which of these four positions correspond to the minimum, we need to evaluate the conic sections given by Equation (18), resulting in the decision regions depicted in Figure 8a. This position of the minimum allows us to easily compute the minimum of the horizontal illuminance, and thus the uniformity.
- To determine for a given number of luminaries the spacings for which the illumination constraints are satisfied, we proposed the bounds Equations (19) and (21). These bounds provide a range for the spacing between luminaries, from which we can find an expression for the minimum number of luminaries (Equation (22)) required to meet the illumination constraints.
- The resulting rules of thumb were tested for a wide variety of system parameters, as illustrated in Table 3.

These guidelines will help a design engineer to construct a combined illumination and positioning or communication system that meets the illumination standard.

Author Contributions: Research, J.M.M.; Writing, J.M.M.; Editing, H.S.; Supervising, H.S.; Funding, H.S. All authors have read and agreed to the published version of the manuscript.

Funding: José Miguel Menéndez acknowledges the National Secretariat of Higher Education, Science, Technology and Innovation of Ecuador (SENESCYT) for their financial support. This work is partially funded by the EOS grant 30452698 from the Belgian Research Councils FWO and FNRS. Further, it is funded by the Flemish Government (AI Research Program).

Institutional Review Board Statement: Not applicable.

Informed Consent Statement: Not applicable.

Data Availability Statement: Not applicable.

Conflicts of Interest: The authors declare no conflict of interest.

References

1. Cheng, Y.-K.; Cheng, K.W.E. General study for using LED to replace traditional lighting devices. In Proceedings of the 2006 2nd International Conference on Power Electronics Systems and Applications, Hong Kong, China, 12–14 November 2006; pp. 173–177.
2. Hou, Y.; Xiao, S.; Bi, M.; Xue, Y.; Pan, W.; Hu, W. Single LED beacon-based 3-D indoor positioning using off-the-shelf devices. *IEEE Photonics J.* **2016**, *8*, 1–11. [CrossRef]
3. Wang, T.Q.; Sekercioglu, Y.A.; Neild, A.; Armstrong, J. Position accuracy of time-of-arrival based ranging using visible light with application in indoor localization systems. *J. Light. Technol.* **2013**, *31*, 3302–3308. [CrossRef]
4. Yang, S.H.; Jeong, E.M.; Kim, D.R.; Kim, H.S.; Son, Y.H.; Han, S.K. Indoor three-dimensional location estimation based on LED visible light communication. *Electron. Lett.* **2013**, *49*, 54–56. [CrossRef]
5. Menendez, J.; Steendam, H. Influence of the aperture-based receiver orientation on RSS-based VLP performance. In Proceedings of the 2017 International Conference on Indoor Positioning and Indoor Navigation, Sapporo, Japan, 18–21 September 2017.
6. Stevens, N.; Steendam, H. Influence of transmitter and receiver orientation on the channel gain for RSS Ranging-based VLP. In Proceedings of the 2018 11th International Symposium on Communication Systems, Networks & Digital Signal Processing (CSNDSP'18), Budapest, Hungary, 18–20 July 2018.
7. Cincotta, S.; He, C.; Neild, A.; Armstrong, J. High angular resolution visible light positioning using a quadrant photodiode angular diversity aperture receiver (QADA). *Opt. Express* **2018**, *26*, 9230–9242. [CrossRef] [PubMed]
8. Sahin, A.; Eroglu, Y.S.; Guvenc, I.; Pala, N.; Yuksel, M. Accuracy of AOA-based and RSS-based 3D localization for visible light communications. In Proceedings of the V2015 IEEE 82nd Vehicular Technology Conference (VTC2015-Fall), Boston, MA, USA, 6–9 September 2015; pp. 1–5.
9. Jung, S.-Y.; Hann, S.; Park, C.-S. TDOA-based optical wireless indoor localization using LED ceiling lamps. *IEEE Trans. Consum. Electron.* **2011**, *57*, 1592–1597. [CrossRef]
10. Nadeem, U.; Hassan, N.U.; Pasha, M.A.; Yuen, C. Highly accurate 3D wireless indoor positioning system using white LED lights. *Electron. Lett.* **2014**, *50*, 828–830. [CrossRef]

11. Hijikata, S.; Terabayashi, K.; Umeda, K. A simple indoor self-localization system using infrared LEDs. In Proceedings of the 2009 Sixth International Conference on Networked Sensing Systems (INSS), Pittsburgh, PA, USA, 17–19 June 2009; pp. 1–7.
12. Steendam, H.; Wang, T.; Armstrong, J. Theoretical lower bound on VLC-based indoor positioning using received signal strength measurements and an aperture-based receiver. *J. Light. Technol.* **2017**, *in press*. [CrossRef]
13. Cassarly, W.; Hayford, M. Illumination optimization: The revolution has begun. *Int. Opt. Des. Conf.* **2002**, *4832*, 258–269.
14. Davenport, T.; Hough, T.; Cassarly, W. Optimization for illumination systems: The next level of design. *Photon Manag.* **2004**, *5456*, 81–90.
15. Jacobson, B.; Gengelbach, R. Lens for uniform LED illumination: An example of automated optimization using Monte Carlo ray-tracing of an LED source. *Nonimaging Opt. Maximum Effic. Light Transf. VI* **2001**, *4446*, 121–128.
16. Bastiaens, S. Towards Centimetre-Order Indoor Localisation with RSS-Based Visible Light Positioning. Ph.D. Thesis, University of Ghent, Ghent, Belgium, 2022.
17. Ghassemlooy, Z.; Popoola, W.; Rajbhandari, S. *Optical Wireless Communications: System and Channel Modellin with Matlab®*; CRC Press: Boca Raton, FL, USA, 2012.
18. Komine, T.; Nakagawa, M. Fundamental analysis for visible-light communication system using LED lights. *IEEE Trans. Consum. Electron.* **2004**, *50*, 100–107. [CrossRef]
19. Normung, D. *Light and Lighting–Lighting of Work Places–Part 1: Indoor Work Places*; DIN EN 12464-1; DIN: Berlin, Germany, 2011.

Article

Two-Stage Link Loss Optimization of Divergent Gaussian Beams for Narrow Field-of-View Receivers in Line-of-Sight Indoor Downlink Optical Wireless Communication (Invited)

Xinda Yan [1], Yuzhe Wang [1], Chao Li [2,*], Fan Li [3,*], Zizheng Cao [2] and Eduward Tangdiongga [1]

[1] Eindhoven Hendrik Casimir Institute, Eindhoven University of Technology,
     5612 AZ Eindhoven, The Netherlands; x.yan@tue.nl (X.Y.); y.wang1@tue.nl (Y.W.); e.tangdiongga@tue.nl (E.T.)
[2] Peng Cheng Laboratory, Shenzhen 518055, China; caozzh@pcl.ac.cn
[3] School of Electronics and Information Technology, Sun Yat-Sen University, Guangzhou 510275, China
*   Correspondence: lich03@pcl.ac.cn (C.L.); fan39@mail.sysu.edu.cn (F.L.)

Abstract: The predominant focus of research in high-speed optical wireless communication (OWC) lies in line-of-sight (LOS) links with narrow infrared beams. However, the implementation of precise tracking and steering necessitates delicate active devices, thereby presenting a formidable challenge in establishing a cost-effective wireless transmission. Other than using none-line-of-sight (NLOS) links with excessive link losses and multi-path distortions, the simplification of the tracking and steering process can be alternatively achieved through the utilization of divergent optical beams in LOS. This paper addresses the issue by relaxing the stringent link budget associated with divergent Gaussian-shaped optical beams and narrow field-of-view (FOV) receivers in LOS OWC through the independent optimization of geometrical path loss and fiber coupling loss. More importantly, the geometrical path loss is effectively mitigated by modifying the transverse intensity distribution of the optical beam using manipulations of multi-mode fibers (MMFs) in an all-fiber configuration. In addition, the sufficiently excited higher order modes (HOMs) of MMFs enable a homogenized distribution of received optical powers (ROPs) within the coverage area, which facilitates the mobility of end-users. Comparative analysis against back-to-back links without free-space transmission demonstrates the proposed scheme's ability to achieve low power penalties. With the minimized link losses, experimental results demonstrate a 10 Gbps error-free (BER < 10^{-13}) LOS OWC downlink transmission at 2.5 m over an angular range of $10° \times 10°$ without using any optical pre-amplifications at a typical PIN receiver. The proposed scheme provides a simple and low-cost solution for high-speed and short-range indoor wireless applications.

Keywords: indoor OWC; divergent Gaussian beam; link loss; MMF

Citation: Yan, X.; Wang, Y.; Li, C.; Li, F.; Cao, Z.; Tangdiongga, E. Two-Stage Link Loss Optimization of Divergent Gaussian Beams for Narrow Field-of-View Receivers in Line-of-Sight Indoor Downlink Optical Wireless Communication (Invited). *Photonics* **2023**, *10*, 815. https://doi.org/10.3390/photonics10070815

Received: 16 June 2023
Revised: 10 July 2023
Accepted: 10 July 2023
Published: 13 July 2023

1. Introduction

Over the past five decades, the capacity of radio frequency (RF) wireless communications has experienced exponential growth, surpassing a million-fold increase. Currently, more than 85% of internet traffic is generated within indoor environments [1], resulting in an overwhelming demand that exceeds the capacity of the RF spectrum. As a prospective solution to alleviate this congestion, the utilization of higher-frequency regions within the electromagnetic spectrum for indoor wireless communication has garnered considerable anticipation [2].

The visible and infrared (IR) light spectrum offers a vast range of over 300 THz of unlicensed bandwidth, presenting an abundant resource. Due to the limited penetration capability of optical beams through opaque obstacles, the confinement of light within rooms or compartments provides an inexhaustible supply of bandwidth resources through wavelength reuse [3]. These distinctive advantages over RF-based counterparts have sparked widespread interest in optical wireless communication (OWC) within the academic community. In the context of indoor OWC, whether operating in the visible or IR band, cost-effectiveness is a crucial consideration, emphasizing the need to avoid complex devices and high computational requirements. While the visible band combines communication and illumination with a constrained modulation bandwidth, the 1550 nm window in the IR band offers eye-safety up to 10 dBm optical power. Furthermore, it is compatible with existing fiber-optic networks and well-suited for high-speed wireless transmission. Presently, considerable research efforts have focused on single-device exclusive links in line-of-sight (LOS) OWC with narrow optical beams, achieving an impressive transmission rate of 112 Gbps per beam [4]. Two-dimensional beam-steering for narrow beams is commonly achieved through actively controlled elements, such as microelectromechanical systems (MEMS) [5] and liquid-crystal spatial light modulators (LC-SLMs) [6]. Nevertheless, precise tracking and steering techniques are essential to eliminate any misalignment between transceivers. A localization accuracy of $0.038°$ and a wide field of view (FOV) of $70° \times 70°$ have been experimentally demonstrated by utilizing light detection and ranging (LiDAR) [7,8]. Nonetheless, the integration of LiDAR with OWC introduces complexity and high costs. To address the need for cost-effective and high-speed indoor wireless downlinks, Koonen et al. proposed a passive and compact two-dimensional arrayed waveguide grating router (AWGR) that enables equivalent infrared (IR) beam-steering without any moving parts [9]. By discretely tuning the wavelength, a slightly divergent optical beam scans a relatively large coverage area. However, achieving seamless coverage without interference between spatially adjacent users and considering the variance of received optical power (ROP) within the coverage area due to a Gaussian-shaped optical beam pose challenges. Alternatively, increasing the divergent angle of the optical beam can potentially cover the same area as AWGRs without any additional operations, albeit at the expense of increased link losses. Moreover, employing a ground glass-based diffuser to achieve this angular expansion naturally transforms the Gaussian-shaped beam generated by lasers into a flat-top beam [10]. Therefore, a more divergent beam holds the potential advantage of a homogenized intensity distribution compared to AWGRs. In this study, as elaborated by [11], sensitive and deterministic multi-mode fibers (MMFs) are manipulated within an all-fiber configuration to modify the transverse intensity distribution of the laser source, which plays a crucial role in optimizing either geometrical link loss or ROP homogenization. Notably, offset launch techniques flatten the averaged intensity distribution of speckle patterns when higher-order modes (HOMs) are predominantly excited [12,13].

This paper assumes an ideal Gaussian distribution for the divergent optical beam, with the transverse intensity distribution along the propagation axis characterized by geometric optics. The validity of these two assumptions is experimentally verified. Our analysis reveals two major types of optical power loss in divergent optical beams, namely geometrical path loss and fiber coupling loss, with an optical fiber coupled collimator serving as the light-gathering device for photodiodes. Simulation results indicate that geometrical path loss and fiber coupling loss are interdependent, resulting in an unacceptably high total link loss in short-range OWC applications. To address the stringent link budget in such scenarios of intensity modulation and direct detection (IM-DD), the independent optimization of geometrical path loss and fiber coupling loss is pursued. Specifically, the fiber coupling loss is mitigated through the use of a receiving collimator with adjustable focus. Essentially, this minimization can be achieved by altering the coupling distance between the receiving lenses and the optical fiber to compensate for focus shifts. As a result, the fiber coupling loss is eliminated, allowing for the separate optimization of these two significant loss factors. Conversely, the complete elimination of geometrical path loss in divergent beams is not feasible. Nonetheless, the transverse intensity distribution of the Gaussian beam can be modified using MMFs. By iteratively applying controlled perturbations to the MMF, adaptive shaping of the divergent optical beam can be achieved based on the specific locations of portable devices. Furthermore, the introduction of offset launch techniques leads to a homogeneous and optimized ROP. Consequently, the significant variation in

ROP experienced by portable devices within the coverage area is mitigated. It is worth noting that the proposed receiving structure with a narrow FOV effectively suppresses ambient light and multi-path distortion. In addition, the mobility of portable devices can be enhanced through the deployment of receivers with a larger FOV.

The remainder of this paper is organized as follows. Section 2 describes the basic formulas of Gaussian beams and parameters of optics used in simulations and experiments. In Section 3, the propagation and reception of the Gaussian beam is simulated. Section 4 evaluates the bit error rate (BER) performance of the proposed scheme in IM-DD using non-return-to-zero on-off keying (NRZ-OOK) modulation over the coverage area, and conclusions and discussions are presented in Sections 5 and 6, respectively.

2. Basic Formulas of Gaussian Beams and Parameters of Optics

By considering a slowly varying envelope (SVE) of the electric field, the Gaussian beam emerges as an analytical solution to the paraxial Helmholtz Equation [14]. In the context of indoor OWC, the transverse intensity distribution of single-mode fibers (SMFs) is presumed to adhere to an ideal Gaussian profile, with this Gaussian shape being preserved along its propagation axis in indoor environments [15]:

$$I(r,z) = I_0 exp\left(\frac{-2r^2}{\omega^2(z)}\right),\tag{1}$$

where I_0 is the peak intensity, r is the transverse distance concerning the propagation axis Z, and z is the axial distance. The expression indicates that the transverse intensity distribution of a Gaussian beam is circular and symmetric with no obvious boundaries and weakens with increasing transverse distance. More precisely, its intensity drops to $1/e^2$ of the peak intensity I_0 when $r = \omega(z)$, and $\omega(z)$ is typically referred to as the beam radius. The peak intensity I_0 can be expressed as:

$$I_0 = \frac{2P_o}{\pi\omega^2(z)},\tag{2}$$

where P_o is the total optical power emitted to the free space. Due to diffraction, the beam radius $\omega(z)$ keeps varying with z as:

$$\omega(z) = \omega_0\sqrt{1 + \left(\frac{z}{z_R}\right)^2},\tag{3}$$

where ω_0 is the minimum beam radius along the propagation axis named beam waist, and it appears at axial distance $z = 0$. The wavelength and beam waist dependent Rayleigh length can be calculated by Equation (4):

$$z_R = \frac{\pi\omega_0^2}{\lambda},\tag{4}$$

where λ is the operating wavelength of the laser source, and the Rayleigh length represents the ability of optical beams to maintain collimation along their propagation direction. Equation (4) indicates that a larger divergent angle can be obtained by shrinking ω_0, and vice versa. Herein, the target divergent angle of the optical beam is obtained by focusing a collimated beam.

In order to analyze the propagation characteristics of the divergent Gaussian beam within the context of short-range indoor OWC, a comprehensive set of simulation and experimental results is provided in the subsequent sections. The optical parameters employed in both the simulations and experiments remain consistent, and they are detailed in Table 1. Specifically, the emitting collimator holds a fixed focal length, while the receiving collimator features an adjustable focal length.

Table 1. Parameters of the optics used in simulations and experiments.

Optics	Size [1]	FL [2]	CA [3]	Beam Waist	NA [4]
Fixed Col. [5]	24 mm	37.20 mm	X	3.5 mm	0.24
Lens [6]	25 mm	20.00 mm	22.5 mm	X	0.54
Zoom Col. [7]	1.2-inch	Adjustable	20.5 mm	X	0.25

[1] Outer Diameter; [2] Focal length; [3] Clear Aperture; [4] Numerical Aperture; [5] Collimator with a fixed focal length; [6] Focusing lens; [7] Collimator with a adjustable focal length.

3. Simulation Results of Gaussian Beams

After a laser-fed SMF passes through the emitting collimator, the beam radius evolution of a Gaussian beam over 200 m is shown in Figure 1. At this range, the Gaussian beam gradually approaches at $y = kx$, where $k = \omega_0/z_R$. In this case, the Gaussian beam can be seen as a point source at $z = 0$ with a half-divergent angle $\theta = \omega_0/z_R$ in rad.

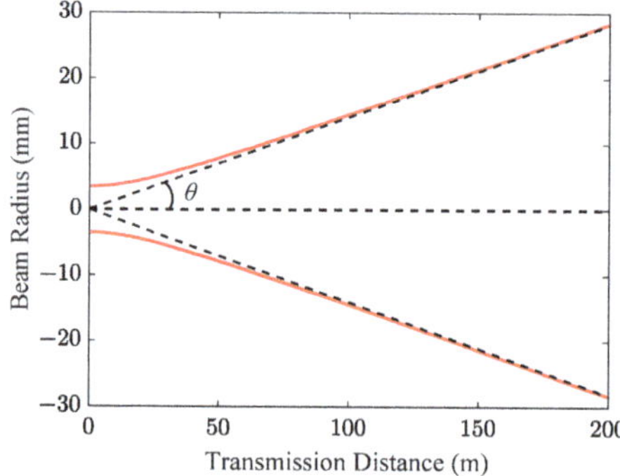

Figure 1. Beam radius evolution of the collimated beam within 200 m.

Figure 1 depicts the passage of the collimated beam through the divergent lens, and the corresponding θ is 0.14 mrad. In contrast to transmission distances spanning hundreds of meters, indoor OWC only requires a few meters. Figure 2 depicts the focusing process of the collimated optical beam after passing through the divergent lens. The emitting collimator's output facet is positioned at $z = 0$, while the thin lens marked by an arrow, neglecting its thickness, is situated at $z = 40$ mm. Since the separation between the emitting collimator and the divergent lens is constrained to several centimeters, the optical beam retains its collimation. As a matter of fact, its collimation distance is sufficient to extend a few meters away, and this characteristic accounts for the negligible geometrical path loss observed in indoor OWC employing narrow optical beams.

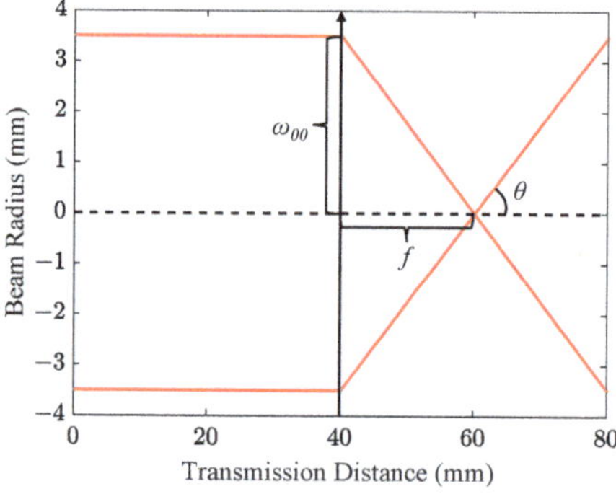

Figure 2. The focusing process of a collimated optical beam.

The divergence of the optical beam can be adjusted by lenses. As the optical beam encounters a lens, the beam waist of this lens is expressed as:

$$\omega_{01} = \frac{\omega_{00} f}{z_R},$$ (5)

where ω_{00} and ω_{01} are the beam waist of the emitting collimator and the divergent lens with a focal length of f, respectively. z_R is the Rayleigh length of the emitting collimator. After the optical beam reaches its beam waist of 2.81 um as calculated by Equation (5), it diverges at the same rate as focusing, and the corresponding half-divergent angle θ is 9.92°, which equals the divergent angle calculated by $atan(\omega_{00}/f)$. As a result, the beam radius evolution, as determined by Equation (3), can be effectively substituted with a simplified geometric optics approach.

Following free-space transmission, the optical beam is collected by the receiving collimator. During this stage, geometrical path loss arises when the diameter of the coverage area exceeds that of the receiving collimator's clear aperture (CA). To evaluate the geometrical path loss, it is critical to establish the receiving model of divergent Gaussian beams. At a target transmission distance z, the optical power within transverse distance r of a Gaussian beam is the integral of Equation (1) from 0 to r:

$$P(r,z) = \int_0^r I(r,z)dr = P_o\left(1 - exp\left(\frac{-2r^2}{\omega^2(z)}\right)\right),$$ (6)

where $\omega(z)$ is the radius of the coverage area at a transmission distance z. Assuming the receiving collimator with a CA of φ_c is located at a transverse distance r, and the intensity distribution between the two circles with an on-axis dot and tangent to the receiving collimator is uniform [16], then the optical power captured by the receiving collimator can be derived from Equation (6) as:

$$P_{colli}(r,z,\varphi_c) = \begin{cases} P_o\left(1 - exp\left(\frac{-\varphi_c^2}{2\omega^2(z)}\right)\right), r = 0 \\ \dfrac{P_o\varphi_c\left(exp\left(\frac{-2\left(r-\frac{\varphi_c}{2}\right)^2}{\omega^2(z)}\right) - exp\left(\frac{-2\left(r+\frac{\varphi_c}{2}\right)^2}{\omega^2(z)}\right)\right)}{8r}, else \end{cases}$$ (7)

For a more intuitive look, the transverse distance r is represented by receiving angle $atan(r/z)$ in the rest of the paper.

Figure 3a shows the geometrical path loss versus varying receiving angles within the coverage area at several short-range free-space transmission distances. Due to the Gaussian-shaped laser beam, the geometrical path loss scales up with the receiving angle. Regardless of transmission distances, the difference in geometrical path loss between the center (receiving angle = 0°) and the boundaries (receiving angle = ± 10°) of the coverage area is fixed at approximately 8.7 dB. In addition, the geometrical path loss is increased with the transmission distance at any receiving angle. Roughly, doubling the transmission distance will increase the geometrical path loss by 6 dB. Thereby, the geometrical path loss is one of the major link losses in divergent Gaussian beams.

The other significant link loss in the case of divergent optical beams is the fiber coupling loss. As depicted in Figure 4, an optical fiber is precisely positioned at the nominal focal length (f) of the receiving collimator. There is no focus shift (Δf) of the receiving collimator when its incident optical beam is collimated, and a divergent optical beam introduces a focus shift of the receiving collimator along the propagation direction.

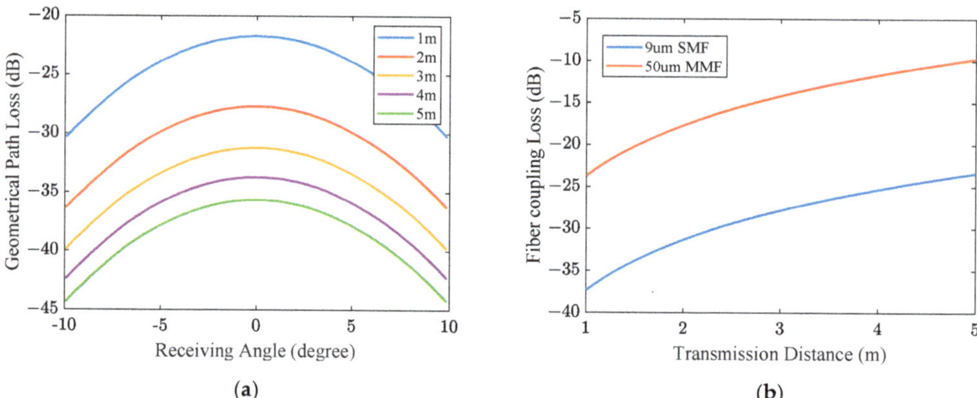

Figure 3. Two major types of optical link losses in divergent Gaussian beams. (**a**) Geometrical path loss versus receiving angles at different transmission distances; (**b**) Fiber coupling losses versus transmission distances using different fiber core diameters.

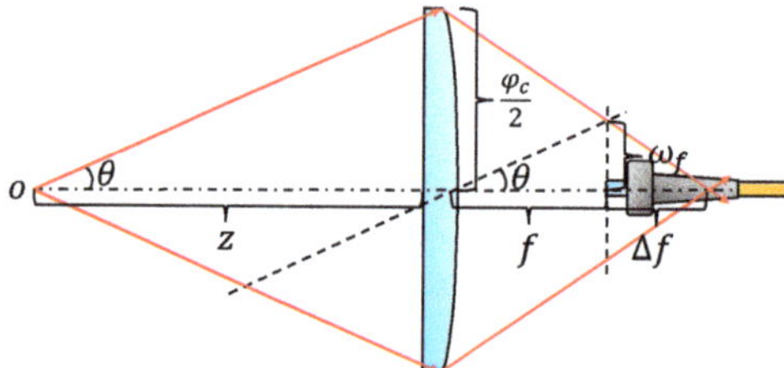

Figure 4. Fiber coupling of the receiving collimator.

According to geometric optics shown in Figure 4, the beam radius at the nominal focal length of the receiving collimator ω_f is:

$$\omega_f = \frac{f \varphi_c}{2z},$$ (8)

and the focus shift of the receiving collimator Δf equals:

$$\Delta f = \frac{f \omega_f}{\varphi_c/2 - \omega_f},$$ (9)

substituting Equation (8) into Equation (9), we have:

$$\Delta f = \frac{f^2}{z - f},$$ (10)

in OWC applications, where $z >> f$, the first-order term of f can be safely omitted.

Figure 5a shows the focus shift of the receiving collimator when its focus length is set to the same as the emitting collimator (37.2 mm). When the transmission distance of the optical beam is 1 m to 5 m, the focus shift of the receiving collimator decreased from 1.43 mm to 0.28 mm monotonically. Figure 5b shows the beam radius at the nominal focal length of the receiving collimator. At the same transmission range, the beam radius decreased from 381 um to 76 um. Figure 3b illustrates the fiber coupling loss when using optical fibers with 2 typical core diameters without considering their numerical aperture (NA) limitation. It is evident that a greater transmission distance corresponds to a reduced fiber coupling loss, which stands in contrast to the geometrical path loss. Thus, optimizing the link loss requires a careful consideration of the tradeoff between geometrical path loss and fiber coupling loss.

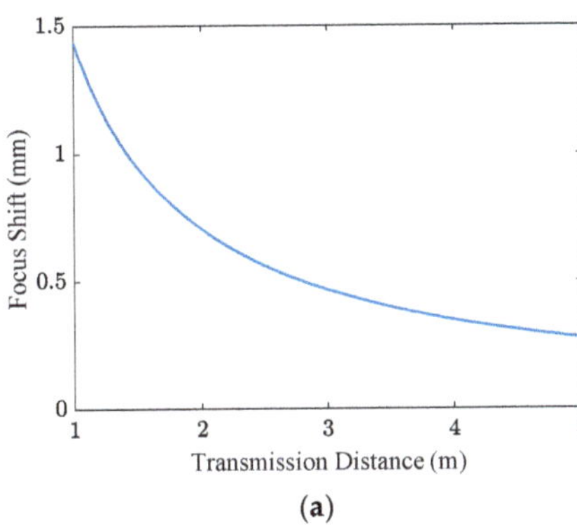
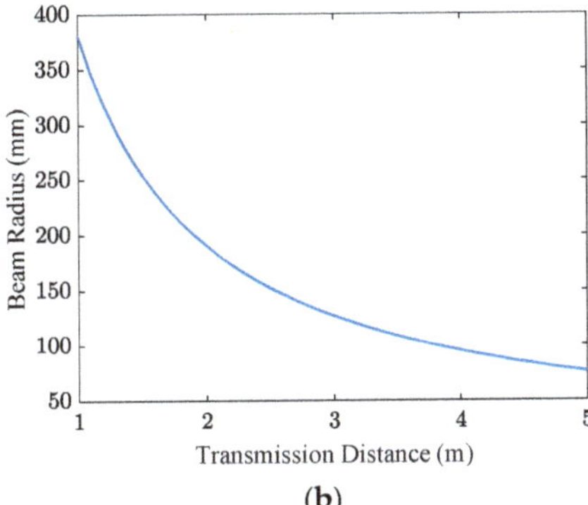

Figure 5. Focus shift of the receiving collimator and beam radius at the focus of the receiving collimator. (**a**) Focus shift versus transmission distances; (**b**) Beam radius at the focal length of the receiving collimator versus transmission distances.

The total link loss of the proposed system is the product of α and β, where α and β presents the geometrical term and the fiber coupling term, respectively. Recall the uniformity of optical intensity within the area of the receiving collimator, the total link loss in dB can be estimated by:

$$\eta = 10log_{10}(\alpha\beta) = 10log_{10}\left(\frac{(\varphi_c/2)^2}{R_{beam}^2}\frac{R_{core}^2}{\omega_f^2}\right), \tag{11}$$

where R_{beam} is the beam radius at the receiving plane, and R_{core} is the core radius of the coupling optical fiber, substituting Equation (8) into Equation (11), we have:

$$\eta = 20log_{10}\left(\frac{R_{core}}{f\theta}\right), \tag{12}$$

where the unit of θ is rad. With the objective of achieving a larger divergent angle θ, the total link loss can be alleviated by either reducing the focal length of the receiving collimator or increasing the radius of the fiber core coupled to the receiving collimator. Nevertheless, a smaller focal length generally implies a collimator with a reduced CA, thereby leading to an increased geometrical path loss. Instead of optimizing Equation (12), a more effective approach to mitigating the total link loss involves compensating for the focus shift and the selection of a receiving collimator featuring a larger CA. Then, the total link loss becomes:

$$\eta = 10log_{10}\alpha = 20log_{10}\left(\frac{\varphi_c}{2z\theta}\right) \tag{13}$$

In addition, the core radius of the coupling fiber should be larger than the practical beam waist of the receiving collimator. As a result, the fiber coupling loss is eliminated, and the achievable ROP is much increased.

4. Experimental Results

4.1. Validity of the Gaussian Propagation Model

Figure 6 shows the experimental setup for measuring the ROPs of the divergent Gaussian beam. In the experiment, the half-divergent angle of the Gaussian beam is 10° and the transmission distance is 2.5 m. At the receiver, a zoomable receiving collimator is fixed on a laterally fixed rail, and the ROPs of different receiver angles are obtained by translating the receiving collimator along the rail axis. Figure 7 showcases a comparison between the numerically simulated transverse ROP distribution and the experimentally measured discrete ROP values at intervals of 1°. The theoretically predicted curve and the experimentally measured curve with linear interpolation exhibit the same trend, with the ROP variance between them not exceeding 1 dB at any receiving angle. The result verifies the validity of employing geometric optics and uniform reception to describe the propagation of Gaussian beams in terms of intensity within short-range indoor OWC. Since the simulation only considers geometrical path loss, it can be inferred that the fiber coupling loss is effectively eliminated through the compensation of focus shifts. It is important to note that due to the narrow FOV of the receiving structure, the orientation of the collimator needs to be adjusted accordingly to align transceivers at varying transverse distances, potentially introducing ROP errors. Furthermore, due to constraints in the length of the rail, only positive transverse distances along the lateral axis were measured. Nevertheless, the spatial intensity distribution of Gaussian beams has a circular symmetry. However, the inhomogeneity of the ROP is as high as 10 dB.

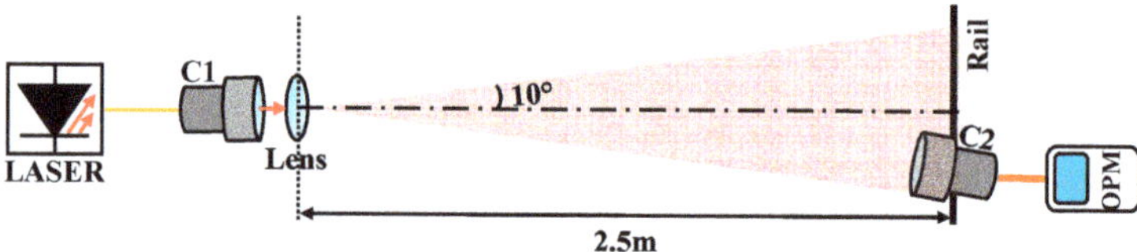

Figure 6. Experimental setup for measuring ROP distribution of a divergent Gaussian beam. C1: Emitting collimator; C2: Receiving collimator; OPM: Optical power meter.

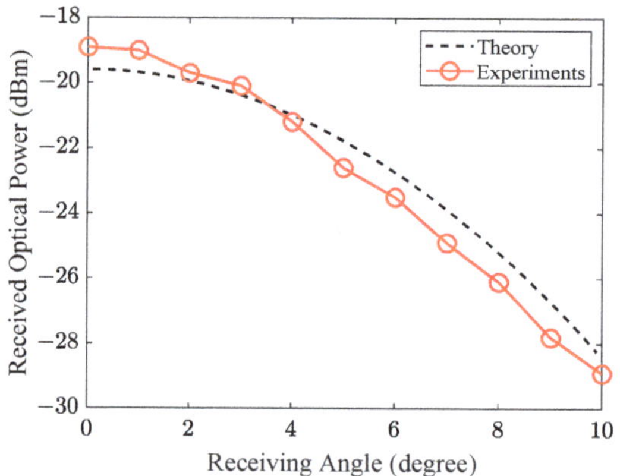

Figure 7. Theoretical and experimental ROP distribution of Gaussian beams.

4.2. MMF-Based Optical Beam-Shaping

Altering the spatial intensity distribution of a Gaussian beam offers an intuitive approach to addressing inhomogeneities. In this context, a segment of MMF is introduced between the SMF pigtailed laser source and the emitting collimator. By intentionally perturbing the transmission matrix of the MMF, the optical beam can be adaptively focused on any desired position within the coverage area. To assess the bit error rate (BER) performance across the coverage area, an experimental setup, as depicted in Figure 8, is implemented utilizing the MMF-based optical transmitter.

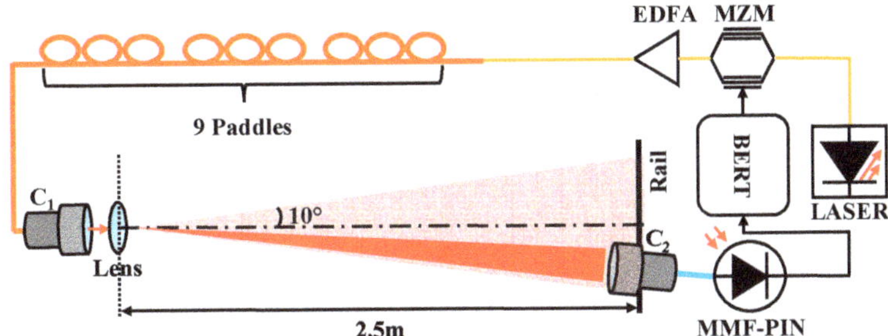

Figure 8. The MMF-based optical transmitter in OWC. BERT: Bit-error tester; MZM: Mach–Zehnder modulator; EDFA: Erbium-doped fiber amplifier; C1: Emitting collimator; C2: Receiving collimator; MMF-PIN: multimode-fiber PIN.

Initially, a Mach–Zehnder modulator (MZM, Fujitsu, FTM7937EZ) modulates the 10 Gbps baseband signal generated by a bit-error tester (BERT, 10 Gbps multi-channel BER tester, Luceo) onto the 1550 nm optical beam emitted by the laser source (Santec, MLS-2100). The modulated optical signal is then amplified through an Erbium-Doped Fiber Amplifier (EDFA, Amonics, AEDFA-BO-13) to 10 dBm. Next, the signal is directed into a 5 m long multi-mode patch cable (OM1 62.5/125 μm graded-index, Thorlabs, GIF625) with a radial offset distance of around 30 μm. The MMF is carefully wound around three three-paddle polarization controllers (nine control units in total, Thorlabs, FPC560). The optimized ROP is obtained by sequentially adjusting the rotation angle of each paddle in iterations. Then, the modulated optical signal is emitted into free space via the emitting collimator (Thorlabs, F810FC-1550) and the divergent lens (Thorlabs, AL2520M-C). After 2.5 m free-space transmission, the optical beam covers an area of 0.61 m², and a receiving collimator (Thorlabs, C40FC-C) is applied to capture the incoming light and then project it into another multi-mode patch cable (OM4 50/125 um graded-index, Thorlabs, GIF50E). The light coupled by the OM4 multi-mode patch cable is directly input into an MMF-PIN device (Thorlabs, DXM12DF), enabling optical-to-electrical (O/E) conversion. The received electrical signal is then compared to the transmitted electrical signal using the BERT, allowing for the counting of error bits. The optimization process for MMFs is iterative, requiring the measured ROP at each iteration to be transmitted back to the transmitter side until the desired target ROP is achieved. The optimized ROPs from 0 to 10° (1° interval) at a transmission distance of 2.5 m are sequentially measured.

Figure 9a shows the great homogenizing ability of MMF-based optical transmitters, and the optimized ROP variance within the coverage area is less than 2 dB. In contrast, SMF-based transmitters exhibit a significant ROP variance of 10 dB, owing to the Gaussian-shaped intensity distribution. Furthermore, at a 0 degree receiving angle, a notable ROP gain of 5.9 dB is observed, with the gain further increasing as the receiving angle expands. At a receiving angle of 10°, the ROP gain reaches 12.9 dB. This enhanced ROP substantially reduces the sensitivity requirements of direct detection receivers. Using a simple PIN-TIA photodiode as the O/E device without any optical pre-amplification, the BER performance of a 10 Gbps LOS OWC system through 2.5 m free-space transmission (Receiving angle = 8°) and an optical back-to-back link without free-space transmission are compared in Figure 9b. Remarkably, the MMF-based beam-shaping approach has a negligible ROP penalty (<0.2 dB), and the penalty may mainly caused by optical distortion at the edge of receiving lenses. We can conclude that the proposed scheme enables a 10 Gbps error-free transmission over an angular coverage of 10° × 10° at 2.5 m.

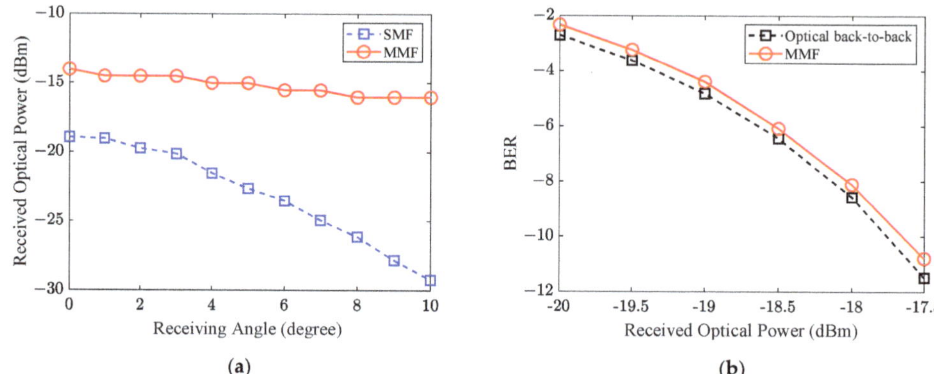

Figure 9. Performance comparison. (**a**) Comparison of ROP distribution between SMF and MMF; (**b**) Comparison of BER performance between optical back-to-back and MMF.

5. Conclusions

In the context of LOS OWC employing Gaussian divergent beams and narrow FOV receivers, a comparison is presented between the theoretical and experimental distributions of ROPs. The small discrepancy observed between the simulation and experimental results verifies the validity of geometric optics and uniform reception principles. Guided by these fundamental principles, the analysis focuses on the two major link losses inherent in the proposed system. Experimental results demonstrate the successful elimination of fiber coupling loss and the effective mitigation of geometrical path loss. Furthermore, a homogeneous ROP distribution, deviating from the Gaussian profile, is attained by manipulating MMFs, facilitating the mobility of end-users in indoor OWC. More importantly, the proposed scheme leads to a reduction in the complexity of tracking and steering operations.

6. Discussions

In future endeavors, the further optimization of link losses can be pursued by minimizing the divergent angle of the optical beam after the localization process. Leveraging classical iterative optimization algorithms and sufficient number of control units, multi-user access can be achieved utilizing a single fixed laser source. In addition, the enhanced mobility of portable devices can be achieved utilizing receivers with a larger FOV. However, it is important to acknowledge that the manipulation of MMFs requires iterations based on feedback loops, resulting in time-consuming optimization. To shorten the optimization time, the manipulation of MMFs should exploit their inherent properties to expedite the process. Lastly, the perturbations of MMFs are introduced by mechanical forces in the proposed scheme, and the inertia of these moving parts will limit the achievable rate of MMF optimizations.

Author Contributions: Conceptualization, Z.C.; methodology, X.Y.; software, Y.W.; validation, X.Y., Y.W.; data curation, X.Y.; writing—original draft preparation, X.Y.; writing—review and editing, E.T., Z.C., F.L. and C.L.; visualization, X.Y.; supervision, E.T.; project administration, Z.C., F.L. and C.L.; funding acquisition, E.T. All authors have read and agreed to the published version of the manuscript.

Funding: This research has been financed by Dutch Research Council NWO Gravitation Nanophotonics (Grant No. 024.002.033).

Institutional Review Board Statement: Not applicable.

Informed Consent Statement: Not applicable.

Data Availability Statement: Not applicable.

Conflicts of Interest: The authors declare no conflict of interest.

References

1. Koonen, T.; Mekonnen, K.; Cao, Z.; Huijskens, F.; Pham, N.Q.; Tangdiongga, E. Ultra-high-capacity Wireless Communication by means of Steered Narrow Optical Beams. *Phil. Trans. R. Soc. A* **2020**, *378*, 32114920. [CrossRef] [PubMed]
2. Khalighi, M.A.; Uysal, M. Survey on Free Space Communication: A Communication Theory Perspective. *IEEE Commun. Surv. Tutor.* **2014**, *16*, 2231–2258. [CrossRef]
3. Ghassemlooy, Z.; Arnon, S.; Uysal, M.; Xu, Z.; Cheng, J. Emerging Optical Wireless Communications-Advances and Challenges. *IEEE J. Sel. Areas Commun.* **2015**, *33*, 1738–1749. [CrossRef]
4. Gomez-Agis, F.; Van der Heide, S.P.; Okonkwo, C.M.; Tangdiongga, E.; Koonen, A.M.J. 112 Gbit/s transmission in a 2D Beam Steering AWG-based Optical Wireless Communication System. In Proceedings of the ECOC2017, Göteborg, Sweden, 17–21 September 2017.
5. Tuantranont, A.; Bright, V.M.; Zhang, J.; Zhang, W.; Neff, J.A.; Lee, Y.C. Optical Beam Steering using MEMS-Controllable Microlens Array. *Sens. Actuator A Phys.* **2001**, *91*, 363–372. [CrossRef]
6. Feng, F.; White, I.H.; Wilkinson, T.D. Free Space Communications with Beam Steering a Two-Electrode Tapered Laser Diode Using Liquid-Crystal SLM. *J. Light. Technol.* **2013**, *31*, 2001–2007. [CrossRef]
7. Li, Z.; Li, Y.; Zang, Z.; Han, Y.; Wu, L.; Li, M.; Li, Q.; Fu, H.Y. LiDAR Integrated IR OWC System with the Abilities of User Localization and High-Speed Data Transmission. *Opt. Express.* **2022**, *30*, 20796–20808. [CrossRef] [PubMed]
8. Zhang, X.; Kwon, K.; Henriksson, J.; Luo, J.; Wu, M.C. A Large-Scale Microelectromechanical-Systems-based Silicon Photonics LiDAR. *Nature* **2022**, *603*, 253–258. [CrossRef] [PubMed]
9. Koonen, T.; Gomez-Agis, F.; Huijskens, F.; Mekonnen, K.A.; Cao, Z.; Tangdiongga, E. High-Capacity Optical Wireless Communication Using Two-Dimensional IR Beam Steering. *J. Light. Technol.* **2018**, *36*, 4486–4493. [CrossRef]
10. Jiang, S.; Tan, Y.; Peng, Y.; Zhao, J. Tunable Optical Diffusers Based on the UV/Ozone-Assisted Self-Wrinkling of Thermal-Cured Polymer Films. *Sensors* **2021**, *21*, 5820. [CrossRef] [PubMed]
11. Li, C.; Zhang, Y.; Yan, X.; Wang, Y.; Zhang, X.; Cui, J.; Zhu, L.; Li, J.; Li, Z.; Yu, S.; et al. Adaptive Beamforming for Optical Wireless Communication via Fiber Modal Control. *arXiv* **2023**, arXiv:2304.11112.
12. Zhang, B.; Zhang, Z.; Feng, Q.; Lin, C.; Ding, Y. A Reference-Defining Criterion for Light Focusing through Scattering Media based on Circular Gaussian Distribution of Speckle Background Intensity. *Sci. Rep.* **2018**, *8*, 6385. [CrossRef] [PubMed]
13. Raddatz, L.; White, I.H.; Cunningham, D.G.; Nowell, M.C. An Experimental and Theoretical Study of the Offset Launch Technique for the Enhancement of the Bandwidth of Multimode Fiber Links. *J. Light. Technol.* **1998**, *16*, 324–331. [CrossRef]
14. Kogelnik, H.; Li, T. Laser Beams and Resonators. *Appl. Opt.* **1966**, *5*, 1550–1567. [CrossRef] [PubMed]
15. Paschotta, R. *Field Guide to Lasers*; SPIE Press: Bellingham, WA, USA, 2008; pp. 18–19.
16. Kolev, D.R.; Wakamori, K.; Matsumoto, M. Transmission Analysis of OFDM-Based Services Over Line-of-Sight Indoor Infrared Laser Wireless Links. *J. Light. Technol.* **2012**, *30*, 3727–3735. [CrossRef]

 *photonics*

Article

Long-Distance, Real-Time LED Display-Camera Communication System Based on LED Point Clustering and Lightweight Image Processing

Jingwen Li [1,2,3,4,†], Chuhan Pan [1,2,3,4,†], Junxing Pan [1,2,3,4,†], Jiajun Lin [1,2,3,4], Mingli Lu [5], Zoe Lin Jiang [6] and Junbin Fang [1,2,3,4,*]

[1] Guangdong Provincial Key Laboratory of Optical Fiber Sensing and Communications, Guangzhou 510632, China
[2] Guangdong Provincial Engineering Technology Research Center on Visible Light Communication, Guangzhou 510632, China
[3] Guangzhou Municipal Key Laboratory of Engineering Technology on Visible Light Communication, Guangzhou 510632, China
[4] Department of Optoelectronic Engineering, Jinan University, Guangzhou 510632, China
[5] Academic Affairs Office of Beijing Vocational College of Agriculture, Beijing 102442, China
[6] School of Computer Science and Technology, Harbin Institute of Technology, Shenzhen 518055, China
* Correspondence: tjunbinfang@jnu.edu.cn
† These authors contributed equally to this work.

Abstract: LED displays can be used to realize the dual functions of display and communication simultaneously. However, existing LED display-based visible light communication (VLC) systems suffer due to their short transmission distance and are not practical. A long-distance, real-time display-camera communication (DCC) system is proposed in this paper. First, a LED-DCC point clustering scheme is proposed to increase the transmission distance by clustering multiple adjacent LED display points for improving the quality of the VLC signal captured by an image sensor. Then, a lightweight, back-forth, fast image processing algorithm is proposed to reduce the introduced additional computational complexity caused by point clustering and enhance the reliability of information extraction from the real-time captured images/video frames. The experimental system was implemented with a 2.2-inch 16×16 point LED display and the CMOS camera on the smartphone. Experimental results show that the proposed system can achieve a maximum data transmission distance of 7 m under a bit error rate (BER) of 0.5, which is about 9 times that of the previous LED-DCC system, and can achieve a data transmission distance of 175 cm under the 7% forward error correction (FEC) limit, which is about 12 times that of the previous LED-DCC system. Additionally, the decoding latency for extracting information from each video frame is only 13.26 ms, which guarantees real-time data reception.

Keywords: LED display; visible light communication (VLC); display-camera communication (DCC); clustered LED points; lightweight image processing; CMOS image sensor

Citation: Li, J.; Pan, C.; Pan, J.; Lin, J.; Lu, M.; Jiang, Z.L.; Fang, J. Long-Distance, Real-Time LED Display-Camera Communication System Based on LED Point Clustering and Lightweight Image Processing. *Photonics* **2022**, *9*, 721. https://doi.org/10.3390/photonics9100721

Received: 4 August 2022
Accepted: 28 September 2022
Published: 3 October 2022

Publisher's Note: MDPI stays neutral with regard to jurisdictional claims in published maps and institutional affiliations.

1. Introduction

As a new display technology, LED displays are widely used in public as information displays, cluster displays, outdoor media, and in other fields, including financial information displays of the stock exchange, passenger guidance information displays at ports and stations, dynamic information displays for airport flights, road traffic information displays, information releases for large exhibitions, command centers, etc. [1]. Modulating the flicker frequency of the LED display's light-emitting unit element (i.e., display pixel) above the flicker fusion threshold can enable the LED display to display normally while emitting high-speed visible light communication (VLC) signals that are imperceptible to the human eye for covert information transmission, thus providing the LED display with the dual-use function of displaying and broadcasting information simultaneously [2–5]. At the same time, the VLC signal emitted by the LED display can be received by the complementary metal oxide semiconductor (CMOS) image sensor, which can recover the transmitted information. At present, many portable mobile electronic devices (i.e., smartphones) are equipped with high-resolution CMOS cameras, which can easily build up an LED display-camera communication (LED-DCC) system and realize some novel industry applications [4]; for example, real-time road information broadcasting and reception based on LED traffic lights, invisible advertising pushing based on airport flight dynamic displays, etc., which can provide mobile value-added service functions for LED displays.

Unlike conventional optical camera communication (OCC) [6], the optical signal transmitting element of LED-DCC is the LED display pixel point, which is constrained by display conditions and has the characteristics of a small, single-point, light-emitting area and low-signal power, limiting the communication performance of LED-DCC. Based on the potential application prospects of DCC, there has been some research carried out on DCC systems that make a trade-off between the visual experience and communication performance, including the data rate and bit error rate (BER) [7]. In 2016, Nguyen, V. et al. proposed a spatially adaptive embedding scheme, TextureCode, to achieve flicker-free communication by exploiting the low sensitivity of the human visual system in the texture-rich region, and also proposed a TextureCode-Hybrid scheme, which is a mix of the HiLight and TextureCode schemes under plain and high-texture blocks, achieving a higher transmission rate [8]. RainBar [9] features a code locator detection and localization scheme to allow flexible frame synchronization and accurate code extraction. ChromaCode [10] proposed an outcome-based adaptive embedding scheme, which adapts to both pixel lightness and frame texture. In 2019, RU codes and vRU codes [11] were proposed to combine 2D barcodes with images and videos for unobtrusive DCC at high data rates, based on the properties of the human visual system and the concept of a complementary framework. In the DaViD system [12], Xu, J. et al. addressed the spatial synchronization problem by utilizing localization patterns for detecting the modulation area and a separate optimization of columns and rows for data resampling. Based on that, clean data frames can be reconstructed by using a slight temporal oversampling. In 2021, Ryu et al. proposed a DCC method based on spatial frequency modulation by hiding data bits through the modulation of the high spatial frequency on the discrete cosine domain, and "0" and "1" bits are embedded in the transition between blurred and sharpened frames [13]. In 2022, Klein, J. et al. proposed a frame recovery method based on differential modulation to solve the synchronization problem for the invisible DCC system, in which the original display frames are recovered from asynchronous recordings in the receiver [4]. To avoid flicker and the resulting interference with the displayed content, these DCC systems use different modulation schemes for the raw pixel intensities of images and videos. Due to the distortion from the interframe interference problem [9], the schemes for embedding data in video content are more complex with regard to the design of channel coding and demodulation, which leads to higher computational complexity and difficulty in real-time communication.

In 2021, a real-time DCC system based on LED displays and smartphones with an Alternate Bit-flipping Repeat Coding (ABRC) scheme for the synchronization problem between LED displays and the smartphones' cameras, and a fast image processing algorithm to decrease the computational complexity of image processing were proposed [14]. The previous work modulates a single pixel on a 16×16 point LED display, which can achieve a data transmission rate of 30 bps, and the data decoding latency on the Android smartphone for data extraction is only 11.83 ms. However, the maximum data transmission distance is only 80 cm under a BER of 0.5 and the data transmission distance is 15 cm under the 7% forward error correction (FEC) limit, limiting its practical use. The main reason for the short communication distance is that the real-time processing capability of the fast image processing algorithm cannot support the real-time information extraction of larger area pixel points, which limits the quality of the received visible light signal. To increase the data transmission distance, it is important to enhance the quality of the received VLC signals, i.e., either increasing the DCC light-emitting area on the transmitter side or increasing the resolution of the image sensor on the receiver side in the LED-DCC system. However, both solutions will multiply the computational latency for image processing several times and will eventually degrade the real-time data transmission performance of the LED-DCC system.

In this paper, we propose a long-distance, real-time DCC system based on LED point clustering and lightweight image processing. First, we propose a LED-DCC point clustering scheme, which uses multiple LED display points to cluster together to increase the light-emitting area for sending information, as well as the data transmission distance. Then, to solve the problem of processing latency introduced by point clustering and improve the reliability of information extraction, a lightweight, back-forth, fast image processing algorithm is proposed, which can quickly realize the high-precision positioning and data extraction of the LED-DCC area using the adaptive scanning method and variable step lengths.

The proposed LED-DCC system has been experimentally verified and demonstrated on a 2.2-inch 16×16 point LED display with a refresh rate of 150 fps and on a commercial Android smartphone with a camera image sensor with a resolution of 3840×2160. Experimental results show that the proposed LED-DCC system can reach the maximum data transmission distance of 7 m under a BER of 0.5, which is about 9 times that of the previous LED-DCC system, and can reach a data transmission distance of 175 cm under the 7% FEC limit, which is about 12 times that of the previous LED-DCC system. Additionally, the data decoding latency caused by extracting information from each video frame is only 13.26 ms, which is similar to the previous LED-DCC system [14], even though the proposed system needs to process many more pixels (about 10 times) than the previous LED-DCC system. Therefore, the proposed system has advantages not only in data transmission distance, but also in data transmission rate and system reliability.

The remainder of this paper is structured as follows: Section 2 presents a detailed description of the proposed LED-DCC system, including the system architecture, the proposed LED-DCC point clustering scheme, and the proposed lightweight, back-forth, fast image processing algorithm. The experimental results and discussion are provided in Section 3. Finally, conclusions and future works are presented in Section 4.

2. The Proposed Long-Distance, Real-Time LED-DCC System

The schematic diagram of the proposed long-distance, real-time LED-DCC system is shown in Figure 1a. The system can be divided into a transmitter side with the LED display with VLC function and a receiver side with the smartphone's CMOS camera used as a photoelectric sensor array.

On the transmitter side, the hidden data is encoded via the ABRC scheme [14] that replicates the original information bits multiple times in an alternating flip for synchronization between the transmitter and the receiver. To increase the LED-DCC light-emitting area for communication, as well as the transmission distance, each data bit in the encoded data frame is inserted in several nearby pixels of the video frame to be displayed on the LED display. Since each LED display point is modulated by the value of each pixel, each hidden data bit can be carried by multiple clustered LED-DCC points at the same time. Finally, the high-speed, modulated LED display points broadcast high-speed visible light signals at a baud rate, the same as the refresh rate of the LED display, which is imperceptible to human eyes.

On the receiver side, the modulated signal transmitted through the VLC channel is captured by the smartphone's CMOS camera. To solve the problem of processing latency introduced by point clustering and improve the reliability of information extraction, a lightweight, back-forth, fast image processing algorithm is proposed to quickly locate the LED-DCC display area from the high-resolution video frame and extract the transmitted data bits in a real-time mode. Finally, the extracted data frame is decoded with the ABRC scheme to recover the hidden data.

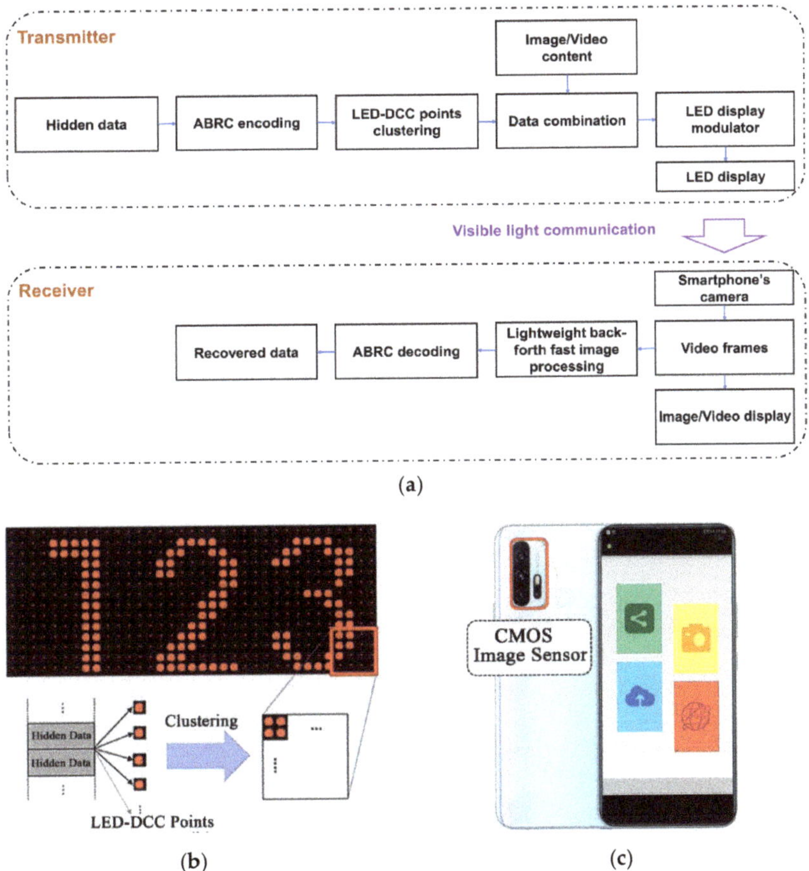

Figure 1. (a) The overall block diagram of the proposed long-distance, real-time LED-DCC system. (b) LED displays for displaying and communication. (c) Smartphone's CMOS camera for receiving VLC signals.

In the long-distance, real-time LED-DCC system, two key technologies are proposed: a LED-DCC point clustering scheme for improving the system communication performance and a lightweight, back-forth, fast image processing algorithm for high-precision, real-time data reception.

2.1. LED-DCC Point Clustering Scheme

In the previous work, the maximum achievable distance at which the VLC signal broadcast by an LED display point could be received and recovered was only 15 cm under the 7% FEC limit. The data transmission distance is limited by the quality of the received VLC signal. Increasing the light-emitting area or the brightness of the LED-DCC point can help improve the visible signal quality captured by the image sensor. Increasing the brightness of LED display points requires the support of hardware, e.g., the driving circuit and the LED display element. Furthermore, increasing the brightness of LED display points may not only change the brightness of the display screen, but also the other display parameters, such as contrast and/or sharpness, as well as the viewing experience.

Therefore, to solve the problem of the short transmission distance in the LED-DCC systems, the LED-DCC point clustering scheme is proposed to increase the light-emitting area by making use of a large number of pixels in the LED display, improving data transmission distance and system reliability.

As shown in Figure 2, each data bit in the codeword encoded by the ABRC scheme is synthesized with N adjacent LED display points. Therefore, the value of the data bit is carried by N LED display points. Since the N LED display points are adjacent, they can be processed as a clustered point group.

$$\text{LED}_i(x, y) = bt \times p_i(x, y), \ i \in [1, N] \tag{1}$$

where $p_i(x, y)$ is the value of the i-th pixel located at the coordinates (x, y) in the light-emitting area of N adjacent pixels in each video frame (frame$_j$), N is the total number of pixels in the clustered group, and b_t is the spread-spectrum code chip that encodes each original data bit based on the ABRC scheme. b_t is combined with $p_i(x, y)$, $i \in [1, N]$ in each frame$_j$ of the video. Finally, the hidden data is broadcasted by the LED-DCC clustering points (LED$_i(x, y)$), $i \in [1, N]$) located at the coordinates (x, y) in a 2D point array at the same baud rate as the refresh rate of the LED display while avoiding flicker perceived by the human eye.

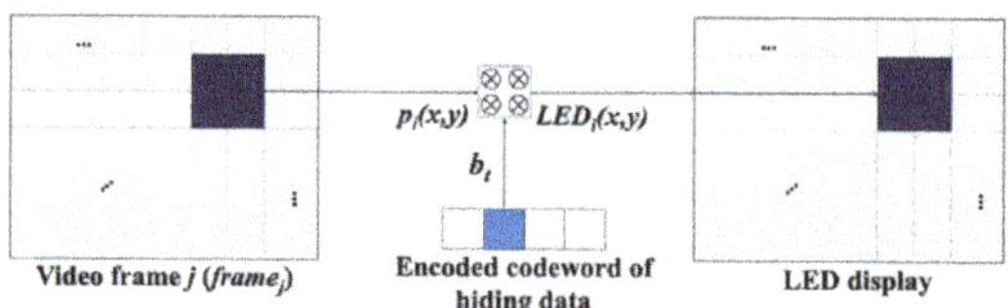

Figure 2. The LED-DCC point clustering scheme.

2.2. Lightweight, Back-Forth, Fast Image Processing Algorithm

As described in the proposed schemes in Section 1, the LED-DCC point clustering scheme and high-resolution image sensor can effectively improve the quality of the received VLC signals and support long-distance LED-DCC. The high resolution, e.g., 3840 × 2160, brings more photoelectric imaging pixels to support reliable long-distance communication. However, it also brings a higher data processing capacity; therefore, there are higher requirements for the real-time image processing speed of a smartphone's software and hardware. The extraction of the LED light-emitting area in the captured image is the main factor affecting the image processing speed. To reduce the latency of data decoding, a fast image processing algorithm [14] can detect the contour of the LED light-emitting area and segment it from the captured video image in real time. However, the algorithm just supports the real-time LED-DCC data reception at a short transmission distance due to its limited real-time processing capability for larger area pixel points. For the proposed LED-DCC system, there is a high processing latency introduced by the high-resolution image and the enlarged light-emitting area containing the clustered points. In this paper, we propose a lightweight, back-forth, fast image processing algorithm that uses variable step lengths to search back and forth to quickly locate the LED light-emitting area from the high-precision video frame and an adaptive binarization method to convert the pixel value of the LED-DCC point into a bit and then recover the transmitted data through the ABRC scheme. With the proposed lightweight, back-forth, fast image processing algorithm, the problem of processing latency introduced by LED-DCC point clustering is solved, and the reliability of the LED-DCC system can be improved. Therefore, the proposed LED-DCC system is able to support long-distance, real-time data reception.

Figure 3 shows an example of processing a captured image frame via a lightweight, back-forth, fast image processing algorithm. Due to the CMOS camera of the smartphone being set to work in underexposure mode with a fast exposure time and small aperture, the brightness value of pixels in most areas in the captured image frame, except for in the LED light-emitting area, is zero. That is, except for outside the LED display area, the sum of pixel values in each row (column) is close to zero. The video frame captured by the camera in the YUV format is processed in real time by the proposed lightweight, back-forth, fast image processing algorithm, which performs two steps of LED light-emitting area detection and LED-DCC data bit extraction. The specific demodulation procedure can be divided into the following two steps:

- LED light-emitting area detection. After the brightness (Y) data of each video frame is extracted as a gray image, the method of pixel sampling is used with the initial sampling step length l_{init} to perform vertical integration processing on the Y value of the pixels with a constant distance, and to quickly detect the fuzzy left or right boundary of the LED light-emitting area. To avoid the error caused by sampling, the left or right boundary of the blur is taken as the center, and the left or right side is separated by $2 \times l_{init}$. Then, the algorithm adjusts the step length to a smaller step length (l_{small}) and performs vertical integration processing on the pixels in this nearby area to determine the precise boundaries of the LED light-emitting area. Similarly, the precise top and bottom boundaries can be determined by horizontal integration with variable sampling step lengths. As shown in Figure 3, the red dashed lines indicate the precise left and top boundaries of the LED light-emitting area detected. Appropriately increasing l_{init} can improve the positioning speed of the LED display area, and reducing l_{small} can improve the positioning accuracy of the LED display area. In our system, the value of l_{init} is set as 5 and the value of l_{small} is set as 1.
- LED-DCC data bit extraction. Once the vertical and horizontal boundaries of the LED lighting-emitting areas are detected, the pixel coordinates of the corners of the LED-DCC emitting areas are easily obtained. Thus, the coordinates of the LED-DCC light-emitting area are quickly detected through the relative position offset based on the coordinates of the pixels in the upper left and lower right corners, and the y values of all pixels in the LED-DCC area are integrated. Then, the symbol value of the LED-DCC clustered points is determined by the Sauvola-based adaptive binarization method [15], converting the y value of the LED-DCC clustered points into the bit. Finally, ABRC decoding is conducted to recover the transmitted data.

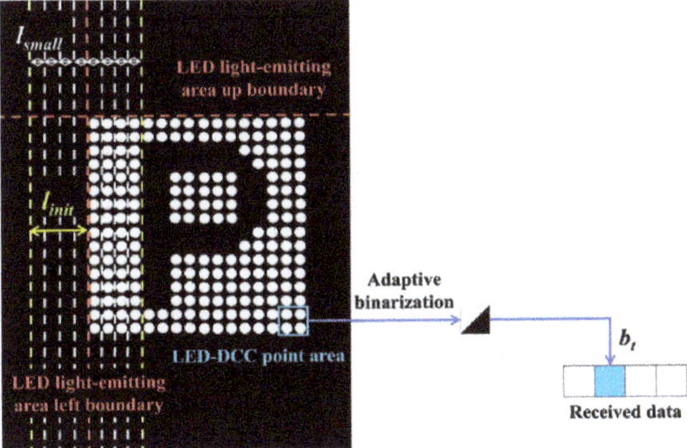

Figure 3. The demodulation procedure of lightweight, back-forth, fast image processing algorithm.

3. Experimental Results

An experimental system was implemented on a 2.2-inch 16×16 point LED display and on a demodulator APP for an Android smartphone, and a series of experiments were conducted to verify the performance of the proposed long-distance LED-DCC system, including data transmission distance, data transmission rate, and data decoding latency, which are critical for real-time data reception.

3.1. Experimental System

As shown in Figure 4a, the experimental system mainly consists of an LED display that is 4×4 cm^2 for transmitting hidden data while displaying image/video, and the smartphone's CMOS camera with a capture frame rate of 30 fps for VLC signal capturing and data reception. As shown in Figure 4b, the experimental system was tested in a dim environment to avoid interference from other light sources, and the main hardware configurations of the experimental system are listed in Table 1.

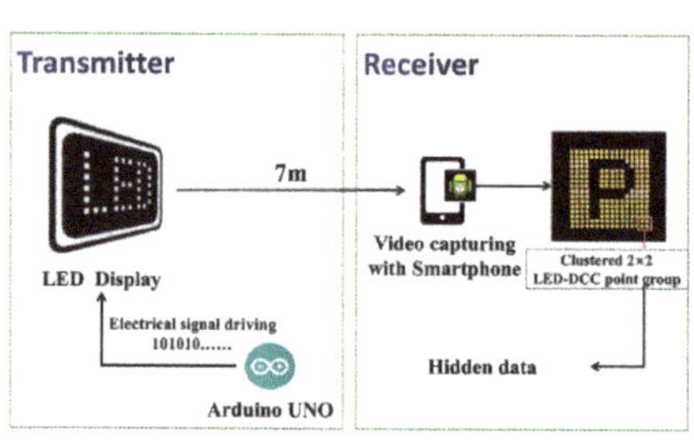

(**a**)

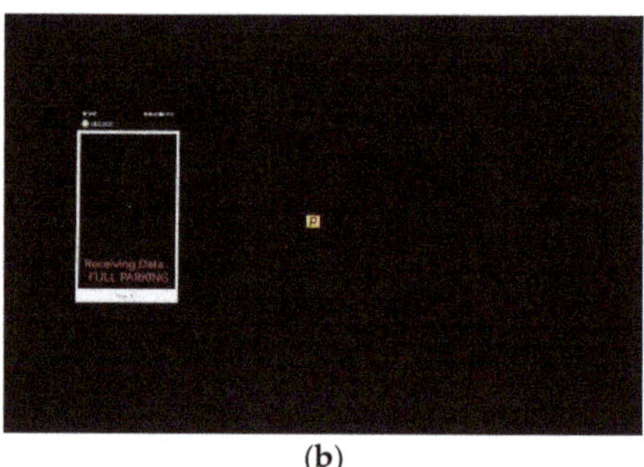

(**b**)

Figure 4. (**a**) The schematic diagram and (**b**) photography of the long-distance, real-time LED-DCC system experimental setup.

Table 1. The experimental system hardware configuration.

	Hardware	Configuration
LED-DCC Transmitter	MCU	ATmega328P-PU
	Clock Speed	16 MHz
	LED Display Panel Resolution	16×16
	LED Display Panel Size	4×4 cm^2 (2.2-inch)
	Area of a LED-DCC Point	3.14 mm^2
	Area of Clustered 2×2 LED-DCC Point	12.56 mm^2
	LED Display Panel Refresh Rate	150 fps
	LED Display Panel Driver	74HC595 8-bit Shift Register
LED-DCC Receiver	Smartphone Model	HUAWEI P30
	Operating System	Android 10
	Processor	Kirin 980 @ 2.6 GHz
	Camera Resolution	3840×2160
	Capture Frame Rate	30 fps

The Arduino microcontroller is used to encode the hidden data via the ABRC scheme and embed it in the VLC signal. The modulated VLC signal is then emitted from an LED display driven by the integrated 74HC595 8-bit Shift Register. In our experiment, the LED display with 16×16 LED display points is controlled to display a "parking" pattern, and four adjacent LED display points at the bottom right corner of the LED display are selected as a clustered 2×2 LED-DCC point group, whose area is 12.56 mm^2.

On the receiver side, an 8-megapixel HUAWEI P30 is employed as a receiver, which captures modulated VLC signals via its CMOS camera and recovers the hidden data with the demodulator APP. The lightweight, back-forth, fast image processing algorithm and the ABRC decoding information are integrated into the demodulator APP of the smartphone. In addition, the exposure mode of the smartphone's CMOS camera is fixed in the demodulator APP to an ISO value of 50 and an exposure time of $1/150$ s in the experimental setup.

3.2. Data Transmission Performance

In our demonstration, the LED display is controlled to display the pattern of the parking sign and broadcast the packets. One packet consists of an 8-bit start frame delimiter (SFD) and a character string "FULL PARKING" as the content of the hidden data, as shown in Figure 5. After ASCII encoding, the total length of the data frame is 104 bits. Then, the original information bit is repeatedly encoded by alternate bit-flipping five times according to the ABRC coding scheme, and thus, the final length of the encoded data frame is 520 symbols. The encoded data frame was repeatedly sent 100 times and compared with the received data of the smartphone's APP to measure the BER of the transmission. The BER was used to evaluate the channel capacity, and therefore, we did not utilize any error correction coding in the data transmission experiments.

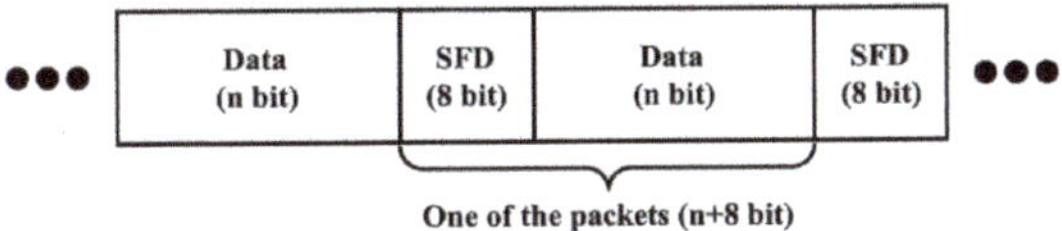

Figure 5. The structure of the packet.

As shown in Figure 6, the BER performance and the data transmission rate with the data transmission distance are evaluated. The black dashed line shows the 7% FEC limit which corresponds to a BER of 3.8×10^{-3}. When the data transmission distance between the LED display and the smartphone's CMOS camera is increased from 10 cm to 700 cm, the measured BER reaches 0.5, which means that the channel capacity reaches 0. As a result, the experimental proposed LED-DCC system can reach a maximum transmission distance of 700 cm, at which the usable capacity is close to 0, demonstrating the superiority of the proposed long-distance LED-DCC system over the existing LED-DCC systems in terms of data transmission distance. Furthermore, when the data transmission distance is within 175 cm, the proposed system can still achieve successful data transmission under a BER of less than 3.8×10^{-3} and a data transmission rate of up to 30 bps, meaning that the proposed system is robust within 175 cm. Note that in this experiment, a clustered 2×2 LED-DCC point is used as a data transmission channel. An LED display can be viewed as a multi-parallel array of emitters with a high pixel count; thus, it is possible to achieve high data rates of several Mbit/s, despite being limited by the low capture frame rate of CMOS cameras.

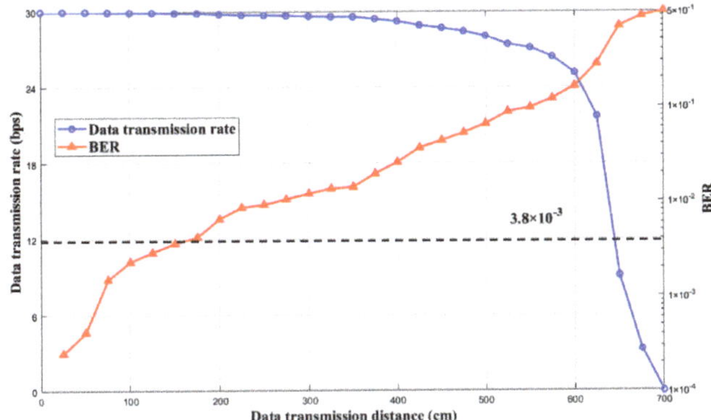

Figure 6. The data transmission performance of the proposed LED-DCC system.

To verify the effect of the proposed LED-DCC point clustering scheme and the lightweight, back-forth, fast image processing algorithm, the data transmission distance of the proposed system is compared with that of the previous LED-DCC system [14] and a 2×2 point clustered LED-DCC system, i.e., the previous LED-DCC system using 2×2 clustered LED-DCC points. The experimental results are shown in Figure 7.

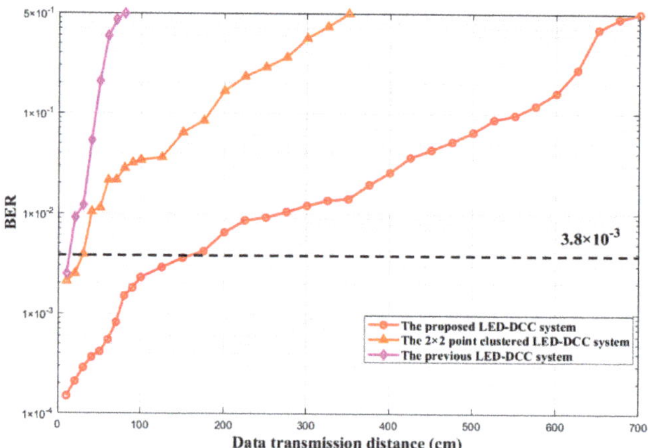

Figure 7. The BER performance of different LED-DCC systems versus data transmission distance.

As shown in Figure 7, the BER increases dramatically with the increment of the data transmission distance. When the BER of the LED-DCC system is higher than 3.8×10^{-3}, it may not ensure successful data transmission. The previous LED-DCC system can achieve a maximum transmission distance of 15 cm, whereas that of the 2×2 point clustered LED-DCC system is 40 cm, which is about 3 times that of the previous LED-DCC system under the BER less than 3.8×10^{-3}. This is attributed to the LED-DCC point clustering scheme since it improves the power of transmitted VLC signals by enlarging the light-emitting area in the LED-DCC system. Furthermore, due to the proposed lightweight, back-forth, fast image processing algorithm, which enables high-precision LED light-emitting area detection and, therefore, increases the signal detection on the receiver's side, the proposed LED-DCC system can achieve a maximum distance of 175 cm under the 7% FEC limit, which is about 4 times that of the 2×2 point clustered LED-DCC system. Therefore, the proposed LED-DCC system has a longer data transmission distance and higher data transmission rate.

Based on basic principles of trigonometric and optical measurements, it is speculated that the 2×2 point clustered LED-DCC system should only double the maximum data transmission distance compared with the previous LED-DCC system. However, the experimental results show that the data transmission distance achieved by the proposed LED-DCC system is about 12 times that of the previous LED-DCC system under the 7% FEC limit, which clearly demonstrates that the LED-DCC point clustering scheme and the lightweight, back-forth, fast image processing algorithm effectively improve the data transmission distance, as analyzed in Section 2. It is anticipated that a much longer distance can be achieved if more adjacent LED display points are clustered and the proposed image processing algorithm is utilized. However, considering the characteristics of the optical wireless channel [16,17], the performance of the system would degrade with the decline in channel capacity as the distance increases, which is challenging in practical applications.

3.3. Data Decoding Latency

Data decoding latency is another key factor for the performance of the LED-DCC system, especially for real-time data reception. In the above experiments, the data decoding latency of each frame for the data reception of the proposed LED-DCC system and the previous LED-DCC system were measured. As shown in Table 2, the proposed LED-DCC system decodes a single video frame with an average data decoding latency of only about 13.26 ms, which still supports real-time data reception at a capture frame rate of 30 fps. Therefore, the proposed LED-DCC system can support real-time data reception.

Table 2. The average data decoding latency for data reception.

	The Previous LED-DCC System [14]	The Proposed LED-DCC System
LED light-emitting area detection	10.34 ms	13.07 ms
LED-DCC data bit extraction	1.49 ms	0.19 ms
Total	11.83 ms	13.26 ms

Note that although the proposed LED-DCC system has a higher processing latency than the previous LED-DCC system due to a larger light-emitting area in the images and more precise image processing, the data decoding latency does not increase significantly and becomes a bottleneck for real-time data reception, which verifies the efficiency of the proposed image processing algorithm.

4. Conclusions

In this paper, a long-distance, real-time DCC system based on LED point clustering and lightweight image processing is proposed. The proposed LED-DCC point clustering scheme has solved the problem of a short transmission distance in existing LED-DCC systems. Meanwhile, this paper also proposes the lightweight, back-forth, fast image processing algorithm to solve the problem of processing latency introduced by point clustering and to improve the reliability of information extraction. The experimental setup is implemented on a 2.2-inch 16 × 16 point LED display and a CMOS camera on a smartphone with a resolution of 3840 × 2160, and an Android smartphone demodulator APP was engineered. The experiment results show that the maximum data transmission distance of the proposed LED-DCC system can reach 7 m under a BER of 0.5, which is about 9 times that of the previous LED-DCC system, and can reach a data transmission distance of 175 cm under the 7% FEC limit, which is about 12 times that of the previous LED-DCC system. Additionally, the data decoding latency caused by extracting information from each video frame is only 13.26 ms, which guarantees real-time data reception.

In the LED-DCC systems, the main limitation of the current maximum transmission distance and the transmission rate is the low frame rate and resolution of the image sensor. However, it is possible to achieve high data rates in the range of several Mbit/s by using MIMO or using more than one color channel for data modulation, e.g., the R, G, and B channels, and achieve a longer transmission distance by using a high-resolution image sensor. Future works should be directed towards improving the channel capacity of the LED-DCC system while testing the performance under outdoor conditions, promoting the expansion of more potential application scenarios of LED-DCC.

Author Contributions: Conceptualization, J.L. (Jingwen Li) and C.P.; methodology, J.L. (Jingwen Li); software, C.P. and J.P.; validation, C.P.; formal analysis, J.L. (Jingwen Li) and J.P.; investigation, C.P. and J.L. (Jiajun Lin); resources, M.L., J.F. and Z.L.J.; data curation, J.P.; writing—original draft preparation, J.L. and C.P.; writing—review and editing, J.L. (Jingwen Li), C.P., J.P., J.L. (Jiajun Lin), M.L., J.F. and Z.L.J.; visualization, J.P. and J.L. (Jiajun Lin); supervision, M.L., J.F. and Z.L.J.; project administration, J.F.; funding acquisition, J.F. All authors contributed to the critical reading and writing of the manuscript. All authors have read and agreed to the published version of the manuscript.

Funding: This work was funded by the National Key Research and Development Program of China (No. 2018YFB1801900, No. 2019YFE0123600), National Natural Science Foundation of China (No. 62171202), Guangdong Provincial Postgraduate Education Innovation Project (No. 2019SFKC08), European Union's Horizon 2020 Research and Innovation Programme under the Marie SklodowskaCurie grant agreement No. 872172 (TESTBED2 project: www.testbed2.org (accessed on 1 September 2022)), HKU-SCF FinTech Academy and themebased research funding (T35-710/20-R) of RGC from the HK

Photonics **2022**, 9, 721

Government, and National Innovation and Entrepreneurship Training Program for Undergraduate (No. 202010559036).

Institutional Review Board Statement: Not applicable.

Informed Consent Statement: Not applicable.

Data Availability Statement: The data that supports the findings of this study are available within the article.

Conflicts of Interest: The authors declare no conflict of interest.

References

1. Kasilingam, G. A Survey of Light Emitting Diode (LED) Display Board. *Indian J. Sci. Technol.* **2013**, *7*, 185–188. [CrossRef]
2. Goto, Y.; Takai, I.; Yamazato, T.; Okada, H.; Fujii, T.; Kawahito, S.; Arai, S.; Yendo, T.; Kamakura, K. A New Automotive VLC System Using Optical Communication Image Sensor. *IEEE Photonics J.* **2016**, *8*, 1–17. [CrossRef]
3. Karunatilaka, D.; Zafar, F.; Kalavally, V.; Parthiban, R. LED Based Indoor Visible Light Communications: State of the Art. *IEEE Commun. Surv. Tutor.* **2015**, *17*, 1649–1678. [CrossRef]
4. Klein, J.; Xu, J.; Brauers, C.; Jochims, J.; Kays, R. Investigations on Temporal Sampling and Patternless Frame Recovery for Asynchronous Display-Camera Communication. *IEEE Trans. Circuits Syst. Video Technol.* **2022**, *32*, 4004–4015. [CrossRef]
5. Wang, J.; Huang, W.; Xu, Z. Demonstration of a Covert Camera-Screen Communication System. In Proceedings of the International Wireless Communications and Mobile Computing Conference, IWCMC, Valencia, Spain, 26–30 June 2017; pp. 910–915. [CrossRef]
6. Le, N.T.; Hossain, M.A.; Jang, Y.M. A Survey of Design and Implementation for Optical Camera Communication. *Signal Process. Image Commun.* **2017**, *53*, 95–109. [CrossRef]
7. Liu, W.; Xu, Z. Some Practical Constraints and Solutions for Optical Camera Communication. *Philos. Trans. R. Soc. A Math. Phys. Eng. Sci.* **2020**, *378*, 20190191. [CrossRef]
8. Nguyen, V.; Tang, Y.; Ashok, A.; Gruteser, M.; Dana, K.; Hu, W.; Wengrowski, E.; Mandayam, N. High-Rate Flicker-Free Screen-Camera Communication with Spatially Adaptive Embedding. In Proceedings of the IEEE INFOCOM 2016—The 35th Annual IEEE International Conference on Computer Communications, San Francisko, CA, USA, 10–14 April 2016; IEEE: New York, NY, USA, 2016; pp. 1–9.
9. Wang, Q.; Zhou, M.; Ren, K.; Lei, T.; Li, J.; Wang, Z. Rain Bar: Robust Application-Driven Visual Communication Using Color Barcodes. In Proceedings of the 2015 IEEE 35th International Conference on Distributed Computing Systems, Columbus, OH, USA, 29 June–2 July 2015; IEEE: New York, NY, USA, 2015; pp. 537–546.
10. Zhang, K.; Wu, C.; Yang, C.; Zhao, Y.; Huang, K.; Peng, C.; Liu, Y.; Yang, Z. ChromaCode. In Proceedings of the 24th Annual International Conference on Mobile Computing and Networking, New Delhi, India, 29 October–2 November 2018; ACM: New York, NY, USA, 2018; pp. 575–590.
11. Chen, C.; Huang, W.; Zhang, L.; Mow, W.H. Robust and Unobtrusive Display-to-Camera Communications via Blue Channel Embedding. *IEEE Trans. Image Process.* **2019**, *28*, 156–169. [CrossRef]
12. Xu, J.; Klein, J.; Brauers, C.; Kays, R. Transmitter Design and Synchronization Concepts for DaViD Display Camera Communication. In Proceedings of the 2019 28th Wireless and Optical Communications Conference (WOCC), Beijing, China, 9–10 May 2019; pp. 2–6. [CrossRef]
13. Ryu, S.-S.; Lee, K.-H.; Bae, J.-S.; Kim, J.-O. Spatial Frequency Modulation for Display-Camera Communication. In Proceedings of the 2021 International Conference on Electronics, Information and Communication (ICEIC), Jeju, South Korea, 31 January–3 February 2021; IEEE: New York, NY, USA, 2021; Volume 2015, pp. 1–5.
14. Bao, X.; Pan, J.; Cai, Z.; Li, J.; Huang, X.; Chen, R.; Fang, J. Real-Time Display Camera Communication System Based on LED Displays and Smartphones. *Opt. Express* **2021**, *29*, 23558–23568. [CrossRef] [PubMed]
15. ZBar. ZBar Bar Code Reader. 2011. Available online: http://zbar.sourceforge.net/ (accessed on 15 July 2011).
16. Al-Kinani, A.; Wang, C.-X.; Zhou, L.; Zhang, W. Optical Wireless Communication Channel Measurements and Models. *IEEE Commun. Surv. Tutor.* **2018**, *20*, 1939–1962. [CrossRef]
17. Basha, M.; Sibley, M.J.N.; Mather, P.J.; Makama, A. Design and Implementation of a Tuned Analog Front-End for Extending VLC Transmission Range. In Proceedings of the 2018 International Conference on Computing, Electronics & Communications Engineering (iCCECE), Southend, UK, 6–17 August 2018; IEEE: New York, NY, USA, 2018; Volume 20, pp. 173–177.

Review

AI-Enabled Intelligent Visible Light Communications: Challenges, Progress, and Future

Jianyang Shi [1,2,3], Wenqing Niu [1], Yinaer Ha [1], Zengyi Xu [1], Ziwei Li [1,4], Shaohua Yu [4] and Nan Chi [1,2,3,*]

1   Key Laboratory for Information Science of Electromagnetic Waves (MoE), Fudan University, Shanghai 200433, China; jy_shi@fudan.edu.cn (J.S.); 21110720066@m.fudan.edu.cn (W.N.); 19210720063@fudan.edu.cn (Y.H.); 21110720075@m.fudan.edu.cn (Z.X.); lizw@fudan.edu.cn (Z.L.)
2   Shanghai Engineering Research Center of Low-Earth-Orbit Satellite Communication and Applications, Shanghai 200433, China
3   Shanghai Collaborative Innovation Center of Low-Earth-Orbit Satellite Communication Technology, Shanghai 200433, China
4   Peng Cheng Laboratory, Shenzhen 518055, China; shaohuayu@fudan.edu.cn
*   Correspondence: nanchi@fudan.edu.cn

Abstract: Visible light communication (VLC) is a highly promising complement to conventional wireless communication for local-area networking in future 6G. However, the extra electro-optical and photoelectric conversions in VLC systems usually introduce exceeding complexity to communication channels, in particular severe nonlinearities. Artificial intelligence (AI) techniques are investigated to overcome the unique challenges in VLC, whereas considerable obstacles are found in practical VLC systems applied with intelligent learning approaches. In this paper, we present a comprehensive study of the intelligent physical and network layer technologies for AI-empowered intelligent VLC (IVLC). We first depict a full model of the visible light channel and discuss its main challenges. The advantages and disadvantages of machine learning in VLC are discussed and analyzed by simulation. We then present a detailed overview of advances in intelligent physical layers, including optimal coding, channel emulator, MIMO, channel equalization, and optimal decision. Finally, we envision the prospects of IVLC in both the intelligent physical and network layers. This article lays out a roadmap for developing machine learning-based intelligent visible light communication in 6G.

Keywords: visible light communication; artificial intelligence; machine learning; physical layer; network layer

Citation: Shi, J.; Niu, W.; Ha, Y.; Xu, Z.; Li, Z.; Yu, S.; Chi, N. AI-Enabled Intelligent Visible Light Communications: Challenges, Progress, and Future. *Photonics* **2022**, *9*, 529. https://doi.org/10.3390/photonics9080529

Received: 10 July 2022
Accepted: 26 July 2022
Published: 29 July 2022

1. Introduction

As 5G's commercialization progresses, the number of 5G base stations worldwide has surpassed one million. This marks the beginning of globally competitive future-oriented research on 6G networks. According to several research reports [1–3], it is widely assumed that 6G communication will go beyond the current wireless spectrum and shift towards higher frequencies. The millimeter-wave and terahertz spectrum have long been the research focus academically and industrially, except that the equipment is of extremely high cost. Recently, the spectrum of light, i.e., visible and infrared light, provides a potential supplement for 6G. During the last decade, visible light communication is being cast in the spotlight by 6G researchers as a green, energy-efficient, high-speed communication method [4].

Visible light communication transmits (VLC) signals in a spectrum range of 400–800 Thz, which owns a very different physical property compared with both conventional wireless transmission and optical communication. Communication with visible light provides benefits of electromagnetic interference resistance, vast spectrum resources, and high-speed transmission capabilities. Moreover, it can be equipped with common lighting systems to allow simultaneous illumination and communication. Furthermore, the short wavelength of light source allows for the creation of super-compact cells, which are ideal for 6G communication. Nevertheless, signal communication at such a small wavelength poses critical challenges to transmitting and receiving devices. Semiconductor materials with wide bandgaps must be employed to achieve such high-frequency photons [5]. The extra electro-optical and photoelectric conversions compared to wireless communications introduce undesirable nonlinear distortions and hinder the high-speed transmission in visible light communications [6,7]. Traditional algorithms and strategies can help to mitigate the specific negative influence from visible light to its communication performance [8,9]. However, these algorithms cannot offset the performance difference between VLC applications and their existing counterparts. Thankfully, artificial intelligence (AI) has become a critical component of the 6G network [10]. It is expected to be the optimal solution for enabling visible light communication.

Machine learning (ML) has emerged to be the most popular technique for prediction, classification, and pattern identification, and has shown great success in data mining, image recognition, and other areas in the last decade. The recent development of AI processing units further accelerates the advancement of the more powerful deep neural networks (DNN). Many machine learning techniques have been successfully implemented in the fields of optical communication [11] and wireless communication [12]. However, the machine learning algorithm also has their own set of drawbacks, such as high computational complexity, long training times, and poor generalization. In the more complicated visible light communications, these issues will be amplified. In the more complicated visible light communications, these issues will be amplified. Therefore, machine learning should be wisely adopted to the visible light communication scenario, in the case that it may not be a viable solution.

Nowadays, wireless networks have progressed from software-defined radio (SDR) and cognitive radio (CR) [13] to AI-powered intelligent radio (IR) [10]. Visible light communication, as a communication method sprouting from 6G, aims to skip the first two stages and go directly to the IR stage. To accomplish this leap, we need to build the framework of intelligent visible light communication (IVLC). IVLC will be a broad concept covering both the intelligent physical layer and the intelligent network layer (including the traditional data link layer and network layer). As we have seen, 6G is still in its early stages of development, and 6G-based IVLC is in an even more preliminary stage. Therefore, the intelligent physical layer, which is more different from traditional wireless, could be the core breakthrough point in forthcoming years.

In this paper, we will introduce the concepts of the intelligent physical layer and the intelligent network layer of IVLC. The underlying physical layer will be given great consideration. Among the existing machine learning algorithms, there are four main categories according to the purpose of implementation: regression [14], classification [15], clustering [16], and dimensionality reduction [17]. However, in IVLC, especially in the physical layer, the existing machine learning algorithm is not designed to achieve the above functions. For this reason, we redefine the categories of machine learning techniques in the physical layer of the IVLC based on the communication system framework, including optimal coding, channel emulator, MIMO, channel equalization, and optimal decision. As seen in Section 3, such categorization intersects with the traditional ML applications, which facilitates readers who are interested in investigating intelligent visible light communication. Each module of the communication framework is featured with unique issues and thus requires specially-designed machine learning algorithms. We will go through the major obstacles of visible light communication and discuss how AI-empowered IVLC could overcome them. It is possible that the newly emerging intelligent visible light communication may play a key role in the 6G communication network, enabling worldwide smart connectivity and the construction of air–space–ground–sea integrated networks.

2. Statues and Challenges of VLC

Visible light communication introduces special nonlinearities due to the additional electro-optical conversion, which can significantly impair communication performance. Due to the spontaneous radiation properties of LEDs, visible light signals can only be directly modulated in communication. This means that changes in signal amplitude will directly affect the carrier concentration, and thus the recombination of electrons and holes [18]. In this section, we present the complexity of visible channels in terms of physical channels and modulation formats, and afterward, show the superiority of machine learning and the attendant costs.

2.1. Visible Light Communication E2E Channel

The VLC end-to-end channel $\mathcal{H}(x)$ includes a digital-to-analog converter (DAC, in arbitrary waveform generator), electronic amplifier (EA), bias tee, LED, transmission channel, receiver, and analog-to-digital converter (ADC, in oscilloscope), as shown in Figure 1. The entire transmission link contains electrical voltage signals, current signals, and optical signals, as well as the conversion between them. However, due to the complexity of the visible light channel, research now focuses on the LED emitter and the transmission channel, which are (d), (e), and (f) of Figure 1.

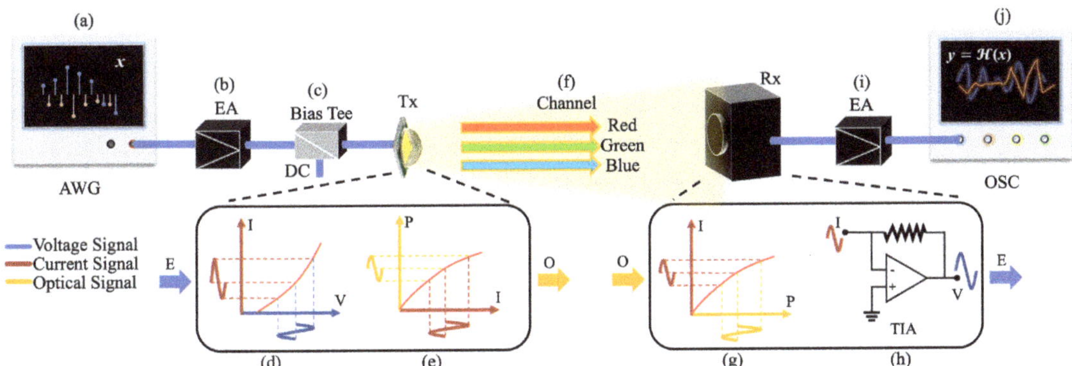

Figure 1. Overview of VLC end-to-end channel; (**a**) DAC, (**b**) EA, (**c**) bias tee, (**d**) V-I, (**e**) I-P, (**f**) transmission channel, (**g**) P-I, (**h**) transimpedance amplifier (TIA), (**i**) EA and (**j**) ADC.

From the communication point of view, the most primitive LED transmission model is the frequency domain model, which is given as [19]:

$$H(\omega) = e^{-\frac{\omega}{\omega_c}} \tag{1}$$

where ω_c is a fitted coefficient. This model mainly represents not only the frequency-selective fading phenomenon of the visible light channel, but also the inter-symbol interference (ISI) and linear memory effects. The high-frequency fading of the exp (exponential) fits well with the limited bandwidth of visible light communication. Therefore, when only linear channels are considered, the expression form of exp is also applicable to the underwater visible light channel [20].

However, a linear model alone cannot describe a channel that is as complex as visible light communication. The first consideration is the V-I transfer model of VLC [21], which extends from the solid state power amplifier (SSPA) model [22]. Similarly, there are equivalent circuit models [23] to equate the V-I transfer curve and frequency response of VLC. Such a model implies that the nonlinear term is only amplitude-dependent, independent of frequency and time, and independent of adjacent symbols.

If we want to take all the above factors into account, a simple way is to equate the overall linearity and nonlinearity to the Volterra series [24,25], as a black-box model. A second-order Volterra expansion can be expressed as [25]:

$$y(m) = \sum_{i=0}^{N-1} h(i)x(m-i) + \sum_{i=0}^{L-1}\sum_{j=k}^{L-1} w(ij)x(m-i)(m-j) \tag{2}$$

where h and w are the linear and nonlinear weights, and N and L are the tap numbers of linearity and nonlinearity. When the signal bandwidth is small (14 MHz bandwidth) and the signal amplitude is small (40 mA DC bias), the second-order Volterra is proven to have better similarity [24]. High-speed visible light communication, however, often requires a higher signal-to-noise ratio (SNR) and modulation bandwidth to meet the 6G transmission rate requirements. Higher-order Volterra series may be expected to fit well, but would introduce an exponential increase in computational complexity.

None of the above-mentioned channel models from the communication dimension actually take the substance of photoelectric and electro-optical conversion into account in computation. They simply treat it as a black box. Since LED is a GaN wide-bandgap semiconductor material based on a multi-quantum wells (QWs) structure, the carrier rate equation can be modeled for LED from semiconductor material considerations [26–29]. The model is named as the ABC model, where A, B, and C represent Shockley–Read–Hall (SRH, nonradiative) recombination, radiative recombination, and Auger recombination (nonradiative), respectively. Considering both recombination and leakage of carriers, the recombination rate R can be expressed as follows [28]:

$$R = An + Bn^2 + Cn^3 + f(n) \tag{3}$$

$$f(n) \approx an + bn^2 + cn^3 + dn^4 \ldots \tag{4}$$

where A, B, and C represent the SRH, radiative, and Auger recombination coefficient. n is the minority carrier concentration. $f(n)$ is the carrier leakage term, which has been expanded into the Taylor series. The carrier lifetime τ, which determines the modulation bandwidth, is given as [18]:

$$\tau = \frac{n}{R} \approx \frac{n}{An + Bn^2 + Cn^3 + an + bn^2 + cn^3 + dn^4 \ldots} \tag{5}$$

The carrier density is determined by the effective injected current density, which is expressed as [27]:

$$\frac{\Delta n}{\Delta R} = -R + \frac{\eta_{inj}J}{q_e} \tag{6}$$

$$J = \frac{I}{w_{active}} \tag{7}$$

where η_{inj} is the injection efficiency, J is the current density, q_e is the elementary charge, I is the current intensity, and w_{active} is the thickness of the effective active region. Then, the optical output power is given by [18]:

$$P = V_{active}E_{phot}\eta_{EQE}Bn^2 \tag{8}$$

here, V_{active} is the volume of the active layer, E_{phot} is the photon energy and η_{EQE} is the external quantum efficiency (EQE).

As can be seen here, since the initial transmission signal is in the form of a voltage; it first goes through a nonlinear V-I conversion [21]. Then, the relationship between current and carriers is a dynamic nonlinear relationship. Moreover, at higher currents, there will be an efficiency droop [28]. It is also easy to understand that the effective radiative carriers have only second-order terms and the total number of carriers has higher-order terms.

When the current increases, the number of carriers increases, and the effective light-emitting carrier ratio P_{ratio} will first increase and then decrease.

$$P_{ratio} = \frac{Bn^2}{An + Bn^2 + Cn^3 + an + bn^2 + cn^3 + dn^4 \dots} \tag{9}$$

Because of the aforesaid dynamic nonlinear equations, all of the carrier-related publications mentioned above use a tiny signal and low bandwidth assumption in their derivation. The carrier density is minimal enough to ignore the higher-order terms due to the low current density caused by the tiny signal. If the signal has a limited bandwidth, the pulse duration is sufficient to make the left side of Equation (6) equal to zero. Abandoning the higher-order and differential terms would greatly simplify the modeling of visible light channels, but it also poses the problem of not being able to satisfy the modeling of high-speed VLC. In later work based on both the carrier rate equation and an equivalent discrete-time circuit modeling [7,30], the same assumptions were required, and the experimental verification of the channel modeling was performed with only 2 MHz [30].

In addition to the large signals and high bandwidth that make visible light channel modeling difficult [19,21], another challenge is transmission channel modeling [31–33]. Much work has focused on indoor visible light transmission modeling, such as multipath impulse-response analysis [31], ray-tracing methods based on channel impulse response [32], and photon-based statistical modeling [33]. However, this is only one aspect of visible light communication applications. When transmitting in an outdoor environment, the effects of atmospheric turbulence in visible wavelengths must also be modeled. Water environment modeling is also essential while broadcasting underwater.

Furthermore, as illustrated in Figure 1, numerous modules introduce nonlinearities. For example, there are high-power electrical amplifiers at the transmitter side, which can have large nonlinearities at high currents. The linear dynamic range of the receiver PIN is usually smaller than the LED, and too much optical power can cause saturation of the PIN, which is serious especially when using APD detectors. Therefore, the nonlinear modeling of the driver circuit and the receiver based on different detector implementations are also very important.

Because of the additional optoelectronic and electro-optical conversion, as well as the rest of the nonlinear modules, the VLC end-to-end channel is extremely complex, as summarized in Table 1. This also presents a significant problem for VLC's high-speed connectivity.

Table 1. Challenges of visible light communication E2E channel.

Challenges	Reasons	References
Optoelectronic and electro-optical conversion	Introduces additional nonlinearity	[26–29]
Large signals	Brings the device into the nonlinear region	[21]
Wide bandwidth	Introduces severe ISI	[19]
Different transmission channel modeling	Diverse application scenarios, such as indoor, underwater	[31–33]

2.2. Modulation Format in VLC

Because of the explained limitations in commercially available LED light sources, LED-based VLC system typically presents extremely limited bandwidth (several MHz). Apart from developing LEDs with novel structure and the optimization of the driving circuits, using advanced modulation formats is also an alternative for high-speed VLC systems. In this section, we will introduce several common modulation technologies in the VLC system.

On-off keying (OOK) as the most basic modulation format in a communication system, uses the "on" and "off" state of the carrier to transmit the binary information "1" and "0". The advantages of OOK modulation are simple implementation and low cost. In

2001, early research on LED-based VLC system applies an OOK non-return to zero (NRZ) modulation [34]. With the development of equalization technology, a 662-Mbit/s VLC link based on a single blue LED using OOK-NRZ modulation has been demonstrated [35].

Pulse amplitude modulation (PAM) is a one-dimensional (1-D) multilevel modulation. Compared with OOK, the spectral efficiency of PAM is less restricted. In [36], based on Volterra decision feedback equalization (DFE), a 1.1-Gbit/s white LED-based VLC system is experimentally demonstrated. Investigation on the comparison of the performance of PAM with different orders has also been reported in [37]. Experimental results indicate that through a three-tap pre-equalizer, a data rate of 2 Gbit/s is achieved.

For the VLC system, there is strong noise at the low-frequency components. Although this noisy spectrum can be avoided through up-conversion, the in-phase (I) and quadrature (Q) channels are not fully utilized if using 1-D modulation such as OOK or PAM. Carrierless amplitude-phase modulation (CAP) as a variant of QAM can not only avoid the low-frequency noise but also demonstrate a fuller utilization of the I and Q channels. It uses a pair of orthogonal Hilbert filters for up-conversion instead of subcarrier. CAP has been widely used in VLC systems due to its merits of low complexity and high spectral efficiency. The early demonstration of CAP modulation in VLC systems has been reported in 2012, in which a 1.1-Gbit/s 23-cm free space transmission is realized [38]. Multiband CAP has also been proposed for multi-user application; through flexible bit allocation, a VLC system with the spectral efficiency of 4.85 bit/s/Hz is demonstrated [39].

When using the above modulation technologies, equalizers with several taps are required to mitigate the ISI because of the bandwidth limitation effect in VLC system. If there is strong ISI, the taps of the equalizer will increase rapidly. Alternatively, multicarrier modulation technologies such as orthogonal frequency division multiplexing (OFDM), discrete multitone (DMT), and discrete Fourier transform spread (DFTS)-OFDM are possible to avoid ISI.

The OFDM signal is generated as follows: First, the transmitting sequence in the frequency domain is divided into parallel subchannels. Then, the time-domain symbols in a slot are the inverse fast Fourier transform (IFFT) of the frequency-domain symbols from each subcarrier. After adding CP and parallel to serial conversion, a complex-valued OFDM signal is generated. However, the transmitting signal is restricted to real value in VLC system. Therefore, an extra up-conversion is required for complex-to-real conversion. In [40], a 3-Gbit/s OFDM VLC system based on bit loading and power loading is demonstrated, indicating that OFDM has great potential of combining with adaptive bit- and power-allocation algorithms.

DMT is similar to OFDM, except that it uses Hermitian symmetry before IFFT, so that the signal after IFFT is real-valued. The step of up-conversion is not needed. In [41], using the maximum ratio combination a 2.3-Gbit/s underwater DMT VLC system is realized. Additionally, adaptive bit- and power-allocation algorithms can also be applied for DMT. It is reported that using a bit-loading and power-loading scheme, the data rate of the underwater VLC system based on Si-substrate LED has achieved 3.37 Gbit/s.

Although OFDM and DMT offer desirable resistance to ISI and flexible bit and power allocation, they are faced with a high peak-to-average power ratio (PAPR). DFTS-OFDM is proposed to mitigate the problem. The difference is that DFTS-OFDM employs an extra FFT operation between the serial-to-parallel conversion and IFFT. In [42], the authors have proven that by employing DFTS-OFDM the PAPR can be significantly reduced.

The complementary cumulative distribution function (CCDF) of the transmitted signal with different modulation formats is illustrated in Figure 2. The results indicate that the PAPR of OOK-NRZ is the lowest, followed by PAM 4. while the PAPR of CAP and PAPR of DFTS-OFDM are higher but exhibit similar performance. Obviously, OFDM has the highest PAPR. As a result, signals using different modulation may experience different nonlinear channel responses, which further aggravates the complexity.

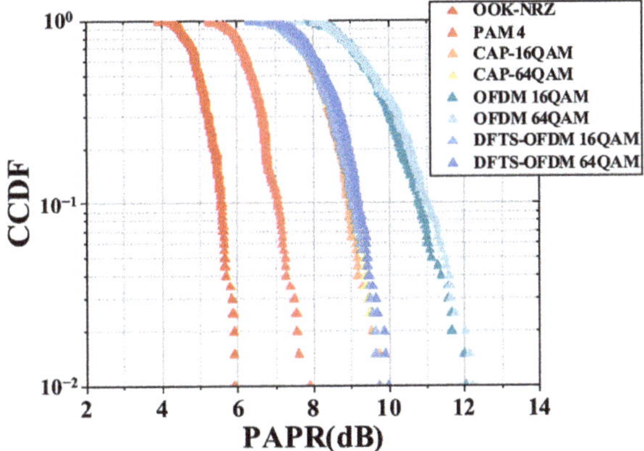

Figure 2. PAPR of different modulation formats in VLC.

2.3. Advantages and Disadvantages of ML in VLC

To show the performance benefits and costs of machine learning in visible light communication, we simply construct a visible light simulation model based on [19,25,30], as shown in Figure 3. The second-order Volterra series is used to replace the one single tap time-discretization, which represents the memory rate equation. The conversion curve of voltage, current, and optical power is used to represent the memoryless optical transform. *exp* is used as the overall channel frequency response. It should be emphasized that this is only a simplified simulation model; some parameters are determined by some communication experimental data. This simulation channel is not the focus of this article, but it is enough to illustrate the characteristics of machine learning.

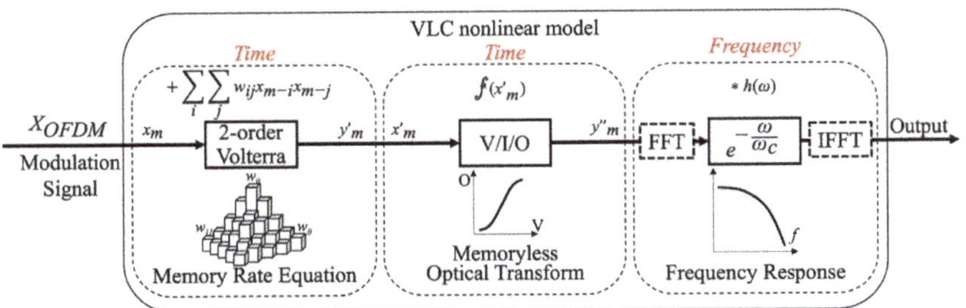

Figure 3. Block diagram of VLC simulation model.

In order to visualize the performance of machine learning, we chose to compare it in the field of channel equalization. Figure 4 shows the performance difference between no channel equalization, traditional nonlinear equalization, and machine learning. We used the least mean square (LMS)-based second-order Volterra algorithm as a representative of the traditional nonlinear equalizer. Both linear and nonlinear taps were set to 31. A one-layer hidden-layer multilayer perceptron (MLP) was used as a representative of machine learning. The size of the input layer was 31 and the size of the hidden layer was 128. As demonstrated in the figure, both nonlinear algorithms can have a good performance improvement. MLP outperforms the Volterra algorithm at various SNR. However, at low SNR, the enhancement is not as much, which is because this is an additive noise-limited system at this point. At high SNR, the MLP's ability to compensate for nonlinearities is even more evident.

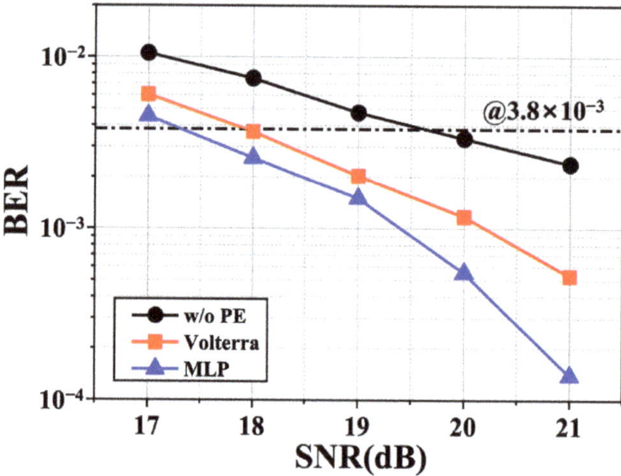

Figure 4. BER performance versus SNR.

As indicated by the above results, machine learning does improve performance. However, we also need to consider its drawbacks. The first is the computational complexity, as shown in Table 2. In this simulation, the trainable parameters of MLP reach 4225, while it is 527 for the Volterra algorithm. MLP has a much higher computational complexity than Volterra, which will be more obvious in actual systems where nonlinearities are more severe. This also means that machine learning algorithms need more convergence time.

Table 2. Computational complexity of MLP and Volterra.

Algorithm	Input Layer	1st Weight Layer	2nd Weight Layer	Drd Weight Layer	Trainable Parameters
MLP (general)	N	W_1	W_2	W_D	$(N+1) \times W_1 + \sum_{i=2}^{D} (W_{i-1}+1) \times W_i$
Volterra (general)	$N_{lin}, N_{non\text{-}lin}$	/	/	/	$N_{lin} + \frac{N_{non\text{-}lin} \times (N_{non\text{-}lin}+1)}{2}$

Another problem with machine learning is generalizability. Figure 5 shows some results of generalizability studies on the Volterra algorithm and MLP. It can be seen that different training data lengths affect the performance of the equalizer. If the trained model is used for other random seed-generated data, the performance is degraded by a specific degree. This degree of degradation is relatively less in the Volterra algorithm. As mentioned above, there are many different modulation formats in visible light communication. For this reason, we also tried to use the model trained based on OFDM signals for CAP signal recovery. In this respect, one can see the more serious generalizability problems of MLP. While machine learning has better performance, data-driven learning can cause it to learn features that do not belong to the channel, for example, the data stream itself. The issues mentioned above can significantly slow down the application and development of machine learning in intelligent visible light communication. Much work should be carried out to address these aspects.

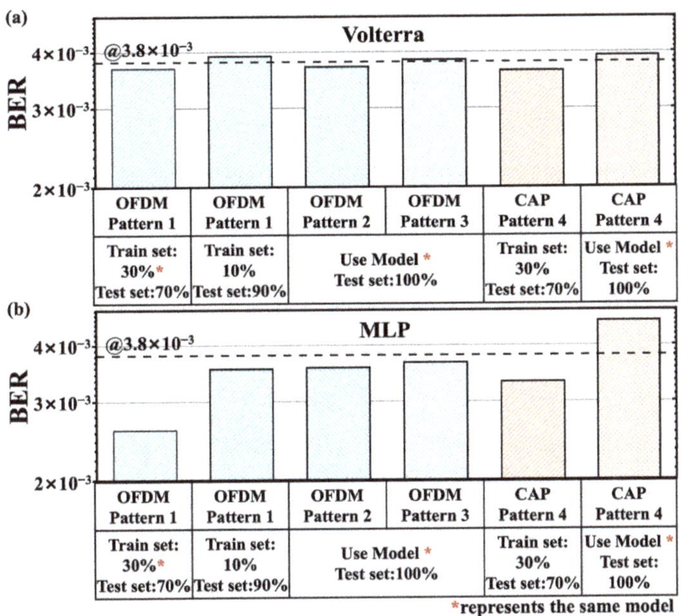

Figure 5. Generalizability of (**a**) Volterra and (**b**) MLP for different cases; * represents the same model.

3. Machine Learning in Physical Layer of IVLC

In this section, we will present some applications of machine learning in the physical layer of intelligent visible light communications, such as channel emulator, channel equalization, optimal decision, MIMO, and optimal coding, as shown in Figure 6.

3.1. Channel Emulator

As discussed in Section 2, the end-to-end channel for visible optical communication is exceptionally complex. In the transmission model, for example, in atmospheric environments, gas molecules and aerosol particles in the atmosphere absorb and scatter light radiation in the near-infrared band, resulting in a loss of signal received power. In addition, the change of atmospheric turbulence causes severe distortion to the optical signals. For another example, in the underwater environment, the attenuation of underwater light depends on the wavelength, where the attenuation of the signal increases with frequency. Moreover, there are other propagation effects such as temperature fluctuations, salinity, scattering, dispersion, and beam steering. For underwater VLC applications whose bandwidth is not too high (tens of MHz), the power attenuation with frequency can be approximately modeled as a linear relationship, allowing the modeling of underwater VLC multipath channels using compressive sensing (CS) method [43]. Traditional methods for high-speed point-to-point VLC cannot support accurate VLC end-to-end channel modeling, but machine learning is able to simulate the complicated nonlinear dynamics of VLC channels [44]. In massive multiple-input multiple-output (m-MIMO) VLC, the machine learning-based methods enable accurate estimation of the channel matrix [45].

3.1.1. TTHNet

Conducting an experimental transmission test in an underwater environment is costly, but there is no accurate analytic model as a reference for underwater high-speed VLC. In order to reduce the cost of testing underwater VLC systems, a machine learning method is needed to model the underwater channel. The two-tributaries heterogeneous neural network (TTHnet) uses a convolutional neural network (CNN) for modeling the linearity of the underwater VLC channel and a two-layer MLP with a hollow layer for modeling the nonlinearity of the underwater VLC channel [44]. The two-branch heterogeneous structure makes full use of the CNN's shared parameters, thus reducing the system complexity. At the same time, it utilizes the MLP's extremely strong nonlinear fitting capability to fit the nonlinearity in the channel. Experiments show that the channel modeled by TTHnet is extremely close to the real channel, and the average spectrum mismatch is only 36.2% of the MLP-based channel emulator and 44.3% of the CNN-based channel emulator.

3.1.2. FFDNet

Since the modulation bandwidth of a single LED is limited, the use of m-MIMO LED and PD arrays are expected to substantially increase the capacity and transmission rate of VLC systems. However, due to the complexity of VLC channels, it is extremely difficult to estimate the m-MIMO channel matrix, which requires deep learning methods. Fast and flexible denoising convolutional neural network (FFDnet) is used for channel estimation in millimeter-wave communication recently [46,47], which is also applicable in VLC [45]. As an image denoising tool using machine learning, FFDnet is able to recover the input noisy channel matrix into an almost noiseless channel matrix. Compared with the minimum mean square error (MMSE) method, the FFDnet has a stronger denoising effect, which can increase the peak signal-to-noise ratio (PSNR) of the recovered channel matrix image. Unlike the nonlinear channel modeling in point-to-point high-speed VLC links, the channel matrix is treated as an image and processed using machine learning methods of image processing, which is of great importance in channel estimation of m-MIMO-VLC channels.

3.1.3. Conclusions

The channel capacity determines the upper bound of the communication system rate, and therefore, the accuracy of the channel estimation determines the communication efficiency of the actual system. Complex VLC channels should be accurately predicted thanks to the widespread use of powerful ML techniques in channel estimation. ML algorithms will guide IVLC to break through its own bottlenecks and complete the comprehensive integration of high-speed communication and large-scale heterogeneous networking to achieve technical solutions for next-generation communication.

3.2. Channel Equalization

Channel equalization techniques generally estimate the transfer function of communication channels and try to remove the channel distortion by an adaptive filter [48]. However, the common equalizers with linear adaptive algorithms become powerless in the field of high-speed VLC, because of the intrinsically limited modulation bandwidth of LEDs [49] and nonlinear distortion introduced by photoelectric devices and VLC channels. Recently, ML-based equalizers, such as artificial neural networks (ANN) [50], etc., have been developed for VLC systems. ML-based equalizers have shown outstanding equalizing performance, especially on modeling nonlinear phenomena, by adopting neural-network-based algorithms. Despite this, challenges such as massive computational complexity, slow convergence speed, and relatively poor generalization still prevent the further practical application of ML-based equalizers for VLC systems. Therefore, researchers have developed many variants, as presented next, to overcome those challenges.

3.2.1. Pre-Equalization GK-DNN

Conventionally, one would replace postequalization with pre-equalization to reduce the computational complexity and power consumption at the receiver side. Research works such as a weighted lookup table (WLUT), etc., have been proposed to mitigate the nonlinear distortion in VLC systems [51]. However, LUT-based pre-equalization methods suffer from a massive increase in computational complexity when dealing with high-order and high-ISI communication scenarios. Therefore, researchers have come up with ML-based pre-equalization methods in the field of VLC systems to provide a new way of solving computational problems of LUTs.

In [52], a pre-equalization method, namely Gaussian kernel-aided deep neural network pre-distortion (GK-DNN-PD), is proposed for a high-order modulated high-speed VLC system. GK-DNN-PD outperforms the LUT-PD in terms of memory depth (MD) and the required training dataset, which leads to lower computational complexity. The experimental results show a 1.56 dB Q-factor gain compared with LUT-PD.

The proposed GK-DNN-PD method consists of two phases: the training phase and the communication testing phase. In the training phase, the received signal, which is not pre-distorted, will be linearly equalized, giving us the label sets of the GK-DNN channel estimator. Then, the clean transmitted signal with certain MD would be the feature sets. Then, the GK-DNN channel estimator will be trained to obtain the weight and bias of the estimator. Next in the communication testing phase, the weight and bias obtained in the first phase would be used to pre-distort the clean signal that is to be transmitted. Specifically, the difference between the clean signal and the output of the GK-DNN channel estimator is also considered, in addition to the weight and bias during the pre-distortion progress. Additionally, clipping operation is also adopted to reduce the peak to PAPR, which consequently reduces the nonlinear degradation.

Moreover, an NN-based pre-equalizer is proposed in [53] to mitigate the semiconductor optical amplifier (SOA) pattern effect for 50G PON, confirming the feasibility of NN-based pre-equalizer in intensity modulation and direct detection (IM/DD) system.

3.2.2. Postequalization GK-DNN

Since the conventional nonlinear postequalization methods based on the Volterra series suffer from a massive increase in computational complexity when dealing with high-order nonlinearity, researchers have turned to the ML for new inspirations. However, the time-consuming training progress of most ML-based postequalizers limits its actual application. To accelerate the training processing and greatly relieve the computational complexity of the equalizer at the receiver side, researchers have proposed the Gaussian kernel-aided deep neural network (GK-DNN) [54] in the field of VLC systems.

Compared to the classical MLP, the major unique feature of GK-DNN is that the input data would go through a functional mapping that is based on Gaussian function, namely the Gaussian kernel, which maps the windowed input data to a nonlinear space to reduce the number of iterations and time consumption of the fitting progress. The researchers believe that the adjacent symbols' influence towards the central (or current) one is in accordance with Gaussian distribution, hence the mapping operation would accelerate the training processing. The expression of the Gaussian kernel is given in [54]. It should be noted that the scope-controlling parameter of the Gaussian kernel would greatly affect the equalization performance of GK-DNN. Generally, the larger the parameter is, the faster the training process would be. However, there is a trade-off between the training process acceleration and equalization performance. Therefore, the Gaussian kernel parameter selection is vital to obtain the best performance. Moreover, the selection of the number of hidden layer nodes is equivalently important, which directly decides the computational complexity of the equalizer. According to the experimental results in [54], the GK-DNN equalizer could efficiently realize the postequalization in the VLC system with the aid of Gaussian kernel, which reduces the iteration epochs of the neural network by 47.06%.

3.2.3. Postequalization FSDNN

The frequency-slicing deep neural network (FSDNN) is a variant application of DNN that could be used in a high-speed VLC system [55]. It has the characteristics of processing high and low frequency respectively to decrease computation complexity by 11.15% compared to the traditional MLP when it comes to the equalization performance in VLC system.

In order to solve the nonlinear frequency spectrum fading issue of the received signal after going through the VLC channel, DNN is introduced as an outstanding postequalizer to equalize linear and nonlinear distortion. However, the DNN structure must be complex enough, which means that more layers and nodes are needed and computation complexity improves to handle complicated linear and nonlinear distortions. For the expectation to release the pressure of DNN, it is worth noticing that high and low domain frequency suffer different degrees of fading. The high-frequency spectrum suffers more serious amplitude attenuation, while the low-frequency spectrum suffers less fading in the received signal in VLC system, so complex MLP structure is unnecessary for the low-frequency domain. Therefore, the received signal can be separated into high-frequency and low-frequency domains and processed, respectively, using a DNN equalizer with different complexity.

The received wide-band signal is split into two narrow-band parts in the frequency domain. Its frequency spectrum is separated into two sub-bands using a low-pass filter and a high-pass filter. Then, the two sub-band signals are respectively fed into two MLPs to train individually. The main factors of the two-MLP network should be tested artificially and adjusted to optimal values, including the number of layers, nodes in every layer, taps, and epochs. Once the MLP is finished training and the weight values are fixed, the sum of the output signal from two MLPs is the equalized and recovered signals.

3.2.4. Postequalization TFDNet

The commonly used ML-based equalizers in VLC systems often aim at fitting the waveform of the transmitting signal, which is a time-domain-serial signal. It is expected that the well-learned received signal should have the same spectrum as the transmitted one. However, waveform-fitting ML equalizers would sometimes cause the spectrum difference between the equalized signal and the original one. This suggests that we should take both time- and frequency-domain information into consideration to obtain a better equalization performance.

A novel postequalizer, namely joint time-frequency deep neural network (TFDNet), is reported in [56] to compensate for the nonlinear distortions in the VLC system. TFDNet could reveal comprehensive information of nonstationary signals received in the VLC system by considering both time and frequency domain information simultaneously. TFDNet can be divided into three main procedures: (1) the received one-dimensional (1D, time domain) signal goes through a short-time Fourier transformation (STFT) operation and would be transferred into a two-dimensional (2D, time-frequency domain) signal, which is a matrix and could be denoted as Y; (2) then, the obtained STFT matrix Y is fed into the NN to be trained. The labels could always be obtained by manipulating the original transmitting signal. If we assume that each row of Y represents a certain frequency component, then Y would be fed into the following network column by column; (3) finally, after the NN finishes the training progress, the reconstructed transmitting signal could be obtained by carrying out the inverse STFT (ISTFT) operation, where the analysis window must satisfy the COLA constraint [57]. Experimental results in [56] also confirm that the proposed TFDNet could resist severe nonlinear distortions and achieve a 0.1 Gbps and 0.2 Gbps data rate gain for VLC system compared to other nonlinear compensators such as Volterra and DNN.

3.2.5. Postequalization DBMLP

To further improve the utility of NN equalizers, researchers had proposed a modified double-branch multilayer perceptron (DBMLP) postequilibrium algorithm [44] to further reduce the consumption of energy and computational resources. DBMLP reconstructed the MLP postequalization algorithm using the structure of the Volterra series postequalization algorithm as a template. DBMLP combines the advantages of linear adaptive filters and MLP, which can improve the BER performance of the algorithm while reducing the complexity of the algorithm by 74.1%. The core structure of DBMLP is two branches of linear and nonlinear ones. In the DBMLP structure, a CNN with a convolutional layer and a dense-layer structure to simulate the linear distortion in the signal bandwidth is the first branch. In addition, a hollow MLP with an airlift layer and two dense-layer structures to simulate the nonlinear distortion outside the signal bandwidth is the second branch. The nonlinearity of the output of the first branch is corrected by the output of the second branch, and the hollow layer can ignore the effect of the intermediate signal on the signals on both sides.

To further reduce power consumption and complexity, a pruning algorithm based on DBMLP is proposed [58]. The algorithm performs the operation of pruning by setting the smaller absolute value of weights of the connections to be pruned to 0 based on sparsity. The weights of the linear branch are not prunable while the nonlinear ones are prunable. The experimental results confirm the superiority of this approach.

3.2.6. Post-Equalization PCVNN

To improve the SNR in Underwater Visible Light Communication (UVLC) system, high LED power must be encouraged due to the LED's incoherent characteristic and the water medium's considerable attenuation. The nonlinearity grows more severe as the signal amplitude increases. Consequently, symbols on the outside of the constellation sustain a more nonlinear distortion than those on the inside. Based on complex-valued neural network (CVNN) [59], an adaptive partition equalizer (PCVNN) [60] has been presented, which reduces the complexity and has superior performance.

In PCVNN, the constellation is segmented into two areas by a proper threshold to distinguish between large-amplitude signals and small-amplitude signals. Then, the large- and small-amplitude signals are fed into two complex-valued neural networks. Finally, a fully connected neural network is then used to combine the signals into a complete one. Since large and small signals experience different nonlinear impairments, such a network structure can recover the signal more accurately and can greatly reduce the complexity of the model for small signals. The final experimental results also verified this conjecture [60]. PCVNN achieves up to 56.1% computational complexity reduction compared with the standard CVNN at the same performance.

3.2.7. Postequalization LSTM-Equalizer

High-speed VLC is limited by inherent nonlinear effects. Linear equalizers with limited taps seem powerless, and the Volterra series schemes suffer from high computational complexity when the high-order taps are required. With the rise of ML in solving nonlinear problems, long short-term memory (LSTM) networks are studied for VLC systems.

In [61], researchers proposed a memory-controlled LSTM NN equalizer for both linear and nonlinear compensation, which outperforms the conventional Volterra-based and FIR-based equalizers. LSTM carries out channel equalization as a pattern classifier where the output of LSTM cells is activated by a specially designed function. Training data with high priority would be assigned by LSTM to the latest training sequence. The proposed LSTM equalizer in [61] contains an input layer, a logical hidden layer with long and short-term memory, a classification layer, and an output layer with a merge node. A standard LSTM cell structure is used for long/short-term memory links. Moreover, a batch random resequencing procedure is adopted to control the memory effect.

Recently, the variants of LSTM have also drawn the attention of researchers because the simple LSTMs have a slow convergence speed. This is because the LSTM unit's inner parameters prolong the training period. A convolution-enhanced LSTM (CE-LSTM) equalizer, which extracts the features by using a convolutional layer, is proposed in [62] to shrink the complexity of the LSTM network and speed up the convergence progress. The experimental results also confirmed the feasibility of the proposed CE-LSTM equalizer.

3.2.8. Postequalization MPANN

Although the ML-based equalizers for mitigating both the linear and nonlinear distortions in VLC systems have been booming recently, the computational complexity is still a problem that needs to be further solved. Therefore, an ML-based equalizer with relatively optimal equalization performance while still maintaining a low complexity is needed in the field of VLC. One promising way is to greatly relieve the equalizer's complexity by moderately sacrificing partial performance.

Researchers have developed a simplified ML-based equalizer, namely the memory-polynomial artificial neural network (MPANN) [63], to prune the network structure and still maintain similar equalization performance as MLP or other NNs. Likewise, the input data to be fed into MPANN could be obtained by windowing the received time-serial signal. The length of the window is usually called the memory length, which also represents the dimensions of the features. The major characteristic of MPANN is that its input layer, namely the memory-polynomial layer (MP layer), would expand the input features by one certain function, which is memory polynomial expansion. In addition, the Gaussian, Fourier basis, and other trigonometric polynomials (e.g., Legendre, Chebyshev, etc.) could be the function in the input layer. It is believed that the demanded nodes of the modified NN structure could be significantly decreased if one could provide a prior knowledge of the nonlinear model. Therefore, the memory polynomial expansion is adopted to map the input features to higher dimensional data space. Then the output pattern of the MP layer is multiplied by the corresponding weights and fed into the following hidden layer of the NN. A regular activating (ReLU) and weighting process are conducted in the hidden layer and back propagation (BP) algorithm is utilized to update the parameters. Then, finally, the output layer is utilized to output the equalized symbol. The experimental results confirmed that the MPANN could achieve the same equalization performance as the regular MLPs and only requires less than a quarter of the complexity [63].

3.2.9. Conclusions

As can be seen from the above presentation, the application of neural networks in channel equalization has become more than a simple application. The integration of neural networks with communication systems is starting to emerge. Figure 7 illustrates the existing neural network channel equalization in VLC. Different branches of neural networks are beginning to emerge, and many more choose to extract communication-specific features from the input data. Beyond that, fast development of computational power resources make it promising to implement ML-based modules in the field of VLC. ML-based methods with powerful nonlinear phenomenon modeling ability open a new gate to solving the inherent nonlinear problems in VLC system. However, further optimization and improvement would be needed for those ML-based equalizers in terms of computational complexity, convergence speed, and generalization. Table 3 compares the equalizers mentioned above.

Table 3. Summarization of machine learning algorithms for channel equalization.

Equalizers	GK-DNN	FSDNN	TFDNet	MPANN	DBMLP	PCVNN	LSTM
Main types of NN	MLP	MLP	MLP	MLP	MLP	MLP	RNN
Number of hidden layers	2	1	1	1	1	1	1
Activation function	ReLU	ReLU	ReLU	ReLU	Tanh	ReLU	Tanh, Sigmoid
Optimizer	Adagrad	Adam	Adam	Adam	Adam	Adam	Adam
Complexity	Moderate	Low	High	Low	High	Low	High
Convergence speed	Fast	Moderate	Moderate	Moderate	Slow	Slow	Slow
Pre-equ.	✓						
Post-equ.	✓	✓	✓	✓	✓	✓	✓
Deployment location	Waveform	Waveform	Waveform	Waveform	Waveform	Symbol	Symbol

3.3. Optimal Decision

After channel equalization, a constellation decision is required to recover the original data. The most common decision scheme is based on the Euclidean distance between the received symbols and the standard constellation points, because the decision scheme is supposed to have the best performance in the additive Gaussian white noise (AWGN) channel. However, as mentioned in Section 2, the VLC channel is not a simple AWGN channel, but has nonuniform noise distribution and nonlinear effect. As a result, the distortion of the constellation diagram may not exhibit a uniform Gaussian distribution around the standard constellation points. On the contrary, the constellation clusters may exhibit deviation or exhibit distortion highly relevant to the signal power. Apparently, the conventional decision is not the optimal decision scheme, and novel algorithms that take the statistical characteristics into account are required. Machine-learning-based decision schemes have been widely investigated in visible light communication systems and great improvement has been reported. These schemes can be divided into two categories: classification and clustering.

3.3.1. K-Means

K-means is a common unsupervised ML algorithm, which is used for spherical clustering. The main idea of K-means is to update the centroids of the constellation clusters, and the class of the new input data depends on the Euclidean distance between the input data and the dynamic centroids of the constellation clusters. It works especially well when the constellation clusters exhibit overall deviation. K-means has been applied in multiband carrierless amplitude and phase (CAP) VLC systems; experimental results indicate that for each sub-band, a decision based on K-means achieves a 1.6–2.5 dB Q-factor gain compared to a conventional scheme [64]. Moreover, if the deviation is known at the transmitting end, K-means-based predistortion is also proposed [65]. However, if the constellation clusters are not spherical, the performance of K-means will be decreased.

3.3.2. DBSCAN

Apart from the power-relevant nonlinear effect, a random jitter in the time domain is also a detrimental factor in VLC systems. Decision schemes based on Euclidean distance or K-means ignore the chronological order of the received sequence, so these methods are not suitable to deal with random jitter. Meanwhile, density-based spatial clustering of applications with noise (DBSCAN), as one of the clustering unsupervised ML algorithms, can divide clusters according to density, and thus has great potential in mitigating the impairment from random jitter. In [66], a DBSCAN-based decision scheme is demonstrated in VLC systems. The received one-dimensional sequence with random jitter is converted into a two-dimensional sequence with the time-axis. The key point of applying DBSCAN in the VLC system is the normalization of amplitude and the time-axis, because it is closely relevant to the density. Experimental results prove that the sequence with random jitter can still be divided into the appropriate cluster according to the density of the two-dimensional sequence.

3.3.3. GMM

The Gaussian mixture model (GMM) refers to decomposing a complex probability density function into several Gaussian probability density functions. In brief, GMM can make use of the linear combination of several single Gaussian probability density functions, so that the model can fit a more complex probability distribution that cannot be described by a single Gaussian function. Theoretically, if GMM contains enough Gaussian probability density functions and the weight is set reasonably, the model can fit samples with an arbitrary distribution. In a low-order modulation VLC system, the clustering algorithm deals with nonlinear problems in the constellation decision of the received signal. However, in a high-order modulation VLC system, the nonlinear effect is more obvious. When it comes to strong nonlinearity, the constellation points on the outer ring may not be a regular circular distribution. In [67], GMM is used to cluster the observation vectors formed by continuous symbols to obtain the distribution relationship between continuous symbols. The traditional soft-decision or hard-decision algorithm will directly remove the correlation between symbols, leading to linear and nonlinear damage, which will result in the lack of information leading to system performance degradation. When the correlation between adjacent symbols is taken into account, the GMM system achieves 1 to 1.5dB sensitivity improvement. The more continuous symbols are considered, the more obviously performance improves.

3.3.4. SVM

The support vector machine (SVM), as one of the classical supervised ML algorithms, is usually used for classification. Through a small amount of training data, SVM can find the optimum classification plane between two clusters, and the classifier model only depends on several support vectors. If the data are not linearly separable, a kernel trick can be used for nonlinear classification. Although SVM was originally used for binary classification, multiclass SVM strategies such as one-versus-one (OVO) and one-versus-all (OVA) have been proposed. In [68], SVM-based detection is proposed and demonstrated in VLC systems. Experimental results indicate that the SVM-based scheme has a 35% increase compared with the conventional decision scheme when there is a strong nonlinear effect. In [69], a constellation decision based on SVM is investigated in integrated optical fiber and VLC systems when there is random phase rotation.

3.3.5. ANN

The artificial neural network (ANN) is an alternative ML algorithm for classification. The input of the ANN-based classifier is not limited to the in-phase (I) and quadrature (Q) components of the present point, whereas the I/Q components of the adjacent points in the time domain can also serve as features. The output of the ANN-based classifier is the estimated label. In this context, an ANN-based classifier serves as the nonlinear mapping process. In [70], ANN is used for the 8-color-shift keying (8-CSK) decision in an RGB-LED VLC system, and other ML algorithms are investigated for comparison.

3.3.6. Conclusions

Existing research has proven the feasibility of ML-based decision schemes in VLC systems. The ML-based decision schemes are summarized in Table 4. The two clustering algorithms are unsupervised, and the computational complexity is low. However, the accuracy is usually lower than supervised algorithms. In supervised algorithms, SVM has fewer computational complexity than GMM and ANN. However, the computational complexity of SVM and GMM will increase when the modulation order becomes higher. ANN is supposed to have better performance as more features can be applied for nonlinear mapping. In the future, through pruning and prior channel knowledge, the complexity of ML-based decision schemes can be further reduced.

Table 4. Summarization of machine learning algorithms for optimal decision.

Algorithms	Supervision	Computational Complexity	Application
K-means	N	Low	Low nonlinearity
DBSCAN	N	Low	Time varying
GMM	Y	High	Moderate nonlinearity, ISI
SVM	Y	Moderate	Moderate nonlinearity
ANN	Y	High	High nonlinearity

3.4. MIMO

The multiple-input multiple-output (MIMO) technique has been developing in the field of VLC recently. Imaging MIMO and nonimaging MIMO are the two main types of MIMO-VLC systems [71]. The channel matrix is diagonal and of full rank for imaging MIMO scenarios; thus, a strict alignment is required between every LED and corresponding PD. In the nonimaging MIMO scenario, the signals will leak into each other and generate interchannel interference (ICI). Hence, to separate the mixed signals, algorithms are required. Conventional methods based on successive interference cancellation (SIC) rely on the power proportionality of transmitters and require transmitting diversity occasionally. Moreover, the interference is most likely to be nonlinear in the MIMO-VLC system due to the optoelectrical devices. Therefore, researchers have turned to the lately booming ML for new inspiration to compensate for the ICI and nonlinearity simultaneously.

3.4.1. ICA

Researchers have proposed an ML-based method in [72], namely joint IQ independent component analysis (ICA), to settle spatial multiplexing problems in VLC MIMO system and enhance spectral efficiency (SE).

The model in [72] is a 2×2 MIMO-VLC system where superposed signals are generated. A 16-quadrature amplitude modulation (16-QAM) signal is transmitted at the Tx1 and a quadrature phase-shift keying (QPSK) is transmitted at the Tx2. As the power ratio of two Txs—namely the scaling factor—changes, the superposed constellation (SC) could be different. Blind source separation (BSS) using ICA could be adopted to separate and recover the two independent data streams from two Txs. ICA assumes that the subcomponents that compose the mixed observed signal are non-Gaussian and are statistically independent of each other. Moreover, the observed mixed signal is assumed to be the linear combination of the source signals. The proposed 2×2 MIMO-VLC model with different SCs just meets those assumptions mentioned above. If we assume the source signal matrix as s, the observed matrix as x, and the mixing matrix as A, where x = As, then we can obtain the recovered signal by searching for the unmixing matrix W that can linearly transform x (which is whitened) so that the estimated subcomponents are independent of each other: s = Wx. The goal of ICA is to find the unmixing matrix W which is approximately equal to the A-1. It should be noted that the mixing matrix A is of full rank. Two mixed time-domain signals in MIMO CAP-modulated system could not be separated since they share the same pulse-shaping filter pairs and consequently lost the mutual independence features at every time slot. Therefore, the joint IQ ICA method deals with the SCs at the receiver side.

3.4.2. MIMO-MBNN

MIMO-MBNN is featured in a hybrid structure that combines both two linear and one nonlinear equalizer to improve the received signal quality. It further removes the nonlinear loss using a NN network, which enables this equalizer to work within high nonlinearity. Although DNN could provide a powerful fitting function, its training consumes considerable computation compared with LMS and Volterra. Previous work [73] has shown that MIMO-MBNN outperforms SISO-DNN and SISO-LMS in operation range (2.33 times the area) and refreshed the record (2.1Gbps within 7% FEC) of communication rate in SR-MIMO (single receiver MIMO) VLC, demonstrating a '1 + 1 > 2' effect.

Using the SR-MIMO system in [73] as an example: the system faces both ISI and ICI; therefore, the equalizer has to remove them both. Firstly, the MIMO signal is arranged in fixed length to train the equalizer. The two linear branches deal with the linear ISI within the corresponding single channel. In the meantime, a nonlinear branch imports the training vector from both channels and uses an NN to fit the nonlinear function including the ISI and ICI. The NN outputs an R^2 vector; each element is specified for a single channel. Next, the output from a linear branch is mixed with the corresponding output from the nonlinear branch. The mixed result is the final output. As a supervised learning process, the output result is compared with the label. An Adam optimizer updates the weight in both linear and nonlinear branches. This combined structure successfully utilizes the strength of the linear equalizer and nonlinear NN network and avoids their weakness. Underwater VLC or long-haul optical communication systems could be benefited from this algorithm that improves their robustness against nonlinear loss and power jittering.

3.4.3. Joint Spatial and Temporal ANN Equalizer

Conventionally, MIMO decoding algorithm and compensator-like decision feedback equalizer (DFE) are adopted in MIMO-VLC receivers to compensate the spatial crosstalk and remove the ISI step by step. The proposal of the vertical Bell Labs layered space time (V-BLAST) paves the way for joint spatial and temporal equalization in MIMO systems [74].

Considering the inherent nonlinear feature of MIMO-VLC systems, researchers have proposed the joint spatial and temporal ANN equalizer in [75] for both imaging and nonimaging MIMO-VLC links. The structure of the joint spatial and temporal ANN equalizer is similar to a matrix DFE structure. The data structure being fed to ANN contains the received signal vector with a feedforward delay line and the estimated signal vector with a feedback delay line. As for the input layer of the ANN, it contains two dimensions where one is spatial and the other is a temporal dimension. The number of input nodes is slightly adjusted to the oversampling factor of the fractionally spaced equalization scheme. According to the experimental results in [75], the proposed ANN-based joint spatial and temporal equalization scheme could outperform the traditional DFE and is able to compensate for the nonlinear channel distortion and cross-talks. Additionally, the proposed method could have better performance when the channel is ill-conditioned.

3.4.4. Adaptive ANN Equalizer

Spatial complexity is a major obstacle for machine learning in VLC MIMO applications. In SR-MIMO-VLC, crosstalk between channels will lead to a significant increase in the size of the data-driven neural network [73]. The traditional MIMO-LMS can obtain the channel matrix more efficiently, although it cannot take into account the nonlinear impairment. Two adaptive ANN (AANN) equalizers are proposed to combine ANN and MIMO-LMS with an adaptive parameter [76]. The spatial complexity of the AANN can be less than 10% of MIMO-MBNN.

An adaptive algorithm determines different algorithmic processes by the power ratio of two transmitted signals. When the power ratio is close to one, the SISO ANN can be used to equalize the two signals. However, when the power ratio is out of balance, it is necessary to use MIMO-ANN algorithms, such as L-DBMLP-L (combination of MIMO-LMS, DBMLP, and MIMO-LMS) or one hidden layer MBNN (OHL MBNN). This constitutes two AANN algorithms, namely ADP L/DBMLP-L and ADP MIMO ANN. The proposed AANN can achieve the same transmission performance, but with lower spatial complexity [76].

3.4.5. Conclusions

In summary, the ML-based ICI and ISI cancellation methods for the MIMO-VLC system are expected to be promising due to the rapid development of computational power. ML methods such as DNNs can model the nonlinear phenomenon of VLC systems where the conventional linear methods become powerless. The future research trend for ML in high-speed MIMO-VLC systems is still mainly about spatial and temporal joint equalization, as well as compensating nonlinear effects.

3.5. Optimal Coding

In general, to design a communication system is to split the system into several independent concatenated modules. These modules will realize the functionalities such as source/channel coding/decoding, modulation/demodulation, pre- and postequalization. The optimization of each module is carried out independently, either based on data-driven statistics features or based on mathematical models. However, the optimization of a single module cannot guarantee the overall optimization of the end-to-end communication of the entire physical layer. An intriguing approach is the end-to-end joint optimization for the physical layer [77]. The methodology is to treat physical-layer communication as an end-to-end signal-reconstruction problem, and to apply the concept of autoencoder to represent the physical-layer communication modules (the transmitter, the channel, and the receiver) by one deep neural network. Autoencoder is an unsupervised deep learning algorithm. The goal of the autoencoder is to find an optimized representation at its intermediate layer. This intermediate representation is robust to channel perturbations, allowing the output to be reconstructed with minimal error.

The end-to-end learning method extended its applications in MIMO and OFDM cases [78–80] with satisfying results. The autoencoder establishes a unified physical-layer framework that can be used in complex communications scenarios, and can obtain a lower bit error rate than the classic method through learning with lower computational complexity. Most of these autoencoder approaches are optimized on the symbol-wise categorical cross entropy [80]. However, the communication system is defined by the bit error rate. Novel approaches developed in [80,81] can optimize the bitwise mutual information between the input and output. Apart from that, VLC systems should consider more instability of complicated nonlinear distortions when searching for the optimal coding scheme. Considering recent works of VLC-based autoencoders, substantial efforts are still demanded to develop practical solutions in the wireless optical system.

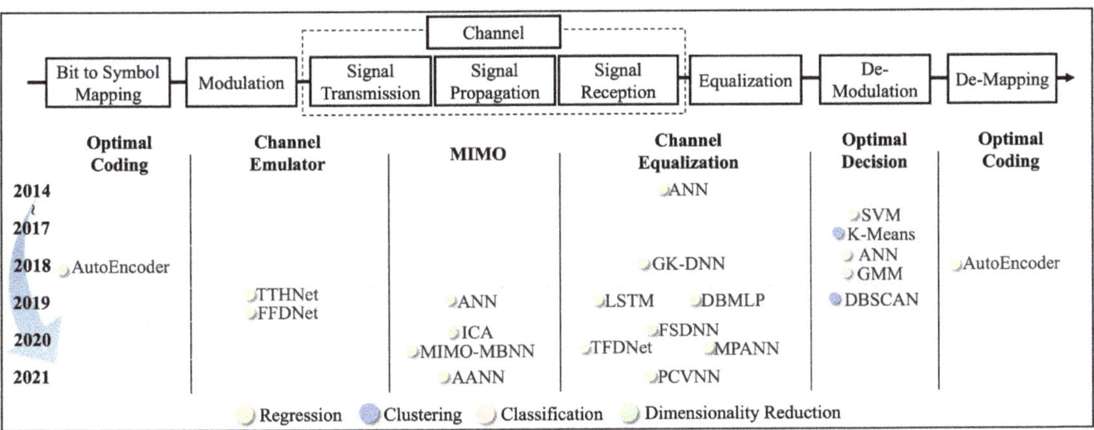

Figure 6. Overview of machine learning algorithms in the intelligent physical layer; Optimal coding: AutoEncoder [82–86]; Channel emulator: TTHnet [44], FFDNet [45]; MIMO: ANN [75], ICA [72], MIMO-MBNN [73], AANN [76]; Channel Equalization: ANN [50], GK-DNN [52,54], LSTM [61], DBMLP [44], FSDNN [55], TFDNet [56], MPANN [63], PCVNN [60]; Optimal decision: SVM [68], K-means [64], ANN [70], GMM [67], DBSCAN [66].

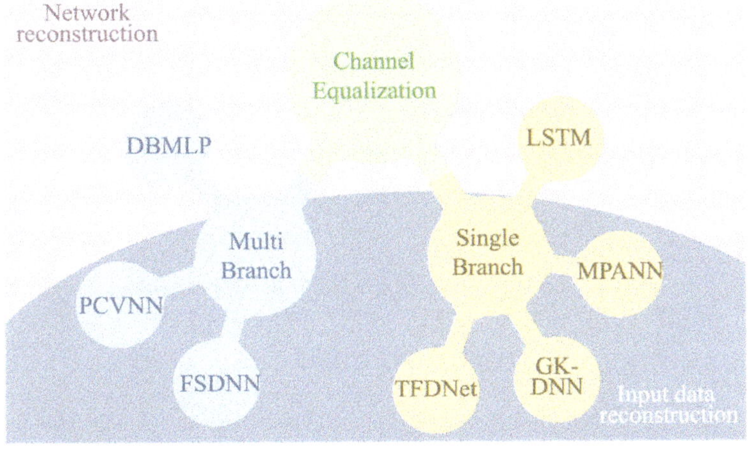

Figure 7. Channel equalization in VLC systems.

3.5.1. VLC-Based Autoencoder

End-to-end learning of the transmitter and receiver for communications over a visible light channel was first proposed in [82]. The transmitter and receiver are mapped to the encoder part and decoder part of the autoencoder structure. Symbol-level precoding at the encoder input transfers the original symbols to one-hot vectors. To accurately reconstruct the input from the received signal, the decoder works as a classifier. The transmission channel is built with two fixed weighted layers considering the color crosstalk of optical antennas and additive noise. Then, the categorical cross-entropy function is used to evaluate the loss between the input and the output probability vector of the transmitted symbols. Gradients of symbol errors direct the updating of the encoder and decoder layers through the backward-propagation process. The average symbol error rate performance reflects the superiority of the proposed closed-loop optimized transceiver to a minimum distance maximizing modulation scheme [83]. Apart from the photodiodes-based autoencoder [82,84], in [85] a convolutional autoencoder structure is proposed for image sensor communication systems. The system utilizes spatially separated LED arrays to convey an OOK-modulated signal and an optical image sensor as the receiver. Convolutional layers are implemented to overcome irradiance spread and lens blur induced by the sensor. The 2D convolution operations of the proposed autoencoder bring performance gain for image-decoding strategies in the simulated ISC systems. Besides the above works on decreasing error rates, the method in [86] takes flicker and illumination levels into account and tries to address real-life application constraints. However, all these works stay with numerical validations. As discussed in Section 2.1, it should be noticed that the channel estimation work is still in a naive stage. Experimental validation of the aforementioned end-to-end communication schemes will suffer performance degeneration due to the unconsidered dynamic nonlinear impairments of devices. To the best of our knowledge, there is still no experimental demonstration of an autoencoder-based VLC system.

3.5.2. Fiber/Wireless-Based Autoencoder

The parameters of the fiber channel or wireless channel are very important for end-to-end learning, so that the back-propagation algorithm of the neural network can be effectively calculated. In an optical fiber communication system, the channel is governed by the nonlinear Schrödinger equation (NLSE). Therefore, in order to apply autoencoder in optical fiber communication, an appropriate equivalent deep neural network of NLSE has to be established [87–90]. The experimental results demonstrated that the input information was mapped to a set of robust transmitted waveforms via autoencoder and detected with a measured BER under FEC threshold in intensity-modulation direct-detection (IM/DD) optical-fiber systems [87,88]. The successful demonstrations in both wireless communication and optical fiber systems clearly validate that end-to-end learning can be a promising technology to fundamentally reconsider communication-system design [77,87].

3.5.3. Conclusions

Above all, the only difference between VLC, fiber, and wireless-based autoencoders lies in the channel. While the output signals of fiber or wireless can be analytically modeled with prior domain knowledge and experience, predicting the outputs of a VLC channel by the mathematically convenient models is very hard, if not impossible. Hence, applying data-driven ML in VLC systems appears to be a reasonable direction. However, even the most recent works have not been practically implemented. Future works should pay attention to dealing with more realistic issues such as the low-frequency interference problem or dynamic nonlinear distortions in the VLC systems. Moreover, the current method relies on cost-prohibitive computational and temporal resources. Meta-learning-enabled online training will speed up the application of end-to-end communication links in the 6G architecture.

4. Future Trend of ML in IVLC

Over the last decade, visible light communication has achieved rapid advances both in technologies and in applications. The fundamental impetus to this achievement is not only the progress in communication technologies including coding, modulation, and signal processing, but also the rapid development of optoelectronic chips and devices. The bandwidth of the state-of-art VLC devices can exceed 1 GHz [91], nearly 100 times larger than that of 10 years ago. However, compared with the devices used in optical fiber communication, which usually boast nearly 100 GHz bandwidth, the bandwidth for VLC is too small to afford roughly equivalent data rates with optical fiber communication. Thus, the integration of wired/wireless communication in future ultrahigh-data-rate 6G networks could still be a challenge. Coherent light sources, i.e., lasers, are supposed to have larger bandwidths. Semipolar and nonpolar LEDs are reported to increase the modulation bandwidth. Microstructure, microcavity, and plasmon may be promising new approaches to enable ultrahigh data rates [92]. The intrinsic diversified characteristics of VLC devices require AI technologies to understand the device model, optimize the transmission link, and manipulate the whole network effectively.

The future development of IVLC in the intelligent physical layer and intelligent network layer will be presented in this part, as shown in Figure 8. We strive to provide readers with some insight.

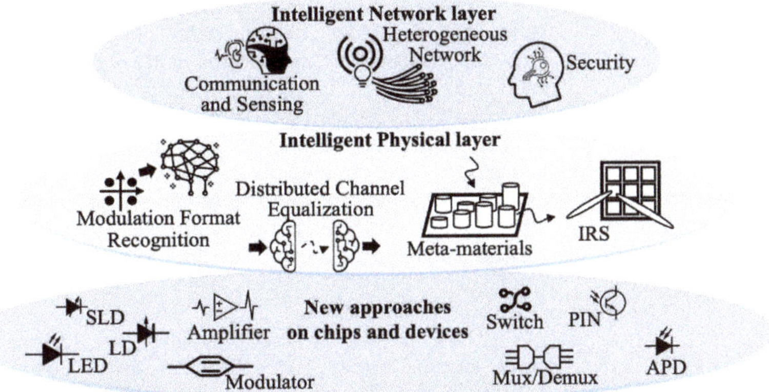

Figure 8. High-level view of intelligent physical layer and intelligent network layer of IVLC.

4.1. Intelligent Physical Layer

In this section, we will go through how electromagnetism and information theory can be better applied to AI-driven visible light communication. We will then present the possible forms of distributed learning of machine learning at the intelligent physical layer. Finally, a future application of machine learning in IVLC will be presented.

4.1.1. Fundamental Electromagnetism Theory and Frontiers in Optical Physics

The frequency bands available for next-generation wireless communication evolve to higher frequency bands such as millimeter wave, terahertz, and visible light [93]. In this case, the network spectrum and resources are unprecedentedly plentiful. Efforts are underway to sense the space electromagnetic information, to govern the spectrum allocation and beamforming by combining the fundamental electromagnetism theory with the conventional information theory. Intelligent Reflecting Surfaces (IRS) is a promising research direction in mm-waves and terahertz waves [94]. In visible light communication, we can imagine the optical phased array antenna. The feature size of the photonic circuits is far larger than that of the microelectronic circuits. Metamaterials, metasurfaces, and metalenses enable manipulation of the propagation, polarization, amplitude, and phase of the injected light at a deep subwavelength scale. This kind of optical artificially structured material is ultracompact in a scale of tens or hundreds of nanometers. On one hand, AI technologies can be a useful tool to design such kinds of materials by using optical reverse design. On the other hand, metamaterial can be a powerful way to realize optical neural networks and optical computing. All of these ideas are related to optical physics and could be promising research avenues towards 6G.

4.1.2. Distributed Channel Equalization

As mentioned above, existing research on machine learning in VLC has mainly focused on the standalone receiver or transmitter side. However, the available computational resources of the terminals and the central office are apparently different, so that leaving the computational complexity at the receiving or transmitting end only is not a reasonable solution. Therefore, ML-based distributed channel equalization at both the transmitting and receiving end is worthy of investigation. In [65], predistortion based on K-means is proposed to mitigate nonlinear impairment in VLC systems. A spatiotemporal neural network-based predistortion equalizer has also been proposed in RF communication systems in order to compensate for the nonlinearity caused by the power amplifier [95]. CNN has also been proposed to be applied for behavioral modeling and digital predistortion, and the model's coefficients can be significantly reduced [96]. Both NN-based distortion and postequalization do promise great performance.

However, an essential premise for pre-equalization is that the channel information state (CSI) is known to the transmitter. The CSI is usually obtained by training sequence. Unfortunately, the VLC channel is usually not static, and there is a mismatch between the estimated CSI and the actual CSI, as the statistic characteristics of the transmitted signal may be changed after pre-equalization or predistortion. The entire channel equalization should be computed simultaneously at the transmitter and receiver side in a distributed manner, taking into account the computational resources of both the receiver and transmitter as well as performance optimization. The CSI mismatch can be compensated since the training sequence is available at the receiving end. Therefore, distributed channel-equalization methods are required in the future VLC system.

4.1.3. Modulation Format Recognition

With the continuous improvement of the transmission capacity, the future VLC networks must be a mixture of multiple transmission rates and multiple modulation formats, thus modulation format recognition (MFR) will become an essential part of the overall communication system. In the early years, traditional machine learning methods, such as decision trees and SVM [97], were applied in VLC format recognition. The drawback of these solutions is that they rely on manual feature extraction, which leads to a lack of flexibility and portability.

Recently, deep learning is widely used in pattern recognition because of its ability to mine useful feature information at a deeper level. DNN is able to extract deeper layers of the signal by stacking layers of hidden layers [98]. However, as the number of hidden layers in DNN increases, the structural complexity increases. Therefore, it is especially important to design the network structure of DNNs rationally. CNN performs local feature extraction of information by setting appropriately sized convolutional kernels, and subsequently achieves classification through fully connected layers [99,100]. The feature-extraction part is mainly composed of convolutional and pooling layers, and the recognition and classification part is the same as the fully connected layer of the BP neural network. Other schemes that combine format recognition with deep learning, such as RNN [101], PNN [102], etc., have been gradually proposed.

It can be foreseen that due to the complexity of visible communication channels and the diversity of modulation formats, there will be an urgent need for machine learning-based modulation format technology. AI-driven MFR will effectively improve the recognition rate and accuracy, accelerating the pace of visible light communication applications.

4.2. Intelligent Network Layer

In this section, communication-aware integrated networks, as well as heterogeneous networks, will be introduced first. Finally, since visible light communication has the characteristics of both wireless and optical communication, network security will be introduced.

4.2.1. Converged Communication and Sensing

The convergence of sensing and communication network has become one of the leading trends in 6G technology and services [93]. The 6G network is expected to be a fusion of mobile communication, sensing, and intelligent computing. Such a converged network refers to a system that has the capabilities of target positioning (ranging, speed measurement, angle measurement), target imaging, target recognition, and target tracking. The development of higher-frequency bands such as millimeter waves, terahertz, and visible light will have more and more overlaps with traditional sensing frequency bands. Wireless communication and wireless sensing show more and more similarities in system design, signal processing, and data processing. Therefore, the AI-propelled VLC technologies will be found to be efficient in sensing technologies. A pragmatic solution is to allow communication and sensing in the same frequency spectrum. Research endeavors must be devoted to the solution of avoiding interference, and improving spectrum utilization. The codesign of the sensing and communication waveform may be aided by AI tools. The functionalities of communication and sensing are obtained based on software and hardware resource sharing or information sharing, which can effectively improve spectrum efficiency, hardware efficiency, and information-processing efficiency.

4.2.2. Heterogeneous Network

6G networks are becoming more and more heterogeneous, composed of different access standards and various network deployment methods. Therefore, it is necessary to consider the indoor and outdoor heterogeneous networking and interconnection issues of visible light communication networks and other communication technologies, for instance, the power line communications, optical fiber access networks, and mobile communications in radio frequency and even higher-frequency bands. Within this complex architecture of the future network, it is difficult for traditional models and algorithms to provide efficient and reliable technical support. AI-driven network technologies have made a series of progress in wireless communication to handle interference coordination and resource scheduling (including power allocation, channel allocation, and access control), which will reduce future wireless resource management costs and improve service quality.

The traditional network structure is network-centric with quite limited flexibility. The users are passive nodes, and the cell is generally preset to a fixed shape according to the transmission scheme and does not change with the communication traffic. Distributed artificial intelligence can implement a fully user-centric network architecture. By taking advantage of the diversity of user locations and service requirements, virtual amorphous communities can be constructed to provide better services for each user. AI can be used to understand users and perform user prediction, inference, and big data analysis. Moreover, AI can realize self-organizing network operations and management by network edge computing, and eventually form global closed-loop optimization.

4.2.3. Security Network

The characteristics of optical networks and wireless communications are combined in VLC systems. As a result, VLC suffers more complex cyber security challenges. When VLC becomes the infrastructure of future communication services, it will be subjected to additional attacks [103].

The first is the jamming attack in visible light communication. Since VLC is an LOS channel, it is highly susceptible to interference by external interference sources. Brief jamming attacks can cause enormous volumes of data to be incorrect or leaked due to the high transmission rates. ML can identify the presence and level of interference by learning interference signals targeting the physical layer in optical communications, as an example of deep Q-learning (DQN) [103].

The second is the cyber-physical attack in visible light communication. VLC is vulnerable to external intervention because it runs in an open environment. Injecting damaging signals into unprotected visible communication equipment can have a variety of negative consequences for communication [104]. By learning the characteristics of different devices or watermarks carried by the transmission, ML can be used to identify illicit transmission devices in VLC systems, as an example of TF-FSN [105].

The third is eavesdropping in visible light communication. VLC channels are vulnerable to unauthorized terminals because of their broadcast nature. Eavesdropping in public places compromises the privacy of legitimate users. Eavesdroppers' capacity to infer information over the channel is reduced by a smart beam anti-eavesdropping system based on deep reinforcement learning (DRL) [106]. The suggested intelligent beamforming method based on DRL may also make full use of complicated and high-dimensional structure information, improving network security for users.

VLC has some fixed security properties; however, entities communicating under the same space are vulnerable to discovery and information theft. Future research will focus on how to successfully use machine learning to counter attacks in the visible light field.

5. Conclusions

In this paper, we have provided a comprehensive study on AI-driven intelligent visible light communication. The major challenges in visible light communication are addressed, as well as the specific contribution of AI-enabled IVLC in overcoming these challenges.

Nonlinearity introduced by VLC due to additional electro-optical conversion is one of the major challenges that can significantly impair communication performance. Because of the spontaneous radiation properties of LEDs, visible light signals can only be directly modulated in communication. This means that changes in signal amplitude will directly affect the carrier concentration and thus the recombination of electrons and holes. This also increased the difficulty of modeling the visible light communication E2E channel. The improvement of machine learning for visible light communication performance is demonstrated in the paper, but again, its drawbacks of high computational complexity and low generalizability are presented.

Detailed applications in the intelligent physical layer of IVLC are categorized into five scenarios based on the communication system framework: optimal coding, channel emulator, MIMO, channel equalization, and optimal decision. For each of these categories, detailed elaboration is given to the state of the art. Among these technologies, autoencoder has the potential to revolutionize the existing physical layer communication architecture as a means of optimizing end-to-end communication.

Finally, we envisage the prospect of the intelligent physical layer and intelligent network layer. As AI continues to integrate in optical physics and electromagnetism area, the derived IRS, metamaterials, metasurfaces, metalenses, and optical computing will further drive the development of IVLC. In particular, optical computing and optical neural networks will be the focus of development. As IVLC may be deployed in 6G on a large scale, electricity-based digital signal processing will consume a lot of energy. The use of optical computing will greatly reduce consumption. Distributed channel equalization combines the existing communication system and the multilayer mechanism of neural networks, which will be an effective means of rapid deployment of IVLC. At the network layer, the main role of AI will be to reduce human intervention. It can achieve better resource scheduling and security through intelligent learning.

Hopefully, it will be a thorough investigation of intelligent visible light communication and serve as a practical guide for large-scale deployment of visible light communications in future 6G networks.

Author Contributions: Conceptualization, J.S. and N.C.; methodology, J.S. and N.C.; software, J.S.; validation, W.N.; formal analysis, J.S. and W.N.; data curation, J.S.; writing—original draft preparation, J.S., W.N., Y.H. and Z.X.; writing—review and editing, Z.L. and N.C.; visualization, J.S., W.N. and Z.X.; supervision, N.C.; project administration, J.S. and N.C.; funding acquisition, S.Y. and N.C. All authors have read and agreed to the published version of the manuscript.

Funding: This research was funded by the Natural Science Foundation of China Project (No. 61925104, No. 62031011), the Major Key Project of PCL (PCL2021A14), China Postdoctoral Science Foundation (2021M700025), and China National Postdoctoral Program for Innovative Talents (BX2021082).

Institutional Review Board Statement: Not applicable.

Informed Consent Statement: Not applicable.

Data Availability Statement: The data that support the findings of this study are available from the corresponding author upon reasonable request.

Conflicts of Interest: The authors declare no conflict of interest.

References

1. You, X.; Wang, C.-X.; Huang, J.; Gao, X.; Zhang, Z.; Wang, M.; Huang, Y.; Zhang, C.; Jiang, Y.; Wang, J. Towards 6G wireless communication networks: Vision, enabling technologies, and new paradigm shifts. *Sci. China Inf. Sci.* **2021**, *64*, 110301. [CrossRef]
2. Latva-aho, M.; Leppänen, K.; Clazzer, F.; Munari, A. *Key Drivers and Research Challenges for 6G Ubiquitous Wireless Intelligence*; University of Oulu: Oulu, Finland, 2020.
3. Zong, B.; Fan, C.; Wang, X.; Duan, X.; Wang, B.; Wang, J. 6G technologies: Key drivers, core requirements, system architectures, and enabling technologies. *IEEE Veh. Technol. Mag.* **2019**, *14*, 18–27. [CrossRef]
4. Chi, N.; Haas, H.; Kavehrad, M.; Little, T.D.; Huang, X.-L. Visible light communications: Demand factors, benefits and opportunities [Guest Editorial]. *IEEE Wirel. Commun.* **2015**, *22*, 5–7. [CrossRef]
5. Akasaki, I.; Amano, H.; Kito, M.; Hiramatsu, K. Photoluminescence of Mg-doped p-type GaN and electroluminescence of GaN pn junction LED. *J. Lumin.* **1991**, *48*, 666–670. [CrossRef]
6. Neokosmidis, I.; Kamalakis, T.; Walewski, J.W.; Inan, B.; Sphicopoulos, T. Impact of nonlinear LED transfer function on discrete multitone modulation: Analytical approach. *J. Lightwave Technol.* **2009**, *27*, 4970–4978. [CrossRef]
7. Linnartz, J.-P.M.G.; Deng, X.; Alexeev, A.; Mardanikorani, S. Wireless Communication over an LED Channel. *IEEE Commun. Mag.* **2020**, *58*, 77–82. [CrossRef]
8. Wang, Y.; Tao, L.; Huang, X.; Shi, J.; Chi, N. 8-Gb/s RGBY LED-based WDM VLC system employing high-order CAP modulation and hybrid post equalizer. *IEEE Photonics J.* **2015**, *7*, 7904507.
9. Ying, K.; Qian, H.; Baxley, R.J.; Yao, S. Joint optimization of precoder and equalizer in MIMO-VLC systems. *IEEE J. Sel. Areas Commun.* **2015**, *33*, 1949–1958. [CrossRef]
10. Letaief, K.B.; Chen, W.; Shi, Y.; Zhang, J.; Zhang, Y.-J.A. The roadmap to 6G: AI empowered wireless networks. *IEEE Commun. Mag.* **2019**, *57*, 84–90. [CrossRef]
11. Khan, F.N.; Lu, C.; Lau, A.P.T. Machine learning methods for optical communication systems. In Proceedings of the Signal Processing in Photonic Communications 2017, New Orleans, LA, USA, 24–27 July 2017.
12. Zhu, G.; Liu, D.; Du, Y.; You, C.; Zhang, J.; Huang, K. Toward an intelligent edge: Wireless communication meets machine learning. *IEEE Commun. Mag.* **2020**, *58*, 19–25. [CrossRef]
13. Mitola, J.; Maguire, G.Q. Cognitive radio: Making software radios more personal. *IEEE Pers. Commun.* **1999**, *6*, 13–18. [CrossRef]
14. Hosmer, D.W., Jr.; Lemeshow, S.; Sturdivant, R.X. *Applied Logistic Regression*; John Wiley & Sons: Hoboken, NJ, USA, 2013; Volume 398.
15. Breiman, L.; Friedman, J.H.; Olshen, R.A.; Stone, C.J. *Classification and Regression Trees*; Routledge: London, UK, 2017.
16. Ester, M.; Kriegel, H.-P.; Sander, J.; Xu, X. A density-based algorithm for discovering clusters in large spatial databases with noise. In Proceedings of the Second International Conference on Knowledge Discovery and Data Mining (KDD-96), Portland, OR, USA, 2–4 August 1996; pp. 226–231.
17. Roweis, S.T.; Saul, L.K. Nonlinear dimensionality reduction by locally linear embedding. *Science* **2000**, *290*, 2323–2326. [CrossRef]
18. Windisch, R.; Knobloch, A.; Kuijk, M.; Rooman, C.; Dutta, B.; Kiesel, P.; Borghs, G.; Dohler, G.; Heremans, P. Large-signal-modulation of high-efficiency light-emitting diodes for optical communication. *IEEE J. Quantum Electron.* **2000**, *36*, 1445–1453. [CrossRef]
19. Le Minh, H.; O'Brien, D.; Faulkner, G.; Zeng, L.; Lee, K.; Jung, D.; Oh, Y.; Won, E.T. 100-Mb/s NRZ visible light communications using a postequalized white LED. *IEEE Photonics Technol. Lett.* **2009**, *21*, 1063–1065. [CrossRef]
20. Kaushal, H.; Kaddoum, G. Underwater optical wireless communication. *IEEE Access* **2016**, *4*, 1518–1547. [CrossRef]
21. Elgala, H.; Mesleh, R.; Haas, H. An LED model for intensity-modulated optical communication systems. *IEEE Photonics Technol. Lett.* **2010**, *22*, 835–837. [CrossRef]

22. Costa, E.; Pupolin, S. M-QAM-OFDM system performance in the presence of a nonlinear amplifier and phase noise. *IEEE Trans. Commun.* **2002**, *50*, 462–472. [CrossRef]
23. Lin, R.-L.; Chen, Y.-F. Equivalent circuit model of light-emitting-diode for system analyses of lighting drivers. In Proceedings of the 2009 IEEE Industry Applications Society Annual Meeting, Houston, TX, USA, 4–8 October 2009; pp. 1–5.
24. Kamalakis, T.; Walewski, J.W.; Ntogari, G.; Mileounis, G. Empirical Volterra-series modeling of commercial light-emitting diodes. *J. Lightwave Technol.* **2011**, *29*, 2146–2155. [CrossRef]
25. Ying, K.; Yu, Z.; Baxley, R.J.; Qian, H.; Chang, G.-K.; Zhou, G.T. Nonlinear distortion mitigation in visible light communications. *IEEE Wirel. Commun.* **2015**, *22*, 36–45. [CrossRef]
26. Schwarz, U.T.; Braun, H.; Kojima, K.; Kawakami, Y.; Nagahama, S.; Mukai, T. Interplay of built-in potential and piezoelectric field on carrier recombination in green light emitting InGaN quantum wells. *Appl. Phys. Lett.* **2007**, *91*, 123503. [CrossRef]
27. Schwarz, U.T. Emission of biased green quantum wells in time and wavelength domain. In Proceedings of the Gallium Nitride Materials and Devices IV (International Society for Optics and Photonics—SPIE), San Jose, CA, USA, 26–29 January 2009; p. 72161U.
28. Dai, Q.; Shan, Q.; Wang, J.; Chhajed, S.; Cho, J.; Schubert, E.F.; Crawford, M.H.; Koleske, D.D.; Kim, M.-H.; Park, Y. Carrier recombination mechanisms and efficiency droop in GaInN/GaN light-emitting diodes. *Appl. Phys. Lett.* **2010**, *97*, 133507. [CrossRef]
29. McKendry, J.J.; Massoubre, D.; Zhang, S.; Rae, B.R.; Green, R.P.; Gu, E.; Henderson, R.K.; Kelly, A.; Dawson, M.D. Visible-light communications using a CMOS-controlled micro-light-emitting-diode array. *J. Lightwave Technol.* **2011**, *30*, 61–67. [CrossRef]
30. Deng, X.; Mardanikorani, S.; Wu, Y.; Arulandu, K.; Chen, B.; Khalid, A.M.; Linnartz, J.-P.M. Mitigating LED nonlinearity to enhance visible light communications. *IEEE Trans. Commun.* **2018**, *66*, 5593–5607. [CrossRef]
31. Barry, J.R.; Kahn, J.M.; Krause, W.J.; Lee, E.A.; Messerschmitt, D.G. Simulation of multipath impulse response for indoor wireless optical channels. *IEEE J. Sel. Areas Commun.* **1993**, *11*, 367–379. [CrossRef]
32. Lopez-Hernandez, F.J.; Perez-Jimenez, R.; Santamaria, A. Ray-tracing algorithms for fast calculation of the channel impulse response on diffuse IR wireless indoor channels. *Opt. Eng.* **2000**, *39*, 2775–2780.
33. Zhang, M.; Zhang, Y.; Yuan, X.; Zhang, J. Mathematic models for a ray tracing method and its applications in wireless optical communications. *Opt. Express* **2010**, *18*, 18431–18437. [CrossRef]
34. Tanaka, Y.; Komine, T.; Haruyama, S.; Nakagawa, M. Indoor visible communication utilizing plural white LEDs as lighting. In Proceedings of the 12th IEEE International Symposium on Personal, Indoor and Mobile Radio Communications (PIMRC 2001) (Cat. No. 01TH8598), San Diego, CA, USA, 30 September–3 October 2001; pp. 81–85.
35. Fujimoto, N.; Yamamoto, S. The fastest visible light transmissions of 662 Mb/s by a blue LED, 600 Mb/s by a red LED, and 520 Mb/s by a green LED based on simple OOK-NRZ modulation of a commercially available RGB-type white LED using pre-emphasis and post-equalizing techniques. In Proceedings of the 2014 The European Conference on Optical Communication (ECOC), Cannes, France, 21–25 September 2014; pp. 1–3.
36. Stepniak, G.; Maksymiuk, L.; Siuzdak, J. 1.1 GBIT/S white lighting LED-based visible light link with pulse amplitude modulation and Volterra DFE equalization. *Microw. Opt. Technol. Lett.* **2015**, *57*, 1620–1622. [CrossRef]
37. Li, X.; Bamiedakis, N.; Guo, X.; McKendry, J.; Xie, E.; Ferreira, R.; Gu, E.; Dawson, M.; Penty, R.; White, I. Wireless visible light communications employing feed-forward pre-equalization and PAM-4 modulation. *J. Lightwave Technol.* **2016**, *34*, 2049–2055. [CrossRef]
38. Wu, F.; Lin, C.; Wei, C.; Chen, C.; Huang, H.; Ho, C. 1.1-Gb/s white-LED-based visible light communication employing carrier-less amplitude and phase modulation. *IEEE Photonics Technol. Lett.* **2012**, *24*, 1730–1732. [CrossRef]
39. Haigh, P.A.; Burton, A.; Werfli, K.; Le Minh, H.; Bentley, E.; Chvojka, P.; Popoola, W.O.; Papakonstantinou, I.; Zvanovec, S. A multi-CAP visible-light communications system with 4.85-b/s/Hz spectral efficiency. *IEEE J. Sel. Areas Commun.* **2015**, *33*, 1771–1779. [CrossRef]
40. Chun, H.; Manousiadis, P.; Rajbhandari, S.; Vithanage, D.A.; Faulkner, G.; Tsonev, D.; McKendry, J.J.D.; Videv, S.; Xie, E.; Gu, E. Visible Light Communication Using a Blue GaN μ LED and Fluorescent Polymer Color Converter. *IEEE Photonics Technol. Lett.* **2014**, *26*, 2035–2038. [CrossRef]
41. Wang, F.; Liu, Y.; Jiang, F.; Chi, N. High speed underwater visible light communication system based on LED employing maximum ratio combination with multi-PIN reception. *Opt. Commun.* **2018**, *425*, 106–112. [CrossRef]
42. Ryu, S.-B.; Choi, J.-H.; Bok, J.; Lee, H.; Ryu, H.-G. High power efficiency and low nonlinear distortion for wireless visible light communication. In Proceedings of the 2011 4th IFIP International Conference on New Technologies, Mobility and Security, Paris, France, 7–10 February 2011; pp. 1–5.
43. Ma, X.; Yang, F.; Liu, S.; Song, J. Channel estimation for wideband underwater visible light communication: A compressive sensing perspective. *Opt. Express* **2018**, *26*, 311–321. [CrossRef]
44. Zhao, Y.; Zou, P.; Yu, W.; Chi, N. Two tributaries heterogeneous neural network based channel emulator for underwater visible light communication systems. *Opt. Express* **2019**, *27*, 22532–22541. [CrossRef]
45. Gao, Z.; Wang, Y.; Liu, X.; Zhou, F.; Wong, K.-K. FFDNet-based channel estimation for massive MIMO visible light communication systems. *IEEE Wirel. Commun. Lett.* **2019**, *9*, 340–343. [CrossRef]
46. He, H.; Wen, C.-K.; Jin, S.; Li, G.Y. Deep learning-based channel estimation for beamspace mmWave massive MIMO systems. *IEEE Wirel. Commun. Lett.* **2018**, *7*, 852–855. [CrossRef]

47. Jin, Y.; Zhang, J.; Jin, S.; Ai, B. Channel estimation for cell-free mmWave massive MIMO through deep learning. *IEEE Trans. Veh. Technol.* **2019**, *68*, 10325–10329. [CrossRef]

48. Burse, K.; Yadav, R.N.; Shrivastava, S. Channel equalization using neural networks: A review. *IEEE Trans. Syst. Man Cybern. Part C (Appl. Rev.)* **2010**, *40*, 352–357. [CrossRef]

49. Le Minh, H.; O'Brien, D.; Faulkner, G.; Zeng, L.; Lee, K.; Jung, D.; Oh, Y. High-speed visible light communications using multiple-resonant equalization. *IEEE Photonics Technol. Lett.* **2008**, *20*, 1243–1245. [CrossRef]

50. Haigh, P.A.; Ghassemlooy, Z.; Rajbhandari, S.; Papakonstantinou, I.; Popoola, W. Visible light communications: 170 Mb/s using an artificial neural network equalizer in a low bandwidth white light configuration. *J. Lightwave Technol.* **2014**, *32*, 1807–1813. [CrossRef]

51. Liang, S.; Jiang, Z.; Qiao, L.; Lu, X.; Chi, N. Faster-than-Nyquist precoded CAP modulation visible light communication system based on nonlinear weighted look-up table predistortion. *IEEE Photonics J.* **2018**, *10*, 7900709. [CrossRef]

52. Zhao, Y.; Zou, P.; Shi, M.; Chi, N. Nonlinear predistortion scheme based on Gaussian kernel-aided deep neural networks channel estimator for visible light communication system. *Opt. Eng.* **2019**, *58*, 116108. [CrossRef]

53. Xue, L.; Yi, L.; Lin, R.; Huang, L.; Chen, J. SOA pattern effect mitigation by neural network based pre-equalizer for 50G PON. *Opt. Express* **2021**, *29*, 24714–24722. [CrossRef]

54. Chi, N.; Zhao, Y.; Shi, M.; Zou, P.; Lu, X. Gaussian kernel-aided deep neural network equalizer utilized in underwater PAM8 visible light communication system. *Opt. Express* **2018**, *26*, 26700–26712. [CrossRef]

55. Chi, N.; Hu, F.; Li, G.; Wang, C.; Niu, W. AI based on frequency slicing deep neural network for underwater visible light communication. *Sci. China Inf. Sci.* **2020**, *63*, 160303. [CrossRef]

56. Chen, H.; Zhao, Y.; Hu, F.; Chi, N. Nonlinear Resilient Learning Method Based on Joint Time-Frequency Image Analysis in Underwater Visible Light Communication. *IEEE Photonics J.* **2020**, *12*, 7901610. [CrossRef]

57. Griffin, D.; Lim, J. Signal estimation from modified short-time Fourier transform. *IEEE Trans. Acoust. Speech Signal Process.* **1984**, *32*, 236–243. [CrossRef]

58. Zhao, Y.; Chi, N. Partial pruning strategy for a dual-branch multilayer perceptron-based post-equalizer in underwater visible light communication systems. *Opt. Express* **2020**, *28*, 15562–15572. [CrossRef]

59. Zhao, Z.; Vuran, M.C.; Guo, F.; Scott, S.D. Deep-waveform: A learned OFDM receiver based on deep complex-valued convolutional networks. *IEEE J. Sel. Areas Commun.* **2021**, *39*, 2407–2420. [CrossRef]

60. Chen, H.; Niu, W.; Zhao, Y.; Zhang, J.; Chi, N.; Li, Z. Adaptive deep-learning equalizer based on constellation partitioning scheme with reduced computational complexity in UVLC system. *Opt. Express* **2021**, *29*, 21773–21782. [CrossRef]

61. Lu, X.; Lu, C.; Yu, W.; Qiao, L.; Liang, S.; Lau, A.P.T.; Chi, N. Memory-controlled deep LSTM neural network post-equalizer used in high-speed PAM VLC system. *Opt. Express* **2019**, *27*, 7822–7833. [CrossRef] [PubMed]

62. Li, Z.; Hu, F.; Li, G.; Zou, P.; Wang, C.; Chi, N. Convolution-enhanced LSTM neural network post-equalizer used in probabilistic shaped underwater VLC system. In Proceedings of the 2020 IEEE International Conference on Signal Processing, Communications and Computing (ICSPCC), Macau, China, 21–24 August 2020; pp. 1–5.

63. Hu, F.; Holguin-Lerma, J.A.; Mao, Y.; Zou, P.; Shen, C.; Ng, T.K.; Ooi, B.S.; Chi, N. Demonstration of a low-complexity memory-polynomial-aided neural network equalizer for CAP visible-light communication with superluminescent diode. *Opto-Electron. Adv.* **2020**, *3*, 200009. [CrossRef]

64. Lu, X.; Wang, K.; Qiao, L.; Zhou, W.; Wang, Y.; Chi, N. Nonlinear compensation of multi-CAP VLC system employing clustering algorithm based perception decision. *IEEE Photonics J.* **2017**, *9*, 7906509. [CrossRef]

65. Lu, X.; Zhao, M.; Qiao, L.; Chi, N. Non-linear compensation of multi-CAP VLC system employing pre-distortion base on clustering of machine learning. In Proceedings of the Optical Fiber Communication Conference (Optical Society of America), San Diego, CA, USA, 11–15 March 2018; p. M2K.1.

66. Lu, X.; Qiao, L.; Zhou, Y.; Yu, W.; Chi, N. An IQ-Time 3-dimensional post-equalization algorithm based on DBSCAN of machine learning in CAP VLC system. *Opt. Commun.* **2019**, *430*, 299–303. [CrossRef]

67. Lu, F.; Peng, P.-C.; Liu, S.; Xu, M.; Shen, S.; Chang, G.-K. Integration of multivariate gaussian mixture model for enhanced pam-4 decoding employing basis expansion. In Proceedings of the Optical Fiber Communication Conference (Optical Society of America), San Diego, CA, USA, 11–15 March 2018; p. M2F.1.

68. Yuan, Y.; Zhang, M.; Luo, P.; Ghassemlooy, Z.; Lang, L.; Wang, D.; Zhang, B.; Han, D. SVM-based detection in visible light communications. *Optik* **2017**, *151*, 55–64. [CrossRef]

69. Niu, W.; Ha, Y.; Chi, N. Support vector machine based machine learning method for GS 8QAM constellation classification in seamless integrated fiber and visible light communication system. *Sci. China Inf. Sci.* **2020**, *63*, 202306. [CrossRef]

70. Li, J.; Guan, W. The optical barcode detection and recognition method based on visible light communication using machine learning. *Appl. Sci.* **2018**, *8*, 2425. [CrossRef]

71. Wang, Y.; Chi, N. Demonstration of high-speed 2×2 non-imaging MIMO Nyquist single carrier visible light communication with frequency domain equalization. *J. Lightwave Technol.* **2014**, *32*, 2087–2093. [CrossRef]

72. Wang, Z.; Han, S.; Chi, N. Performance enhancement based on machine learning scheme for space multiplexing 2×2 MIMO-VLC system employing joint IQ independent component analysis. *Opt. Commun.* **2020**, *458*, 124733. [CrossRef]

73. Zou, P.; Zhao, Y.; Hu, F.; Chi, N. Enhanced performance of MIMO multi-branch hybrid neural network in single receiver MIMO visible light communication system. *Opt. Express* **2020**, *28*, 28017–28032. [CrossRef]

74. Wolniansky, P.W.; Foschini, G.J.; Golden, G.D.; Valenzuela, R.A. V-BLAST: An architecture for realizing very high data rates over the rich-scattering wireless channel. In Proceedings of the 1998 URSI International Symposium on Signals, Systems, and Electronics (Cat. No. 98EX167), Pisa, Italy, 2 October 1998; pp. 295–300.

75. Rajbhandari, S.; Chun, H.; Faulkner, G.; Haas, H.; Xie, E.; McKendry, J.J.; Herrnsdorf, J.; Gu, E.; Dawson, M.D.; O'Brien, D. Neural network-based joint spatial and temporal equalization for MIMO-VLC system. *IEEE Photonics Technol. Lett.* **2019**, *31*, 821–824. [CrossRef]

76. Zhao, Y.; Zou, P.; He, Z.; Li, Z.; Chi, N. Low spatial complexity adaptive artificial neural network post-equalization algorithms in MIMO visible light communication systems. *Opt. Express* **2021**, *29*, 32728–32738. [CrossRef]

77. O'shea, T.; Hoydis, J. An introduction to deep learning for the physical layer. *IEEE Trans. Cogn. Commun. Netw.* **2017**, *3*, 563–575. [CrossRef]

78. Dörner, S.; Cammerer, S.; Hoydis, J.; Ten Brink, S. Deep learning based communication over the air. *IEEE J. Sel. Top. Signal Process.* **2017**, *12*, 132–143. [CrossRef]

79. Felix, A.; Cammerer, S.; Dörner, S.; Hoydis, J.; Ten Brink, S. OFDM-autoencoder for end-to-end learning of communications systems. In Proceedings of the 2018 IEEE 19th International Workshop on Signal Processing Advances in Wireless Communications (SPAWC), Kalamata, Greece, 25–28 June 2018; pp. 1–5.

80. Cammerer, S.; Aoudia, F.A.; Dörner, S.; Stark, M.; Hoydis, J.; Ten Brink, S. Trainable communication systems: Concepts and prototype. *IEEE Trans. Commun.* **2020**, *68*, 5489–5503. [CrossRef]

81. Balevi, E.; Andrews, J.G. Autoencoder-based error correction coding for one-bit quantization. *IEEE Trans. Commun.* **2020**, *68*, 3440–3451. [CrossRef]

82. Lee, H.; Lee, I.; Lee, S.H. Deep learning based transceiver design for multi-colored VLC systems. *Opt. Express* **2018**, *26*, 6222–6238. [CrossRef]

83. Liang, X.; Yuan, M.; Wang, J.; Ding, Z.; Jiang, M.; Zhao, C. Constellation design enhancement for color-shift keying modulation of quadrichromatic LEDs in visible light communications. *J. Lightwave Technol.* **2017**, *35*, 3650–3663. [CrossRef]

84. Hao, L.; Wang, D.; Cheng, W.; Li, J.; Ma, A. Performance enhancement of ACO-OFDM-based VLC systems using a hybrid autoencoder scheme. *Opt. Commun.* **2019**, *442*, 110–116. [CrossRef]

85. Lee, H.; Lee, S.H.; Quek, T.Q.S.; Lee, I. Deep Learning Framework for Wireless Systems: Applications to Optical Wireless Communications. *IEEE Commun. Mag.* **2019**, *57*, 35–41. [CrossRef]

86. Ulkar, M.G.; Baykas, T.; Pusane, A.E. VLCnet: Deep learning based end-to-end visible light communication system. *J. Lightwave Technol.* **2020**, *38*, 5937–5948. [CrossRef]

87. Karanov, B.; Chagnon, M.; Thouin, F.; Eriksson, T.A.; Bülow, H.; Lavery, D.; Bayvel, P.; Schmalen, L. End-to-end deep learning of optical fiber communications. *J. Lightwave Technol.* **2018**, *36*, 4843–4855. [CrossRef]

88. Chagnon, M.; Karanov, B.; Schmalen, L. Experimental demonstration of a dispersion tolerant end-to-end deep learning-based IM-DD transmission system. In Proceedings of the 2018 European Conference on Optical Communication (ECOC), Roma, Italy, 23–27 September 2018; pp. 1–3.

89. Häger, C.; Pfister, H.D. Wideband time-domain digital backpropagation via subband processing and deep learning. In Proceedings of the 2018 European Conference on Optical Communication (ECOC), Roma, Italy, 23–27 September 2018; pp. 1–3.

90. Häger, C.; Pfister, H.D. Physics-based deep learning for fiber-optic communication systems. *IEEE J. Sel. Areas Commun.* **2020**, *39*, 280–294. [CrossRef]

91. Xie, E.; Bian, R.; He, X.; Islim, M.S.; Chen, C.; McKendry, J.J.; Gu, E.; Haas, H.; Dawson, M.D. Over 10 Gbps VLC for long-distance applications using a GaN-based series-biased micro-LED array. *IEEE Photonics Technol. Lett.* **2020**, *32*, 499–502. [CrossRef]

92. Rajbhandari, S.; McKendry, J.J.; Herrnsdorf, J.; Chun, H.; Faulkner, G.; Haas, H.; Watson, I.M.; O'Brien, D.; Dawson, M.D. A review of gallium nitride LEDs for multi-gigabit-per-second visible light data communications. *Semicond. Sci. Technol.* **2017**, *32*, 023001. [CrossRef]

93. Polese, M.; Jornet, J.M.; Melodia, T.; Zorzi, M. Toward End-to-End, Full-Stack 6G Terahertz Networks. *IEEE Commun. Mag.* **2020**, *58*, 48–54. [CrossRef]

94. Pan, C.; Ren, H.; Wang, K.; Kolb, J.F.; Elkashlan, M.; Chen, M.; Di Renzo, M.; Hao, Y.; Wang, J.; Swindlehurst, A.L. Reconfigurable intelligent surfaces for 6G systems: Principles, applications, and research directions. *IEEE Commun. Mag.* **2021**, *59*, 14–20. [CrossRef]

95. Rawat, M.; Ghannouchi, F.M. Distributed spatiotemporal neural network for nonlinear dynamic transmitter modeling and adaptive digital predistortion. *IEEE Trans. Instrum. Meas.* **2011**, *61*, 595–608. [CrossRef]

96. Hu, X.; Liu, Z.; Yu, X.; Zhao, Y.; Chen, W.; Hu, B.; Du, X.; Li, X.; Helaoui, M.; Wang, W. Convolutional neural network for behavioral modeling and predistortion of wideband power amplifiers. *IEEE Trans. Neural Netw. Learn. Syst.* 2021; Early Access, 1–15. [CrossRef]

97. Fehske, A.; Gaeddert, J.; Reed, J.H. A new approach to signal classification using spectral correlation and neural networks. In Proceedings of the First IEEE International Symposium on New Frontiers in Dynamic Spectrum Access Networks, 2005 (DySPAN 2005), Baltimore, MD, USA, 8–11 November 2005; pp. 144–150.

98. Khan, F.N.; Zhong, K.; Zhou, X.; Al-Arashi, W.H.; Yu, C.; Lu, C.; Lau, A.P.T. Joint OSNR monitoring and modulation format identification in digital coherent receivers using deep neural networks. *Opt. Express* **2017**, *25*, 17767–17776. [CrossRef] [PubMed]

99. Xu, W.; Wang, Y.; Wang, F.; Chen, X. PSK/QAM modulation recognition by convolutional neural network. In Proceedings of the 2017 IEEE/CIC International Conference on Communications in China (ICCC), Qingdao, China, 22–24 October 2017; pp. 1–5.
100. Yashashwi, K.; Sethi, A.; Chaporkar, P. A learnable distortion correction module for modulation recognition. *IEEE Wirel. Commun. Lett.* **2018**, *8*, 77–80. [CrossRef]
101. Tang, Y.; Huang, Y.; Wu, Z.; Meng, H.; Xu, M.; Cai, L. Question detection from acoustic features using recurrent neural network with gated recurrent unit. In Proceedings of the 2016 IEEE International Conference on Acoustics, Speech and Signal Processing (ICASSP), Shanghai, China, 20–25 March 2016; pp. 6125–6129.
102. Liao, X.; Li, B.; Yang, B. A novel classification and identification scheme of emitter signals based on ward's clustering and probabilistic neural networks with correlation analysis. *Comput. Intell. Neurosci.* **2018**, *2018*, 1458962. [CrossRef] [PubMed]
103. Natalino, C.; Schiano, M.; Di Giglio, A.; Wosinska, L.; Furdek, M. Field demonstration of machine-learning-aided detection and identification of jamming attacks in optical networks. In Proceedings of the 2018 European Conference on Optical Communication (ECOC), Roma, Italy, 23–27 September 2018; pp. 1–3.
104. Li, Y.; Hua, N.; Yu, Y.; Luo, Q.; Zheng, X. Light source and trail recognition via optical spectrum feature analysis for optical network security. *IEEE Commun. Lett.* **2018**, *22*, 982–985. [CrossRef]
105. Liu, W.; Li, X.; Huang, Z.; Wang, X. Transmitter Fingerprinting for VLC Systems via Deep Feature Separation Network. *IEEE Photonics J.* **2021**, *13*, 7300407. [CrossRef]
106. Xiao, L.; Sheng, G.; Liu, S.; Dai, H.; Peng, M.; Song, J. Deep reinforcement learning-enabled secure visible light communication against eavesdropping. *IEEE Trans. Commun.* **2019**, *67*, 6994–7005. [CrossRef]

Review

Advances in Visible Light Communication Technologies and Applications

Zuhang Geng [1,2], **Faisal Nadeem Khan** [3,*], **Xun Guan** [3] and **Yuhan Dong** [1,2]

1   Tsinghua Shenzhen International Graduate School, Tsinghua University, Beijing 100084, China
2   Department of Electronic Engineering, Tsinghua University, Beijing 100084, China
3   Tsinghua-Berkeley Shenzhen Institute, Tsinghua Shenzhen International Graduate School,
    Tsinghua University, Beijing 100084, China
*   Correspondence: faisal.khan@sz.tsinghua.edu.cn

Abstract: With the development of light emitting diode (LED) lighting technology and its wide applications, visible light communication (VLC) technology has also seen significant advancements. VLC is regarded as a supplementary technology to radio frequency (RF) due to its unregulated spectrum and extraordinarily high communication rates. In this paper, the advantages, architectures, key technologies, application scenarios and machine learning (ML) applications of VLC are reviewed and summarized.

Keywords: visible light communication; machine learning; non-orthogonal multiple access; orthogonal frequency division multiplexing

Citation: Geng, Z.; Khan, F.N.; Guan, X.; Dong, Y. Advances in Visible Light Communication Technologies and Applications. *Photonics* **2022**, *9*, 893. https://doi.org/10.3390/photonics9120893

Received: 20 October 2022
Accepted: 18 November 2022
Published: 23 November 2022

Publisher's Note: MDPI stays neutral with regard to jurisdictional claims in published maps and institutional affiliations.

1. Introduction

In recent years, the number of mobile devices has increased exponentially. However, the existing radio frequency (RF) might not fully satisfy the communication requirements in some scenarios due to its limited bandwidth, where intense competition for available radio resources leads to degradation of communication performance.

To solve the contradiction between limited bandwidth and the increased number of mobile devices, visible light communication (VLC) becomes a potential supplementary technology to RF. As shown in Figure 1, the wavelengh range of visible light is from 380 nm to 780 nm and thus, it has abundant bandwidth. As a result, VLC devices are capable of reaching much greater transmission speeds than RF devices. Furthermore, the spectrum of VLC is unregulated and it results in lower costs.

Another superiority of VLC over RF is the use of light emitting diode (LED) for data transmission. Residential LEDs have much higher power efficiency and longer lifespan than traditional incandescent light bulbs. VLC is regarded as a green communication technology with low energy consumption as it could satisfy the requirements of both illumination and communication simultaneously [1]. The wide employment of LED further reduces the cost of devices in VLC networks.

Due to the superiorities mentioned above, researches in VLC has exponentially increased in both theoretical techniques and practical applications. Several surveys in the past few years have been noted as well. For instance, H. Haas et al. review key advancements in physical layer techniques which significantly improve the transmission speeds of LEDs and discuss the challenges in achieving Tbps LiFi systems in [2]. L. E. M. Matheus et al. [3] discuss the main characteristics and future directions of VLC-based applications and the research platforms of VLC. An overview of optimization techniques to improve the performance of VLC networks is presented in [4] and it mainly focuses on how to apply the new technologies for RF networks to VLC networks. Additionally, the key performance metrics and recent achievements of hybrid LiFi and WiFi networks (HLWNets) are presented in [5], which focuses on network architectures, cell deployments, multiple access and modulation schemes, illumination requirements and backhaul. A. Al-Kinani et al. [6] investigate the main optical channel characteristics and present a comprehensive overview of optical wireless communication (OWC) channel models. Furthermore, the channel models are compared in terms of computation complexity and accuracy. J. Luo et al. [7] discuss VLC-based indoor positioning and propose a novel taxonomy method. The VLC usage in vehicular communication applications is reviewed in [8,9], while the technology shortcomings and challenges are also presented. M. A. Arfaoui et al. [10] summarize the relevant technologies from the perspective of physical layer security (PLS) and point out the future research directions for PLS-VLC systems. Table 1 summarizes the surveys mentioned above and demonstrates the topics of the existing surveys and this paper.

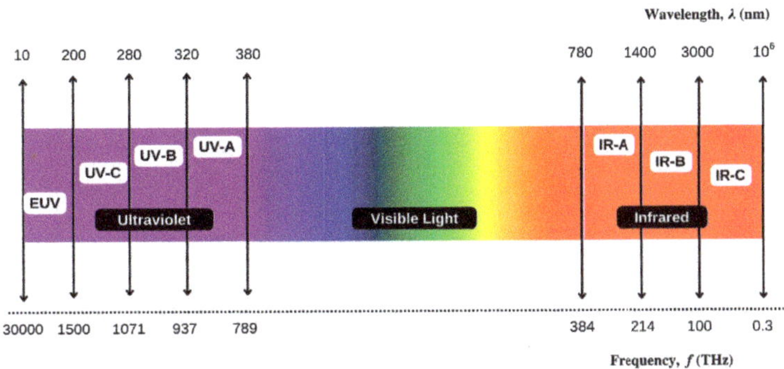

Figure 1. The optical spectrum.

Table 1. Surveys on Visible Light Communication.

Content Explored		H. Haas [2]	L. E. M. Matheus [3]	M. Obeed [4]	X. Wu [5]	A. Al-Kinani [6]	J. Luo [7]	A. -M. Căilean [8]	A. Memedi [9]	M. A. Arfaoui [10]	This Survey
Superiorities of VLC		✓	✓	✓	✓	✓	✓	✓	✓	✓	✓
Architecture	Transmitter	✓	✓	✓	✓	✓	✓	✓	✓	✓	✓
Architecture	Receiver	✓	✓	✓	✓	✓	✓	✓	✓	✓	✓
Architecture	Transmission distance		✓	✓	✓	✓		✓			✓
Channel modeling		✓	✓	✓		✓		✓	✓	✓	✓
Light modulation	OOK, PWM, PPM		✓	✓	✓		✓				✓
Light modulation	OFDM		✓	✓	✓		✓				✓
Physical layer security	Keyless security techniques			✓	✓					✓	✓
Physical layer security	Key-based security techniques			✓	✓					✓	✓
Physical layer security	NOMA		✓	✓	✓					✓	✓
Machine learning											✓
VLC applications	Indoor communication	✓	✓	✓	✓	✓					✓
VLC applications	Positioning		✓		✓		✓	✓			✓
VLC applications	Vehicular communication		✓					✓	✓		✓
VLC applications	Underwater communication		✓			✓					✓

However, most of the previous articles focus mainly on one of the research areas in VLC, which is not comprehensive enough. Additionally, to the best of our knowledge, none of the previous surveys deals with machine learning (ML). In this paper, we present a comprehensive survey on VLC as well as investigating the latest research progress of VLC and its applications in various emerging fields. The remainder of this paper is organized as follows. The architecture of VLC systems is presented in Section 2, while in Section 3, several key technologies of VLC are discussed, including channel modeling, light modulation, PLS and non-orthogonal multiple access (NOMA). Additionally, ML algorithms applied in VLC systems are discussed as well. Section 4 provides the latest advances in VLC applications. Then, Section 5 concludes the paper.

2. Architecture of VLC Systems

Intensity modulation with direct detection (IM/DD) is usually adopted in VLC systems and LEDs are usually utilized as transmitters [11]. The whole architecture of a VLC system is shown in Figure 2. After the modulator and pulse shaper, the transmitted information is loaded on an electrical signal, which is going to be converted into an optical signal at the transmitter. At the receiver side, the photodetector generates an electrical signal according to the intensity of the received optical signal. The inherent interference caused by the ambient light and the multi-path problem will lead to the reduction of transmission performance. As a result, bandpass filters and amplifiers are adopted to restore the time-domain signals. Then, the original information is recovered by the demodulator.

The visible light signals might be scattered by small particles in atmospheric or underwater channels due to its relatively low wavelengths. To solve this problem, a lens is adopted to collect and focus the received beam onto the photodiode (PD) at the receiver side [12,13]. Furthermore, multi-hop VLC is proposed to extend the communication range and it is regarded as a solution to the main issues including attenuation, scattering and divergence [14–16].

The transmission distance and data rate are limited by the channel conditions, light sources and photodetectors. In atmospheric channels, the transmission distance varies from few meters to few kilometers [16–20]. In contrast, the transmission distance is usually restricted to 500 meters in underwater channels as a result of the significant attenuation [21–23].

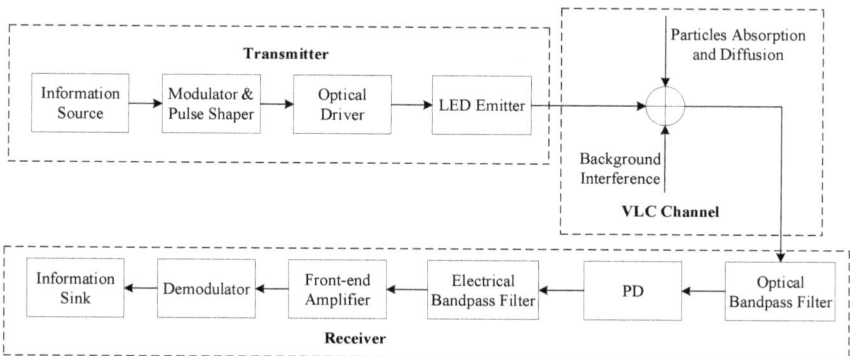

Figure 2. Architecture of a VLC system.

3. Key Technologies for VLC Systems

Considering the fact that the optical channel suffers from various noise and interference effects, techniques in physical and media access control (MAC) layers are proposed to ensure the desired communication quality. For example, research on channel characterization could help to reduce the influences caused by the noise and interference. Light modulation helps to utilize light intensity to convey data bits. Furthermore, research in physical layer security ensures the communication security of VLC systems. In the MAC layer, multiple access (MA) is adopted in many VLC scenarios to support multiple devices connected simultaneously and NOMA significantly improves the resource efficiency compared to traditional schemes. Furthermore, ML is proposed as a supplement of traditional algorithms in VLC systems.

3.1. Channel Characterization

In some VLC scenarios, the complex link and the dynamically changing channel state lead to challenges in signal transmission and system design. To solve this problem, research in VLC channels is required to obtain real-time channel status. In VLC systems, geometrical optics is usually adopted for channel modeling due to the short wavelength of visible light [8] and the channel variation is mainly caused by the translation and rotation of the receiver, which is different from conventional RF channels. The indoor VLC propagation scenario including both line of sight (LOS) and non-LOS (NLOS) paths is shown in Figure 3.

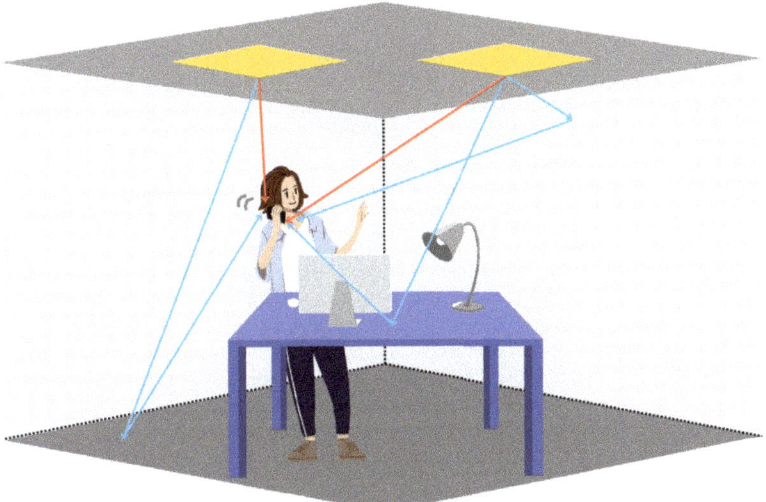

Figure 3. Indoor VLC propagation scenario.

As a key feature of VLC channel, the instantaneous impulse has been researched in [24,25]. However, VLC channels become time-varying with the movement or rotation of the receiver. As a result, methods for dynamic channel modeling are required to obtain the instantaneous channel state information (CSI), which is proved to be crucial to improving system throughput in [26,27].

J. Chen et al. proposed a movement–rotation (MR) correlation function to measure VLC channel variations for LOS channel in [28]. The MR correlation function could be utilized to measure VLC channel variations without time dependence as the receiver moves and rotates. Simulation results show that receiver rotations usually result in small channel fluctuations, while receiver movements lead to large-scale channel variations. The authors utilize the MR correlation function to approximate the correlation function of VLC channel gain and analyze the system performances for varying VLC channels, which is significant for the design and analysis of adaptive data transmission.

Another channel model for VLC is presented in [29], where X. Zhu et al. propose a novel three-dimensional (3D) space-time non-stationary geometry-based stochastic model (GBSM) for indoor multiple input multiple output (MIMO) VLC channels. This is one of the first VLC GBSM to support 3D translational and rotational motions, special radiation patterns of LEDs and space-time-frequency non-stationarity. The authors investigate several key statistical properties including channel DC gain, received power, channel 3dB bandwidth, space-time-frequency correlation function (STFCF) and root mean square (RMS) delay spread. The proposed GBSM is a better fit to the measured data than existing models, which confirms the accuracy and practicality of it.

It could be seen that the latest studies mainly focus on the dynamic changes of the channel state based on existing mathematical functions or models. It plays an important role in analyzing the dynamic characteristics of the channel, reducing the influence caused by noise and interference and designing new VLC systems.

3.2. Light Modulation

As mentioned above, one of the most significant differences between VLC and RF is that the transmitted data have to be encoded in the optical signals with different intensities. There are two main limitations for light modulation in VLC. Firstly, dimmer circuits are equipped on light bulbs to provide the ability to control light intensity. Secondly, no human-perceivable fluctuations are allowed in the modulated light waves in order to prevent serious detrimental physiological damage to humans [30].

The traditional light modulation schemes include on-off keying (OOK), pulse width modulation (PWM), pulse position modulation (PPM) and orthogonal frequency division multiplexing (OFDM). OOK mudulation turns the LED on and off to transmit the data bits 1 and 0, respectively. It is easy to implement yet it suffers from flickering and limited data rate. PPM loads the transmitted data on the position of the pulse while the length of the pulse corresponds to the value of the signal in PWM. As a result, PWM and PPM are capable of transmitting data without the variation of the pulse intensity.

Recently, several PWM schemes have been proposed for VLC systems. In order to increase the bandwidth usages for VLC, M. A. S. Sejan et al. [31] propose multilevel PWM (MPWM). The suggested modulation technique can impose additional bits by adopting pulse height and width modifications simultaneously. However, the increase of pulse height and width levels leads to higher BER performance than traditional schemes.

In previous studies, rectangular waves are usually assumed as the underlying received signal waveforms. However, uninterrupted on/off switching of electronic elements leads to non-rectangular waves, which is demonstrated in [32]. K. Yan et al. [33] propose a new precise PPM-VLC received-signal model considering the non-rectangular waves. Based on the PPM-VLC received-signal model, a new PPM-VLC demodulation scheme is proposed. Furthermore, a package-template of an L-PPM symbol is constructed and the new modulator evaluates the similarity between the package-template and the received signal waveform to recover the transmitted information. Therefore, it solves the problem of dynamic threshold adjustment, which implies better performance than a traditional demodulator.

Additionally, the four-level pulse amplitude modulation (PAM4) format has been extensively researched and experimentally proven to be an appealing technique to boost spectral efficiency. However, optical multipath interference (MPI) noises extremely decrease the transmission quality of IM/DD systems [34–37]. C. Huang et al. propose two algorithms for MPI noise elimination in [38] by removing the fluctuation of MPI-impaired PAM4 signals and estimating the MPI noise, respectively. Without altering the current IM/DD system design, these algorithms could be implemented in the receiver digital signal processing (DSP) module directly. Furthermore, the simulation and experimental results demonstrate their capacity for suppressing the MPI noise for IM/DD transmission systems with high-speed and high-order modulation formats.

Given that single-carrier modulation schemes suffer from high inter-symbol interference (ISI), OFDM is adopted in VLC with the advantage of high spectral efficiency and characteristics for resisting ISI and multipath fading [39]. However, the traditional OFDM time-domain signal in RF is complex-valued and bipolar, which is contradictory to IM/DD systems [11]. Generally, Hermitian symmetry is adopted to generate real-valued time-domain signals. In order to obtain unipolar signals, several optical OFDM (O-OFDM) schemes have been proposed, which are summarized in Table 2. The most popular O-OFDM schemes are asymmetrically clipped optical OFDM (ACO-OFDM) and direct-current (DC) biased optical OFDM (DCO-OFDM). ACO-OFDM only occupies odd subcarriers and clips the negative part of time-domain signals with no information lost [40]. Whereas, in DCO-OFDM, a DC bias is added to make the time-domain signal nonnegative [41]. Another O-OFDM scheme named pulse-amplitude-modulated discrete multitone (PAM-DMT) only utilizes the imaginary part of subcarriers and clips the negative part of time-domain signals as well as ACO-OFDM [42]. Based on the O-OFDM schemes mentioned above, several variants are proposed in succession. For instance, hybrid ACO-OFDM (HACO-OFDM) combines both ACO-OFDM and PAM-DMT [43], while layered ACO-OFDM (LACO-OFDM) transmits layers of ACO-OFDM simultaneously and recovers the transmitted data iteratively [44]. LACO-OFDM has much higher spectral efficiency and lower peak-to-average power ratio (PAPR) than ACO-OFDM and HACO-OFDM.

Table 2. Summary of Optical OFDM Schemes in VLC.

Work	O-OFDM Scheme	Utilized Spectral Resource	Signal Processing	Features
J. Armstrong et al. [40]	ACO-OFDM	Odd subcarriers	Clipping operation	Low complexity, low spectral efficiency and high power efficiency
J. B. Carruthers et al. [41]	DCO-OFDM	All subcarriers	Adding a DC bias	Low complexity, high spectral efficiency and low power efficiency
S. C. J. Lee, et al. [42]	PAM-DMT	The imaginary part of subcarriers	Clipping operation	Low complexity, low spectral efficiency and high power efficiency
B. Ranjha et al. [43]	HACO-OFDM	Odd subcarriers and the imaginary part of even subcarriers	Clipping operation	Higher spectral efficiency than ACO-OFDM
Q. Wang et al. [44]	LACO-OFDM	Layers of the half of the remained subcarriers	Clipping operation	Higher spectral efficiency than HACO-OFDM and an iterative receiver with higher complexity
R. Bai et al. [45]	AAO-OFDM	All subcarriers	Clipping operation and absolute operation	Higher spectral efficiency than ACO-OFDM
R. Bai et al. [46]	ALACO-OFDM	All subcarriers	Clipping operation and absolute operation	Higher spectral efficiency than AAO-OFDM and an iterative receiver with higher complexity

Asymmetrically clipped absolute value optical OFDM (AAO-OFDM) is proposed in [45], it utilizes two streams to send ACO-OFDM signal and absolute value optical OFDM (AVO-OFDM) signal, respectively. The signs of the AVO-OFDM signals are modulated to the frequency-domain symbols of ACO-OFDM. As a result, signals of AVO-OFDM do not require any DC bias to generate unipolar values. Furthermore, in [46], absolute value layered asymmetrically clipped optical OFDM (ALACO-OFDM) is proposed, and it reaches higher spectral efficiency by transmitting ACO-OFDM signals in the first L layers and absolute value optical OFDM (AVO-OFDM) signals on the remaining subcarriers simultaneously. Taking uncoded bit-error-ratio (BER) and achievable information rate into account, R. Bai et al. [46] designed two optimal optical power allocation schemes, respectively. Additionally, analysis shows that ALACO-OFDM achieves a higher information rate at moderate to high signal-to-noise ratios (SNRs) and has lower PAPR than ACO-, AAO- and LACO-OFDM.

Furthermore, signals are modulated using the indices of a medium in the index modulation (IM) approach to further enhance the spectral efficiency or power efficiency of O-OFDM systems [47]. By utilizing discrete Hartley transform (DHT), X. -Y. Xu et al. [48] propose a new O-OFDM-IM scheme with lower complexity and higher spectral efficiency than traditional O-OFDM schemes based on discrete Fourier transform (DFT). It is superior in SNR performance as well. However, the PAPR performance of this scheme is not satisfying compared with its traditional counterparts sometimes.

In this section, we discussed several modulation schemes, which generate unipolar real-valued signals and meet the requirements of IM/DD systems. OOK, PWM and PPM are simple to realize while OFDM could reduce ISI and is suitable for MIMO. The best scheme should be chosen according to the specific scenario.

3.3. Physical Layer Security

Considering the fact that light does not penetrate through walls, VLC has higher security than RF systems. However, unlike fiber-optic systems, security problems might occur in VLC systems owing to their open and broadcast nature [49]. PLS techniques have been fully studied and applied in RF systems [4,50]. The two main categories of PLS techniques are keyless security techniques and key-based security techniques. As a result of the differences between RF and VLC, PLS technologies developed for RF systems may not be directly applicable to VLC systems.

Keyless security techniques usually utilize the randomness of noise, channels and different resources to enhance the security of networks [50]. N. Su et al. [51] propose a novel spatial constellation design technique based on a multi-user generalized space shift keying for indoor multi-user MIMO-VLC (MU-MIMO-VLC) scenario to enhance the PLS. The authors adjust the transmission power of each transmitting LED by adjusting the CSI of legitimate users, which optimizes the received signal constellation for legitimate users in terms of BER and only generates interference for the eavesdroppers. Simulation results demonstrate that the BER of an eavesdropper declines significantly and the confidentiality improvement depends on the relative position between users.

Y. M. Al-Moliki et al. [52] propose a chaos-based physical-layer encryption method for OFDM-based VLC schemes. By using the position-sensitive and real-valued CSI of the VLC channel, a chaotic key creation approach is developed to create the secret key. However, based on floating-point arithmetic, chaotic systems have a significant resource and latency overhead, which is inappropriate for resource-limited VLC devices and high-data-rate VLC systems. Furthermore, in Y. M. Al-Moliki et al. [53] propose a key-based lightweight channel-independent (LCI) physical-layer encryption method, which generates dynamic keys and ciphertexts with the random nature of the input data and applies phase encryption of OFDM symbols in the frequency domain. The proposed method is suitable for resource-restricted scenarios and has relatively low complexity. Another lightweight cipher scheme for VLC systems is proposed in [54]. The proposed approach secures the underlying OFDM signals using straightforward substitution and phase shuffling techniques with low computational complexity and latency.

The key-based security techniques mentioned above are superior to keyless security techniques in design complexity. However, keyless security techniques could provide better security than key-based security techniques. Through advanced coding techniques at the physical layer, PLS has the potential to take advantage of elements of the environment around it [55], which can contribute to realizing the robust end-to-end security and satisfying the requirements of the next generation networks.

3.4. NOMA

In VLC systems, traditional MA techniques, including frequency division multiple access (FDMA), time division multiple access (TDMA) and code division multiple access (CDMA), suffer from low resource efficiency [56]. To solve this problem, orthogonal frequency division multiple access (OFDMA) is proposed to reuse subcarrier resources. H. Marshoud et al. [57] propose NOMA in order to further enhance resource efficiency and system capacity in VLC systems, aiming at increasing the throughput, reducing the latency and improving the fairness and connectivity. Multiple users' signals are superimposed in the power domain and each user could utilize the entire time and frequency resources. In NOMA, users with poor channel conditions are assigned more signal power, while users with good channel conditions corresponds to less power. The transmitted information is recovered by successive interference cancellation (SIC) at the receiver side.

Considering uniformly distributed users, L. Yin et al. [56] derive the distribution function of the channel gain in a closed form. Additionally, the performance of NOMA is evaluated and compared with orthogonal multiple access (OMA) and a closed-form expression of the ergodic sum rate gain of NOMA over OMA is derived. In [58], the BER performance in a downlink NOMA-VLC network is analyzed and an exact, simple and generic analytic expression is derived for the BER performance, which is the first work to study the BER performance of NOMA-based systems. The performance of a hybrid NOMA-VLC-RF system with imperfect CSI and uniformly distributed users is evaluated in [59]. The authors derive closed-form expressions for the corresponding average sum-rate and average energy efficiency, which coincide with the simulation results. M. Le-Tran et al. [60] analyze the NOMA performance in a downlink VLC system with an optical backhauled link and derive the closed-form expressions of the user outage probability, the sum throughput, the average BER and the energy efficiency with guaranteed transmission rates. The theoretical and simulation results imply that NOMA systems provide significant performance gains for high-rate optical backhaul links compared with OMA systems at medium to high SNR ranges. The first non-OFDM-based NOMA scheme for VLC with arbitrary modulation order in multiple access and broadcast channel is proposed in [61], which is appropriate for high-SNR regimes and requires lower computational complexity than the existing schemes.

In NOMA, all time and frequency resources are shared by all users with different power. It offers a higher quality of service and better capacity for resisting interference, and it is regarded as a promising technique.

3.5. Machine Learning

In the past few years, ML has attracted intense interest of researchers and has been regarded as a potential technology to solve the various challenges in wireless communication systems. ML performs well in resources allocation, channel equalization, estimation and modeling [62]. As a result, the applications in VLC are exponentially increasingly proposed. For instance, M. Najla et al. [63] study the selection between RF and VLC bands for device-to-device (D2D) communication, assuming the condition that sudden drops in channel quality occur in VLC links. A deep neural network (DNN)-based framework to select RF or VLC for D2D pairs is proposed to obtain an initial band selection decision. The authors further present a low-complexity heuristic algorithm to improve the accuracy of the band selection and simulation results show the close-to-optimal performance of the proposed algorithm.

To overcome the deterioration in communication and positioning performance resulting from the long transmission distance in vehicular VLC (V-VLC), J. He et al. [64] propose and experimentally demonstrate an ML-assisted image sensor-based visible-light-based positioning (VLP) scheme. At the transmitter side, a new coding method is adopted to increase the data rate and adjust the LED lighting power according to the ambient light intensity. At the receiver side, a convolutional neural network (CNN) and an artificial neural network (ANN) are used for decoding and vehicle positioning, respectively, which help to realize long-distance communication, high-accuracy positioning and LED dimming simultaneously.

The issue of dynamically deploying unmanned aerial vehicles (UAVs) in VLC for improving the energy efficiency of UAV-enabled networks is investigated in [65]. In order to jointly maximize the usage of UAVs, user association and power efficiency while satisfying user lighting and communication needs, an approach is developed to address this issue by combining the ML framework of gated recurrent units (GRUs) with CNNs. The proposed method can forecast future light distributions by modeling long-term historical illumination distributions, which significantly reduces power consumption when compared to traditional methods, according to simulation results.

Collaborative constellation (CC) design is useful for enhancing performance while drastically lowering the total optical power in MIMO VLC systems. M. Le-Tran et al. [66] propose CCNet, a DL-based constellation design technique for MIMO VLC systems with CC that can drastically reduce complexity while preserving near-optimal performance when compared to previous schemes. CCNet is initially trained offline to decrease the mean square error (MSE) and the ordinary CSI is effectively preprocessed to further enhance the performance of CCNet.

In order to enable efficient resource management in VLC, Z.-Y. Wu et al. [67] proposed a data-driven ML-based approach to forecast LOS link outages and minimize severe signal degradation. Furthermore, a predictor is designed to learn the channel variation patterns and predict LOS link outages and recoveries by utilizing a deep recurrent neural network (RNN) with long-short-term-memory (LSTM) units. For both uplink and downlink, the proposed predictor achieves a 91% hit rate for outages and an 83% hit rate for signal recoveries when predicting the channel in the next one second. As a result, the development of effective resource management strategies in VLC networks could be greatly aided by this predictor.

Table 3 summarizes part of the latest research advances of ML in VLC systems. Research on ML has been growing exponentially recently and it will provide significant support for the advancement of VLC technology.

Table 3. Summary of ML applications in VLC.

Work	Application Scenarios	Neural Network	Advantages
Mehyar Najla et al. [63]	The selection between RF and VLC bands for D2D communication	DNN	Close-to-optimal performance
Jing He et al. [64]	Long-distance transmission in V-VLC	CNN and ANN	Long-distance communication, high-accuracy positioning and LED dimming
Yining Wang et al. [65]	Dynamically deploying UAVs in VLC	CNN	Future light distribution forecast and low power consumption
Manh Le-Tran et al. [66]	Collaborative constellation design	DNN	Low complexity and near-optimal performance
Zi-Yang Wu et al. [67]	Prediction of LOS link outage	deep LSTM-based RNNs	High hit rate for signal outages and recoveries

4. VLC Applications

VLC has a wide range of applications in various fields. Thanks to the characteristics of VLC, it could achieve many functions that cannot be realized by RF. For instance, VLC performs much better than RF in scenarios that require a high data rate, such as underwater high-speed video communication. Additionally, VLC is capable of meeting the demand for illumination and communication simultaneously, while RF is not applicable evidently.

In this section, we will discuss the research advances in the past few years that mainly focus on indoor communication, positioning, vehicular communication applications and underwater communication.

4.1. Indoor Communication

With the wide application of LED bulbs, more and more attention has been paid to the research of indoor VLC systems. M. A. Arfaoui et al. [68] propose realistic and measurement-based channel models for indoor VLC systems. The modified truncated Laplace (MTL) model and the modified Beta (MB) model are designed for stationary users, while the sum of modified truncated Gaussian (SMTG) model and the sum of modified Beta (SMB) model are proposed for mobile users.

Another significant issue is the orientation variation of terminals in VLC. A. A. Purwita et al. [69] propose a random process model for user equipment (UE) orientation variation and present how it affects the optical channel conditions. The results demonstrate that the blockage and diffuse connection have considerable consequences, particularly when the UE is situated far from an access point (AP).

In terms of system implementation, C. -H. Yeh et al. [70] achieve a data rate of 1.7 to 2.3 Gbps with a communication distance of 1 to 4 m. The illumination is extremely low and is set to be 6.9 to 136.1 lux. Two blue and two green LEDs are utilized and a 4×4 color-polarization-multiplexing method is proposed. Additionally, to maintain the improved signal performance, the measured BER of each LED is lower than the forward error correction objective of 3.8×10^{-3}.

4.2. Positioning

The Global Positioning System (GPS) has been widely used to provide real-time positioning and navigation. However, signals transmitted by satellites are usually weakened by obstacles, which reduces the accuracy of indoor positioning [71,72]. WiFi and Bluetooth positioning systems are usually adopted to improve the performance of indoor positioning in the previous works [73,74]. In recent years, positioning systems utilizing a visible light signal rather than RF are proposed. A typical prototype of an LED-based indoor positioning system is shown in Figure 4 where several LEDs are utilized as transmitters and provide lighting. Received signal strength (RSS), time-of-arrival (TOA), time difference of arrival (TDOA) or angle of arrival (AOA) information at the receiver side could be utilized to evaluate the localization. Moreover, the position coordinates could be obtained directly from captured images if the PDs are replaced by cameras [75]. The general architecture of a simplex VLC positioning system is presented in Figure 5.

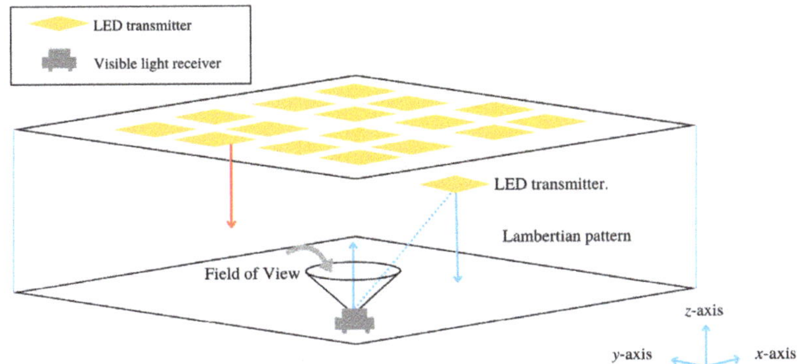

Figure 4. Illustration of the VLC-based positioning system.

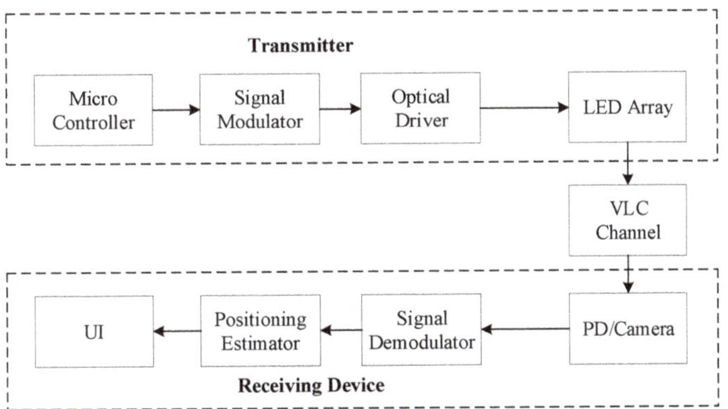

Figure 5. The general architecture of a simplex VLC positioning system.

B. Zhou et al. [76] propose a new VLC localization algorithm with the assumption of unknown LED emitting power, UE position and UE orientation. The joint optimization of all unknown parameters is adopted and a successive linear least square (SLLS)-based VLP algorithm is proposed. The authors derive the closed-form Cramer–Rao lower bound (CRLB) on each unknown parameter and analyze the performance limits of the proposed algorithm. Based on improved hybrid bat algorithm (IHBA), Y. Chen et al. [77] propose an indoor VLC 3D positioning system. The simulation results demonstrate higher positioning accuracy and shorter convergence time of IHBA compared with the existing VLC 3D positioning algorithm.

A position estimation DNN (PE-DNN)-aided receiver is proposed in [78], which utilizes the received pilot signals to extract the feature of the channel impulse response (CIR). Then, the coordinates are obtained from the CIR and an LED and a PD is sufficient to achieve centimeter-level positioning accuracy. Furthermore, it could achieve information transmission and positioning simultaneously, which ensures the compatibility and practicality of this VLC system.

Indoor positioning with VLC is regarded as a potential complement to GPS and it is capable of providing more accurate localization. Nevertheless, obstacles and reflection components might reduce the accuracy of positioning and these issues remain to be studied.

4.3. Vehicular Communication

Transportation systems have recently been developing by leaps and bounds. Intelligent transportation systems (ITS) are discussed in [79] and information exchange between vehicles and with infrastructure is indispensable for achieving ITS, where RF communication plays an important role. However, with the wide range of LED adoption, V-VLC could be realized by utilizing the LED-equipped lighting modules and transportation infrastructure [8].

Figure 6 presents a typical V-VLC traffic scenario, which includes infrastructure to vehicle communication, head-to-tail, tail-to-head, head-to-head and tail-to-tail communication. The building blocks of a generic V-VLC system are shown in Figure 7. Encoder, modulator, LED driver and optical transmitter front-end are equipped at the transmitter side, while optical receiver front-end, demodulator and decoder are equipped at the receiver side, correspondingly. The optical channel is interfered by other light sources, weather conditions and reflections.

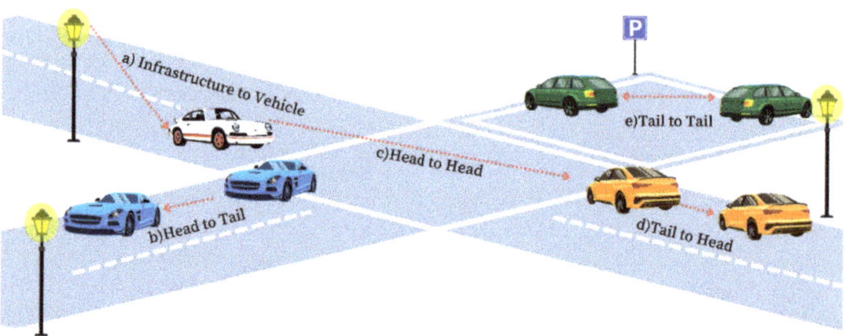

Figure 6. A typical V-VLC traffic scenario.

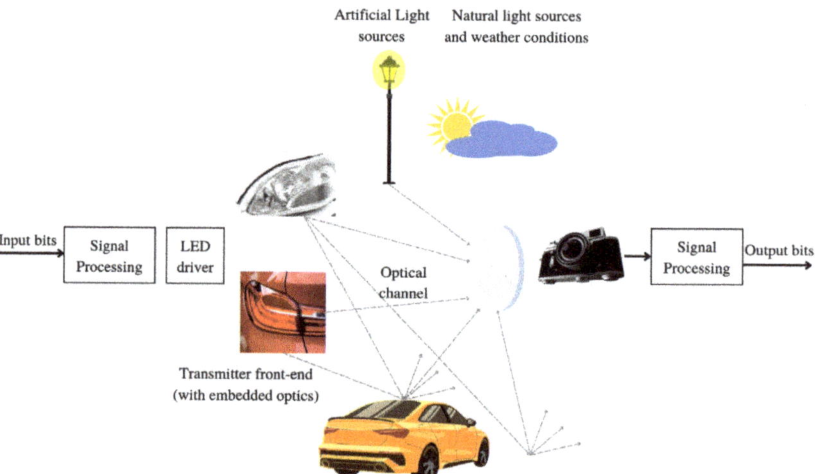

Figure 7. Building blocks of a generic V-VLC system.

Platooning is a key scenario for autonomous driving, where vehicles utilize vehicle-to-vehicle (V2V) communication and distance sensors to automatically adjust their position. It has gained strong interest in realizing illumination, data transmission and range-finding simultaneously with the automotive lighting. In [80], a system named visible light communication rangefinder (VLCR) is proposed. The V2V distance is estimated by utilizing the phase-shift between the original signal and the received signal, where the Doppler effect is found to be neglected. Experimental results show that the range-finding function could work at up to 25 m and the system is able to support a 500 kbps link with a BER below 10^{-6} and a transmission distance of up to 30 m.

Based on GNU Radio, M. S. Amjad et al. [81] introduce a flexible IEEE 802.11 compliant system for outdoor V-VLC and a high-power LED headlight is adopted to support communication distances beyond 75 m even in broad daylight. The authors also study the impact of optics alignment on the receiver's performance and analyze how the daylight influences the PD noise floor.

In [82], a vehicular MIMO VLC system based on two commercial headlights and a self-designed PIN array is presented and the strategy for selecting the best MIMO demultiplexing scheme by analyzing the rank and type of the channel matrix is discussed. Additionally, the authors proposed a modified pilot-aided phase recovery method based on polynomial curve fitting (PCF) and a record-breaking data rate of 3.08 Gbps at a 2 m indoor transmission link is realized. Furthermore, the overall data rates reach 336 Mbps and 362 Mbps in the day and at night, respectively, when the transmission distance is extended to 100 m, which are the highest transmission data rates of a vehicular MIMO VLC system for a 100 m transmission distance until now.

A major disadvantage of VLC is that the headlights and taillights are not capable of communicating directly with side-to-side vehicles, which demonstrates the lack of preventing blind-spot oversight [83]. Moreover, stringent latency and reliability are required for the security of vehicle driving in V-VLC and the performance of V-VLC is influenced by the outdoor environment, which bring challenges for V-VLC applications.

4.4. Underwater Communication

Underwater wireless communication (UWC) refers to data transmission via wireless carriers, i.e., RF waves, acoustic waves, and optical waves in unguided water environments. The characteristics of the underwater channel present many challenges compared to traditional wired and wireless communications through the atmosphere. Higher transmission bandwidth and data rates make underwater optical wireless communication (UOWC) more suitable than RF and acoustic counterparts for UWC systems [12]. Additionally, the turbulence of underwater environments results in changes in water density and salinity, which may reduce the performance of UOWC systems. It has been proved in [84] that flicker index and BER are significantly influenced by average temperature, average salinity concentration, temperature–salinity gradient ratio, temperature dissipation rate and energy dissipation rate.

Many researchers focus on improving transmission data rate and extending communication distance in UOWC experimental systems. For instance, X. Yang et al. [21] utilize the arrays consisting of series-connected monochromatic LEDs to reach a data rate of 130 Mbps over a 7 m underwater channel in LED-to-LED UOWC systems. Moreover, X. Chen et al. [22] realized a data rate of 500 Mbps with a transmission distance up to 150 m by combining partial response shaping and trellis-coded modulation (TCM) technology for the first time. Furthermore, H. Zhou et al. [23] propose a new mathematical model of UOWC. It could be used for water quality measurement and the proposed system could reach a 50 m link with the data rate of 80 Mbps.

With the exponentially increasing research in VLC, it will be utilized as a complement to RF in more and more scenarios to improve the performance of communication systems. However, challenges including flickering, dimming, noise and interference remain a significant impediment to the widespread adoption of VLC, which requires further research.

5. Conclusions

In this work, we thoroughly investigate the literature on VLC in recent years. With a high data rate and unlicensed spectrum, VLC is regarded as an excellent alternative to RF to satisfy the increasing demand for wireless resources. Related techniques are researched in physical and MAC layers to reduce the impact of interference and ensure desired communication performance. Additionally, ML has been adopted in VLC systems and the related research has increased exponentially, which shows a new direction for VLC research.

VLC has a wide range of applications in many short-range communication scenarios, such as indoor communication, indoor positioning, vehicular communication and underwater communication. However, there are still many issues and challenges in the application of VLC technology in multiple scenarios, such as flickering, long-distance transmission and interference. Many promising VLC technologies are not yet well developed, and this research area needs further investigation.

Author Contributions: Conceptualization, Z.G., F.N.K., X.G. and Y.D.; writing—original draft preparation, Z.G.; writing—review and editing, F.N.K., X.G. and Y.D.; supervision, F.N.K., X.G. and Y.D.; project administration, F.N.K.; funding acquisition, F.N.K., X.G. and Y.D. All authors have read and agreed to the published version of the manuscript.

Funding: The work was supported in part by the GuangDong Basic and Applied Basic Research Foundation under Grant 2022A1515010209 and Shenzhen Natural Science Foundation under Grant JCYJ20200109143016563. Faisal Nadeem Khan would like to acknowledge the support of Tsinghua Shenzhen International Graduate School and Tsinghua–Berkeley Shenzhen Institute under Scientific Research Startup Fund (Project number: 01010600001(CD2022004C)).

Institutional Review Board Statement: Not applicable.

Informed Consent Statement: Not applicable.

Data Availability Statement: Not applicable.

Conflicts of Interest: The authors declare no conflict of interest.

Abbreviations

The following abbreviations are used in this manuscript:

VLC	Visible light communication
RF	Radio frequency
ML	Machine learning
LED	Light emitting diode
OWC	Optical wireless communication
MAC	Media access control
PLS	Physical layer security
MA	multiple access
NOMA	Non-orthogonal multiple access
IM/DD	Intensity modulation with direct detection
LOS	Line of sight
NLOS	Non-line of sight
CSI	Channel state information
MR	Movement-rotation
GBSM	Geometry-based stochastic model
MIMO	Multiple input multiple output
STFCF	Space-time-frequency correlation function
RMS	Root mean square
OOK	On-off keying

PWM	Pulse width modulation
PPM	Pulse position modulation
OFDM	Orthogonal frequency division multiplexing
PAM4	4-level pulse amplitude modulation
MPI	Multipath interference
DSP	Digital signal processing
SNR	Signal-to-noise ratio
IM	Index modulation
DHT	Discrete Hartley transform
DFT	Discrete Fourier transform
LCI	Lightweight channel-independent
SIC	Successive interference cancellation
OMA	Orthogonal multiple access
DNN	Deep neural network
CNN	Convolutional neural network
ANN	Artificial neural network
UAV	Unmanned aerial vehicle
GRU	Gated recurrent unit
CC	Collaborative constellation
MSE	Mean square error
RNN	Recurrent neural network
LSTM	Long-short-term-memory
GPS	Global positioning system
RSS	Received signal strength
TOA	Time-of-arrival
TDOA	Time difference of arrival
AOA	Angle of arrival
PD	Photodiode
UE	User equipment
VLP	Visible light-based positioning
CIR	Channel impulse response
ITS	Intelligent transportation systems
V-VLC	Vehicular VLC
D2D	Device-to-device
V2V	Vehicle-to-vehicle
PCF	Polynomial curve fitting

References

1. Li, X.; Huo, Y.; Zhang, R.; Hanzo, L. User-Centric Visible Light Communications for Energy-Efficient Scalable Video Streaming. *IEEE Trans. Green Commun. Netw.* **2017**, *1*, 59–67. [CrossRef]
2. Haas, H. Opportunities and Challenges of Future LiFi. In Proceedings of the 2019 IEEE Photonics Conference (IPC), San Antonio, TX, USA, 29 September–3 October 2019; pp. 1–12.
3. Matheus, L.E.M.; Vieira, A.B.; Vieira, L.F.M.; Vieira, M.A.M.; Gnawali, O. Visible Light Communication: Concepts, Applications and Challenges. *IEEE Commun. Surv. Tutor.* **2019**, *21*, 3204–3237. [CrossRef]
4. Obeed, M.; Salhab, A.M.; Alouini, M.-S.; Zummo, S.A. On Optimizing VLC Networks for Downlink Multi-User Transmission: A Survey. *IEEE Commun. Surv. Tutor.* **2019**, *21*, 2947–2976. [CrossRef]
5. Wu, X.; Soltani, M.D.; Zhou, L.; Safari, M.; Haas, H. Hybrid LiFi and WiFi Networks: A Survey. *IEEE Commun. Surv. Tutor.* **2021**, *23*, 1398–1420. [CrossRef]
6. Al-Kinani, A.; Wang, C.-X.; Zhou, L.; Zhang, W. Optical Wireless Communication Channel Measurements and Models. *IEEE Commun. Surv. Tutor.* **2018**, *20*, 1939–1962. [CrossRef]
7. Luo, J.; Fan, L.; Li, H. Indoor Positioning Systems Based on Visible Light Communication: State of the Art. *IEEE Commun. Surv. Tutor.* **2017**, *19*, 2871–2893. [CrossRef]
8. Căilean, A.-M.; Dimian, M. Current Challenges for Visible Light Communications Usage in Vehicle Applications: A Survey. *IEEE Commun. Surv. Tutor.* **2017**, *19*, 2681–2703. [CrossRef]
9. Memedi, A.; Dressler, F. Vehicular Visible Light Communications: A Survey. *IEEE Commun. Surv. Tutor.* **2021**, *23*, 161–181. [CrossRef]
10. Arfaoui, M.A.; Soltani, M.D.; Tavakkolnia, I.; Ghrayeb, A.; Safari, M.; Assi, C.M.; Haas, H. Physical layer security for visible light communication systems: A survey. *IEEE Commun. Surv. Tutor.* **2020**, *22*, 1887–1908. [CrossRef]

11. Gancarz, J.; Elgala, H.; Little, T.D.C. Impact of lighting requirements on VLC systems. *IEEE Commun. Mag.* **2013**, *12*, 34–41. [CrossRef]

12. Zeng, Z.; Fu, S.; Zhang, H.; Dong, Y.; Cheng, J. A Survey of Underwater Optical Wireless Communications. *IEEE Commun. Surv. Tutor.* **2017**, *1*, 204–238. [CrossRef]

13. Khalighi, M.A.; Uysal, M. Survey on Free Space Optical Communication: A Communication Theory Perspective. *IEEE Commun. Surv. Tutor.* **2014**, *4*, 2231–2258. [CrossRef]

14. Kazemi, H.; Safari, M.; Haas, H. A Wireless Optical Backhaul Solution for Optical Attocell Networks. *IEEE Trans. Wireless Commun.* **2019**, *2*, 807–823. [CrossRef]

15. Eldeeb, H.B.; Yanmaz, E.; Uysal, M. MAC Layer Performance of Multi-Hop Vehicular VLC Networks with CSMA/CA. In Proceedings of the 12th International Symposium on Communication Systems, Networks and Digital Signal Processing (CSNDSP), Online, 20–22 July 2020 .

16. Shakil Sejan, M.A.; Habibur Rahman, M.; Chung, W.-Y. Optical OFDM Modulation in Multi-hop VLC for Long Distance Data Transmission Over 30 meters. In Proceedings of the IEEE Photonics Conference (IPC), Vancouver, BC, Canada, 28 September– 1 October 2020.

17. Bian, R.; Tavakkolnia, I.; Haas, H. 15.73 Gb/s Visible Light Communication with Off-the-Shelf LEDs. *J. Light. Technol.* **2019**, *10*, 2418–2424. [CrossRef]

18. Lee, C.; Islim, M.S.; Das, S.; Spark, A.; Videv, S.; Rudy, P.; Raring, J. 26 Gbit/s LiFi System with Laser-Based White Light Transmitter. *J. Light. Technol.* **2022**, *5*, 1432–1439. [CrossRef]

19. Fahs, B.; Chowdhury, A.J.; Hella, M.M. A 12-m 2.5-Gb/s Lighting Compatible Integrated Receiver for OOK Visible Light Communication Links. *J. Light. Technol.* **2016**, *16*, 3768–3775. [CrossRef]

20. Wang, L.; Wang, X.; Kang, J.; Yue, C.P. A 75-Mb/s RGB PAM-4 Visible Light Communication Transceiver System with Pre- and Post-Equalization. *J. Light. Technol.* **2021**, *5*, 1381–1390. [CrossRef]

21. Yang, X.; Tong, Z.; Zhang, H.; Zhang, Y.; Dai, Y.; Zhang, C.; Xu, J. 7-M/130-Mbps LED-to-LED Underwater Wireless Optical Communication Based on Arrays of Series-Connected LEDs and a Coaxial Lens Group. *J. Light. Technol.* **2022**, *17*, 5901–5909. [CrossRef]

22. Chen, X.; Yang, X.; Tong, Z.; Dai, Y.; Li, X.; Zhao, M.; Xu, J. 150 m/500 Mbps Underwater Wireless Optical Communication Enabled by Sensitive Detection and the Combination of Receiver-Side Partial Response Shaping and TCM Technology. *J. Light. Technol.* **2021**, *14*, 4614–4621. [CrossRef]

23. Zhou, H.; Zhang, M.; Wang, X.; Ren, X. Design and Implementation of More Than 50m Real-Time Underwater Wireless Optical Communication System. *J. Light. Technol.* **2022**, *12*, 3654–3668. [CrossRef]

24. Al-Kinani, A.; Sun, J.; Wang, C.-X.; Zhang, W.; Ge, X.; Haas, H. A 2-D Non-Stationary GBSM for Vehicular Visible Light Communication Channels. *Trans. Wirel. Commun.* **2018**, *12*, 7981–7992 [CrossRef]

25. Jani, M.; Garg, P.; Gupta, A. Performance Analysis of a Mixed Cooperative PLC–VLC System for Indoor Communication Systems. *IEEE Syst. J.* **2020**, *1*, 469–476. [CrossRef]

26. Dehghani Soltani, M.; Wu, X.; Safari, M.; Haas, H. Bidirectional User Throughput Maximization Based on Feedback Reduction in LiFi Networks. *IEEE Trans. Commun.* **2018**, *7*, 3172–3186. [CrossRef]

27. Feng, S.; Zhang, R.; Li, X.; Wang, Q.; Hanzo, L. Dynamic Throughput Maximization for the User-Centric Visible Light Downlink in the Face of Practical Considerations. *Trans. Wirel. Commun.* **2018**, *8*, 5001–5015. [CrossRef]

28. Chen, J.; Tavakkolnia, I.; Chen, C.; Wang, Z.; Haas, H. The Movement-Rotation (MR) Correlation Function and Coherence Distance of VLC Channels. *J. Light. Technol.* **2020**, *24*, 6759–6770. [CrossRef]

29. Zhu, X.; Wang, C.-X.; Huang, J.; Chen, M.; Haas, H. A Novel 3D Non-Stationary Channel Model for 6G Indoor Visible Light Communication Systems. *Trans. Wirel. Commun.* **2022**. [CrossRef]

30. Roberts, R.D.; Rajagopal, S.; Lim, S.-K. IEEE 802.15.7 physical layer summary. In Proceedings of the 2011 IEEE GLOBECOM Workshops (GC Wkshps), Houston, TX, USA, 5–9 December 2011; pp. 772–776.

31. Sejan, M.A.S.; Naik, R.P.; Lee, B.-G.; Chung, W.-Y. A Bandwidth Efficient Hybrid Multilevel Pulse Width Modulation for Visible Light Communication System: Experimental and Theoretical Evaluation. *IEEE Open J. Commun. Soc.* **2022**, 1991–2004. [CrossRef]

32. Chitturi, M.V.; Benekohal, R.F.; Girianna, M. Turn-on and turn-off characteristics of incandescent and light-emitting diode signal modules. *ITE J.* **2005**, *69*, 69–73.

33. Yan, K.; Li, Z.; Cheng, M.; Wu, H.-C. QoS Analysis and Signal Characteristics for Short-Range Visible-Light Communications. *IEEE Trans. Veh. Technol.* **2021**, *7*, 6726–6734. [CrossRef]

34. Zhong, K.; Zhou, X.; Gui, T.; Tao, L.; Gao, Y.; Chen, W.; Lu, C. Experimental study of PAM-4 CAP-16 and DMT for 100 Gb/s short reach optical transmission systems. *Opt. Exp.* **2015**, *2*, 1176–1189. [CrossRef] [PubMed]

35. Okuda, S.; Yamatoya, T.; Yamaguchi, T.; Azuma, Y.; Tanaka, Y. High-power low-modulating-voltage 1.5 μm-band CWDM uncooled EMLs for 800 Gb/s (53.125 Gbaud-PAM4) transceivers. In Proceedings of the 2021 Optical Fiber Communications Conference and Exhibition (OFC), San Francisco, CA, USA, 6–10 June 2021.

36. Rahman, T.; Calabrò, S.; Stojanovic, N.; Hossein, M.S.B.; Wei, J.; Kuschnerov, M.; Xie, C. Blind adaptation of partial response equalizers for 200 Gb/s per lane IM/DD systems. In Proceedings of the 2021 European Conference on Optical Communication (ECOC), Bordeaux, France, 13–16 September 2021; pp. 1–3.

37. Yamamoto, S.; Taniguchi, H.; Nakamura, M.; Kisaka, Y. O-Band 10-km PAM transmission using nonlinear-spectrum-shaping encoder and transition-likelihood-based decoder with symbol- and likelihood-domain feedbacks. In Proceedings of the 2021 European Conference on Optical Communication (ECOC), Bordeaux, France, 13–16 September 2021; pp. 1–4.

38. Huang, C.; Song, H.; Dai, L.; Cheng, M.; Yang, Q.; Tang, M.; Deng, L. Optical Multipath Interference Mitigation for High-Speed PAM4 IMDD Transmission System. *J. Light. Technol.* **2022**, *16*, 5490–5501. [CrossRef]

39. Elgala, H.; Mesleh, R.; Haas, H.; Pricope, B. OFDM visible light wireless communication based on white LEDs. In Proceedings of the 2007 IEEE 65th Vehicular Technology Conference—VTC2007-Spring, Dublin, Ireland, 22–25 April 2007; pp. 2185–2189.

40. Armstrong, J. OFDM for optical communications. *J. Light. Technol.* **2009**, *3*, 189–204. [CrossRef]

41. Carruthers, J.B.; Kahn, J.M. Multiple-subcarrier modulation for nondirected wireless infrared communication. *IEEE J. Sel. Areas Commun.* **1996**, *3*, 538–546. [CrossRef]

42. Lee, S.C.J.; Randel, S.; Breyer, F.; Koonen, A.M.J. PAM-DMT for intensity-modulated and direct-detection optical communication systems. *IEEE Photon. Technol. Lett.* **2009**, *23*, 1749–1751. [CrossRef]

43. Ranjha, B.; Kavehrad, M. Hybrid asymmetrically clipped OFDM-based IM/DD optical wireless system. *J. Opt. Commun. Netw.* **2014**, *4*, 387–396. [CrossRef]

44. Wang, Q.; Qian, C.; Guo, X.; Wang, Z.; Cunningham, D.G.; White, I.H. Layered ACO-OFDM for intensity-modulated direct-detection optical wireless transmission. *Opt. Express* **2015**, *9*, 12382–12393. [CrossRef]

45. Bai, R.; Wang, Q.; Wang, Z. Asymmetrically clipped absolute value optical OFDM for intensity-modulated direct-detection systems. *J. Lightw. Technol.* **2017**, *17*, 3680–3691. [CrossRef]

46. Bai, R.; Hranilovic, S. Absolute Value Layered ACO-OFDM for Intensity-Modulated Optical Wireless Channels. *IEEE Trans. Commun.* **2020**, *11*, 7098–7110. [CrossRef]

47. Basar, E.; Wen, M.; Mesleh, R.; Di Renzo, M.; Xiao, Y.; Haas, H. Index Modulation Techniques for Next-Generation Wireless Networks. *IEEE Access* **2017**, *5*, 16693–16746. [CrossRef]

48. Xu, X.-Y.; Zhang, Q.; Yue, D.-W. Orthogonal Frequency Division Multiplexing with Index Modulation Based on Discrete Hartley Transform in Visible Light Communications. *IEEE Photonics J.* **2022**, *3*, 1–10. [CrossRef]

49. Classen, J.; Chen, J.; Steinmetzer, D.; Hollick, M.; Knightly, E. The spy next door: Eavesdropping on high throughput visible light communications. In Proceedings of the 2nd International Workshop on Visible Light Communications Systems, Paris, France, 11 September 2015; pp. 9–14.

50. Yang, N.; Wang, L.; Geraci, G.; Elkashlan, M.; Yuan, J.; Di Renzo, M. Safeguarding 5G wireless communication networks using physical layer security. *IEEE Commun. Mag.* **2015**, *4*, 20–27. [CrossRef]

51. Su, N.; Panayirci, E.; Koca, M.; Yesilkaya, A.; Poor, H.V.; Haas, H. Physical Layer Security for Multi-User MIMO Visible Light Communication Systems with Generalized Space Shift Keying. *IEEE Trans. Commun.* **2021**, *4*, 2585–2598. [CrossRef]

52. Al-Moliki, Y.M.; Alresheedi, M.T.; Al-Harthi, Y. Chaos-based physical-layer encryption for OFDM-based VLC schemes with robustness against Known/chosen plaintext attacks. *IET Optoelectron.* **2019**, *13*, 124–133. [CrossRef]

53. Al-Moliki, Y.M.; Alresheedi, M.T.; Al-Harthi, Y.; Alqahtani, A.H. Robust Lightweight-Channel-Independent OFDM-Based Encryption Method for VLC-IoT Networks. *IEEE Internet Things J.* **2022**, *6*, 4661–4676. [CrossRef]

54. Melki, R.; Noura, H.N.; Chehab, A. Efficient & Secure Physical Layer Cipher Scheme for VLC Systems. In Proceedings of the 2019 IEEE 90th Vehicular Technology Conference (VTC2019-Fall), Honolulu, HI, USA, 22–25 September 2019; pp. 1–6.

55. Shiu, Y.-S.; Chang, S.Y.; Wu, H.-C.; Huang, S.C.-H.; Chen, H.-H. Physical layer security in wireless networks: A tutorial. *IEEE Trans. Wirel. Commun.* **2011**, *2*, 66–74. [CrossRef]

56. Yin, L.; Popoola, W.O.; Wu, X.; Haas, H. Performance Evaluation of Non-Orthogonal Multiple Access in Visible Light Communication. *IEEE Trans. Commun.* **2016**, *12*, 5162–5175. [CrossRef]

57. Marshoud, H.; Kapinas, V.M.; Karagiannidis, G.K.; Muhaidat, S. Non-orthogonal multiple access for visible light communications. *IEEE Photon. Technol. Lett.* **2016**, *1*, 51–54. [CrossRef]

58. Marshoud, H.; Sofotasios, P.C.; Muhaidat, S.; Karagiannidis, G.K.; Sharif, B.S. Error performance of NOMA VLC systems. In Proceedings of the 2017 IEEE International Conference on Communications (ICC), Paris, France, 21–25 May 2017; pp. 1–16.

59. Hammadi, A.A.; Sofotasios, P.C.; Muhaidat, S.; Al-Qutayri, M.; Elgala, H. Non-orthogonal multiple access for hybrid VLC-RF networks with imperfect channel state information. *IEEE Trans. Veh. Technol.* **2020**, *5*, 4238–4253. [CrossRef]

60. Le-Tran, M.; Vu, T.-H.; Kim, S. Performance Analysis of Optical Backhauled Cooperative NOMA Visible Light Communication. *IEEE Trans. Veh. Technol.* **2021**, *12*, 12932–12945. [CrossRef]

61. Uday, T.; Kumar, A.; Natarajan, L. NOMA for Multiple Access Channel and Broadcast Channel in Indoor VLC. *IEEE Wirel. Commun. Lett.* **2021**, *3*, 609–613. [CrossRef]

62. Naser, S.; Bariah, L.; Muhaidat, S.; Sofotasios, P.C.; Al-Qutayri, M.; Damiani, E.; Debbah, M. Toward Federated-Learning-Enabled Visible Light Communication in 6G Systems. *IEEE Wirel. Commun.* **2022**, *29*, 48–56. [CrossRef]

63. Najla, M.; Mach, P.; Becvar, Z. Deep Learning for Selection Between RF and VLC Bands in Device-to-Device Communication. *IEEE Wirel. Commun. Lett.* **2020**, *10*, 1763–1767. [CrossRef]

64. He, J.; Zhou, B. A Deep Learning-Assisted Visible Light Positioning Scheme for Vehicles with Image Sensor. *IEEE Photonics J.* **2022**, *4*, 1–7. [CrossRef]

65. Wang, Y.; Chen, M.; Yang, Z.; Luo, T.; Saad, W. Deep Learning for Optimal Deployment of UAVs with Visible Light Communications. *IEEE Trans. Wirel. Commun.* **2020**, *11*, 7049–7063. [CrossRef]

66. Le-Tran, M.; Kim, S. Deep Learning-Based Collaborative Constellation Design for Visible Light Communication. *IEEE Commun. Lett.* **2020**, *11*, 2522–2526. [CrossRef]
67. Wu, Z.-Y.; Ismail, M.; Serpedin,E.; Wang, J. Efficient Prediction of Link Outage in Mobile Optical Wireless Communications. *IEEE Trans. Wirel. Commun.* **2021**, *2*, 882–896. [CrossRef]
68. Arfaoui, M.A.; Soltani, M.D.; Tavakkolnia, I.; Ghrayeb, A.; Assi, C.M.; Safari, M.; Haas, H. Measurements-Based Channel Models for Indoor LiFi Systems. *IEEE Trans. Wirel. Commun.* **2021**, *2*, 827–842. [CrossRef]
69. Purwita, A.A.; Soltani, M.D.; Safari, M.; Haas, H. Terminal Orientation in OFDM-Based LiFi Systems. *IEEE Trans. Wireless Commun.* **2019**, *8*, 4003–4016. [CrossRef]
70. Yeh, C.H.; Weng, J.H.; Chow, C.W.; Luo, C.M.; Xie, Y.R.; Chen, C.J.; Wu, M.C. 1.7 to 2.3 Gbps OOK LED VLC Transmission Based on 4 × 4 Color-Polarization-Multiplexing at Extremely Low Illumination. *IEEE Photonics J.* **2019**, *4*, 1–6. [CrossRef]
71. Lan, H.; Yu, C.; Zhuang, Y.; Li, Y.; El-Sheimy, N. A novel Kalman filter with state constraint approach for the integration of multiple pedestrian navigation systems. *Micromachines* **2015**, *6*, 926–952. [CrossRef]
72. Zhuang, Y.; El-Sheimy, N. Tightly-coupled integration of WiFi and MEMS sensors on handheld devices for indoor pedestrian navigation. *IEEE Sens. J.* **2016**, *1*, 224–234. [CrossRef]
73. Zhuang, Y.; Syed, Z.; Li, Y.; El-Sheimy, N. Evaluation of two WiFi positioning systems based on autonomous crowdsourcing of handheld devices for indoor navigation. *IEEE Trans. Mobile Comput.* **2016**, *8*, 1982–1995. [CrossRef]
74. Zhuang, Y.; Yang, J.; Li, Y.; Qi, L.; El-Sheimy, N. Smartphone-based indoor localization with Bluetooth low energy beacons. *Sensors* **2016**, *16*, 596. [CrossRef] [PubMed]
75. Zhuang, Y.; Hua, L.; Qi, L.; Yang, J.; Cao, P.; Cao, Y.; Haas, H. A Survey of Positioning Systems Using Visible LED Lights. *IEEE Commun. Surv. Tutor.* **2018**, *3*, 1963–1988. [CrossRef]
76. Zhou, B.; Liu, A.; Lau, V. Joint user location and orientation estimation for visible light communication systems with unknown power emission. *IEEE Trans. Wirel. Commun.* **2019**, *11*, 5181–5195. [CrossRef]
77. Chen, Y.; Zheng, H.; Liu, H.; Han, Z.; Ren, Z. Indoor High Precision Three-Dimensional Positioning System Based on Visible Light Communication Using Improved Hybrid Bat Algorithm. *IEEE Photonics J.* **2020**, *5*, 1–13. [CrossRef]
78. Lin, X.; Zhang, L. Intelligent and Practical Deep Learning Aided Positioning Design for Visible Light Communication Receivers. *IEEE Commun. Lett.* **2020**, *3*, 577–580. [CrossRef]
79. Siegel, J.E.; Erb, D.C.; Sarma, S.E. A survey of the connected vehicle landscape—Architectures enabling technologies applications and development areas. *IEEE Trans. Tell. Transp. Syst.* **2018**, *8*, 2391–2406. [CrossRef]
80. Béchadergue, B.; Chassagne, L.; Guan, H. Simultaneous Visible Light Communication and Distance Measurement Based on the Automotive Lighting. *IEEE Trans. Intell. Veh.* **2019**, *4*, 532–547. [CrossRef]
81. Amjad, M.S.; Tebruegge, C.; Memedi, A.; Kruse, S.; Kress, C.; Scheytt, J.C.; Dressler, F. Towards an IEEE 802.11 Compliant System for Outdoor Vehicular Visible Light Communications. *IEEE Trans. Veh. Technol.* **2021**, *6*, 5749–5761. [CrossRef]
82. Li, G.; Niu, W.; Ha, Y.; Hu, F.; Wang, J.; Yu, X.; Chi, N. Position-Dependent MIMO Demultiplexing Strategy for High-Speed Visible Light Communication in Internet of Vehicles. *IEEE Internet Things J.* **2022**, *13*, 10833–10850. [CrossRef]
83. Rapson, C.J.; Seet, B.-C.; Chong, P.H.J.; Klette, R. Safety Assessment of Radio Frequency and Visible Light Communication for Vehicular Networks. *IEEE Wirel. Commun.* **2020**, *1*, 186–192. [CrossRef]
84. Ata, Y.; Kiasaleh, K. On the performance of wireless optical communication systems in underwater channels. In Proceedings of the 2022 IEEE International Conference on Communications Workshops (ICC Workshops), Seoul, Republic of Korea, 16–20 May 2022; pp. 634–639.

Review

Visible Light Communications: A Survey on Recent High-Capacity Demonstrations and Digital Modulation Techniques

Pedro A. Loureiro [1,2,*], **Fernando P. Guiomar** [1] and **Paulo P. Monteiro** [1,2]

[1] Instituto de Telecomunicações, 3810-193 Aveiro, Portugal; guiomar@av.it.pt (F.P.G.); paulo.monteiro@ua.pt (P.P.M.)

[2] Department of Electronics, Telecommunications and Informatics, University of Aveiro, 3810-193 Aveiro, Portugal

* Correspondence: pedro.a.loureiro@av.it.pt

Abstract: In order to deal with the increasing number of mobile devices and with their demand for Internet services, particularly social media platforms, streaming video, and online gaming, Radio-Frequency (RF) wireless networks have been pushed to their capacity limits. In addition to this, 80% of the total data traffic is carried out by users inside buildings. Therefore, new technologies have started to be considered for indoor wireless communications. Visible Light Communications (VLC) can provide both illumination and communications, appearing as an alternative or complement to RF wireless networks. VLC offers high bandwidth and immunity to interference from electromagnetic sources. This manuscript reviews recent high-capacity VLC demonstrations. The main focus of this work is to present digital-signal-processing techniques used in VLC systems. Different modulation formats are analyzed, which can be divided into two large groups, namely single-carrier and multi-carrier modulation schemes. Finally, some recently proposed capacity-achieving strategies are presented. We discuss how to implement these techniques and how they will be useful for the continued development of VLC systems.

Keywords: 5G and beyond; visible light communications; optical wireless applications; laser diodes; light-emitting diodes

Citation: Loureiro, P.A.; Guiomar, F.P.; Monteiro, P.P. Visible Light Communications: A Survey on Recent High-Capacity Demonstrations and Digital Modulation Techniques. *Photonics* **2023**, *10*, 993. https://doi.org/10.3390/photonics10090993

Received: 31 July 2023
Revised: 23 August 2023
Accepted: 29 August 2023
Published: 30 August 2023

1. Introduction

With the emergence of 5G, a series of new applications have been introduced, such as autonomous driving, communication between objects, and industrial automation. All these applications are achieved due to high data rates, low latency, and ultra-reliable communications. Therefore, it is expected that these requirements will be even higher in the coming years [1]. Furthermore, considering that more than 80% of the total mobile data traffic is generated indoors, it is important to introduce a new technology that works mainly inside buildings and complements Radio-Frequency (RF) networks, which are rapidly becoming highly congested and also limited by electromagnetic interference [2].

In order to tackle this challenge, the possibility of introducing optical wireless communications has emerged as a potential alternative [3]. The optical band includes infrared, visible, and ultraviolet light. The most-common use of light for communications is in fiber optics, which utilizes optical wavelengths, typically infrared, to transmit data over fiber. Moreover, several works have demonstrated wireless infrared communications, known as Free-Space Optics (FSO) [4,5]. However, the performance of FSO communications is highly dependent on the directivity of the optical beam, making it unusable in indoor mobile communications. In contrast, ultraviolet is generally not considered for communications because of the risks introduced. Light emitted at this wavelength can be harmful to human eyes if protection is not used. Furthermore, prolonged exposure can lead to serious health problems, where UVA and UVB are the most dangerous, since they are not absorbed by the Earth's atmosphere. Therefore, Visible Light Communications (VLC) is seen as a potential technology for indoor short-range communications, because optical beams do not need to be very directive to provide good communications, nor are there great risks to human health when exposed to visible light. Moreover, lights are normally used in closed spaces (shopping centers, offices, and houses), where most mobile communications are performed [6].

VLC takes advantage of hundreds of terahertz (430–790 THz) of unlicensed bandwidth in the visible band to perform wireless communications. On the other hand, for the RF band, mainly sub-6 GHz, the frequency allocation is restricted and regulated in each country, making it impossible to explore these bands to provide faster communications [7]. WiFi technology offers two frequency regions to be used by unlicensed devices, 2.4 GHz and 5 GHz, which is quite limited when compared to the hundreds of terahertz available in VLC. Moreover, visible light causes no electromagnetic interference, so it does not affect the performance of other electronic devices [8]. Thus, VLC can be employed in places where sensitive electronic equipment is used, such as hospitals and aircraft, where any significant interference could have tragic effects. Additionally, in this type of communication, the generated signal will be confined within the room, since light cannot penetrate opaque objects, providing secure wireless communications. In contrast, RF communications are characterized by signal propagation for hundreds of meters, passing through walls and other solid surfaces, thus facilitating its intrusive interception [3]. Lastly, this technology offers the possibility of reusing the already implemented lighting infrastructure, thereby reducing installation and operation costs.

This survey focuses mainly on presenting recent high-capacity VLC demonstrations. Throughout the article, several currently used modulation techniques are highlighted and some high-speed VLC systems reported in recent years are presented. The rest of this work is organized as follows. We start in Section 2 by briefly presenting the evolution of VLC standards and the main VLC applications. In Section 3, modulation methods typically used in VLC are addressed. Section 4 presents some state-of-the-art capacity-achieving strategies for VLC indoor applications. Finally, the challenges and future research directions are discussed in Section 5.

2. Progress on Standardization and VLC Applications Scenarios

Recently, the interest in VLC has increased exponentially, as can be easily verified by the number of papers published over the recent years. Figure 1 presents the number of documents published over the years that used the expression "Visible light communication" or the keyword "VLC" within the body or title of the paper. We can see that the number of publications using these expressions in the title reached a halt after 2020, likely influenced by the pandemic, but on the other hand, the number of articles referring to VLC has increased successively every year, demonstrating a clearly growing interest in this topic. This increase is also due to the appearance of the first standards, which validates the potential of this technology and motivates both industry and academia to invest in it.

2.1. VLC Standardization

The VLC standardization process started in 2003 with the creation of the Visible Light Communication Consortium (VLCC) in Japan, aiming at creating the first VLC standard. However, only in 2007, the VLCC proposed the first two standards to the Japan Electronics and Information Technology Industries Association's (JEITA): JEITA CP-1221 and JEITA CP-1222, which introduced the basics of VLC systems [7]. Despite these efforts, none of the referenced standards focus on flickering and dimming mitigation. Therefore, in 2011, the IEEE 802.15.7 standard for the link and physical layer of a VLC system was proposed to address some practical issues associated with VLC systems [9]. First is the integration of the VLC system with the already-standardized wireless communication technology, for example, WiFi. Secondly, this standard solved the interference problem with ambient light sources. Subsequently, mobility issues, such as handover, were properly assessed. Then, Forward-Error-Correction (FEC) schemes were selected in order to improve communication performance. Finally, interference between VLC devices was considered [8]. Broadly speaking, the IEEE 802.15.7 standard is divided into three Physical (PHY) types for VLC: PHY I, PHY II, and PHY III. PHY I works from 11.67 to 266.6 Kbit/s; PHY II operates from 1.25 to 96 Mbit/s; PHY III offers bit rates between 12 and 96 Mbit/s [7]. The first two modes of operation use a single light source, supporting On–Off Keying (OOK) and variable Pulse-Position Modulation (VPPM). On the other hand, PHY III uses multiple optical sources with different wavelengths, introducing the Color-Shift-Keying (CSK)-modulation scheme. For the different modulation format options, there is a trade-off between high bit rates and flickering and dimming mitigation [10].

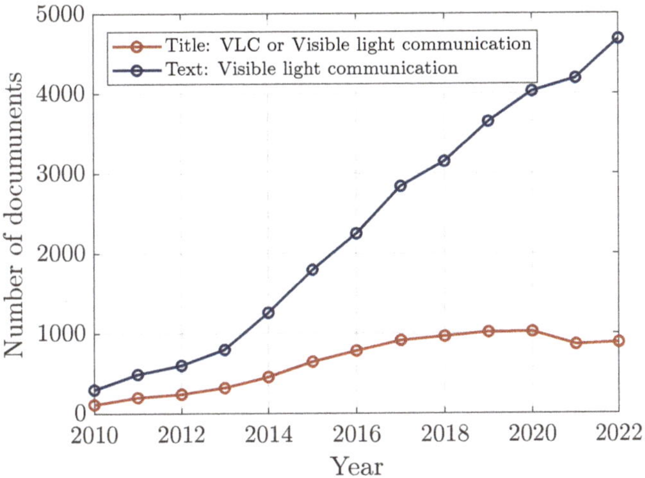

Figure 1. Number of VLC publications over the years. These data were obtained from the Google Scholar search engine.

Recently, the IEEE 802.15.7 standard started to be revisited by a new task group (IEEE P802.15.13) in order to increase the data rate for specialty applications [11]. The main objective of this new standard is to define a PHY and Media Access Control (MAC) layer using wavelengths from 10,000 nm to 190 nm to achieve a multi-gigabit/second optical wireless communication system [12]. The current standard version establishes two options for VLC, the energy-efficient Pulse Modulation (PM)-PHY and the spectrally efficient High-Bandwidth (HB)-PHY. The first option is indicated for low-power applications, such as uplink, the IoT, and Industry 4.0. It offers up to a 200 MHz bandwidth using the low-spectral-efficiency OOK modulation format. Instead, the HB-PHY is based on Orthogonal Frequency Division Multiplexing (OFDM) and offers a bandwidth of up to 1 GHz, also allowing the use of bit loading [11].

However, the IEEE P802.15.13 standard is being designed mainly for industrial applications and is not compatible with existing wireless networks. Therefore, the 802.11 Working Group, dedicated to the development of standards for communications in wireless networks, created the Task Group bb (TGbb) to study the possibility of integrating LiFi into the WiFi standard. 802.11bb intends to introduce a VLC system to the WiFi network, allowing it to address more use cases than IEEE 802.15.7 and IEEE P802.15.13, where just low data rate communications and industrial applications were considered, respectively [13].

These evolutions in standardization have motivated both industry and academia, as mentioned above. While the number of scientific articles easily measures interest in academia, in industry, this can be seen by the appearance of commercial products. Currently, the global leader in VLC technology is pureLiFi. They brought to market the first commercial visible light antennas operating at more than 1 Gbit/s in the downlink direction and 600 Mbit/s in the uplink direction. Their system is compatible with the IEEE 802.11 standard and has been tested in classrooms, hospitals, and real-time sensor monitoring [14]. In turn, other companies started to demonstrate high-speed VLC solutions, such as the Fraunhofer Heinrich Hertz Institute and Oledcomm [15].

2.2. VLC Application Scenarios

Beyond the aforementioned indoor mobile communications, VLC presents a wide range of applications and potentialities. The use of visible light can also be essential for underwater communications, vehicular communications, and indoor localization.

2.2.1. Indoor Wireless Communications

LiFi provides a high-speed bidirectional communication, equivalent to WiFi, but with visible light. Currently, most homes and buildings are equipped with Light-Emitting Diodes (LEDs), which can become LiFi access points, where the lamps are used for both room illumination and communications. Therefore, considering a room with several VLC transmitters, they can be organized in a way to reduce interference, allowing the introduction of coordinated multi-point transmission, which offers the possibility of applying Multiple-Input, Multiple-Output (MIMO) techniques. In the literature, this is the most-explored application, with thousands of works exploring different system approaches in an indoor scenario.

In the early 2000s, the first VLC works began to appear, using the visible light of LEDs to illuminate and carry out communications in indoor scenarios [16,17]. One of the most-common strategies to generate white light relies on the use of a blue LED with a yellow phosphor [18,19]. Although this single-LED approach has captured significant attention mainly owing to its simplicity and low cost, the most-common approach in recent VLC works is based on the use of at least three LEDs with different colors, usually Red, Green, and Blue (RGB), to produce white light [20,21]. Despite the progressive improvement of the LED-based VLC system, the performance is limited by the low bandwidth of the source (typically tens to hundreds of megahertz), not allowing it to exceed 20 Gbit/s. Therefore, more recently, VLC systems using Laser Diodes (LDs) have been proposed to improve the performance of these systems. In LD-based systems, there are some different approaches, but, similar to the implementations with LEDs, most of the works tend to combine the color of multiple transmitters, currently allowing exceeding 40 Gbit/s [22,23]. On the other hand, several modulation formats are explored, always aiming to maximize the system's capacity. In the next section, several modulation techniques used in VLC systems will be presented.

2.2.2. Underwater Wireless Communications

The growing investment and interest in underwater wireless communications is fostered by the increase in underwater human activities, namely oceanography studies, oil exploration, and military warship-to-submarine communication [24]. Traditionally, acoustic waves have been used in these scenarios, being able to support transmission distances in the order of a few kilometers. However, the main drawback of acoustic communications is the slow propagation of sound waves, resulting in a large latency. Moreover, the data rate is limited to tens of Kbit/s due to the strong attenuation of sound in seawater, as verified by some works [25–27]. Alternatively, the use of electromagnetic waves was suggested, initially in the RF band, allowing increasing the capacity of the system in relation to acoustic waves, but on the other hand, the distance is considerably smaller due to the very high attenuation [24]. In order to improve the performance of RF underwater communications, a large-sized antenna is needed, considerably increasing the costs of the system [28]. Therefore, Underwater Optical Communications (UWOC) has been proposed as an alternative to acoustic and RF underwater communication links for short and moderate distances. Visible light has a great potential for this type of communication due to the low absorption of seawater in the blue-green region (400–550 nm) of the visible spectrum, allowing providing data rates up to a few Gbit/s [24,29,30]. Hassan M. Oubei et al. experimentally demonstrated an underwater wireless VLC system with a bit rate of up to 4.8 Gbit/s over a tank of water with 60 cm, achieving a distance of 5.4 m with the help of mirrors. The authors verified that the transmission distance can be increased since the attenuation coefficient at this wavelength is very small [29]. Tsai-Chen Wu et al. decided to experimentally study the performance of these systems using tap water and seawater. A bit rate of 7.2 Gbit/s for a distance of 6.8 m was demonstrated when the light passed through a tank of seawater. On the other hand, as expected, using tap water, the achieved bit rate was 9.2 Gbit/s due to lower attenuation of the light beam [24]. Alternatively, Jianyang Shi et al. experimentally demonstrated a net data rate of 14.6 Gbit/s over 1.2 m of underwater distance using five primary-color LEDs, validating the viability of this alternative approach [30]. For a more-detailed analysis of UWOC systems, several comprehensive surveys can be found in the literature, describing the fundamentals, main research problems, and future directions of this technology [28,31–33].

2.2.3. Vehicular Communications

In recent years, the number of cars on the road has increased exponentially, resulting in a greater number of road accidents, making it one of the main causes of death. Therefore, several government institutions, the automobile industry, and the scientific community have joined efforts to improve road safety. One of the best ways to prevent road accidents is to introduce real-time wireless communications to facilitate the interaction between vehicles and traffic infrastructure [34]. Although some RF-based technologies have been proposed as a solution to this problem, such as WiFi and Bluetooth, they are not ideal due to the very-low-latency synchronization requirements. Therefore, other technologies were studied, with VLC emerging as a promising alternative [34]. VLC can be used in vehicular communication since its environment offers a large number of light sources, such as vehicle lights and traffic lights. Currently, the automobile industry is adopting LEDs as light sources, enabling the introduction of VLC. Therefore, by implementing VLC transmitters/receivers in cars, traffic lights, and streetlights, it will be possible to create a network capable of extracting and exchanging information among multiple users [7]. In this type of scenario, there can be two types of communications, vehicle-to-vehicle and vehicle-to-infrastructure. For example, traffic lights can be used to transmit information about vehicle safety and traffic [6]. In the literature, there are some works demonstrating communication between road infrastructures and moving vehicles [35,36]. In [35], Ning Wang et al. experimentally demonstrated an intelligent transportation system based on VLC. The proposed communication system controls the traffic lights to ensure that large vehicles do not need to perform emergency braking. Furthermore, D. Marabissi et al. presented one of the first experiments using 5G in a VLC system for vehicular communications [36]. In another case, streetlights can offer wireless data communications to cars and pedestrians while also being used to light the streets. Regarding vehicle-to-vehicle communications, this can be used to transmit data between the various vehicles to enhance road safety [37,38]. In this scenario, both headlights and taillights on automobiles can be used as VLC transmitters/receivers to provide reliable communications [6]. Unlike indoor applications, where data rates reach multiple Gbit/s and distances are only a few meters, in this scenario, the distances can exceed one-hundred meters, obviously resulting in lower data rates, also due to interference from other light sources [34]. Nevertheless, the viability of this system for higher distances was demonstrated in [38], where reliable communications were established at 75 m.

2.2.4. Visible Light Positioning with Integrated Communications

Lastly, visible light is seen as a potential solution for indoor localization where the Global Positioning System (GPS) usually fails. GPS is known to be the most-widespread technology for localization applications, owing to its ubiquity and high precision in outdoor environments. However, inside buildings, it has limitations due to the effect of reflections and the difficulty of penetrating walls, offering only an accuracy of several meters, which limits its applicability [3]. Currently, indoor localization based on WiFi is seen as the most-attractive alternative. This technique uses access points to estimate the user's position. However, the main advantage that VLC has over WiFi is the higher number of LEDs when compared to WiFi access points [6]. The higher density of LEDs offers better accuracy. Furthermore, this visible-light-based localization system offers the interesting possibility of integrating wireless communications. This possibility has already been experimentally tested in several works [39–41]. In [41], Kottke et al. experimentally demonstrated LiFi-based positioning and communication with data rates up to 500 Mbit/s and positioning accuracy of more than 7 cm. There are numerous positioning algorithms that allow centimeter accuracy, namely resorting to the Time Of Arrival (TOA) [41], Received Signal Strength (RSS) [42], and Angle Of Arrival (AOA) [43], which are among the most-popular positioning methodologies for VLC systems [44,45].

As exposed above, there are many interesting applications that can use visible light. The application with the greatest potential for use is indoor communications, where it can be used for both lighting and high-speed wireless communications. However, the other mentioned scenarios also present interesting advantages, which could increase the interest in their implementation. Thus, in the next sections, digital modulation techniques typically used in VLC will be presented, which allow reaching higher bit rates in different VLC applications. Recently proposed techniques that could improve the system's capacity will also be highlighted.

3. Modulation Techniques for Visible Light Communications

Contrary to RF communications, in VLC, it is not possible to encode data in the phase and amplitude of the light signal. The information is transmitted using variations in light intensity, and reception is performed by direct detection [46]. This technique is named Intensity-Modulated Direct Detection (IM-DD). Moreover, in VLC, the modulation scheme has to simultaneously ensure high data rate transmission and good lighting quality. Consequently, two factors need to be considered, dimming and flickering [8]. Depending on the activity, different luminosity values have to be considered to enable a good human experience. There are some scenarios where an illuminance in the range of 30–100 lux is sufficient (public places), but others require a higher level of illuminance in the range of 300–1000 lux, such as offices [6]. This means that the considered modulation formats have to support different lighting values without significantly affecting the communications. On the other hand, residual changes of brightness caused by light modulation cannot be detected by the human eye. Therefore, the luminous intensity has to vary at a frequency greater than 200 Hz, as specified by the IEEE 802.15.7 standard [9]. Physical-layer-modulation techniques typically used in VLC can be divided into two groups, single-carrier- and multi-carrier-modulation schemes. Throughout this section, a survey of the most-relevant modulation and signal-processing techniques for high-capacity VLC systems is carried out.

3.1. Single-Carrier Techniques

Single-carrier-modulation techniques have been widely used in VLC systems over the years. The most-widely employed include On–Off Keying (OOK), Pulse Amplitude Modulation (PAM), Carrier-Less Amplitude and Phase Modulation (CAP), Pulse-Position Modulation (PPM), Pulse-Width Modulation (PWM), and Color-Shift Keying (CSK), which we will briefly review in the following.

3.1.1. On–Off Keying and Pulse Amplitude Modulation

The OOK method is the simplest and the easiest to implement, where the data bits "0" and "1" can be transmitted with two different levels of light intensity [46]. Most early VLC works used OOK modulation [47,48]. For example, H. Le Minh et al. applied the OOK modulation format to experimentally demonstrate data transmission at 40 Mbit/s for a link distance of 2 m [47]. On the other hand, in [48], J. Vucic et al. demonstrated an indoor VLC link using white LEDs operating at 125 Mbit/s over a 5 m free-space distance. However, as verified, these works suffered from the limited data rate, which has motivated the development of new modulation techniques, such as pulse-amplitude-modulation methods [8]. PAM is a more-advanced format with higher spectral efficiency, where the data are modulated into the amplitude of the signal. The PAM modulation format has been experimentally demonstrated with bit rates close to 1 Gbit/s using PAM-4 and PAM-8, clearly improving the performance compared to other works using OOK [49,50]. In addition to these works, higher-order modulation formats were considered with bit rates up to 10-Gbit/s using PAM-32 data encoding [51].

3.1.2. Carrier-Less Amplitude and Phase Modulation

A modulation format often used in VLC systems is CAP [19,52–54]. CAP was proposed to be an option to generate a real-value signal. For instance, complex-value signals such as QAM need to be hardware up-converted to an RF frequency at the transmitter in order to directly modulate the VLC transmitter. However, the typical VLC transmitter has a very limited bandwidth, making this approach not recommended [19]. CAP is very similar to this strategy, having both the same spectral characteristics and theoretical performance [19]. It uses two digital filters with an orthogonal impulse response to generate two separate data streams, in-phase and quadrature. The resulting signal is centered at an intermediate frequency. In this way, a real-value signal is generated using a simpler and less-expensive system. In Figure 2, we can see the block diagram of CAP modulation and demodulation. First, the bit stream is encoded and mapped into the constellation, and then, the in-phase ($s_I[n]$)- and quadrature ($s_Q[n]$)-generated signals are separated, taking the real and imaginary parts of the signal. The next step is the orthogonal filters, which are obtained by multiplying, in the time domain, the shaping filters and a sine/cosine:

$$f_I(t) = g(t)\cos(2\pi f_c t), \quad f_Q(t) = g(t)\sin(2\pi f_c t), \tag{1}$$

where $g(t)$ is the impulse response of the shaping filter, for example a square-root-raised-cosine function [54]. The frequency of sine and cosine represents the bandpass frequency of the transmitted signal. After adding both signals, the final step is to convert the signal from digital to analog ($s(t)$). The CAP signal is expressed by:

$$s(t) = s_I(t)f_I(t) - s_Q(t)f_Q(t). \tag{2}$$

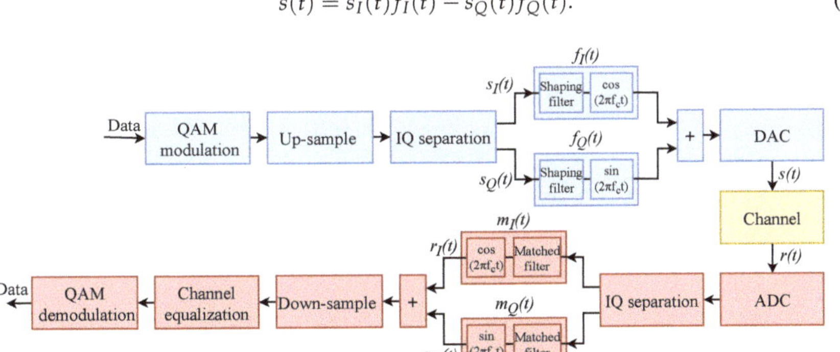

Figure 2. Block diagrams of the CAP modulation and demodulation.

Regarding the CAP demodulation, the inverse operations of the modulation are performed, with only the application of an equalizer, which improves the frequency response and the performance of the system [19]. In [52], the CAP scheme was compared with OFDM in a VLC system. The authors verified that CAP has the potential for low power consumption, low cost, and low complexity. Therefore, it was concluded that the CAP scheme is a good alternative with competitive performance for VLC systems. The main disadvantage of CAP is the poor performance with non-flat frequency channels, with the need to use very complex equalizers, which would reduce its simplicity. In order to solve this problem, Multiband CAP was proposed, where the CAP signal is divided into smaller sub-bands [55]. This alternative was experimentally demonstrated in a VLC system by P. Haigh et al., demonstrating gains over the conventional CAP modulation scheme [56].

3.1.3. Pulse Width/Position Modulation

PWM is an efficient method to modulate the light and control the dimming since the widths of the pulses can be adjusted. Instead, in PPM, the symbol duration is divided into t slots and the pulse is transmitted in one of the t slots, where each position of the pulse represents a different symbol [6]. However, transmitting only one pulse per symbol duration is spectrally inefficient. Therefore, Overlapping PPM (OPPM) and Multipulse PPM (MPPM) were proposed to solve this limitation and transmit more than one pulse per symbol duration [6]. Finally, Variable-PPM (VPPM) is a modulation scheme that controls the dimming of light and, at the same time, enables communications. VPPM has the simplicity and robustness of PPM and can change the dimming by adjusting the pulse width [57].

3.1.4. Color-Shift Keying

Alternatively, in order to overcome the lower data rate and limited dimming support issues of other modulation schemes, the IEEE 802.15.7 standard proposed CSK modulation [9]. It uses multiple optical sources with different colors (wavelengths) [58]. The data are transmitted through the variation of color emitted by RGB VLC transmitters [59]. CSK modulation uses the "Commission Internationale de l'éclairage" (CIE) 1931 color space, which is a graphical representation of all colors perceived by humans, and it is represented in two chromaticity coordinates—x and y—as can be seen in Figure 3 [60]. In Figure 3a–c, an example of the 4CSK, 8CSK, and 16CSK constellations is presented based on the specifications provided by the IEEE 802.15.7 standard. Each symbol represents a different combination of the three colors, resulting in different CIE 1931 coordinates. This approach allows a white color to be produced by joining the three colors, which is the desired color for illumination in indoor and outdoor applications. The main advantage of CSK is that it supports dimming and flickering control. First, by simply varying the driving current of the transmitter, the brightness of the resulting white light is adjusted, while the transmitted power is constant. Therefore, there are no fluctuations in light intensity, reducing potential complications in human health, such as nausea or epilepsy [59]. Over the years, many works have been published using CSK modulation, mainly with the aim of developing algorithms to optimize the constellation points [59,61,62]. Interestingly, different approaches with four LEDs (blue, cyan, yellow, and red) began to appear, different from the three transmitters used in conventional CSK [63]. In this way, it was possible to create square constellations identical to QAM.

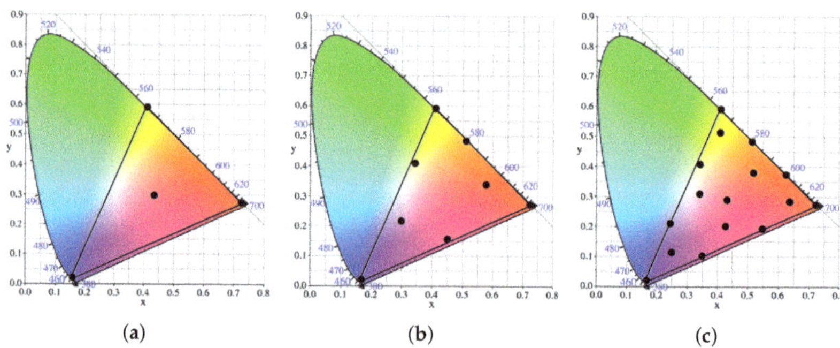

Figure 3. CSK constellations provided by the IEEE 802.15.7 standard: (**a**) 4-CSK; (**b**) 8-CSK; (**c**) 16-CSK.

3.2. Multi-Carrier Techniques

The main limitation of the previous single-carrier modulation methods is that, for higher data rates, the Inter-Symbol Interference (ISI) rises considerably due to the nonlinear frequency response of the VLC transmitters [46]. In order to improve the performance and data rate of band-limited VLC systems, multi-carrier signals can be used. In RF systems, the solution to this problem is to use OFDM. In VLC, OFDM is also frequently used in various works; however, other multi-carrier options can also be considered, as will be described throughout this section.

3.2.1. Orthogonal Frequency Division Multiplexing

An OFDM signal consists of a set of orthogonal sinc-shaped subcarriers in the frequency domain with a minimum inter-subcarrier distance of $\frac{1}{T_s}$, where T_s is the subcarrier symbol period. Figure 4 depicts the diagram of a typical OFDM transmitter and receiver. Firstly, the complex data signal, $X = [X_0 \quad X_1 \quad X_2 \quad \ldots \quad X_{N-1}]$, is generated with a length N, where N is the size of the Inverse Fast Fourier Transform (IFFT) and X_k is a complex value associated with a QAM constellation point. Note that, in an OFDM symbol, each X_k represents the data to be carried on the k-th subcarrier. The output of the Inverse Discrete Fourier Transform (IDFT) is calculated as follows:

$$x[n] = \frac{1}{\sqrt{N}} \sum_{k=0}^{N-1} X_k \exp\left(\frac{j2\pi kn}{N}\right) \quad for \quad 0 \leq n \leq N-1, \tag{3}$$

where x_n is the time domain complex value obtained through the frequency domain M-QAM signal. A Cyclic Prefix (CP) is considered in OFDM signals to avoid ISI, where the last N_{CP} samples are added at the beginning of the OFDM symbol, $x = [x_{N-N_{CP}} \ldots x_{N-1}, x_0 \ldots x_{N-1}]$. Note that, although the use of the CP reduces the data rate due to the introduced redundancy, it allows the equalization at the receiver to be simple. However, to avoid inter-subcarrier interference and preserve the subcarrier orthogonality, time and frequency synchronization are needed [64].

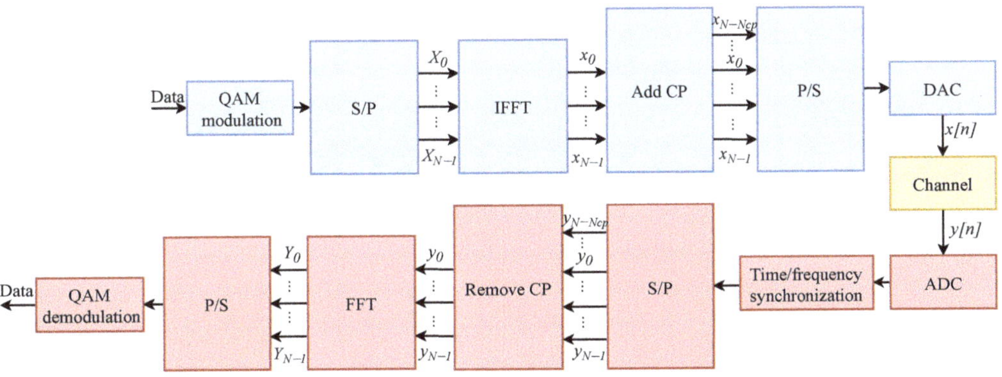

Figure 4. Block diagram of an OFDM communication system with a cyclic prefix.

After transmission through the wireless channel, the CP is removed and the Discrete Fourier Transform (DFT) is performed. Therefore, the received frequency domain signal is represented as follows:

$$Y_k = \frac{1}{\sqrt{N}} \sum_{n=0}^{N-1} y[n] \exp\left(\frac{-j2\pi kn}{N}\right) \quad for \quad 0 \leq k \leq N-1, \tag{4}$$

where $Y = [Y_0 \quad Y_1 \quad Y_2 \quad \ldots \quad Y_{N-1}]$ is the received frequency domain signal and $y[n]$ is the received time domain signal. This process ends with the receiver Digital Signal Processing (DSP), where the QAM symbols' de-mapping is performed.

The OFDM signal can also be used in VLC with some modifications to be compatible with IM-DD, where the signal directly modulates the intensity of the light. OFDM in VLC was proposed for the first time in [65]. In a typical RF system, OFDM is transmitted through an electrical signal, which can be positive or negative. Furthermore, the receiver includes a local oscillator supporting the coherent detection of the transmitted complex signal, unlike VLC systems, which only allow direct detection. The first modification is associated with the fact that the generated signal is a bipolar complex value. Therefore, it is necessary to convert it to a unipolar real signal. Real OFDM signals can be obtained with the Hermitian symmetry constraint on the subcarriers, resulting in a bipolar real-valued signal [66]. Besides that, the two most-used ways to obtain a unipolar signal are Direct Current (DC)-biased Optical OFDM (DCO-OFDM) and Asymmetrically Clipped Optical OFDM (ACO-OFDM). In DCO-OFDM, a positive DC bias is added, making the signal unipolar [65,67]. In contrast, in ACO-OFDM, the signal is clipped to zero and only the positive parts are transmitted [68]. Both the ACO-OFDM and DCO-OFDM methods were compared by Raed Mesleh et al. through simulation tests [69]. In a VLC system, the authors verified that the distortions are more serious for DCO-OFDM, mainly considering high modulation orders, with the LED clipping effect being the most predominant. However, in [70], the authors suggested the use of a third method named Flip-OFDM. In this strategy, the positive and negative parts are separated and transmitted in two consecutive OFDM symbols, with the negative part being flipped. In the same work, Fernando et al. concluded that Flip-OFDM and ACO-OFDM have the same spectral efficiencies and BER performance. A common phenomenon presented in optical OFDM systems is the high Peak-to-Average-Power Ratio (PAPR), and one of the simplest ways to reduce it is to introduce signal clipping [3]. However, clipping introduces significant distortion, which can lead to poor Bit Error Rate (BER) performance. Consequently, some strategies were introduced to mitigate clipping noise in optical OFDM systems [71–73].

The majority of recent VLC works tend to use OFDM [21,74–76]. Liang-Yu Wei et al. experimentally demonstrated a collimated VLC system using tricolor RGB LDs to produce a bit rate of 20.231 Gbit/s over a channel with 1 m. In this work, a downstream OFDM signal was used [74]. Yi-Chien Wu et al. presented a red/green/violet-LD- and yellow-LED-based four-color white-lighting module with high illuminance of 12,800 lux and a Color-Rending Index (CRI) of 60. In this work, the authors experimentally demonstrated a data rate of 20 Gbit/s using OFDM over a transmission link of 0.8 m [75]. Interestingly, Changmin Lee et al. used a DCO-OFDM signal to modulate the light of two LDs in different optical bands, demonstrating an aggregated bit rate of up to 26 Gbit/s [76]. In 2022, Gutema et al. studied the performance of OFDM for optical wireless communications and applied the technique to Wavelength-Division-Multiplexing (WDM)-based visible light communication. They independently modulated three LEDs of different colors (red, green, and blue) for parallel and simultaneous data transmission. The use of OFDM resulted in an aggregate bit rate of 10.81 Gbit/s with a link of 50 cm [21]. Therefore, with the support of the OFDM waveform, these works reached bit rates of more than 10 Gbit/s, clearly surpassing the works with single-carrier modulation formats.

3.2.2. Modified OFDM Waveforms

A multi-carrier modulation format also commonly utilized in various VLC works is Discrete Multitone (DMT) [77–79]. DMT is a variation of OFDM proposed for the Asymmetric Digital Subscriber Line (ADSL). Due to the slowly varying nature of this type of channel, it allows spectral shaping by bit and power loading, to improve the system performance [3]. Xin Zhu et al. experimentally demonstrated a WDM VLC system operating at 10.72 Gbit/s over 1 m indoor free-space transmission. The authors used a single package with Red, Green, Blue, Cyan, and Yellow LEDs (RGBCY). The light of the LEDs was modulated independently using the 64QAM-DMT modulation format [77]. Alternatively, Wei-Chung Wang et al. showed a total data rate of 34.8 Gbit/s over a free-space with a distance of 0.3 m. This was achieved using Red, Green, and Violet (RGV) LDs modulated with the DMT format, resulting in respective WDM data rates of 18/7.2/9.6 Gbit/s [78], thus demonstrating that DMT can also be a competitive alternative to OFDM.

An interesting possibility is to combine different modulation formats, such as CSK and OFDM. CSK only acts on the polarization current of the transmitters, so it is possible to modulate the transmitted light in the same way. This possibility was suggested by Gunawan et al. in their work, allowing them to reach a bit rate of 26.65 Gbit/s with 1.25 m free-space transmission [80]. In addition to this, the use of a hybrid OFDM-PWM scheme was proposed by Tian Zhang et al. for intensity modulation of the VLC transmitter as a possibility to address the issue of the high PAPR of O-OFDM signals [81]. This hybrid scheme combines O-OFDM with PWM, where the OFDM samples are converted into the pulse width of the signal. The authors concluded through simulation and experimental results that hybrid OFDM-PWM had a better BER performance, lower PAPR, higher luminance, and better resilience to the transmitter nonlinearity compared to the original ACO-OFDM scheme. In turn, Ebrahimi et al. extended this study by converting the most-common ACO-OFDM and DCO-OFDM into the PWM and PPM modulation formats [82]. The authors concluded that the hybrid DCO-OFDM-PWM/PPM schemes had a lower PAPR and BER than the ACO-OFDM and ACO-OFDM-PWM/PPM schemes.

In the literature, there are also some lesser-known modified OFDM implementations that are currently being proposed as potential candidates for VLC, despite being initially proposed for RF, such as filtered OFDM and Filter Bank Multi-Carrier (FBMC) [83–85]. Filtered-OFDM, based on the original OFDM, was proposed in 2015 [86]. This method filters the OFDM signal using digital filters. These filters cause no distortion in the pass-band signal while filtering the Out-Of-Band (OOB) part. In this way, the OOB emission is reduced, resulting in a lower inter-sub-band interference, presented in the OFDM waveform. In [83], filtered-OFDM was proposed for VLC. Yanyan Wang et al. verified through simulations that the filtered-OFDM has better BER performance than ACO-OFDM and DCO-OFDM. In contrast to filtered-OFDM, FBMC applies a filter per subcarrier. Therefore, the side lobes of each subcarrier are much weaker, resulting in even lower OOB emission [84]. In [85], FBMC was experimentally demonstrated in a VLC system with a blue LED using a pre-equalization method to improve the modulation bandwidth, achieving a bit rate greater than 2 Gbit/s.

3.2.3. Generalized Frequency Division Multiplexing

Recently, new modulation formats have started to appear as an alternative to OFDM, namely Generalized Frequency Division Multiplexing (GFDM) and Digital Subcarrier Multiplexing (DSCM). Interestingly, they were proposed by two different scientific communities, RF and optical fiber, respectively. The main disadvantages of OFDM are a high PAPR, high OOB emissions, and the requirement to use large CPs, considerably decreasing the spectral efficiency. All these issues have been partially solved in these waveforms [87]. The block diagram of the GFDM transmitter and receiver is shown in Figure 5. Contrary to OFDM, in GFDM signals, multiple pulse-shaped subcarriers with low symbol rates are multiplexed in the frequency domain in order to produce a high bandwidth signal, allowing each subcarrier to be individually modulated having its own bandwidth, pulse shaping, frequency, and CP, avoiding the use of the Fast Fourier Transform (FFT) or IFFT. The transmit signal, $x[n]$, is obtained through the summation of all transmit symbols:

$$x[n] = \sum_{k=0}^{N_{SC}-1} (d_k * g_k)\exp(j2\pi\frac{k}{N_{SC}}n) \quad for \quad 0 \leq n \leq N-1, \tag{5}$$

where d_k is the QAM symbols transmitted on the k-th subcarrier and g_k corresponds to the impulse response of the pulse-shaping filter applied to the k-th subcarrier. The Root-Raised

Cosine (RRC) filter is widely used as a pulse-shaping filter in GFDM. The total number of QAM symbols is $N = N_{SC}M$, where N_{SC} is the number of subcarriers and M is the number of symbols per subcarrier. Furthermore, GFDM takes advantage of the simple equalization of OFDM, adding the flexibility of occupying the desired frequencies and controlling OOB emission. GFDM is a multi-carrier modulation technique that was initially proposed for RF wireless communications in 2009 [88]. As verified for OFDM, the transmitted light signal needs to be unipolar and real. Therefore, the optical GFDM appears with two different approaches, with the same working principle of DCO-OFDM and Flip-OFDM, giving rise to DCO-GFDM and Flip-GFDM [89,90].

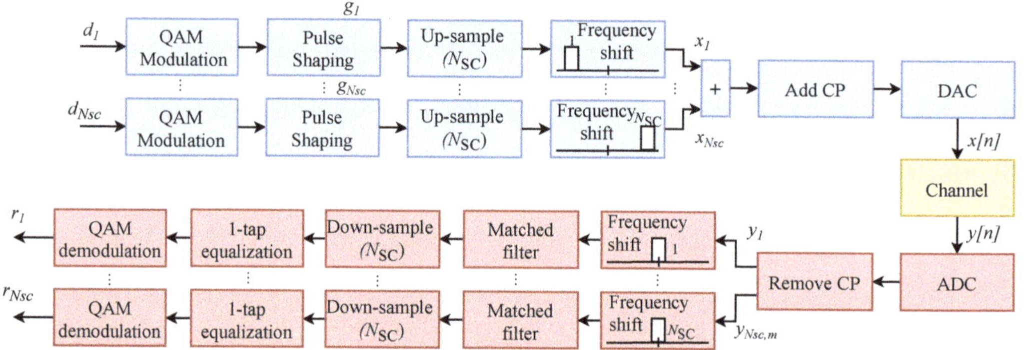

Figure 5. Block diagram of the GFDM transceiver.

In the literature, there are some works that propose the adoption of this waveform in VLC, but contrary to RF-based wireless communications, there are only a few studies on GFDM for VLC [87,89,90]. In [87,89], the authors showed that GFDM and OFDM have a similar BER performance in a VLC system, but the main advantage verified for GFDM was the lower OOB emission. Saengudomlert et al. proposed a Flip-GFDM modulation scheme with dimming support for VLC. Unlike DCO-GFDM, where dimming is adjusted by changing the DC bias, the proposed method avoids signal clipping and provides wider dimming ranges [90].

3.2.4. Digital Subcarrier Multiplexing

On the other hand, DSCM is also a multi-carrier modulation technique, but is more commonly used in fiber optics systems [91–93]. Similar to GFDM, in DSCM, usually, each subcarrier is pulse-shaped using an RRC filter, with a minimum subcarrier spacing of $(1 + \alpha)/\mathrm{Ts}$, where α is the roll-off factor of the pulse-shaping filter, to enable the orthogonality between subcarriers. Furthermore, due to the low OOB emission, no guard bands are needed between adjacent channels, as with OFDM. Despite the many similarities with GFDM, DSCM uses wider subcarriers, so it needs to implement a single-carrier-compatible DSP, which is more complex than the single-tap equalization used in GFDM. In contrast, GFDM needs to use CP to achieve this lower complexity, reducing the spectral efficiency.

DSCM signal modulation and demodulation are presented in Figure 6. After data generation and QAM modulation, the signal is up-sampled. Then, a set of symbols of each subcarrier is pulse-shaped using an RRC filter. Finally, each subcarrier is shifted to its respective central frequency, producing a signal identical to (5). The signal demodulation of each subcarrier is performed individually, each being downconverted to the baseband, filtered with a matched filter, and finally, the QAM symbols are decoded. Recently, record high bit rates were obtained in a VLC system with diffused light using the DSCM waveform [23,94].

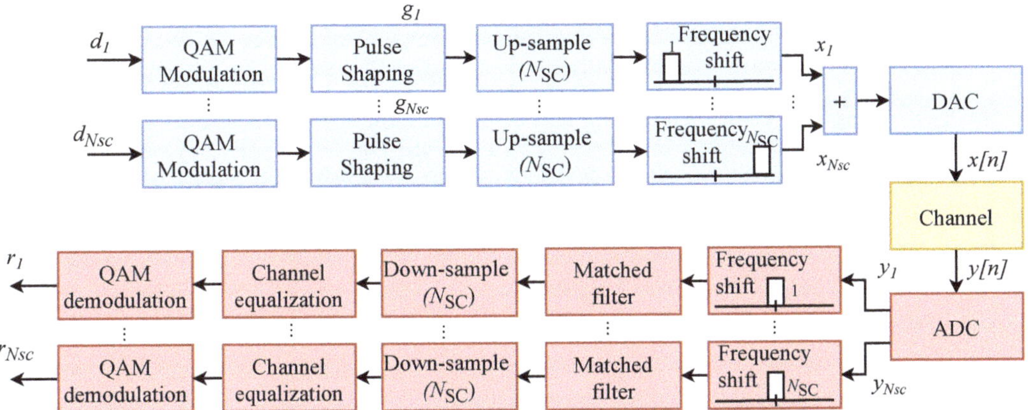

Figure 6. Diagram of the DSCM multiplexing and de-multiplexing.

3.3. VLC MIMO Systems

So far, it has been assumed that the VLC system is composed of a single transmitter and receiver. However, currently, a typical room in a home or office contains several LEDs to ensure sufficient lighting. Therefore, in a visible light communications scenario, MIMO can be implemented due to the multiple transmitters [6]. The introduction of MIMO in VLC allows improving the reliability and bit rate of the system, which is currently limited by the transmitters' bandwidth [95]. However, contrary to RF MIMO systems, which have multiple different channels between the transmitter and the receiver, in VLC systems, this is not verified due to the similarities of the channels. The transmitters and receivers are often confined to a single room, resulting in high channel correlation [6]. Therefore, the angular diversity receiver was introduced in some works to improve the performance of VLC-MIMO systems through the decorrelation of the optical channels [96–99]. An angular diversity receiver is composed of a set of narrow field-of-view detectors that point in different directions [98]. In [96], the authors presented two designs of angle diversity receivers, pyramid receivers and hemispheric receivers. The authors concluded that the proposed receivers outperformed the spatially separated photodiode array. C. Chen et al. demonstrated the operation of a different topology, the generalized angular diversity receiver, which consists of a detector in the middle and multiple inclined detectors around it, in which its inclination is adjustable [97]. Furthermore, some techniques have been also studied to alleviate the channel correlation issue and improve the performance of the VLC-MIMO systems. The proposed VLC-MIMO techniques are based on those used for RF-MIMO systems, where three methods stand out, namely Repetition Coding (RC), Spatial Multiplexing (SMP), and Spatial Modulation (SM) [100].

The main feature of the first method is its simplicity since the same data stream is used in all transmitters, thus allowing increasing the robustness of the system [99,101]. In [101], M. Safari et al. investigated the performance of RC and simple Orthogonal Space–Time Block Codes (OSTBCs), such as the Alamouti scheme. The authors concluded that a multiple-input single-output system with RC outperforms OSTBCs because the signal power from the transmitters constructively adds up at the VLC receiver. Therefore, with this work, it was concluded that the use of OSTBCs is not necessary for VLC. In addition, Reference [99] studied the performance of the RC for a VLC-MIMO system using angular diversity receivers under imperfect channel state information. For different receiver locations and semi-half angles, the analytical results showed that this system has better error performance than a multiple-input, single-output VLC system.

On the other hand, in SMP, the transmitters employ different signals, increasing the spectral efficiency of the system [102,103]. U. Siddiqi et al. proposed an adaptive bit and power loading for a DCO-OFDM VLC MIMO system. The adaptive algorithm chooses between RC and SMP MIMO modes and applies bit- and power-loading methods to improve the data rate for a given target BER [102]. In [103], Guo et al. proposed a novel superposed 64QAM constellation scheme for a 2 × 2 VLC-MIMO system using the SMP scheme. The authors experimentally demonstrated that the proposed superposed constellation combined with the SMP technique in VLC-MIMO systems achieves multiplexing gains, even in highly correlated VLC channels, thus providing better performance than the traditional superposed 64QAM constellation.

Lastly, in SM, only one transmitter is considered at each time slot. In this technique, each VLC transmitter is associated with a particular symbol of the constellation; therefore, whenever it is necessary to transmit that symbol, the corresponding transmitter is activated and the remaining are turned off. In this way, when the receiver receives a certain symbol, it is easy to estimate the transmitter. The main advantage is that, in this way, the information is encoded in two dimensions. In addition to the information encoded in the signal, there is also a modulated signal in space, thus increasing the spectral efficiency of the system. Furthermore, only one transmitter is connected at any one time, thus avoiding cross-channel interference, simplifying receiver complexity [6]. In [100], the authors compared the three MIMO techniques using different 4 × 4 MIMO setups, with different transmitter and receiver positions. The RC technique presents the worst spectral efficiency due to the use of the same signal in all transmitters, but this results in easier system alignment. Contrarily, SMP requires a low channel correlation between the transmitter and the receiver, but provides the highest data rates. In turn, SM provides improved spectral efficiencies even at a low Signal-to-Noise Ratio (SNR) and it works efficiently at high channel correlation.

4. Capacity-Achieving Strategies

As highlighted in the previous section, impressive progress has been made during the last couple of decades in the development of advanced modulation formats and signal-processing techniques to maximize the performance of VLC systems. Notoriously, the adoption of multi-carrier modulation has enabled the efficient exploitation of the available bandwidth, thus optimizing the spectral efficiency of the system.

Following this trend, in this section, we delve in more detail into the recent adoption of advanced multi-carrier modulation techniques that aim at approaching the ultimate limit set by Shannon's capacity. Namely, by adaptively optimizing the allocation of transmitted information over different frequency bands, significant capacity gains can be achieved. Employing traditional QAM formats, this can be achieved through bit- and power-loading techniques. Instead, resorting to capacity-achieving modulation formats such as Probabilistic Constellation Shaping (PCS), it is possible to squeeze out the ultimate spectral efficiency limits of VLC systems.

4.1. Bit Loading and Power Loading

With that problem in mind, some articles suggested the application of adaptive Power Loading (PL) and Bit Loading (BL) in multi-carrier signals for VLC systems. In this way, it was possible to improve the system bit rate and reduce the filtering effect introduced by the channel [104]. In the BL process, the main objective is to adapt the size of the QAM constellations of each subcarrier. In turn, PL applies scalar power ratios to each subcarrier, adjusting the transmitted power. The power ratio between subcarriers emphasizes some subcarriers, at the expense of decreasing the transmitted power in others. In [105], the authors studied the application of both BL and PL for optical fiber systems, obtaining gains when BL and PL were applied independently. However, the best case was found in the joint application of PL and BL, mitigating the filtering penalties. In VLC, R. Bian et al. experimentally demonstrated an RGBY-LED-based system with a bit rate of 15.73 Gbit/s, where each wavelength was modulated using DCO-OFDM with adaptive BL [20].

Some algorithms have been developed to perform a bit-and-power-ratio allocation in order to optimize the bit distribution and the transmitted power of all subcarriers based on the measured SNR. One of the most-well-known algorithms for allocating power across sub-carriers is the water-filling algorithm [106]. Furthermore, other widely used PL/BL approaches are Chow's algorithm [107] and the Levin–Campello algorithm [108,109]. The Levin–Campello algorithm, which was designed to reduce the nonlinearities introduced by ADSL systems, solves both the bit rate maximization and the margin maximization (minimum transmitted power) problems with low complexity. Contrary to the water-filling algorithm, which distributes the power over the subcarriers, the main objective of this algorithm is to allocate bits to each subcarrier. Then, after allocating bits, a power adjustment is made for each sub-carrier.

4.2. Entropy Loading

Despite their simplicity and widespread use in various studies, PL/BL methods have significant limitations, mainly because uniform QAM modulation formats have an integer number of bits per symbol, resulting in an entropy of the constellation equal to:

$$H = \log_2(M), \tag{6}$$

where M is the order of the QAM modulation. Therefore, these methods do not have the flexibility to adapt to the distortions introduced by the channel, and in many cases, the bit rate is not being maximized, thus increasing the gap to Shannon capacity. In order to solve this problem, some works proposed the use of continuous entropies instead of discrete numbers of bits, using PCS [110,111]. The main goal of this method is to adapt the probability distribution function of the QAM constellation, in order to maximize the net bit rate of the system. Usually, in a QAM constellation, all symbols have the same probability, so the constellations have an integer number of bits per symbol. Instead, with PCS, it is possible to continuously adjust the entropy and the average signal power by assigning a probability distribution function [112]. According to Shannon's theory for Additive White Gaussian Noise (AWGN) channels, the Maxwell–Boltzmann distribution requires the minimum signal power to achieve a given bit rate. Therefore, the QAM symbol probabilities can be calculated as follows [110]:

$$P_{x_n} = \frac{\exp(-\lambda|x_n|)}{\sum_{k=1}^{M}\exp(-\lambda|x_k|)}, \tag{7}$$

where $\lambda \geq 0$ is the shaping parameter and x_n is the symbol n in the M-QAM constellation. Therefore, for $\lambda = 0$, the probability the all symbols will be $P_{x_n} = \frac{1}{M}$, resulting in a uniform distribution. For higher λ values, the probability of outer QAM symbols decreases and the probability of inner ones increases, resulting in a lower number of bits per symbol. The source entropy of an M-QAM constellation (number of bits per symbol), H, represents the transmitted information rate and depends on the probability distribution [113]:

$$H = -\sum_{n=1}^{M} P_{x_n}\log_2(P_{x_n}). \tag{8}$$

In Figure 7, we can see four 64QAM constellations for entropies between 3 and 6 bit/symbol. As expected, for lower entropies, the outer constellation symbols start to have a much lower probability compared to the rest. For the Maxwell–Boltzmann distribution, a Distribution Matcher (DM) is needed to convert an input uniformly distributed bit stream into an output non-uniformly distributed symbol sequence. Currently, various options to implement the PCS DMs have been proposed, but the Constant Composition Distribution Matcher (CCDM) is by far the most-used algorithm [114].

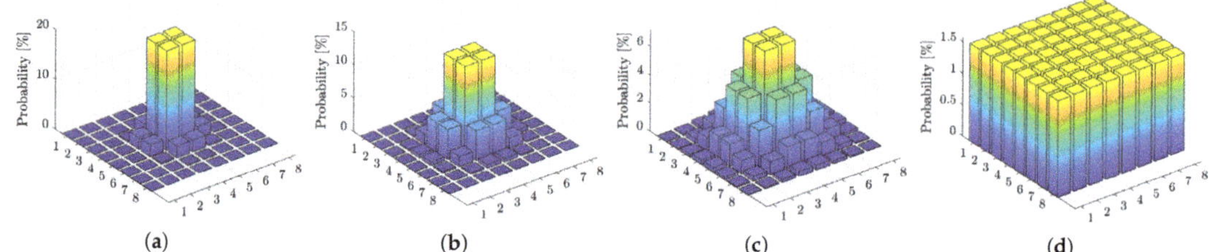

Figure 7. Graphical illustration for PCS in a 64QAM constellation with four different entropies: (**a**) 3 bit/symbol; (**b**) 4 bit/symbol; (**c**) 5 bit/symbol; (**d**) 6 bit/symbol.

In order to take advantage of PCS and multi-carrier signals, some works implemented an Entropy-Loading (EL) method, demonstrating a capacity-achieving solution [21,94,115]. EL is a method that aims to maximize the data rate of the system. The ideal entropy per subcarrier is estimated based on the measured SNR per subcarrier and a performance metric that guarantees an error-free system after FEC. This entropy is estimated through several iterations until it converges to the optimal value. However, considering that the implementation of an ideal DM is complex, it will hardly be possible to converge to the optimal entropy for the measured SNR [94]. The EL method was demonstrated for the first time by Di Che et al. for colored SNR optical channels with band-limited cascaded-Reconfigurable Optical Add–Drop Multiplexers (ROADMs) [116]. The authors demonstrated the advantage of the EL method over the single-carrier PCS in band-limited systems. In VLC systems, the EL scheme was implemented for the first time in 2018 by Xie et al. In their work, the authors experimentally demonstrated a bit rate increment of 26.8% in comparison with OFDM using a bit-loading technique [115]. In turn, using OFDM and PCS together, Gutema et al. achieved a 25% higher transmission rate than the adaptive bit–power-loading algorithm under the same channel conditions [21]. Furthermore, our group experimentally demonstrated an RGB-LD-based VLC system capable of both lighting and high-speed communications. An EL method based on PCS was proposed to maximize the net bit rate of a DSCM signal. A bit rate of 31.2 Gbit/s over 0.90 m of free-space distance was presented [94]. In [23], we extended the first work by presenting a distance-adaptive VLC system. Using again together DSCM and PCS, we experimentally demonstrated a maximum bit rate of 46 Gbit/s at 50 cm, linearly decreasing over distance, down to 26 Gbit/s after 200 cm.

From what has been mentioned above, currently, this modulation technique appears as the main candidate to be used in future VLC systems, as it allows obtaining data rates closer to Shannon's capacity. Moreover, this method allows continuous adaptation to distortions introduced over time in the channel. In a VLC channel, there are usually many lights external to the system (sunlight or indoor lighting) that can affect the performance of the communication link, introducing a variation of the channel characteristics over time. In [112], the authors experimentally demonstrated the application of time-adaptive PCS to change the bit rate according to the conditions of an outdoor FSO channel, adjusting the source entropy of the single-carrier over time. Therefore, in addition to the ability to adjust entropy over frequency (subcarriers), in the EL method, there is also the possibility of making this adjustment over time when there are variations in the channel.

4.3. Geometric Constellation Shaping

In many of the recent works, the use of PCS has been increasingly considered, mainly because it is currently the method that guarantees the best spectral efficiency. However, in addition to this method, which focuses on adjusting the probability of symbols, there is also the possibility of adjusting the geometry of the constellation in order to maximize the data rate of the system. The presented PCS method is ideal for AWGN channels using a Maxwell–Boltzmann probability distribution of symbols. However, for non-AWGN channels, for example, due to the nonlinear response of the VLC transmitter, the adaptation of the constellation geometry can achieve better results. Geometric Constellation Shaping (GCS) is a capacity-achieving technique that uses non-uniformly spaced constellation symbols, and by adjusting the location of the symbols, it is able to closely approximate Shannon's capacity [117]. Recently, the use of autoencoders, based on end-to-end deep learning, has been used to obtain the optimum constellation [118]. Compared to more-traditional methods, it stands out for taking into account the state of the channel at a given moment, with the possibility of adjusting it over time, which can be fundamental in VLC systems. The GCS has been demonstrated for different applications, mainly in fiber optics communications [119–121]. However, in VLC, there are currently few works that use this technique [122,123]. Therefore, the use of this technique, its comparison with other methods (PCS for instance), and its joint use with PCS have to be studied. The performance of future VLC systems can be improved since the system will become even more robust, making it more of a candidate for beyond-5G wireless communications.

5. Summary and Future Work

In the last decade, many VLC articles have been published, and it is expected that it will be a new type of wireless communication present in 6G. However, at the moment, there are a number of key issues. Based on this survey of recent trends and technologies for indoor VLC, we present some challenges that should be studied in the near future. First, there are some commercialization challenges due to the need to make modifications to the lamps, which may diminish the interest of manufacturers. For mobile device manufacturers, integrating new hardware into phones can result in a higher cost, driving up cell phone prices. Therefore, to achieve the goal of integrating VLC in future wireless communications, it is crucial to improve the energy efficiency of current VLC transmitters, as well as their cost to allow their large-scale implementation. Furthermore, it will also be important to clarify the best approach, whether with LEDs, LDs, or even a hybrid solution that uses both possibilities in order to offer good-quality lighting and communications. Moreover, considering that VLC is seen as a good technology to complement RF-based wireless systems, the authors think that the development of a hybrid RF-VLC approach could be a trend in VLC systems, where ways of joining both technologies and how they can cooperate with each other should be studied. This should also be performed without discarding the various applications that VLC can have, including high-speed communications with indoor localization, communication between vehicles, and underwater communications.

Additionally, one of the main problems with indoor VLC is the continuous need for data transmission, while indoor lights may be dimmed or turned off by users. For example, on bright days or before bedtime, light is not needed, but the Internet connection is. For this reason, VLC should be seen, at the moment, as an alternative to WiFi, but not as its replacement, as there will be situations where VLC will not work as a standalone solution. However, it will be interesting to develop strategies to solve this problem. The obvious solution is to use infrared as a backup in these situations, but the performance of non-collimated IR communications is not sufficient. Therefore, the solution will have to go through the use of visible light. Currently, there are some works proposing the operation of the VLC system with the light off [124,125]. The idea of these works is to produce an average brightness of light so low that it cannot be detected by humans, appearing as if the light sources are turned off, but it is possible that the receivers detect these light variations [126]. In [124], Borogovac et al. through simulations concluded that very low light emission is sufficient to maintain data rates of several megabits/second. However, with the system proposed by Tian et al., it was only possible to achieve a data rate of 1.6 Kbit/s at a distance of 10 cm [125]. Therefore, at the moment, there is no clear solution to this problem, and from the authors' perspective, it is important to address this issue in future works.

Regarding the modulation techniques for visible light communications, we currently observe a clear trend for a transition from single-carrier to multi-carrier modulation formats. Several multi-carrier modulation formats were recently presented, each with its advantages and disadvantages, but so far, it is not clear which could be the best approach. OFDM is the accepted modulation technique for WiFi, but for VLC, this is an open research area. Thus, an area of study in the future will be to compare the different formats in a VLC system under different practical scenarios, such as background light interference, since conventional lamps or even sunlight can coexist in the same room as the VLC link.

In summary, VLC is expected to deliver high-speed communications with good energy efficiency and secure communications. However, there is still a way to go to improve the current state of technology and increase its popularity both within industry and the general population.

Author Contributions: Conceptualization, P.A.L., F.P.G. and P.P.M.; writing—original draft preparation, P.A.L.; writing—review and editing, P.A.L., F.P.G. and P.P.M.; supervision, F.P.G. and P.P.M.; project administration, F.P.G. and P.P.M.; funding acquisition, F.P.G. and P.P.M. All authors have read and agreed to the published version of the manuscript.

Funding: This work was partially supported by FCT/MCTES through project OptWire (PTDC/EEI-TEL/2697/2021) and by the H2020-EU.1.3 program through project DIOR (GA ID: 101008280). Pedro A. Loureiro acknowledges the Ph.D. fellowship from FCT (2021.06736.BD).

Institutional Review Board Statement: Not applicable.

Informed Consent Statement: Not applicable.

Data Availability Statement: Not applicable.

Conflicts of Interest: The authors declare no conflict of interest.

Abbreviations

The following abbreviations are used in this manuscript:

ACO-OFDM	Asymmetrically Clipped Optical Orthogonal Frequency Division Multiplexing
ADSL	Asymmetric Digital Subscriber Line
AOA	Angle Of Arrival
AWGN	Additive White Gaussian Noise
BER	Bit Error Rate
BL	Bit Loading
CAP	Carrier-less Amplitude and Phase modulation
CCDM	Constant Composition Distribution Matcher
CIE	Commission Internationale de l'éclairage
CP	Cyclic Prefix
CRI	Color-Rendering Index
CSK	Color-Shift Keying
DCO-OFDM	DC-biased Optical Orthogonal Frequency Division Multiplexing
DM	Distribution Matcher
DMT	Discrete Multitone
DSCM	Digital Subcarrier Multiplexing
DSP	Digital Signal Processing
EL	Entropy Loading
FBMC	Filter Bank Multi-Carrier
FEC	Forward Error Correction
FFT	Fast Fourier Transform
FSO	Free-Space Optics
GCS	Geometric Constellation Shaping
GFDM	Generalized Frequency Division Multiplexing
GPS	Global Positioning System
HB	High Bandwidth

IDFT	Inverse Discrete Fourier Transform
IFFT	Inverse Fast Fourier Transform
IM-DD	Intensity Modulated Direct Detection
ISI	Inter-Symbol Interference
JEITA	Japan Electronics and Information Technology Industries Association
LD	Laser Diode
LED	Light-Emitting Diode
MAC	Media Access Control
MIMO	Multiple-Input, Multiple-Output
MPPM	Multi-Pulse Position Modulation
OFDM	Orthogonal Frequency Division Multiplexing
OOB	Out-Of-Band
OOK	On–Off Keying
OPPM	Overlapping Pulse Position Modulation
OSTBC	Orthogonal Space–Time Block Code
PCS	Probabilistic Constellation Shaping
PHY	Physical
PM	Pulse Modulation
PAM	Pulse Amplitude Modulation
PAPR	Peak-to-Average-Power Ratio
PL	Power Loading
PPM	Pulse Position Modulation
PWM	Pulse Width Modulation
QAM	Quadrature Amplitude Modulation
RC	Repetition Coding
RF	Radio Frequency
RGB	Red, Green, and Blue
RGBCY	Red, Green, Blue, Cyan, and Yellow
RGV	Red, Green, and Violet
ROADM	Reconfigurable Optical Add–Drop Multiplexer
RRC	Root-Raised Cosine
RSS	Received Signal Strength
SM	Spatial Modulation
SMP	Spatial Multiplexing
SNR	Signal-to-Noise Ratio
TGbb	Task Group bb
TOA	Time Of Arrival
UWOC	Underwater Optical Communications
VPPM	Variable-Pulse-Position Modulation
VLC	Visible Light Communications
VLCC	Visible Light Communication Consortium
WDM	Wavelength Division Multiplexing

References

1. Gupta, A.; Jha, R. A Survey of 5G Network: Architecture and Emerging Technologies. *IEEE Access* **2015**, *3*, 1206–1232. [CrossRef]
2. Chun, H.; Gomez, A.; Quintana, C.; Zhang, W.; Faulkner, G.; O'Brien, D. A Wide-Area Coverage 35 Gb/s Visible Light Communications Link for Indoor Wireless Applications. *Sci. Rep.* **2019**, *9*, 4952. [CrossRef]
3. Karunatilaka, D.; Zafar, F.; Kalavally, V.; Parthiban, R. LED Based Indoor Visible Light Communications: State of the Art. *IEEE Commun. Surv. Tutor.* **2015**, *17*, 1649–1678. [CrossRef]
4. Fernandes, M.A.; Monteiro, P.P.; Guiomar, F.P. Free-Space Terabit Optical Interconnects. *J. Lightwave Technol.* **2022**, *40*, 1519–1526. [CrossRef]
5. Guiomar, F.P.; Fernandes, M.A.; Nascimento, J.L.; Rodrigues, V.; Monteiro, P.P. Coherent Free-Space Optical Communications: Opportunities and Challenges. *J. Lightwave Technol.* **2022**, *40*, 3173–3186. [CrossRef]
6. Pathak, P.H.; Feng, X.; Hu, P.; Mohapatra, P. Visible Light Communication, Networking, and Sensing: A Survey, Potential and Challenges. *IEEE Commun. Surv. Tutor.* **2015**, *17*, 2047–2077. [CrossRef]
7. Matheus, L.E.M.; Vieira, A.B.; Vieira, L.F.M.; Vieira, M.A.M.; Gnawali, O. Visible Light Communication: Concepts, Applications and Challenges. *IEEE Commun. Surv. Tutor.* **2019**, *21*, 3204–3237. [CrossRef]

8. Khan, L.U. Visible Light Communication: Applications, Architecture, Standardization and Research Challenges. *Digit. Commun. Netw.* **2017**, *3*, 78–88. [CrossRef]

9. *IEEE Std 802.15.7-2011*; IEEE Standard for Local and Metropolitan Area Networks–Part 15.7: Short-Range Wireless Optical Communication Using Visible Light. IEEE: Piscataway, NJ, USA, 2011; pp. 1–309. [CrossRef]

10. Rajagopal, S.; Roberts, R.D.; Lim, S.K. IEEE 802.15.7 visible light communication: Modulation schemes and dimming support. *IEEE Commun. Mag.* **2012**, *50*, 72–82. [CrossRef]

11. Bober, K.L.; Ackermann, E.; Freund, R.; Jungnickel, V.; Baykas, T.; Lim, S.K. The IEEE 802.15.13 Standard for Optical Wireless Communications in Industry 4.0. In Proceedings of the IECON 2022—48th Annual Conference of the IEEE Industrial Electronics Society, Brussels, Belgium, 17–20 October 2022; pp. 1–6. [CrossRef]

12. Jungnickel, V. Standardization of Li-Fi in IEEE 802.15.13. Fraunhofer HHI. 2018. Available online: https://www.hhi.fraunhofer. de/fileadmin/Departments/PN/MAI/Li-Fi_Standardization_Presentation.pdf (accessed on 30 May 2023).

13. Khorov, E.; Levitsky, I. Current Status and Challenges of Li-Fi: IEEE 802.11bb. *IEEE Commun. Stand. Mag.* **2022**, *6*, 35–41. [CrossRef]

14. Online. Our Solutions. pureLiFi. 2023. Available online: https://www.purelifi.com/our-solutions/ (accessed on 4 July 2023).

15. Chi, N.; Zhou, Y.; Wei, Y.; Hu, F. Visible Light Communication in 6G: Advances, Challenges, and Prospects. *IEEE Veh. Technol. Mag.* **2020**, *15*, 93–102. [CrossRef]

16. Tanaka, Y.; Haruyama, S.; Nakagawa, M. Wireless optical transmissions with white colored LED for wireless home links. In Proceedings of the 11th IEEE International Symposium on Personal Indoor and Mobile Radio Communications, PIMRC 2000, Proceedings (Cat. No.00TH8525), London, UK, 18–21 September 2000; Volume 2, pp. 1325–1329. [CrossRef]

17. Tanaka, Y.; Komine, T.; Haruyama, S.; Nakagawa, M. Indoor visible light data transmission system utilizing white LED lights. *IEICE Trans. Commun.* **2003**, *E86B*, 2440–2454.

18. Vucic, J.; Kottke, C.; Nerreter, S.; Langer, K.D.; Walewski, J.W. 513 Mbit/s Visible Light Communications Link Based on DMT-Modulation of a White LED. *J. Lightwave Technol.* **2010**, *28*, 3512–3518. [CrossRef]

19. Wu, F.M.; Lin, C.T.; Wei, C.C.; Chen, C.W.; Huang, H.T.; Ho, C.H. 1.1-Gb/s White-LED-Based Visible Light Communication Employing Carrier-Less Amplitude and Phase Modulation. *IEEE Photonics Technol. Lett.* **2012**, *24*, 1730–1732. [CrossRef]

20. Bian, R.; Tavakkolnia, I.; Haas, H. 15.73 Gb/s Visible Light Communication with Off-the-Shelf LEDs. *J. Lightwave Technol.* **2019**, *37*, 2418–2424. [CrossRef]

21. Gutema, T.Z.; Haas, H.; Popoola, W.O. WDM Based 10.8 Gbps Visible Light Communication with Probabilistic Shaping. *J. Lightwave Technol.* **2022**, *40*, 5062–5069. [CrossRef]

22. Hu, J.; Hu, F.; Jia, J.; Li, G.; Shi, J.; Zhang, J.; Li, Z.; Chi, N.; Yu, S.; Shen, C. 46.4 Gbps visible light communication system utilizing a compact tricolor laser transmitter. *Opt. Express* **2022**, *30*, 4365–4373. [CrossRef] [PubMed]

23. Loureiro, P.A.; Guiomar, F.P.; Monteiro, P.P. 25G+ Distance-Adaptive Visible Light Communications Enabled by Entropy Loading. In Proceedings of the Optical Fiber Communication Conference (OFC), San Diego, CA, USA, 5–9 March 2023; p. M4F.1. [CrossRef]

24. Wu, T.C.; Chi, Y.C.; Wang, H.Y.; Tsai, C.T.; Lin, G.R. Blue Laser Diode Enables Underwater Communication at 12.4 Gbps. *Sci. Rep.* **2017**, *7*, 40480. [CrossRef]

25. Li, B.; Huang, J.; Zhou, S.; Ball, K.; Stojanovic, M.; Freitag, L.; Willett, P. MIMO-OFDM for High-Rate Underwater Acoustic Communications. *IEEE J. Ocean. Eng.* **2009**, *34*, 634–644. [CrossRef]

26. Song, H.C.; Hodgkiss, W.S. Efficient use of bandwidth for underwater acoustic communication. *J. Acoust. Soc. Am.* **2013**, *134*, 905–908. [CrossRef]

27. Zakharov, Y.V.; Morozov, A.K. OFDM Transmission without Guard Interval in Fast-Varying Underwater Acoustic Channels. *IEEE J. Ocean. Eng.* **2015**, *40*, 144–158. [CrossRef]

28. Ali, M.F.; Jayakody, D.N.K.; Li, Y. Recent Trends in Underwater Visible Light Communication (UVLC) Systems. *IEEE Access* **2022**, *10*, 22169–22225. [CrossRef]

29. Oubei, H.M.; Duran, J.R.; Janjua, B.; Wang, H.Y.; Tsai, C.T.; Chi, Y.C.; Ng, T.K.; Kuo, H.C.; He, J.H.; Alouini, M.S.; et al. 4.8 Gbit/s 16-QAM-OFDM transmission based on compact 450-nm laser for underwater wireless optical communication. *Opt. Express* **2015**, *23*, 23302–23309. [CrossRef]

30. Shi, J.; Zhu, X.; Wang, F.; Zou, P.; Zhou, Y.; Liu, J.; Jiang, F.; Chi, N. Net Data Rate of 14.6 Gbit/s Underwater VLC Utilizing Silicon Substrate Common-Anode Five Primary Colors LED. In Proceedings of the 2019 Optical Fiber Communications Conference and Exhibition (OFC), San Diego, CA, USA, 3–7 March 2019; pp. 1–3.

31. Kaushal, H.; Kaddoum, G. Underwater Optical Wireless Communication. *IEEE Access* **2016**, *4*, 1518–1547. [CrossRef]

32. Zeng, Z.; Fu, S.; Zhang, H.; Dong, Y.; Cheng, J. A Survey of Underwater Optical Wireless Communications. *IEEE Commun. Surv. Tutor.* **2017**, *19*, 204–238. [CrossRef]

33. Spagnolo, G.; Cozzella, L.; Leccese, F. Underwater Optical Wireless Communications: Overview. *Sensors* **2020**, *20*, 2261. [CrossRef] [PubMed]

34. Vappangi, S.; Mani, V.V.; Sellathurai, M. VLC for Vehicular Communications. In *Visible Light Communication: Comprehensive Theory and Applications with MATLAB*; CRC Press: Boca Raton, FL, USA, 2021. [CrossRef]

35. Wang, N.; Qiao, Y.; Wang, W.; Tang, S.; Shen, J. Visible Light Communication based Intelligent Traffic Light System: Designing and Implementation. In Proceedings of the 2018 Asia Communications and Photonics Conference (ACP), Hangzhou, China, 26–29 October 2018; pp. 1–3. [CrossRef]

36. Marabissi, D.; Mucchi, L.; Caputo, S.; Nizzi, F.; Pecorella, T.; Fantacci, R.; Nawaz, T.; Seminara, M.; Catani, J. Experimental Measurements of a Joint 5G-VLC Communication for Future Vehicular Networks. *J. Sens. Actuator Netw.* **2020**, *9*, 32. [CrossRef]
37. Béchadergue, B.; Chassagne, L.; Guan, H. Experimental comparison of pulse-amplitude and spatial modulations for vehicle-to-vehicle visible light communication in platoon configurations. *Opt. Express* **2017**, *25*, 24790–24802. [CrossRef]
38. Avătămăniței, S.A.; Beguni, C.; Căilean, A.M.; Dimian, M.; Popa, V. Evaluation of Misalignment Effect in Vehicle-to-Vehicle Visible Light Communications: Experimental Demonstration of a 75 Meters Link. *Sensors* **2021**, *21*, 3577. [CrossRef]
39. Kouhini, S.M.; Kottke, C.; Ma, Z.; Freund, R.; Jungnickel, V.; Müller, M.; Behnke, D.; Vazquez, M.M.; Linnartz, J.P.M.G. LiFi Positioning for Industry 4.0. *IEEE J. Sel. Top. Quantum Electron.* **2021**, *27*, 1–15. [CrossRef]
40. Kouhini, S.M.; Ma, Z.; Kottke, C.; Mana, S.M.; Freund, R.; Jungnickel, V. LiFi based Positioning for Indoor Scenarios. In Proceedings of the 2021 17th International Symposium on Wireless Communication Systems (ISWCS), Berlin, Germany, 6–9 September 2021; pp. 1–5. [CrossRef]
41. Kottke, C.; Ma, Z.; Kouhini, S.M.; Jungnickel, V. In-building Optical Wireless Positioning Using Time of Flight. In Proceedings of the 2023 Optical Fiber Communications Conference and Exhibition (OFC), San Diego, CA, USA, 5–9 March 2023; pp. 1–3. [CrossRef]
42. Du, P.; Zhang, S.; Chen, C.; Yang, H.; Zhong, W.D.; Zhang, R.; Alphones, A.; Yang, Y. Experimental Demonstration of 3D Visible Light Positioning Using Received Signal Strength with Low-Complexity Trilateration Assisted by Deep Learning Technique. *IEEE Access* **2019**, *7*, 93986–93997. [CrossRef]
43. Hong, C.Y.; Wu, Y.C.; Liu, Y.; Chow, C.W.; Yeh, C.H.; Hsu, K.L.; Lin, D.C.; Liao, X.L.; Lin, K.H.; Chen, Y.Y. Angle-of-Arrival (AOA) Visible Light Positioning (VLP) System Using Solar Cells with Third-Order Regression and Ridge Regression Algorithms. *IEEE Photonics J.* **2020**, *12*, 1–5. [CrossRef]
44. Do, T.H.; Yoo, M. An in-Depth Survey of Visible Light Communication Based Positioning Systems. *Sensors* **2016**, *16*, 678. [CrossRef] [PubMed]
45. Chaudhary, N.; Alves, L.N.; Ghassemblooy, Z. Current Trends on Visible Light Positioning Techniques. In Proceedings of the 2019 2nd West Asian Colloquium on Optical Wireless Communications (WACOWC), Tehran, Iran, 27–28 April 2019; pp. 100–105. [CrossRef]
46. Tsonev, D.; Videv, S.; Haas, H. Light fidelity (Li-Fi): Towards all-optical networking. *Proc. SPIE Int. Soc. Opt. Eng.* **2013**, *9007*, 900702. [CrossRef]
47. Le Minh, H.; O'Brien, D.; Faulkner, G.; Zeng, L.; Lee, K.; Jung, D.; Oh, Y. High-Speed Visible Light Communications Using Multiple-Resonant Equalization. *IEEE Photonics Technol. Lett.* **2008**, *20*, 1243–1245. [CrossRef]
48. Vucic, J.; Kottke, C.; Nerreter, S.; Habel, K.; Buttner, A.; Langer, K.D.; Walewski, J.W. 125 Mbit/s over 5 m wireless distance by use of OOK-Modulated phosphorescent white LEDs. In Proceedings of the 2009 35th European Conference on Optical Communication, Vienna, Austria, 20–24 September 2009; pp. 1–2.
49. Stepniak, G.; Siuzdak, J.; Zwierko, P. Compensation of a VLC Phosphorescent White LED Nonlinearity by Means of Volterra DFE. *IEEE Photonics Technol. Lett.* **2013**, *25*, 1597–1600. [CrossRef]
50. Stepniak, G.; Maksymiuk, L.; Siuzdak, J. 1.1 GBIT/S white lighting LED-based visible light link with pulse amplitude modulation and Volterra DFE equalization. *Microw. Opt. Technol. Lett.* **2015**, *57*, 1620–1622. [CrossRef]
51. Li, X.; Bamiedakis, N.; Wei, J.; McKendry, J.J.D.; Xie, E.; Ferreira, R.; Gu, E.; Dawson, M.D.; Penty, R.V.; White, I.H. µLED-Based Single-Wavelength Bi-directional POF Link with 10 Gb/s Aggregate Data Rate. *J. Lightwave Technol.* **2015**, *33*, 3571–3576. [CrossRef]
52. Wu, F.M.; Lin, C.T.; Wei, C.C.; Chen, C.W.; Chen, Z.Y.; Huang, H.T.; Chi, S. Performance Comparison of OFDM Signal and CAP Signal over High Capacity RGB-LED-Based WDM Visible Light Communication. *IEEE Photonics J.* **2013**, *5*, 7901507. [CrossRef]
53. Wang, Y.; Tao, L.; Huang, X.; Shi, J.; Chi, N. 8-Gb/s RGBY LED-Based WDM VLC System Employing High-Order CAP Modulation and Hybrid Post Equalizer. *IEEE Photonics J.* **2015**, *7*, 1–7. [CrossRef]
54. Khalighi, M.A.; Long, S.; Bourennane, S.; Ghassemlooy, Z. PAM- and CAP-Based Transmission Schemes for Visible-Light Communications. *IEEE Access* **2017**, *5*, 27002–27013. [CrossRef]
55. Olmedo, M.I.; Zuo, T.; Jensen, J.B.; Zhong, Q.; Xu, X.; Popov, S.; Monroy, I.T. Multiband Carrierless Amplitude Phase Modulation for High Capacity Optical Data Links. *J. Lightwave Technol.* **2014**, *32*, 798–804. [CrossRef]
56. Haigh, P.A.; Burton, A.; Werfli, K.; Minh, H.L.; Bentley, E.; Chvojka, P.; Popoola, W.O.; Papakonstantinou, I.; Zvanovec, S. A Multi-CAP Visible-Light Communications System with 4.85-b/s/Hz Spectral Efficiency. *IEEE J. Sel. Areas Commun.* **2015**, *33*, 1771–1779. [CrossRef]
57. Zhao, S.; Xu, J.; Trescases, O. A dimmable LED driver for visible light communication (VLC) based on LLC resonant DC-DC converter operating in burst mode. In Proceedings of the 2013 Twenty-Eighth Annual IEEE Applied Power Electronics Conference and Exposition (APEC), Long Beach, CA, USA, 17–21 March 2013; pp. 2144–2150. [CrossRef]
58. Wu, X.; Soltani, M.D.; Zhou, L.; Safari, M.; Haas, H. Hybrid LiFi and WiFi Networks: A Survey. *arXiv* **2020**, arXiv:2001.04840.
59. Monteiro, E.; Hranilovic, S. Design and Implementation of Color-Shift Keying for Visible Light Communications. *J. Lightwave Technol.* **2014**, *32*, 2053–2060. [CrossRef]
60. CIE. *Commission Internationale de lEclairage Proc*; Cambridge University Press: Cambridge, UK, 1931.
61. Drost, R.J.; Sadler, B.M. Constellation design for color-shift keying using billiards algorithms. In Proceedings of the 2010 IEEE Globecom Workshops, Miami, FL, USA, 6–10 December 2010; pp. 980–984. [CrossRef]

62. Monteiro, E.; Hranilovic, S. Constellation design for color-shift keying using interior point methods. In Proceedings of the 2012 IEEE Globecom Workshops, Anaheim, CA, USA, 3–7 December 2012; pp. 1224–1228. [CrossRef]
63. Singh, R.; O'Farrell, T.; David, J.P.R. An Enhanced Color Shift Keying Modulation Scheme for High-Speed Wireless Visible Light Communications. *J. Lightwave Technol.* **2014**, *32*, 2582–2592. [CrossRef]
64. Gerzaguet, R.; Bartzoudis, N.; Baltar, L.; Berg, V.; Doré, J.B.; Ktenas, D.; Font-Bach, O.; Mestre, X.; Payaro, M.; Färber, M.; et al. The 5G candidate waveform race: A comparison of complexity and performance. *EURASIP J. Wirel. Commun. Netw.* **2017**, *2017*, 13. [CrossRef]
65. Afgani, M.; Haas, H.; Elgala, H.; Knipp, D. Visible light communication using OFDM. In Proceedings of the 2nd International Conference on Testbeds and Research Infrastructures for the Development of Networks and Communities, TRIDENTCOM 2006, Barcelona, Spain, 1–3 March 2006; pp. 6–134. [CrossRef]
66. Armstrong, J. OFDM for Optical Communications. *J. Lightwave Technol.* **2009**, *27*, 189–204. [CrossRef]
67. Elgala, H.; Mesleh, R.; Haas, H.; Pricope, B. OFDM Visible Light Wireless Communication Based on White LEDs. In Proceedings of the 2007 IEEE 65th Vehicular Technology Conference—VTC2007-Spring, Dublin, Ireland, 22–25 April 2007; pp. 2185–2189. [CrossRef]
68. Armstrong, J.; Lowery, A. Power efficient optical OFDM. *Electron. Lett.* **2006**, *42*, 370–372. [CrossRef]
69. Mesleh, R.; Elgala, H.; Haas, H. Performance analysis of indoor OFDM optical wireless communication systems. In Proceedings of the 2012 IEEE Wireless Communications and Networking Conference (WCNC), Paris, France, 1–4 April 2012; pp. 1005–1010. [CrossRef]
70. Fernando, N.; Hong, Y.; Viterbo, E. Flip-OFDM for optical wireless communications. In Proceedings of the 2011 IEEE Information Theory Workshop, Paraty, Brazil, 16–20 October 2011; pp. 5–9. [CrossRef]
71. Elgala, H.; Mesleh, R.; Haas, H. Predistortion in Optical Wireless Transmission Using OFDM. In Proceedings of the 2009 Ninth International Conference on Hybrid Intelligent Systems, Shenyang, China, 12–14 August 2009; Volume 2, pp. 184–189. [CrossRef]
72. He, C.; Armstrong, J. Clipping Noise Mitigation in Optical OFDM Systems. *IEEE Commun. Lett.* **2017**, *21*, 548–551. [CrossRef]
73. He, C.; Lim, Y. Clipping Noise Mitigation in Optical OFDM Using Decision-Directed Signal Reconstruction. *IEEE Access* **2021**, *9*, 115388–115402. [CrossRef]
74. Wei, L.Y.; Hsu, C.W.; Chow, C.W.; Yeh, C.H. 20.231 Gbit/s tricolor red/green/blue laser diode based bidirectional signal remodulation visible-light communication system. *Photonics Res.* **2018**, *6*, 422–426. [CrossRef]
75. Wu, Y.C.; Su, C.Y.; Wang, H.Y.; Cheng, C.H.; Lin, G.R. Miniature R/G/V-LDs+Y-LED Mixed White-Lighting Module with High-Lux and High-CRI for 20-Gbps Li-Fi. In Proceedings of the 2020 Optical Fiber Communications Conference and Exhibition (OFC), San Diego, CA, USA, 8–12 March 2020; pp. 1–3.
76. Lee, C.; Islim, M.S.; Das, S.; Spark, A.; Videv, S.; Rudy, P.; Shah, B.; McLaurin, M.; Haas, H.; Raring, J. 26 Gbit/s LiFi System with Laser-Based White Light Transmitter. *J. Lightwave Technol.* **2022**, *40*, 1432–1439. [CrossRef]
77. Zhu, X.; Wang, F.; Shi, M.; Chi, N.; Liu, J.; Jiang, F. 10.72Gb/s Visible Light Communication System Based on Single Packaged RGBYC LED Utilizing QAM-DMT Modulation with Hardware Pre-Equalization. In Proceedings of the 2018 Optical Fiber Communications Conference and Exposition (OFC), San Diego, CA, USA, 11–15 March 2018; pp. 1–3.
78. Wang, W.C.; Cheng, C.H.; Wang, H.Y.; Lin, G.R. White-light color conversion with red/green/violet laser diodes and yellow light-emitting diode mixing for 34.8 Gbit/s visible lighting communication. *Photonics Res.* **2020**, *8*, 1398–1408. [CrossRef]
79. Cheng, C.H.; Wu, Y.C.; Tsai, C.T.; Wang, H.Y.; Lin, G.R. Four-Color LD+LED Lighting Module for 30-Gbps Visible Wavelength Division Multiplexing Data Transmission. In Proceedings of the 2020 Conference on Lasers and Electro-Optics (CLEO), San Jose, CA, USA, 10–15 May 2020; pp. 1–2.
80. Gunawan, W.H.; Liu, Y.; Chow, C.W.; Chang, Y.H.; Peng, C.W.; Yeh, C.H. Two-Level Laser Diode Color-Shift-Keying Orthogonal-Frequency-Division-Multiplexing (LD-CSK-OFDM) for Optical Wireless Communications (OWC). *J. Lightwave Technol.* **2021**, *39*, 3088–3094. [CrossRef]
81. Zhang, T.; Ghassemlooy, Z.; Rajbhandari, S.; Popoola, W.O.; Guo, S. OFDM-PWM scheme for visible light communications. *Opt. Commun.* **2017**, *385*, 213–218. [CrossRef]
82. Ebrahimi, F.; Ghassemlooy, Z.; Olyaee, S. Investigation of a hybrid OFDM-PWM/PPM visible light communications system. *Opt. Commun.* **2018**, *429*, 65–71. [CrossRef]
83. Wang, Y.; Wei, Y.; Zhang, W.; Wang, C.X. Filtered-OFDM for Visible Light Communications. In Proceedings of the 2018 IEEE/CIC International Conference on Communications in China (ICCC Workshops), Beijing, China, 16–18 August 2018; pp. 227–231. [CrossRef]
84. Jung, S.Y.; Kwon, D.H.; Yang, S.H.; Han, S.K. Inter-cell interference mitigation in multi-cellular visible light communications. *Opt. Express* **2016**, *24*, 8512–8526. [CrossRef]
85. Chen, M.; Cai, Y.; Zhou, J.; Zhou, H.; Liu, Y.; Chen, Q. Bandwidth enhancement with DAC-enabled pre-equalization and real-valued precoding for a FBMC-VLC. *Opt. Lett.* **2022**, *47*, 4826–4829. [CrossRef]
86. Abdoli, J.; Jia, M.; Ma, J. Filtered OFDM: A new waveform for future wireless systems. In Proceedings of the 2015 IEEE 16th International Workshop on Signal Processing Advances in Wireless Communications (SPAWC), Stockholm, Sweden, 28 June–1 July 2015; pp. 66–70. [CrossRef]
87. Kishore, V.; Valluri, S.P.; Vakamulla, V.M.; Sellathurai, M.; Kumar, A.; Ratnarajah, T. Performance Analysis under Double Sided Clipping and Real Time Implementation of DCO-GFDM in VLC Systems. *J. Lightwave Technol.* **2021**, *39*, 33–41. [CrossRef]

88. Fettweis, G.; Krondorf, M.; Bittner, S. GFDM—Generalized Frequency Division Multiplexing. In Proceedings of the VTC Spring 2009—IEEE 69th Vehicular Technology Conference, Barcelona, Spain, 26–29 April 2009; pp. 1–4. [CrossRef]
89. Ahmad, R.; Srivastava, A. Optical GFDM: An improved alternative candidate for indoor visible light communication. *Photonic Netw. Commun.* **2020**, *39*, 152–163. [CrossRef]
90. Saengudomlert, P.; Buddhacharya, S. Unipolar GFDM with Dimming Support for Visible Light Communications. *IEEE Trans. Wirel. Commun.* **2023**, 1. [CrossRef]
91. Guiomar, F.P.; Bertignono, L.; Nespola, A.; Carena, A. Frequency-Domain Hybrid Modulation Formats for High Bit-Rate Flexibility and Nonlinear Robustness. *J. Lightwave Technol.* **2018**, *36*, 4856–4870. [CrossRef]
92. Guiomar, F.P.; Carena, A.; Bosco, G.; Bertignono, L.; Nespola, A.; Poggiolini, P. Nonlinear mitigation on subcarrier-multiplexed PM-16QAM optical systems. *Opt. Express* **2017**, *25*, 4298–4311. [CrossRef]
93. Qiu, M.; Zhuge, Q.; Chagnon, M.; Gao, Y.; Xu, X.; Morsy-Osman, M.; Plant, D.V. Digital subcarrier multiplexing for fiber nonlinearity mitigation in coherent optical communication systems. *Opt. Express* **2014**, *22*, 18770–18777. [CrossRef]
94. Loureiro, P.A.; Silva, V.N.H.; Medeiros, M.C.R.; Guiomar, F.P.; Monteiro, P.P. Entropy loading for capacity maximization of RGB-based visible light communications. *Opt. Express* **2022**, *30*, 36025–36037. [CrossRef]
95. Marshoud, H.; Sofotasios, P.C.; Muhaidat, S.; Karagiannidis, G.K. Multi-user techniques in visible light communications: A survey. In Proceedings of the 2016 International Conference on Advanced Communication Systems and Information Security (ACOSIS), Marrakesh, Morocco, 17–19 October 2016; pp. 1–6. [CrossRef]
96. Nuwanpriya, A.; Ho, S.W.; Chen, C.S. Indoor MIMO Visible Light Communications: Novel Angle Diversity Receivers for Mobile Users. *IEEE J. Sel. Areas Commun.* **2015**, *33*, 1780–1792. [CrossRef]
97. Chen, C.; Zhong, W.D.; Yang, H.; Zhang, S.; Du, P. Reduction of SINR Fluctuation in Indoor Multi-Cell VLC Systems Using Optimized Angle Diversity Receiver. *J. Lightwave Technol.* **2018**, *36*, 3603–3610. [CrossRef]
98. Aljohani, M.K.; Aletri, O.Z.; Alazwary, K.D.; Musa, M.O.I.; El-Gorashi, T.E.H.; Alresheedi, M.T.; Elmirghani, J.M.H. NOMA Visible Light Communication System with Angle Diversity Receivers. In Proceedings of the 2020 22nd International Conference on Transparent Optical Networks (ICTON), Bari, Italy, 19–23 July 2020; pp. 1–5. [CrossRef]
99. Dixit, V.; Kumar, A. Performance analysis of angular diversity receiver based MIMO–VLC system for imperfect CSI. *J. Opt.* **2021**, *23*, 085701. [CrossRef]
100. Fath, T.; Haas, H. Performance Comparison of MIMO Techniques for Optical Wireless Communications in Indoor Environments. *IEEE Trans. Commun.* **2013**, *61*, 733–742. [CrossRef]
101. Safari, M.; Uysal, M. Do We Really Need OSTBCs for Free-Space Optical Communication with Direct Detection? *IEEE Trans. Wirel. Commun.* **2008**, *7*, 4445–4448. [CrossRef]
102. Siddiqi, U.F.; Narmanlioglu, O.; Uysal, M.; Sait, S.M. Joint bit and power loading for adaptive MIMO OFDM VLC systems. *Trans. Emerg. Telecommun. Technol.* **2020**, *31*, e3850. [CrossRef]
103. Guo, X.; Yuan, Y.; Pan, C.; Xiao, J. Interleaved superposed-64QAM-constellation design for spatial multiplexing visible light communication systems. *Opt. Express* **2021**, *29*, 23341–23356. [CrossRef] [PubMed]
104. Mardanikorani, S.; Deng, X.; Linnartz, J.P.M.G. Sub-Carrier Loading Strategies for DCO-OFDM LED Communication. *IEEE Trans. Commun.* **2020**, *68*, 1101–1117. [CrossRef]
105. Brusin, A.M.R.; Guiomar, F.P.; Lorences-Riesgo, A.; Monteiro, P.P.; Carena, A. Enhanced resilience towards ROADM-induced optical filtering using subcarrier multiplexing and optimized bit and power loading. *Opt. Express* **2019**, *27*, 30710–30725. [CrossRef]
106. Che, D.; Shieh, W. Approaching the Capacity of Colored-SNR Optical Channels by Multicarrier Entropy Loading. *J. Lightwave Technol.* **2018**, *36*, 68–78. [CrossRef]
107. Chow, P.; Cioffi, J.; Bingham, J. A practical discrete multitone transceiver loading algorithm for data transmission over spectrally shaped channels. *IEEE Trans. Commun.* **1995**, *43*, 773–775. [CrossRef]
108. Levin, H. A complete and optimal data allocation method for practical discrete multitone systems. In Proceedings of the GLOBECOM'01, IEEE Global Telecommunications Conference (Cat. No.01CH37270), San Antonio, TX, USA, 25–29 November 2001; Volume 1, pp. 369–374. [CrossRef]
109. Campello, J. Practical bit loading for DMT. In Proceedings of the 1999 IEEE International Conference on Communications (Cat. No. 99CH36311), Vancouver, BC, Canada, 6–10 June 1999; Volume 2, pp. 801–805. [CrossRef]
110. Fehenberger, T.; Alvarado, A.; Böcherer, G.; Hanik, N. On Probabilistic Shaping of Quadrature Amplitude Modulation for the Nonlinear Fiber Channel. *J. Lightwave Technol.* **2016**, *34*, 5063–5073. [CrossRef]
111. Cho, J.; Winzer, P.J. Probabilistic Constellation Shaping for Optical Fiber Communications. *J. Lightwave Technol.* **2019**, *37*, 1590–1607. [CrossRef]
112. Guiomar, F.P.; Lorences-Riesgo, A.; Ranzal, D.; Rocco, F.; Sousa, A.N.; Fernandes, M.A.; Brandão, B.T.; Carena, A.; Teixeira, A.L.; Medeiros, M.C.R.; et al. Adaptive Probabilistic Shaped Modulation for High-Capacity Free-Space Optical Links. *J. Lightwave Technol.* **2020**, *38*, 6529–6541. [CrossRef]
113. Che, D.; Shieh, W. Squeezing out the last few bits from band-limited channels with entropy loading. *Opt. Express* **2019**, *27*, 9321–9329. [CrossRef] [PubMed]
114. Schulte, P.; Böcherer, G. Constant Composition Distribution Matching. *IEEE Trans. Inf. Theory* **2016**, *62*, 430–434. [CrossRef]

115. Xie, C.; Chen, Z.; Fu, S.; Liu, W.; He, Z.; Deng, L.; Tang, M.; Liu, D. Achievable information rate enhancement of visible light communication using probabilistically shaped OFDM modulation. *Opt. Express* **2018**, *26*, 367–375. [CrossRef]
116. Che, D.; Shieh, W. Entropy-loading: Multi-carrier constellation-shaping for colored-SNR optical channels. In Proceedings of the 2017 Optical Fiber Communications Conference and Exhibition (OFC), Los Angeles, CA, USA, 19–23 March 2017; pp. 1–3.
117. Qu, Z.; Djordjevic, I.B. On the Probabilistic Shaping and Geometric Shaping in Optical Communication Systems. *IEEE Access* **2019**, *7*, 21454–21464. [CrossRef]
118. Karanov, B.; Chagnon, M.; Thouin, F.; Eriksson, T.A.; Bülow, H.; Lavery, D.; Bayvel, P.; Schmalen, L. End-to-End Deep Learning of Optical Fiber Communications. *J. Lightwave Technol.* **2018**, *36*, 4843–4855. [CrossRef]
119. Chen, B.; Okonkwo, C.; Hafermann, H.; Alvarado, A. Increasing Achievable Information Rates via Geometric Shaping. In Proceedings of the 2018 European Conference on Optical Communication (ECOC), Rome, Italy, 23–27 September 2018; pp. 1–3. [CrossRef]
120. Mirani, A.; Agrell, E.; Karlsson, M. Low-Complexity Geometric Shaping. *J. Lightwave Technol.* **2021**, *39*, 363–371. [CrossRef]
121. Oliveira, B.M.; Neves, M.S.; Guiomar, F.P.; Medeiros, M.C.R.; Monteiro, P.P. End-to-end deep learning of geometric shaping for unamplified coherent systems. *Opt. Express* **2022**, *30*, 41459–41472. [CrossRef]
122. Wu, X.; Chi, N. The phase estimation of geometric shaping 8-QAM modulations based on K-means clustering in underwater visible light communication. *Opt. Commun.* **2019**, *444*, 147–153. [CrossRef]
123. Wei, Y.; Yao, L.; Zhang, H.; Shen, C.; Chi, N.; Shi, J. An Optimal Adaptive Constellation Design Utilizing an Autoencoder-Based Geometric Shaping Model Framework. *Photonics* **2023**, *10*, 809. [CrossRef]
124. Borogovac, T.; Rahaim, M.B.; Tuganbayeva, M.; Little, T.D.C. "Lights-off" visible light communications. In Proceedings of the 2011 IEEE GLOBECOM Workshops (GC Wkshps), Houston, TX, USA, 5–9 December 2011; pp. 797–801. [CrossRef]
125. Tian, Z.; Wright, K.; Zhou, X. The Darklight Rises: Visible Light Communication in the Dark: Demo. In Proceedings of the 22nd Annual International Conference on Mobile Computing and Networking, MobiCom '16, New York, NY, USA, 3–7 October 2016; pp. 495–496. [CrossRef]
126. Beguni, C.; Căilean, A.M.; Avătămăniței, S.A.; Dimian, M. Analysis and Experimental Investigation of the Light Dimming Effect on Automotive Visible Light Communications Performances. *Sensors* **2021**, *21*, 4446. [CrossRef]

Review

Recent Advances in Optical Injection Locking for Visible Light Communication Applications

Xingchen Liu [1,2], Junhui Hu [1], Qijun Bian [1], Shulan Yi [1], Yingnan Ma [1], Jianyang Shi [1,3,4], Ziwei Li [1,3,4,5], Junwen Zhang [1,3,4,5], Nan Chi [1,3,4,5] and Chao Shen [1,3,4,5,*]

[1] Key Laboratory for Information Science of Electromagnetic Waves (MoE), School of Information Science and Technology, Fudan University, Shanghai 200433, China
[2] School of Microelectronics, Fudan University, Shanghai 200433, China
[3] Shanghai Engineering Research Center of Low-Earth-Orbit Satellite Communication and Applications, Shanghai 200433, China
[4] Shanghai Collaborative Innovation Center of Low-Earth-Orbit Satellite Communication Technology, Shanghai 200433, China
[5] Peng Cheng Laboratory, Shenzhen 518055, China
* Correspondence: chaoshen@fudan.edu.cn; Tel.: +86-136-7158-4193

Abstract: The introduction of visible light communication (VLC) technology could increase the capacity of existing wireless communication systems towards 6G networks. In practice, VLC can make good use of lighting system infrastructures to transmit data using light fidelity (Li-Fi). The use of semiconductor light sources, including light-emitting diodes (LEDs) and laser diodes (LDs) are essential to VLC technology because these devices are energy-efficient and have long lifespans. To achieve high-speed VLC links, various technologies have been utilized, including injection locking. Optical injection locking (OIL) is an optical frequency and phase synchronization technique that has been implemented in semiconductor laser systems for performance enhancement. High-performance optoelectronic devices with narrow linewidth, wide tunable emission, large modulation bandwidth and high data transmission rates are desired for advanced VLC. Thus, the features of OIL could be promising for building high-performance VLC systems. In this paper, we present a comprehensive review of the implementation of the injection-locking technique in optical communication systems. The enhancement of characteristics through OIL is elucidated. The applications of OIL in VLC systems are discussed. The prospects of OIL for future VLC systems are evaluated.

Keywords: optical injection locking; visible light communication; laser diode; Li-Fi

Citation: Liu, X.; Hu, J.; Bian, Q.; Yi, S.; Ma, Y.; Shi, J.; Li, Z.; Zhang, J.; Chi, N.; Shen, C. Recent Advances in Optical Injection Locking for Visible Light Communication Applications. *Photonics* **2023**, *10*, 291. https://doi.org/10.3390/photonics10030291

Received: 17 February 2023
Revised: 3 March 2023
Accepted: 7 March 2023
Published: 10 March 2023

1. Introduction

Visible light communication (VLC) is a high-speed communication technique utilizing an unlicensed frequency range of 400–800 THz [1]. Since the development of semiconductor lasers and optical fibers, optical communication systems have been widely used in our daily life [2]. With rapid growing wireless data demands, there is currently a push for environmentally friendly, high-speed and sustainable wireless communication methods in both indoor and outdoor environments. VLC can offer both lighting and communication at the same time, which could be a highly promising complement to conventional radio frequency (RF) wireless communication for high-speed local area networks in future 6G systems [3].

Optical injection locking (OIL) is an optical frequency and phase synchronization technique that is based on photon–photon interactions. Different from multiple section laser diodes, in which lights from different sections affect each other, there is one laser affecting the other in OIL. The latter may occur when external lights are shone into laser cavities [4]. The schematic diagrams of typical OIL schemes are illustrated in Figure 1. A typical external OIL configuration consists of master lasers and slave lasers. The master light can be injected into the slave laser and the output light can be blocked from the master laser via a circulator. A self-injection-locking configuration consists of a reflector through which light can be partially fed back. It is typically well acknowledged that the OIL technique is able to improve the device performance of semiconductor lasers, and there has been a mass of studies with regard to using the technique to build high-quality systems for telecommunication applications [5].

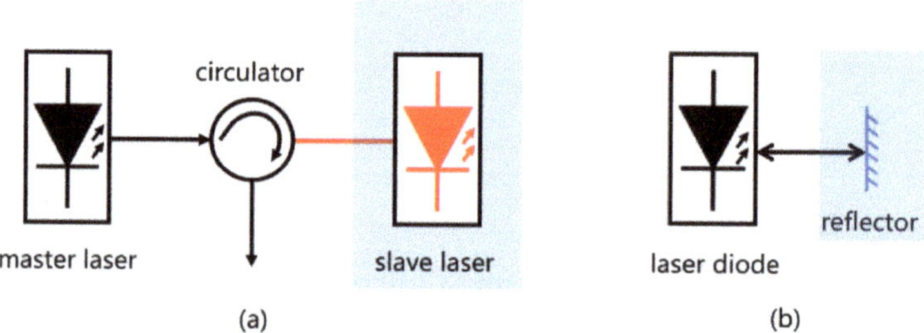

Figure 1. Schematic diagrams of the OIL technique. (**a**) External OIL configuration consists of master lasers and slave lasers. The master light can be injected into the slave laser and the output light can be blocked from the master laser via a circulator. (**b**) Self-injection-locking configuration consists of a reflector through which light can be partially fed back.

In VLC systems, OIL lasers are recently utilized as high-quality light sources for high-bitrate data links. Studies have demonstrated the enhancement effects of the OIL technique on the characteristics of different devices in the visible color regime. Additionally, the application of OIL is now demonstrated in various VLC systems, such as free-space VLC, visible light-based optical fiber communication and underwater wireless optical communication (UWOC).

In this paper, we review the recent advances in OIL and its applications in VLC. Firstly, the enhancement effects of OIL on the characteristics of different semiconductor optoelectronic devices are discussed, which could play vital roles in the future development of VLC systems and LiFi networks. Then, the most recent applications of OIL in VLC systems, including the implementation of OIL in fiber communication and UWOC systems, are analyzed. Finally, we elucidate the expected future trends in high-quality devices, such as ultranarrow-linewidth lasers for coherent optical communication.

2. Applications of Optical Injection Locking (OIL) in Different Fields

In the 1980s, following the development of semiconductor lasers for telecom applications, scientists actively began investigating the applications of OIL in optical communication systems [4]. Over the course of its history, the OIL technique has been utilized in many fields, prior to the development of VLC systems.

OIL has been successfully applied in the infrared wavelength field. Research regarding the effects of OIL on the noise properties of mid-infrared quantum cascade lasers has produced results that indicate that locked slave lasers could operate under reduced-intensity noise levels compared with the free-running operation [6], thereby corroborating the characteristic enhancement ability of the injection-locking technique. This practical approach to achieving low-noise operation for quantum cascade lasers is critical for the majority of applications in gas sensing and absorption spectroscopy. In 2020, a 448 Gb/s four-level pulse amplitude modulation (PAM4) free-space optical communication system with a 600 m free-space link was constructed, which also utilized polarization multiplexing optical injection-locked vertical cavity surface-emitting lasers (VCSELs) [7]. The VCSELs were optically injected with light from distributed feedback (DFB) LDs via the combination of a three-port optical circulator and a polarization controller. The findings from the experiment demonstrated that four 1.55 μm VCSEL transmitters were sufficiently powerful for 448 Gb/s PAM4 signal transmission when using the OIL technique. Additionally, the utilization of an eight-mode self-injection-locked quantum dash laser diode (LD) and a Reed–Solomon encoding technique in a secure PAM4-based free-space optical communication system was presented [8], which could transmit 88 Gbps of data over 555 m optical free-space links.

Moreover, the OIL technique has also been implemented in the generation and transmission of microwave and mmWave signals. Recently, there have been several studies on the generation and transmission of microwave and mmWave signals using the OIL technique. For example, Zhang et al. presented a frequency-modulated microwave generation setup including one master laser and two slave lasers [9]. In such a system, slow and fast perturbations were used to study frequency-modulated and externally locked P1 dynamics. The frequency-modulated continuous-wave generation at 6 GHz was achieved using externally locked lasers, with a comb contrast of up to 42 dB.

In an all-optical Ka-band microwave long-distance dissemination system, which was based on an optoelectronic oscillator (OEO), a single tone with high spectrum purity and low phase noise was excited by an optical injection-locked OEO, thereby achieving the stable phase transmission of mmWave signals [10]. Additionally, self-injection-locked quantum dash LD comb source technology has been used to produce a tunable 50/75 GHz mmWave transmission system in the difficult 1610 nm area [11]. Examples of mmWave generation and transmission systems are shown in Figure 2.

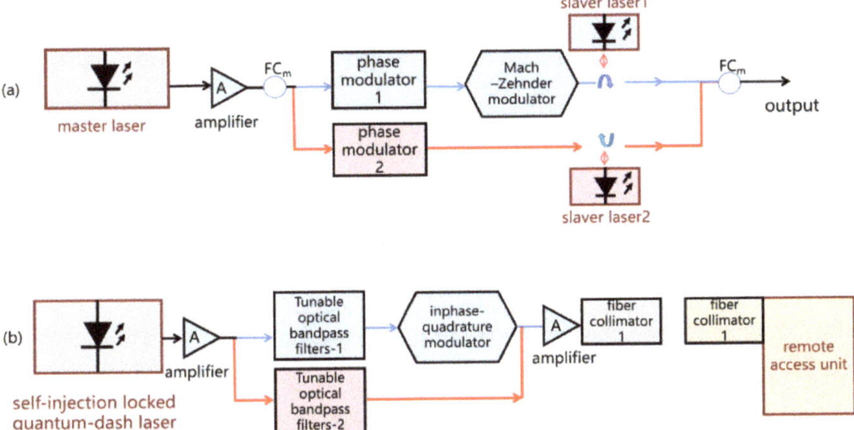

Figure 2. (a) The perturbed P1 dynamics of semiconductor lasers for frequency-modulated continuous-wave generation with external injection locking [8]; (b) A self-injection-locked quantum dash LD comb source, based on a 50/75 GHz mmWave transport system [11].

Furthermore, using a hybrid integrated self-injection-locking DFB laser, a gain-switched optical frequency comb source was proposed, which had eight pure continuous comb lines within 3 dB of the spectral envelope peak and a narrow linewidth of 615 kHz [12]. The carrier to noise ratio and the phase correlation between the comb lines were significantly improved by the self-injection-locking effect that was induced by the silicon nitride mirroring reflector, which could be promising for future radio over fiber and coherent optical communication.

In addition to the generation and transmission of microwaves and mmWaves as well as the production of frequency comb sources, self-injection-locked DFB LDs can be used for the high-sensitivity detection of acoustic emissions via fiber-coil Fabry–Pérot (F-P) interferometer sensors [13] and injection locking GaN blue LDs can be used for laser cooling and the trapping of ytterbium atoms in the field of physics [14].

Overall, OIL has been implemented in numerous applications over recent decades and now plays significant roles in various applications.

3. Enhancement Effects of OIL on the Characteristics of Different Devices

Since GaN-based devices have drawn significant attention for their application in VLC systems, we firstly examined the OIL in violet-blue-green light emitters. GaN-based LDs have demonstrated significantly higher modulation frequencies, greater power and better beam quality in comparison to GaN-based light-emitting diodes (LEDs). These features enable GaN LDs to produce long-distance transmission and open the door for the realization of emerging VLC applications. With the implementation of the OIL technique, the performance of GaN LDs has been further improved. Table 1 summarizes the enhancement effects of OIL on the characteristics of GaN LDs and the comparison with free running LDs.

Table 1. The enhancement effects of OIL on the characteristics of different devices [15–17].

Setup	Pros	Cons
Free-running blue F-P laser diodes	Simple structure	Relatively large emission spectrum linewidth
InGaN/GaN DFB LDs with gratings	Near-single-mode emission and a high side-mode suppression ratio	Fabrication of high-quality DFB grating is challenging
Littrow or Littman external cavity LD systems	A narrower linewidth and a high side-mode suppression ratio	System complexity (i.e., requires dielectric gratings or other wavelength filtering elements)
External cavity semiconductor LD systems	A satisfactory tuning range and high output power	Fine-tuning structure is required
Self-injection-locked LD systems	Good wavelength tunability and high optical power	Additional system complexity

In 2018, researchers from KFUPM and KAUST reported the application of self-injection locking in InGaN/GaN (blue/green) and InGaP/AlGaInP (red) visible light LD systems, in which the free-space optical feedback paths were accomplished using external mirrors [18]. They achieved significant increases of ∼57% (1.53–2.41 GHz) and ∼31% (1.72–2.26 GHz) in the modulation bandwidth and ∼9 (1.0–0.11 nm)- and ∼9 (0.63–0.07 nm)-fold reductions in the spectral linewidths of the green and blue lasers, respectively.

The following year, self-injection-locked green LDs using tunable dual-wavelength systems have been explored [19]. The self-injection-locking scheme was based on an external cavity configuration and utilized either highly or partially reflective mirrors. A tunable longitudinal mode spacing of 0.20–5.96 nm was accomplished, which corresponded to a calculated frequency difference of 0.22–6.51 THz. To further explore the systems, the same group employed the self-injection-locking technique for InGaN/GaN green LDs in an external cavity configuration with a partially reflective mirror using single- and multiwavelength laser systems [20]. The single-stage self-injection-locked laser system was for tunable laser and multiwavelength generation, while the two-stage self-injection-locked laser system was for near-single-wavelength generation. Figure 3 shows the narrow linewidth in single-mode, dual-mode and four-mode self-injection-locked green LDs at ∼525 nm and the two-stage self-injection-locking setup achieved a narrow locked-mode linewidth of ∼34 pm at 524.05 nm.

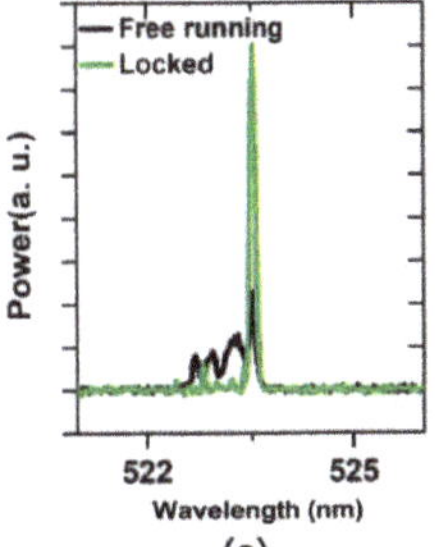
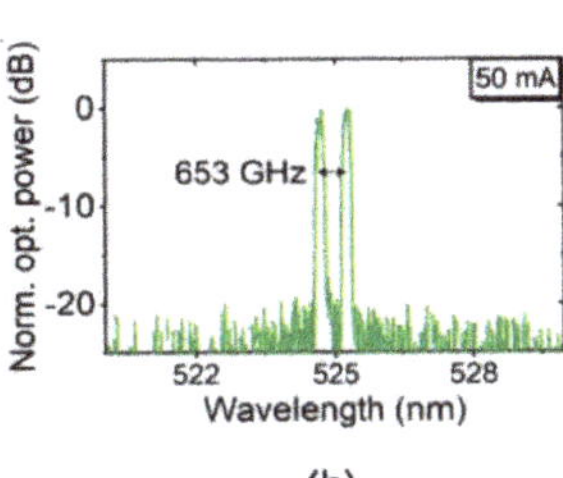
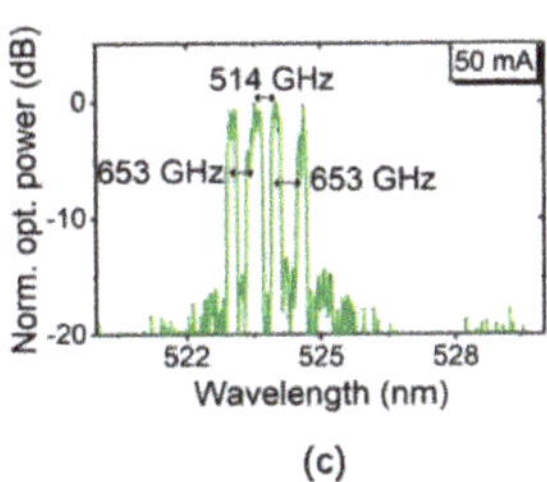

Figure 3. (**a**) The enhancement in the power of free-running (black) and self-injection-locked (green) systems for green LDs [18]; the normalized multiwavelength spectra of self-injection-locked external cavity systems with simultaneous (**b**) dual-and (**c**) four-mode generation, with an injection current (temperature) of 50 mA (40 °C) from a multiwavelength laser system [20].

Later in 2020, a prism-based self-injection-locked seamlessly tunable blue InGaN/GaN LD composite cavity system was presented [21]. The team achieved a clear main peak in the spectrum in contrast to the free-running one. Using an external cavity configuration and the self-injection-locking technique, the robust, simple and compact system achieved a significant enhancement in the wavelength-tuning window of up to 12.11 and 8 nm with ~3 and 14.5 mW optical power via high-reflection and low-reflection configurations, respectively, at just above the threshold current. The measured output power at low, medium, and high injection currents for both high-reflection and low-reflection system configurations and the optical power of a free-running one are presented in Table 2.

Table 2. The measured output power at low, medium and high injection currents for both high-reflection and low-reflection system configurations as well as the optical power of a free-running LD system [21].

Injection Current (mA)	Working Power (mW)		
	Free Running System	High-Reflection System	Low-Reflection System
130	3.4	3	14
260	26	13	23
390	186	93	180

Additionally, in recent years, researchers have managed to utilize VCSELs as injected slave lasers and have found some interesting VCSEL characteristics in the infrared regime. These devices have two orthogonal polarizations of the fundamental transverse mode: the parallelly polarized mode and the orthogonally polarized mode. The mode competition means that one of the modes is the main mode and the other is its subsidiary. However, the situation can reverse under different conditions of the injection signals, which means that the polarizations can be switched by changing the intensity of injections or detuning frequencies [22]. Spikes can be generated by utilizing OIL in slave lasers, which can be further applied in optical spike neural networks. Lu et al. proposed an approach to generate neuron-like spikes for self-injection-locked VCSELs (~1.56 μm) using multifrequency switching [23]. A controllable spiking coding scheme that utilized switching was designed, which reached speeds of up to 1 Gbps experimentally. The affiliation of [24] used optical inputs extracted from digital images as signals that were injected into VCSELs (~1.30 μm), thus achieving all-optical binary convolution.

In summary, the typical implementation of OIL systems is based on packaged LDs and external cavities with reflective setups. These compact systems can obtain considerable decreases in linewidth and increases in output power and modulation bandwidth whilst

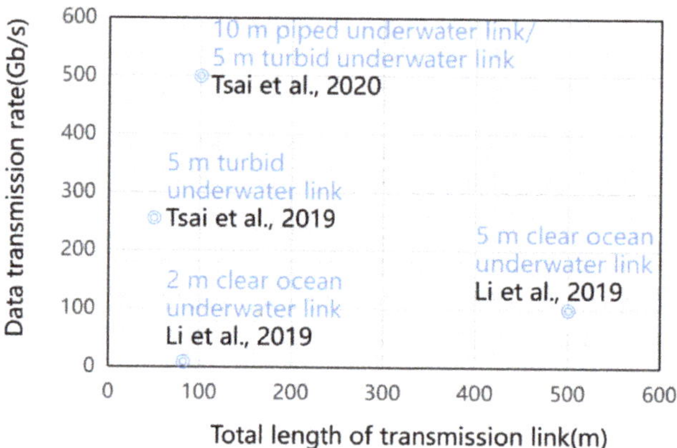

Figure 4. Recent applications of OIL in hybrid free space—UWOC systems. The total length of transmission link (x-axis) contains both underwater portion (blue) and free-space counterpart [33–36].

In 2020, Ming Chi University of Technology established a 500 Gb/s PAM4 free-space optical UWOC convergent system for 100 m free-space transmission using either a 10 m piped underwater link or a 5 m turbid underwater link, which integrated PAM4 modulation with a five-wavelength polarization multiplexing scheme [33]. The critical part of that scheme was the utilization of two-stage OIL and optoelectronic feedback on the LDs. With the two-stage OIL and optoelectronic feedback technique, the -3 dB bandwidths were enhanced by $\sim$10 times compared to a free-running counterpart.

Similarly, another wavelength-division multiplexing PAM4 free-space optical integrated FSO UWOC system was proposed, which had a channel capacity of 100 Gb/s [34]. The system applied 405 nm blue–violet light LDs and 1.7 GHz 450 nm blue light LDs via the two-stage OIL and optoelectronic feedback technique, which were adequately adopted for 100 Gb/s PAM4 signal transmission using a 500 m free-space transmission with a 5 m clear ocean underwater link. Apart from the critical deployment of the two-stage OIL and optoelectronic feedback technique, doublet lenses in the FSO, laser beam reducers and transmissive spatial light modulators were also crucial elements that helped the system to achieve a low bit error rate.

To sum up, with the development of signal modulation and multistage injection locking techniques, practical and high-speed underwater optical wireless links, as shown in Figure 5, which could enable high data rate transmission, are just around the corner.

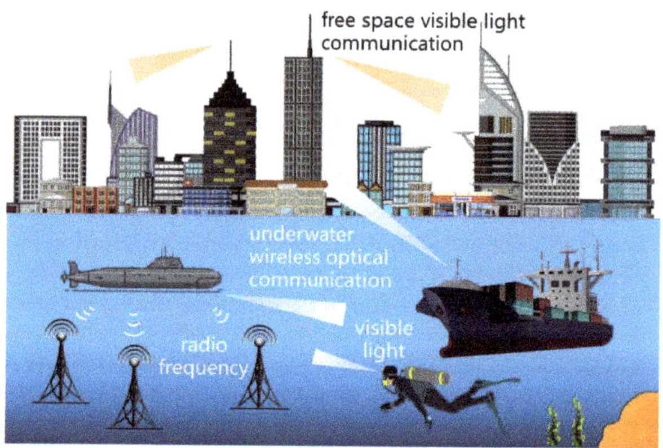

Figure 5. OIL in different VLC applications.

5. Future Trends of OIL in VLC

Current researchers have focused on the key materials and devices, high-speed system technology, networking technology, and the applications of VLC. Towards high performance VLC links, the utilization of OIL technology in VLC systems calls for the development of compact, low cost and manufacturable OIL transmitter on a chip, enabling the on-chip integration of light emitters. The form factor of the existing OIL setups is still large when comparing with commercial transceivers in telecommunication applications. The eventual deployment of VLC systems in practical applications, such as in mobile devices and underwater wireless networks, requires the further scaling down of the device size when OIL is used in VLC transmitter. This could be performed by designing a copackaging scheme of optical components with diode lasers, or on-chip integration of light emitters with passive optical devices. Moreover, future work on developing a high-optical-efficiency OIL is an important research topic. There are various sources of loss in current OIL system that can lead to reduced energy efficiency. Since achieving a low power consumption is one of the key objectives in 6G, the study of OIL technology with high optical efficiency in a visible color regime is another essential point that further research can lay emphasis on.

In addition, coherent optical communication is a promising approach for long-distance VLC, in which the requirements for narrow-linewidth and high-power devices are more demanding. Feedback mechanisms such as OIL can stabilize the frequency and reduce the spectral linewidth of the devices. Thus, future coherent VLC could be developed based on the high-performance visible light emitters using OIL.

Looking beyond, investigations into the implementation of OIL could lead to more advanced high-performance devices and VLC systems in the future.

6. Conclusions

In this review, we elucidated recent advances in the OIL technique, particularly in the visible light color regime. The ability of the OIL approach to enhance the performance of optoelectronic devices is well recognized, and there is a vast array of existing research that has used the technology for optical communication applications. We presented research concerning the enhancement effects of OIL on the characteristics of different devices, which has demonstrated satisfactory increases in output power as well as high SMSR values. We also examined studies on the application of OIL in various VLC systems. The two-stage OIL and optoelectronic feedback technique has turned out to be a salient and practical approach in these intricate real-life situations, thanks to its state-of-art characteristics of high output power and considerable stability.

VLC is an emerging technology that has a bright future, and the development of OIL in VLC could function as a helpful road map for developing greater and more effective VLC networks and systems.

Author Contributions: Writing—original draft preparation, X.L. and Q.B.; visualization, J.H.; review and editing, S.Y. and Y.M.; resources, J.S. and Z.L.; investigation, J.Z.; supervision and project administration, N.C. and C.S. All authors have read and agreed to the published version of the manuscript.

Funding: This research is partially funded by the Natural Science Foundation of China Project, grant number 62274042, 61925104; Natural Science Foundation of Shanghai, grant number 21ZR1406200; Major Key Project of PCL; The joint project of China Mobile Research Institute & X-NET; The Key Research and Development Program of Jiangsu Province (BE2021008-5).

Institutional Review Board Statement: Not applicable.

Informed Consent Statement: Not applicable.

Data Availability Statement: The data of this study are available from the corresponding author upon request.

Conflicts of Interest: The authors declare no conflict of interest. The funders had no role in the design of the study; in the collection, analyses, or interpretation of data; in the writing of the manuscript; or in the decision to publish the results.

References

1. Chi, N.; Zhou, Y.; Wei, Y.; Hu, F. Visible Light Communication in 6G: Advances, Challenges, and Prospects. *IEEE Veh. Technol. Mag.* **2020**, *15*, 93–102. [CrossRef]
2. Sibley, M. *Optical Communications: Components and Systems*; Springer: Cham, Switzerland, 2020.
3. Shen, C.; Ma, C.; Li, D.; Hu, J.; Li, G.; Zou, P.; Zhang, J.; Li, Z.; Chi, N. High-speed visible laser light communication: Devices, systems and applications. In *Broadband Access Communication Technologies XV*; Electr Network; SPIE: Bellingham, WA, USA, 2021.
4. Liu, Z.; Slavik, R. Optical Injection Locking: From Principle to Applications. *J. Light. Technol.* **2019**, *38*, 43–59. [CrossRef]
5. Khan, M.Z.M.; Alkhazraji, E.A.; Ragheb, A.M.; Esmail, M.A.; Tareq, Q.; Fathallah, H.A.; Qureshi, K.K.; Alshebeili, S. Injection-Locked Quantum-Dash Laser in Far L-Band 192 Gbit/s DWDM Transmission. *IEEE Photonics J.* **2020**, *12*, 1504211. [CrossRef]
6. Simos, H.; Bogris, A.; Syvridis, D.; Elsasser, W. Intensity Noise Properties of Mid-Infrared Injection Locked Quantum Cascade Lasers: I. Modeling. *IEEE J. Quantum Electron.* **2013**, *50*, 98–105. [CrossRef]
7. Wu, H.-W.; Lu, H.-H.; Tsai, W.-S.; Huang, Y.-C.; Xie, J.-Y.; Huang, Q.-P.; Tu, S.-C. A 448-Gb/s PAM4 FSO Communication with Polarization-Multiplexing Injection-Locked VCSELs through 600 M Free-Space Link. *IEEE Access* **2020**, *8*, 28859–28866. [CrossRef]
8. Mukherjee, R.; Mallick, K.; Kuiri, B.; Santra, S.; Dutta, B.; Mandal, P.; Patra, A.S. PAM-4 based long-range free-space-optics communication system with self injection locked QD-LD and RS codec. *Opt. Commun.* **2020**, *476*, 126304. [CrossRef]
9. Zhang, L.; Chan, S.-C. Perturbing period-one laser dynamics for frequency-modulated microwave generation with external locking. *Opt. Lett.* **2022**, *47*, 4483–4486. [CrossRef] [PubMed]
10. Zhang, K.; Li, S.; Xie, Z.; Zheng, Z. An All-optical Ka-band Microwave Long-distance Dissemination System Based on an Optoelectronic Oscillator. *IEEE Photonics J.* **2022**, *14*, 5545505. [CrossRef]
11. Khan, M.Z.M. Quantum-Dash Laser-Based Tunable 50/75 GHz mmW Transport System for Future L-Band Networks. *IEEE Photonics Technol. Lett.* **2022**, *34*, 842–845. [CrossRef]
12. Shao, S.; Li, J.; Chen, H.; Yang, S.; Chen, M. Gain-Switched Optical Frequency Comb Source Using a Hybrid Integrated Self-Injection Locking DFB Laser. *IEEE Photonics J.* **2022**, *14*, 6613606. [CrossRef]
13. Karim, F.; Mitul, A.F.; Zhou, B.; Han, M. High-Sensitivity Demodulation of Fiber-Optic Acoustic Emission Sensor Using Self-Injection Locked Diode Laser. *IEEE Photonics J.* **2022**, *14*, 7142610. [CrossRef]
14. Komori, K.; Takasu, Y.; Kumakura, M.; Takahashi, Y.; Yabuzaki, T. Injection-Locking of Blue Laser Diodes and Its Application to the Laser Cooling of Neutral Ytterbium Atoms. *Jpn. J. Appl. Phys.* **2003**, *42*, 5059–5062. [CrossRef]
15. Khan, M.Z.M.; Mukhtar, S.; Holguin-Lerma, J.A.; Alkhazragi, O.; Ashry, I.; Ng, T.K.; Ooi, B.S. Prism-based tunable InGaN/GaN self-injection locked blue laser diode system: Study of temperature, injection ratio, and stability. *J. Nanophotonics* **2020**, *14*, 036001. [CrossRef]
16. Shamim, M.H.M.; Ng, T.K.; Ooi, B.S.; Khan, M.Z.M. Tunable self-injection locked green laser diode. *Opt. Lett.* **2018**, *43*, 4931–4934. [CrossRef]
17. Ding, D.; Lv, X.; Chen, X.; Wang, F.; Zhang, J.; Che, K. Tunable high-power blue external cavity semiconductor laser. *Opt. Laser Technol.* **2017**, *94*, 1–5. [CrossRef]
18. Shamim, M.H.M.; Shemis, M.A.; Shen, C.; Oubei, H.M.; Ng, T.K.; Ooi, B.S.; Khan, M.Z.M. Investigation of Self-Injection Locked Visible Laser Diodes for High Bit-Rate Visible Light Communication. *IEEE Photonics J.* **2018**, *10*, 7905611. [CrossRef]
19. Shamim, M.H.M.; Alkhazragi, O.; Ng, T.K.; Ooi, B.S.; Khan, M.Z.M. Tunable Dual-Wavelength Self-injection Locked InGaN/GaN Green Laser Diode. *IEEE Access* **2019**, *7*, 143324–143330. [CrossRef]
20. Shamim, M.H.M.; Ng, T.K.; Ooi, B.S.; Khan, M.Z.M. Single and Multiple Longitudinal Wavelength Generation in Green Diode Lasers. *IEEE J. Sel. Top. Quantum Electron.* **2019**, *25*, 1501307. [CrossRef]
21. Mukhtar, S.; Ashry, I.; Shen, C.; Ng, T.K.; Ooi, B.S.; Khan, M.Z.M. Blue Laser Diode System with an Enhanced Wavelength Tuning Range. *IEEE Photonics J.* **2020**, *12*, 1502110. [CrossRef]
22. Prucnal, P.R.; Shastri, B.J.; de Lima, T.F.; Nahmias, M.A.; Tait, A.N. Recent progress in semiconductor excitable lasers for photonic spike processing. *Adv. Opt. Photonics* **2016**, *8*, 228–299. [CrossRef]
23. Lu, Y.; Zhang, W.; Fu, B.; He, Z. Frequency-switched photonic spiking neurons. *Opt. Express* **2022**, *30*, 21599. [CrossRef] [PubMed]
24. Zhang, Y.; Robertson, J.; Xiang, S.; Hejda, M.; Bueno, J.; Hurtado, A. All-optical neuromorphic binary convolution with a spiking VCSEL neuron for image gradient magnitudes. *Photonics Res.* **2021**, *9*, B201–B209. [CrossRef]
25. Chen, C.-Y.; Wu, P.-Y.; Lu, H.-H.; Lin, Y.-P.; Chang, C.-H.; Lin, H.-C. A bidirectional lightwave transport system based on PON integration with WDM VLC. *Opt. Fiber Technol.* **2013**, *19*, 405–409. [CrossRef]
26. Huang, X.-H.; Li, C.-Y.; Lu, H.-H.; Chou, C.-R.; Hsia, H.-M.; Chen, Y.-H. A Bidirectional FSO Communication Employing Phase Modulation Scheme and Remotely Injection-Locked DFB LD. *J. Light. Technol.* **2020**, *38*, 5883–5892. [CrossRef]
27. Lu, H.-H.; Li, C.-Y.; Chu, C.-A.; Lu, T.-C.; Chen, B.-R.; Wu, C.-J.; Lin, D.-H. 10 m/25 Gbps LiFi transmission system based on a two-stage injection-locked 680 nm VCSEL transmitter. *Opt. Lett.* **2015**, *40*, 4563–4566. [CrossRef]
28. Mukhtar, S.; Xiaobin, S.; Ashry, I.; Ng, T.K.; Ooi, B.S.; Khan, M.Z.M. Tunable Violet Laser Diode System for Optical Wireless Communication. *IEEE Photonics Technol. Lett.* **2020**, *32*, 546–549. [CrossRef]

29. Shamim, M.; Shemis, M.; Shen, C.; Oubei, H.; Alkhazragi, O.; Ng, T.; Ooi, B.; Khan, M. Analysis of optical injection on red and blue laser diodes for high bit-rate visible light communication. *Opt. Commun.* **2019**, *449*, 79–85. [CrossRef]

30. Lu, H.-H.; Huang, X.-H.; Chen, Y.-T.; Chang, P.-S.; Lin, Y.-Y.; Ko, T.; Liu, C.-X. WDM-VLLC and White-Lighting Ring Networks with Optical Add-Drop Multiplexing Scheme. *J. Light. Technol.* **2022**, *40*, 4196–4205. [CrossRef]

31. Li, C.-Y.; Lu, H.-H.; Chang, C.-H.; Lin, C.-Y.; Wu, P.-Y.; Zheng, J.-R.; Lin, C.-R. Bidirectional hybrid PM-based RoF and VCSEL-based VLLC system. *Opt. Express* **2014**, *22*, 16188–16196. [CrossRef] [PubMed]

32. Ying, C.-L.; Lu, H.-H.; Li, C.-Y.; Chu, C.-A.; Lu, T.-C.; Peng, P.-C. A Bidirectional Hybrid Lightwave Transport System Based on Fiber-IVLLC and Fiber-VLLC Convergences. *IEEE Photonics J.* **2015**, *7*, 7201611. [CrossRef]

33. Tsai, W.-S.; Lu, H.-H.; Wu, H.-W.; Tu, S.-C.; Huang, Y.-C.; Xie, J.-Y.; Huang, Q.-P.; Tsai, S.-E. 500 Gb/s PAM4 FSO-UWOC Convergent System With a R/G/B Five-Wavelength Polarization-Multiplexing Scheme. *IEEE Access* **2020**, *8*, 16913–16921. [CrossRef]

34. Li, C.-Y.; Huang, X.-H.; Lu, H.-H.; Huang, Y.-C.; Huang, Q.-P.; Tu, S.-C. A WDM PAM4 FSO–UWOC Integrated System with a Channel Capacity of 100 Gb/s. *J. Light. Technol.* **2019**, *38*, 1766–1776. [CrossRef]

35. Tsai, W.-S.; Li, C.-Y.; Lu, H.-H.; Lu, Y.-F.; Tu, S.-C.; Huang, Y.-C. 256 Gb/s Four-Channel SDM-Based PAM4 FSO-UWOC Convergent System. *IEEE Photonics J.* **2019**, *11*, 7902008. [CrossRef]

36. Li, C.-Y.; Lu, H.-H.; Wang, Y.-C.; Wang, Z.-H.; Su, C.-W.; Lu, Y.-F.; Tsai, W.-S. An 82-m 9 Gb/s PAM4 FSO-POF-UWOC Convergent System. *IEEE Photonics J.* **2019**, *11*, 7900609. [CrossRef]

Review

A Review of Advanced Transceiver Technologies in Visible Light Communications

Cuiwei He [1] and Chen Chen [2,*]

[1] School of Information Science, Japan Advanced Institute of Science and Technology, Nomi 923-1292, Japan; cuiweihe@jaist.ac.jp

[2] School of Microelectronics and Communication Engineering, Chongqing University, Chongqing 400044, China

* Correspondence: c.chen@cqu.edu.cn

Abstract: Visible Light Communication (VLC) is an emerging technology that utilizes light-emitting diodes (LEDs) for both indoor illumination and wireless communications. It has the potential to enhance the existing WiFi network and connect a large number of high-speed internet users in future smart home environments. Over the past two decades, VLC techniques have made significant strides, resulting in transmission data rates increasing from just a few Mbps to several tens of Gbps. These achievements can be attributed to the development of various transceiver technologies. At the transmitter, LEDs should provide high-quality light for illumination and support wide modulation bandwidths. Meanwhile, at the receiver, optics systems should have functions such as optical filtering, light concentration, and, ideally, a wide field of view (FOV). The photodetector must efficiently convert the optical signal into an electrical signal. Different VLC systems typically consider various transceiver designs. In this paper, we provide a survey of some important emerging technologies used to create advanced optical transceivers in VLC.

Keywords: visible light communication; optical wireless communication; LED; photodetector; fluorescent antenna; optical MIMO

Citation: He, C.; Chen, C. A Review of Advanced Transceiver Technologies in Visible Light Communications. *Photonics* **2023**, *10*, 648. https://doi.org/10.3390/photonics10060648

Received: 28 April 2023
Revised: 29 May 2023
Accepted: 1 June 2023
Published: 3 June 2023

1. Background

Wireless communication using radio frequency (RF) suffers from its limited bandwidth, which may result in a 'spectrum crunch' problem that restricts how fast we can access wireless data. In beyond-5G/6G, a promising solution to increase wireless communication capacity is to use a different part of the electromagnetic spectrum, such as the optical band. This emerging technology is known as optical wireless communication (OWC) and offers the significant advantage of utilizing large amounts of unregulated optical spectrum that is free to use.

OWC is typically classified into different categories based on the transceiver technologies and application scenarios, as depicted in Figure 1. Among the various forms of OWC, visible light communication (VLC) or LiFi has gained considerable research interest in the past two decades. VLC utilizes light-emitting diodes (LEDs) for both indoor illumination and data transmission. The inspiration behind VLC came from the rapid replacement of conventional incandescent and fluorescent light bulbs with LEDs for indoor lighting. In addition to their energy efficiency, LEDs can be directly modulated at very high frequencies, making them ideal for high-speed wireless transmission. Thanks to the development of various critical technologies, the transmission data rate of VLC has increased from a few Mbps to several tens of Gbps, making it a strong candidate for future wireless networks.

In this paper, we review the milestones in VLC's development along with some important transmitter and receiver technologies used in VLC. In particular, one specific focus of this survey is on reviewing the research related to developing fast color converters that not only deliver high-quality illumination but also can enhance the bandwidth performance of the LED transmitter. Additionally, we provide a summary of the research conducted on developing fluorescent antennas at the receiver side. These antennas enable simultaneous optical filtering and light concentration while having a wide field of view (FOV). Moreover, we summarize the use of silicon photomultiplier (SiPM) sensors in VLC for detecting weak-intensity light and also highlight the unique non-linearity problem caused by the device's dead time. In addition, we discuss the implementation of various types of angular diversity receivers in optical multiple-input–multiple-output (MIMO) transmission. Lastly, we outline the recent trends and future challenges in VLC research.

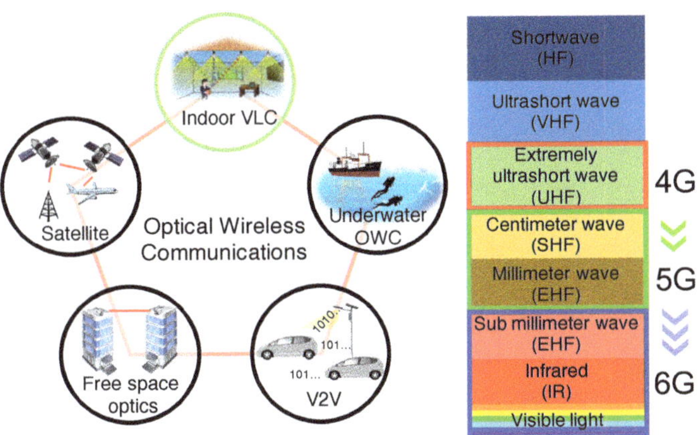

Figure 1. The trend of exploring higher frequency spectrum resources for wireless communications at each generation of wireless network and the possible applications of optical wireless communications.

2. Milestones of VLC Research

Figure 2 shows some key milestones in the development of VLC technologies. Between the 1970s and 1990s, wireless infrared communication research received significant attention [1,2]. In 1997, J. M. Kahn and J. R. Barry published an important paper [3] analyzing the theoretical aspects of infrared communication systems. Many of its theories are also directly relevant to VLC research. Since 2000, energy-efficient LEDs have been rapidly replacing traditional incandescent lighting. The Nobel Prize in 2014 was awarded to three scientists, Isamu Akasaki, Hiroshi Amano, and Shuji Nakamura, for their invention of efficient blue LEDs in the 1980s and 1990s. The invention of blue LEDs has not only enabled bright and energy-saving white light sources but also paved the way for the idea of using LEDs for transmitting data. The current form of VLC, which uses white LEDs for both indoor illumination and data transmission, originated from Nakagawa Lab [4], Japan, in 2002 and has since generated substantial interest worldwide. In 2009, the University of Oxford successfully demonstrated a 100 Mbps VLC transmission link using on–off keying (OOK) modulation [5]. At the same time, VLC using optical orthogonal frequency-division multiplexing (OFDM) modulation also attracted attention. In 2006, J. Armstrong et al. invented a power-efficient optical OFDM modulation method that was later widely considered for both VLC and optical fiber systems [6]. In 2008, the home Gigabit Access Network (OMEGA) project [7–9] was established in Europe and aimed to achieve gigabit data rates for home users via both VLC and RF communications. In 2010, the OMEGA project successfully demonstrated a 513 Mbps VLC transmission link using OFDM modulation with bit loading [10]. In 2011, the term 'LiFi' was first introduced by Prof. Harald Haas during a TEDGlobal talk and has attracted much attention from both the general public and the wireless industry [11]. During the same year, the IEEE 802.15.7 standard was formalized and defined the physical (PHY) and media access control (MAC) layer mechanisms for short-range optical wireless systems [12]. In recent years, the uses of multiplexing techniques, such as MIMO and wavelength-division multiplexing (WDM), to boost the transmission data rate have shown very promising performance. In 2013, the University of Oxford demonstrated a VLC system at 1 Gbps by the use of MIMO [13]. In 2015, Fudan University successfully demonstrated 8 Gbps VLC transmission using WDM with RGBY LEDs [14]. In 2016, the University of Oxford further increased this transmission data rate to 10 Gbps by using WDM and OFDM [15]. In 2019, using off-the-shelf LEDs, this data rate was increased to 15.73 Gbps by the University of Edinburgh [16]. In the same year, Fudan University successfully established an underwater VLC transmission link of 15 Gbps using RGBYC LEDs and WDM [17]. With various new technologies still under development, the VLC research community is aiming to improve the transmission data rate to Tbps using eye-safe lasers [18–21].

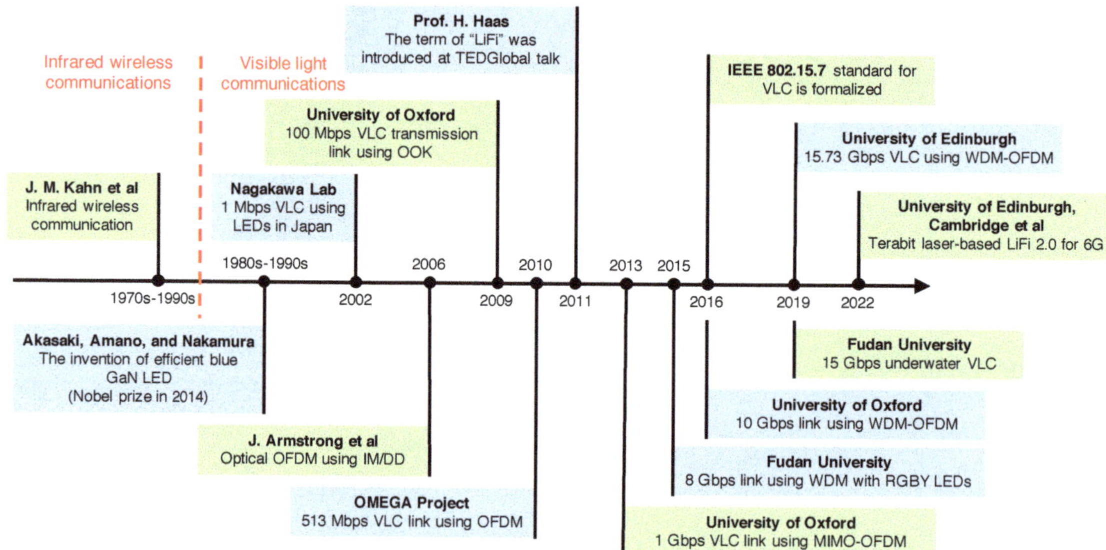

Figure 2. Some major milestones in VLC research.

The transmission data rate is typically regarded as the most important criterion for evaluating the performance of a communication system. Over the past two decades, the data transmission rate of LED-based VLC has increased significantly from only a few Mbps to several tens of Gbps. Based on the type of LED used, a summary of different VLC systems demonstrated to date can be found in Tables 1–4. Using representative work shown in these tables, Figure 3 illustrates how the achieved data transmission rate has been increased over time. Overall, when typical phosphor-coated white LEDs are used, the data transmission rate increases from 1 Mbps to several Gbps. The use of multi-chip RGB LEDs is seen to support the highest data rate thanks to the implementation of WDM to support parallel channels. Moreover, the use of μLEDs enables Gbps data rates via a single transmission link because of their high modulation bandwidths [22]. Recently, the use of organic LEDs (OLEDs) in VLC has also gained a lot of interests since OLEDs have flexible structures and can be potentially manufactured at very low costs. However, due to the high capacitance of OLEDs, the transmission data rate is usually only several Mbps. In a recent study [23], by manufacturing special types of OLEDs for VLC, data transmission rates of more than 1 Gbps were also achieved. The significant improvement in VLC can be attributed to the development of numerous transceiver technologies. In the following sections, some key technologies are reviewed.

Table 1. Phosphor-based white LED-based VLC links.

Year	Transmitter	Receiver	Modulation	Multiplexing	Distance	Data Rate	Ref.
2020	white LED	PIN	DMT	-	1 m	3 Gbps	[24]
2018	white LED	SPAD	OOK	-	1.2 km	2 Mbps	[25]
2015	white LED	PIN	OFDM	-	1.5 m	2 Gbps	[26]
2015	white LED	PIN	OFDM	-	1 m	1.6 Gbps	[27]
2013	white LED	PIN	OFDM	MIMO	1 m	1.1 Gbps	[13]
2012	white LED	PIN	CAP	-	0.23 m	1.1 Gbps	[28]
2012	white LED	APD	DMT	-	0.1 m	1 Gbps	[29]
2010	white LED	APD	DMT	-	0.3 m	513 Mbps	[10]
2009	white LED	PIN	OOK	-	0.1 m	100 Mbps	[5]
2007	white LED	PIN	OFDM	-	0.01 m	100 Mbps	[30]
2006	white LED	PIN	OFDM	-	1 m	16 Kbps	[31]
2002	white LED	PIN	BPSK	-	-	1 Mbps	[32]

Table 2. Multi-chip LED-based VLC links.

Year	Transmitter	Receiver	Modulation	Multiplexing	Distance	Data Rate	Ref.
2021	16 Si-LEDs	PIN	DMT	WDM	1.2 m	24.25 Gbps	[33]
2019	RGBY LEDs	PIN	OFDM	WDM	1.6 m	15.73 Gbps	[16]
2019	RGBYC Si-LEDs	PIN	DMT	WDM	1.2 m	15.17 Gbps	[17]
2017	RGB LEDs	PIN	OFDM	WDM + MIMO	1 m	6.36 Gbps	[34]
2016	RGB LEDs	SPAD	OFDM	WDM	2 m	60 Mbps	[35]
2015	RGBY LEDs	PIN	CAP	WDM	1 m	8 Gbps	[14]
2015	RGB LEDs	PIN	CAP	WDM	1.5 m	4.5 Gbps	[36]
2014	RGB LEDs	APD	OFDM	WDM	0.01 m	4.22 Gbps	[37]
2012	RGB LEDs	APD	OFDM	WDM	0.1 m	3.4 Gbps	[38]
2012	RGB LEDs	APD	DMT	WDM	0.1 m	1.25 Gbps	[39]

Table 3. µLED-based VLC links.

Year	Transmitter	Receiver	Modulation	Multiplexing	Distance	Data Rate	Ref.
2023	µLED	APD	OFDM	-	0.31 m	3.5 Gbps	[40]
2022	µLED	APD	OFDM	WDM	0.25	18.43 Gbps	[41]
2021	µLED	APD	OFDM	-	0.25	4.343 Gbps	[42]
2017	µLED	PIN	OFDM	-	0.275 m	11.12 Gbps	[43]
2016	µLEDs	PIN	OFDM	WDM	1.5 m	10 Gbps	[15]
2016	µLED	APD	PAM-4	MIMO	0.5 m	7.5 Gbps	[44]
2016	µLED	PIN	OFDM	-	0.75 m	5.37 Gbps	[45]
2015	µLED	APD	OFDM	WDM	-	2.3 Gbps	[46]
2014	µLED	PIN	OFDM	-	0.05 m	3 Gbps	[47]
2014	µLED	APD	OFDM	-	0.03m	1.68 Gbps	[48]
2011	µLED	PIN	OOK	-	-	512 Mbps	[49]
2010	µLED	PIN	OOK	-	-	1 Gbps	[50]

Table 4. OLED-based VLC links.

Year	Transmitter	Receiver	Modulation	Multiplexing	Distance	Data Rate	Ref.
2020	OLED	APD	OFDM	-	2 m	1.13 Gbps	[23]
2020	OLED	PIN	OOK	-	0.05 m	2.2 Mbps	[51]
2015	OLED	OPD	OOK	WDM	0.05 m	55 Mbps	[52]
2015	OLED	PIN	OOK	WDM	0.05 m	10Mbps	[53]
2013	OLED	PIN	OOK	-	-	10 Mbps	[54]
2012	OLED	PIN	OOK	-	0.1 m	550 Kbps	[55]

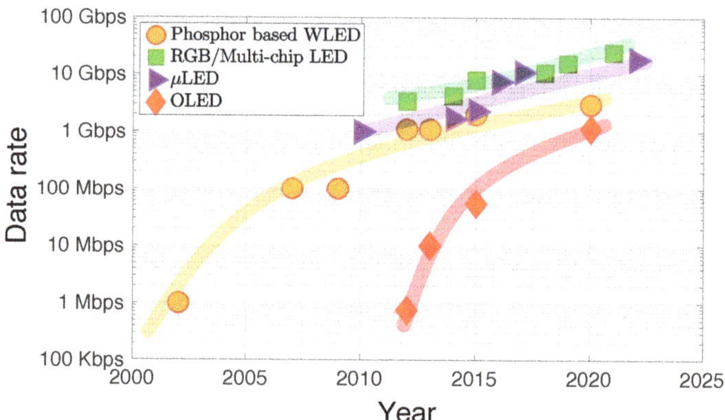

Figure 3. The achieved VLC transmission data rates using different types of LED in the past 20 years. The curved lines illustrate a coarse representation of the data rates achieved over different years.

3. LED Transmitter

As depicted in Figure 3, the choice of the LED transmitter has a substantial influence on the performance of a VLC system. For indoor illumination purposes, the emitted light from the LED needs to be white. There are typically two categories of LEDs used for producing white light: multi-chip RGB LEDs and phosphor-based LEDs. As shown in Figure 4, in an RGB LED, three LED chips emit red, green, and blue light, respectively, to generate white light. Different from a RGB LED, a phosphor-based LED is comprised of a phosphor coating and a single blue LED. The phosphor coating converts a portion of the emitted blue light to yellow, and the combination of blue and yellow light is perceived as white light. Due to their low cost, phosphor-coated white LEDs are the most widely used LED type for illumination purposes.

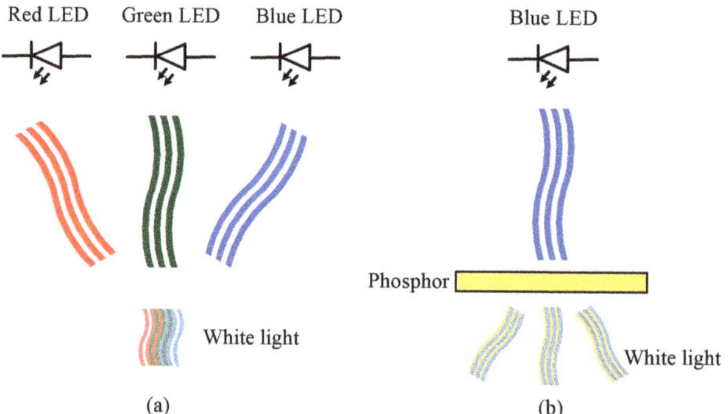

Figure 4. Different types of white LED: (**a**) multi-chip RGB LED (**b**) phosphor-based LED.

3.1. LED Radiation Pattern

One important characteristic of an LED is its radiation pattern, which determines the relative light strength for different emission angles. In most of the VLC research, a Lambertian emitter [3] is used to model the radiation pattern of the LED. The Lambertian order, m, is related to the semi-angle of the LED, $\Phi_{1/2}$, by

$$m = \frac{\ln 2}{\ln(\cos \Phi_{1/2})}. \tag{1}$$

Figure 5 shows the 3-D radiation pattern of the LED with different Lambertian orders. The radiation pattern of the LED is designed for illumination purposes but also significantly influences VLC performance. It can be seen that a higher value of the Lambertian order means the LED is more directional. If a bare photodiode with an area of A is placed at a position with a distance of d from the LED, the optical channel gain is given by

$$h = \frac{(m+1)A}{2\pi d^2} \cos^m(\phi) \cos(\psi) \tag{2}$$

where ϕ is the emergence angle of the light and ψ is the incident angle of the light.

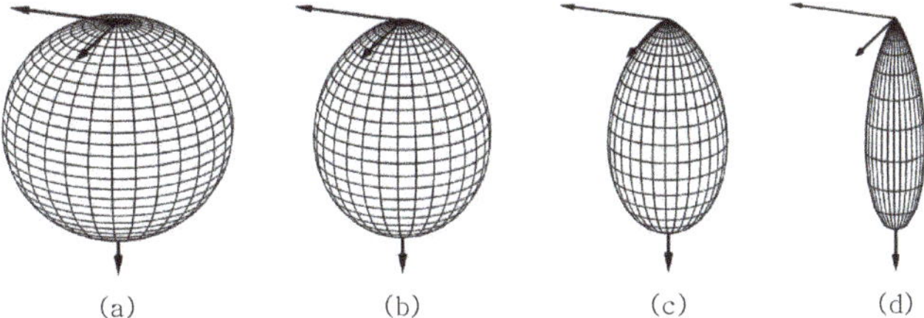

Figure 5. The radiation pattern of LEDs with different Lambertian orders (**a**) $m = 1$ ($\Phi_{1/2} = 60°$), (**b**) $m = 2$ ($\Phi_{1/2} = 45°$), (**c**) $m = 5$ ($\Phi_{1/2} = 30°$), and (**d**) $m = 20$ ($\Phi_{1/2} = 15°$).

In a VLC transmission scenario, the positions of the LED and the photodiode not only affect their distance but also influence both the emergence angle and incident angle of the light. Therefore, unlike most RF systems, the relative position between the transmitter and the receiver has a significant influence on the channel gain and thus affects the transmission performance. Using the Lambertian model, Figure 6 shows an example of how the illuminance is distributed in an office-like environment with a size of 10 m× 6 m. In this example, the vertical distance between the LED luminaires installed on the ceiling and the plane where the illumination level is simulated at 1.75 m, which is a typical distance between the ceiling and a table surface. The Lambertian order is considered to be one, and the positions of the luminaires are shown in Figure 6 using crosses. As expected, the positions directly below the LED luminaires have high illumination levels, e.g., 400 lx–500 lx. In contrast, the illumination levels are relatively low in the corners of the room, e.g., 100 lx. Importantly, it can be clearly seen that the light transmitted from each individual LED luminaire only covers a specific area. This implies that even within a limited space, users may only receive desired signals from a single transmitter without much interferences from other LED luminaires. As a result, the same frequency resources can be reused across different LED luminaires, allowing for the construction of a small indoor cellular system [56–58]. This is normally considered as an another significant advantage of VLC over its RF counterparts.

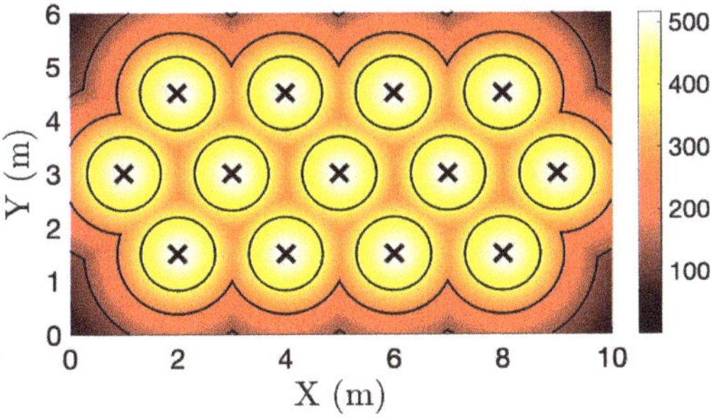

Figure 6. The illumination level distribution (in lx) in a 10 m by 6 m room with 13 LED luminaires.

3.2. Fast Color Converter

Phosphor-based LEDs are currently the most commonly used light source for illumination, making them the popular type of transmitter in VLC research. However, the main limitation when using phosphor-based LEDs as VLC data transmitters is that, although the blue LED can be modulated up to high frequencies, due to the slow photoluminescence (PL) lifetime of the phosphor coating, the generated white light can only be modulated at frequencies up to few megahertz [59]. This is normally known as the 'phosphor bottleneck' problem in VLC and results in a trending research topic on developing new color converters with short PL lifetime to replace the phosphor coating for generating high-quality white light.

A good color converter for VLC applications should possess several important features. Firstly, it should have strong absorption of blue light emitted from Gallium Nitride (GaN) LEDs, typically at 450 nm. Secondly, it should have a high photoluminescence quantum yield (PLQY) to ensure that most absorbed photons lead to the emission of new photons with longer wavelengths rather than non-radiative decay. Thirdly, the generated white light should have good illumination quality, as usually measured by the color rendering index (CRI). Finally, the color converter should have a PL lifetime shorter than the carrier recombination lifetime of the LED so that the bandwidth of the LED is not affected. For both the GaN LED and the color converter, the exponential decay of the processes that lead to emission of photons results in a single-pole response in the frequency domain, given by [60]

$$H(f) = \frac{1}{1 + j2\pi f\tau} \tag{3}$$

where τ is either the PL lifetime of the color converter or the carrier recombination lifetime of the GaN LED. Therefore, the frequency response of the signal power is

$$|H(f)|^2 = H(f)H^*(f) = \frac{1}{1 + (2\pi f\tau)^2}. \tag{4}$$

In this case, the 3 dB bandwidth is given by

$$f_{3\mathrm{dB}} = \frac{1}{2\pi\tau} \tag{5}$$

which is inversely proportional to the PL lifetime. Consequently, a short PL lifetime of the color converter is very desirable for communication purposes.

In the past, several different fast color converters have been developed for VLC applications. In reference [48], the color converter "super yellow", which is a type of conjugated polymer, was combined with a blue μLED to produce high-quality white light. A breakthrough transmission data rate of 1.68 Gbps was achieved at a standard illumination level of 240 lx. Following this, the organic materials group at the University of St Andrews and the optical wireless group at the University of Oxford collaborated to investigate several organic materials for developing fast color converters [61–64]. For example, some studied materials, such as BBEHP-PPV ($\tau = 0.83$ ns), exhibited PL lifetimes of less than one nanosecond [61]. Additionally, by using advanced cascade energy transfer principles, a high-performance color converter that enables the shifting of the emission spectrum from green to red with both a high PLQY and a short PL lifetime was created [64]. In more recent studies, in addition to organic materials, there is also a trend to explore some inorganic perovskite nanocrystals, or quantum dots, for creating fast color converters, and many transmission systems have been demonstrated with very promising performance [65–71].

4. VLC Receiver

In addition to the LED transmitter, the design of the receiver also plays an important role in VLC. As shown in Figure 7, a typical VLC receiver consists of an optical filter, an optical concentrator, and a photodiode. The optical filter selects the wavelength of the light to be detected. The optical concentrator is used to increase the received optical power. The photodiode converts the optical intensity signal into an electrical signal. In many receiver designs, the photodiode is also connected with a transimpedance amplifier (TIA) so that the current signal is converted into a voltage signal.

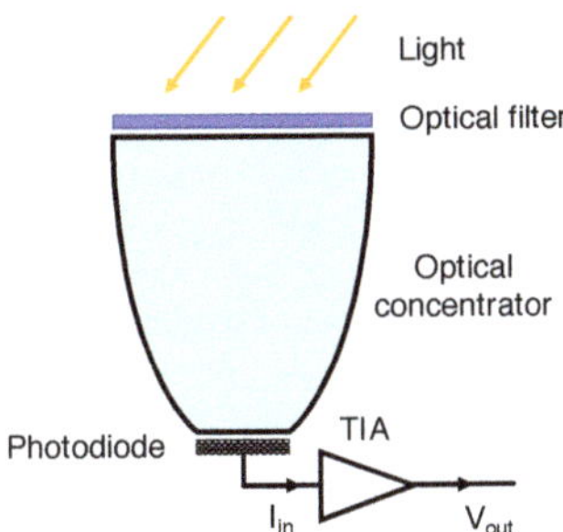

Figure 7. A typical VLC receiver consists of an optical filter, an optical concentrator, and a photodiode.

4.1. Optical Filter

An optical filter can improve the overall performance of the VLC system in several ways. For instance, when using a phosphor-based LED as the data transmitter, a blue filter is usually placed before the photodiode to filter out the yellow light and thus increase the transmission bandwidth. However, in some studies [72,73], it was found that blue filtering does not always improve the transmission performance. The usefulness of blue filtering depends on a range of factors such as the transmission data rate, the frequency range utilized, the effectiveness of the equalizer, the noise level, and even the LED spectrum [72]. This is because if the transmission sampling rate is not too high and very high frequencies are not used, the yellow light still carries information. In this case, using proper equalization techniques, the frequency response caused by both the blue LED and the phosphor can be well equalized and the noise component after equalization is acceptable; the use of blue filtering may not be necessary and may even reduce the performance. In contrast, when the transmission sampling rate becomes high or very high frequencies are utilized, the yellow light carries no information at these frequencies but only introduces shot noise. In this situation, a blue filter can be placed to filter out the yellow light and thus reduce the shot noise. If no blue filtering is used, the equalization can result in significant noise enhancement, which affects the overall transmission performance. Another important function of optical filtering is to effectively separate light of various colors, thus preventing crosstalk between different channels when RGB LEDs or lasers are employed as data transmitters, enabling WDM to achieve high data transmission rates [74].

4.2. Optical Concentrator

In VLC, the received optical power of the light is proportional to the area of the photodiode. However, the capacitance of the photodiode increases when the area of the photodiode increases, and a high capacitance limits the bandwidth of the receiver. To support a high modulation bandwidth, the size of the photodiode needs to be small. Therefore, an optical concentrator is usually placed in front of the photodiode to increase the received optical power for a given photodiode area. The most common concentrators include lenses and compound parabolic concentrators (CPCs). However, these concentrators cannot achieve both high concentration gain and a wide FOV; this is known as the étendue limit [3]. Using typical optical concentrators, the maximum concentration gain, $g(\psi)$, and the FOV, Ψ_{FOV}, are related by

$$g(\psi) = \begin{cases} \dfrac{n^2}{\sin^2 \Psi_{FOV}}, & 0 \leq \psi \leq \Psi_{FOV} \\ 0, & \psi > \Psi_{FOV} \end{cases} \tag{6}$$

where n is the refractive index of the concentrator. When a concentrator and an optical filter are used, the optical channel gain shown in (2) can be adapted to give

$$h = \begin{cases} \dfrac{(m+1)A}{2\pi d^2} \cos^m(\phi) T_s(\psi) g(\psi) \cos(\psi), & 0 \le \psi \le \Psi_{FOV} \\ 0, & \psi > \Psi_{FOV} \end{cases} \tag{7}$$

where $T_s(\psi)$ is the transmission coefficient of the optical filter, which may also vary with the incident angle of the light, especially when narrow-band interference filters are used.

In addition to limiting the receiver's FOV, placing both an optical filter and an optical concentrator in front of the photodiode can also result in a bulky receiver structure, such as the example shown in Figure 7, which is not suitable for small devices such as IoT devices.

4.3. Fluorescent Antenna

To build compact VLC receivers with wide FOVs, a new approach of using optical antennas made of fluorescent materials has been recently studied and shows very promising performance. In addition to their wide FOVs, these antennas are capable of simultaneous optical filtering and light concentration. Figure 8a shows the main physical processes of the light within the fluorescent antenna. When the incident light arrives at the antenna surface, depending on the incident angle of the light, part of the light is reflected back into the air and the rest of the light transmits into the antenna. Within the antenna, the light can pass through the antenna if the wavelength of the light is not within the absorption range of the fluorophore. Alternatively, a photon can be absorbed by the fluorophore. This absorption can be relaxed non-radiatively or it can result in the emission of a photon with longer wavelengths. Since the emitted photons can go in any direction, as shown in Figure 8a, these photons can escape the antenna or be re-absorbed by the fluorophore. At the same time, many photons can be waveguided to the antenna end where a photodiode is placed. Overall, due to these physical processes, a single fluorescent antenna has many functions. First, because the fluorescent materials only absorb light of certain wavelengths, therefore the antennas have the functions of optical filtering. Second, the antennas are designed to have both a cladding layer and a core layer so that many emitted photons from the fluorophores can be trapped within the antenna and guided to the photodiodes. Thus, they are capable of light concentration. Third, since its light concentration principle is based on fluorescence rather than reflection and refraction, it can exceed the étendue limit and achieve both high light concentration gain and a wide FOV. Additionally, since fluorophores with very short PL lifetimes can be selected. The antennas can provide very high transmission bandwidths.

The idea of fluorescent antennas was originally from the study of luminescent solar concentrators (LSCs) [75], and it was then first studied for use in OWCs in [76] for building a wide-FOV receiver. After that, several different antenna structures and fluorescent materials have been explored and tested in different OWC systems. In reference [77], an optical antenna structure made of a fluorescent layer sandwiched between two glass microscope slides was introduced. The fluorescent layer contained an organic fluorescent dye, Coumarin-6 (Cm6), which has strong absorption of 450 nm blue light. The performance of this antenna was studied in a blue LED-based VLC system. In reference [78], this structure was extended to include a second organic fluorescent layer made of 4-(Dicyanomethylene)-2-methyl-6-(4-dimethylaminostyryl)-4H-pyran (DCM) dye, which can absorb green light greater than 500 nm. In this case, this structure can support WDM techniques to boost the transmission data rate when a multi-chip RGB LED is used as the data transmitter. However, the significant drawback of this structure is that it has four wide rectangular edges, and most of the photons that are guided to the antenna edges cannot be detected. To overcome its structure problem, capillary-based antennas were introduced in [79]. A photo of these antennas, which are made of Coumarin-504 (Cm504), Cm6, and DCM, is shown in Figure 8b. In addition to organic materials, some inorganic materials have also been studied. For example, in [80], $CsPbBr_3$ perovskite nanocrystals were considered to make an optical antenna with a polymer-fiber structure. Its performance was demonstrated in an underwater optical wireless transmission link. Another popular approach to designing optical antennas is to use commercially available fluorescent fibers [81–86]. Figure 8c shows one example of a florescent antenna made of two bent fibers that have strong absorption of 450 nm blue light and emission of green light. The benefits of using fluorescent fibers as optical antennas include their flexible structure and the advanced fiber cladding technique that increases photon trap efficiency. In recent studies [87–89], the concept of utilizing a commercially available light-diffusing fiber (LDF) to construct a wide-FOV OWC system was also introduced. In this approach, the end of the LDF fiber is connected to a light source that transmits the signal. Due to the scattering of light in various directions within the fiber and its uniform emission from the fiber surface, this technique enables the use of movable receivers.

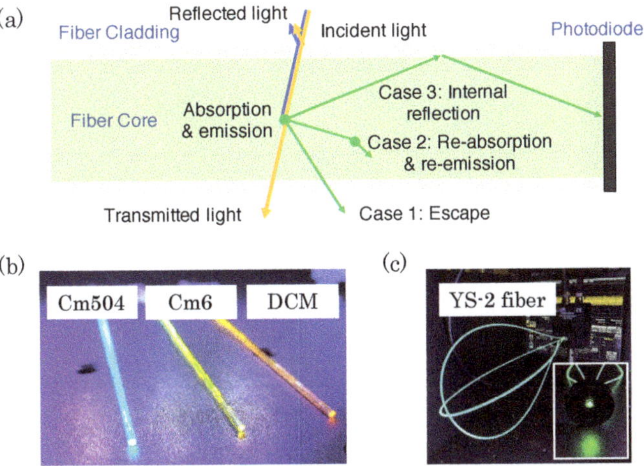

Figure 8. (**a**) The physical processes of the light within the fluorescent antenna, (**b**) three different antenna examples made of organic fluorescent materials [79] , and (**c**) one antenna example using commercially available fluorescent fibers [86].

4.4. Photodetector

In VLC systems, the received light intensity is converted into an electrical signal using a photodetector at the receiver. Two commonly used types of photodiodes are positive–intrinsic–negative (PIN) photodiodes and avalanche photodiodes (APDs). Although the cost of PIN photodiodes is typically much lower than that of APDs, APDs are significantly more sensitive due to the signal amplification resulting from avalanche multiplication. However, the drawback of using an APD is that it produces excess noise [90]. One way to further improve the sensitivity of the receiver is to bias an APD above its breakdown voltage, known as the Geiger mode, to create a single-photon avalanche detector (SPAD) [91,92]. Although SPADs are very sensitive and can detect individual photons, their operational mechanism means that the SPAD needs a period to recover after detecting a single photon. This is usually known as the dead time or the recovery period [91]. Figure 9 shows three simulation trials of the photon detection process with the impact of the SPAD's dead time when different irradiance levels are considered. As observed in Figure 9a, no photons are missed when the irradiance level is low. However, as illustrated in Figure 9b,c, an increase in irradiance results in a rise in both the number of photons arriving at the SPAD and the number of missed photons.

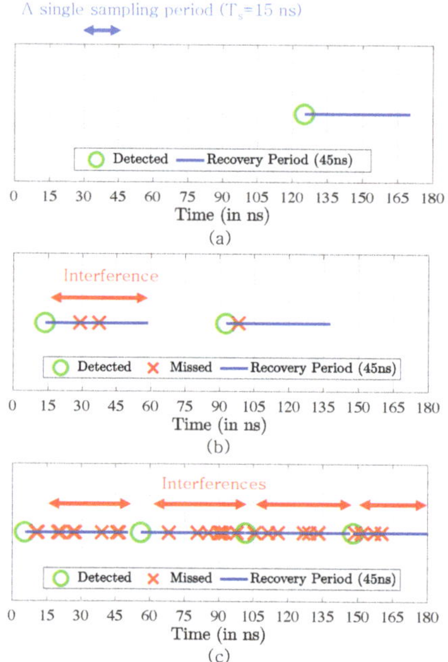

Figure 9. The simulation trials of the photon detection process with the influence of the SPAD's dead time (45 ns) considering a single microcell within a SiPM 30035 chip when the received irradiance levels are (**a**) 1 mW/m^2, (**b**) 10 mW/m^2, and (**c**) 100 mW/m^2. In these figures, the green circles indicate the time of detected photons and the blue lines indicate the associated dead time. The red crosses indicate the missed photons.

In recent studies, the use of a SPAD array sensor, known as SiPM or multi-pixel photon counter (MPPC), in VLC shows very impressive results, particularly when the received light is weak [93–96]. Thanks to advanced silicon fabrication techniques, a single SiPM sensor, only a few millimeters in size, can contain thousands of SPADs. Compared to a single SPAD, the use of a SPAD array ensures that even if some SPADs become inactive after detecting photons, other active SPADs can continue to detect them. This also provides a wide dynamic range of the quantized photon-counting signal to support advanced modulation techniques such as OFDM [97–100]. However, the dead time of individual SPADs still significantly impacts transmission and causes a non-linear response when a SiPM is used to detect light intensity signals. Figure 10 shows the nonlinear relationship between the average photon counting rate and the irradiance at the SiPM with a SiPM 30035 chip manufactured by Onsemi considered. Since SiPMs are non-paralyzable detectors, it also can be seen that the maximum photon counting rate is given by

$$C_{\max} = \frac{N_{\mathrm{SPAD}}}{\tau_{\mathrm{recover}}} \tag{8}$$

where N_{SPAD} represents the number of SPADs within a SiPM and τ_{recover} denotes the dead time. When the transmission sampling rate is low, the nonlinearity caused by the SPAD dead time can cause attenuation in individual signal samples. However, when the transmission sampling rate is high, the dead time can span several symbol periods, as shown in Figure 9b,c, introducing a unique form of ISI. To address the nonlinearity problem caused by SPAD dead time, various pre- and post-equalization techniques have been explored in recent studies [101,102], including using different forms of artificial neural networks [103,104]. As SiPM chip fabrication techniques continue to advance, SiPM sensors are becoming increasingly suitable for low-light-intensity environments, such as underwater OWC [105–107], eye-safe laser-based OWC [108,109], and free space optics (FSO)-based applications for unmanned aerial vehicles (UAVs) [110–112].

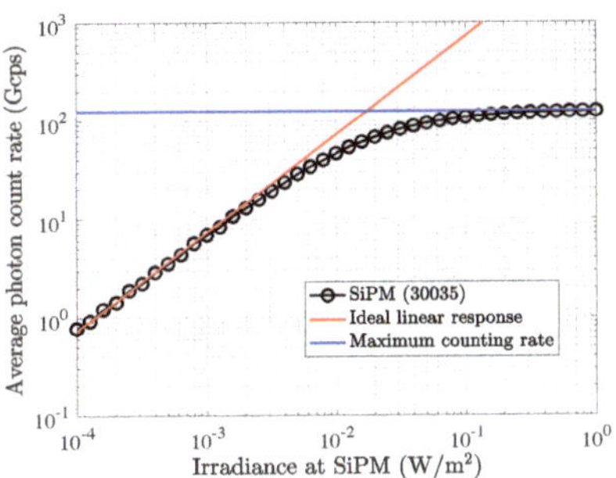

Figure 10. The simulated nonlinear response of the SiPM chip 30035 and the desired linear response.

4.5. Optical MIMO Receiver

In most indoor environments, the illumination is typically provided by several LED luminaires that are spaced at intervals on the ceiling. By using these LED luminaires as data transmitters and multiple photodiodes at the receiver, it is possible to achieve VLC MIMO transmission [113–115]. Several types of VLC MIMO systems are commonly employed, including spatial multiplexing (SMP), optical spatial modulation (OSM), and indoor VLC cellular systems.

An important indoor VLC MIMO configuration utilizes SMP [116–118] in which independent data are transmitted from each of the LED luminaires. Figure 11a depicts the transmitters, L1, L2, L3, and L4, transmitting independent data streams to the user devices, U1 and U2. Multiple photodiodes are used in each user device to receive the transmitted signals, which are then sent to a de-multiplexing module for separation. The independent transmission of data from LEDs means that SMP configuration can support high data rates.

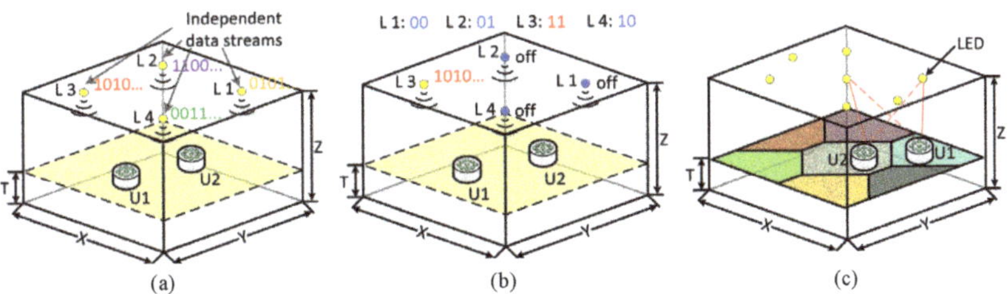

Figure 11. Three typical optical MIMO systems considered in VLC: (**a**) spatial multiplexing, (**b**) spatial modulation, and (**c**) indoor optical cellular system.

Another important VLC MIMO configuration is OSM [119–122], which is illustrated in Figure 11b. In the simplest form of OSM, the transmission is divided into short time slots and only one LED luminaire is active during each time slot. Information is conveyed through both the data symbols modulated on the light intensity and the index of the active LED luminaire. Multi-stream interference (MSI) is avoided in OSM since only one LED is active. However, this is achieved at the expense of a significant reduction in data rate. To enhance the transmission data rate of OSM, generalized OSM can be considered, in which during each time slot, multiple LED luminaires are active to transmit some information [123] or independent data streams [124]. Consequently, more bits of information are modulated into the index of the active LED luminaires and the performance is enhanced via diversity combining or multiplexing.

A more feasible configuration in an indoor environment is a cellular system wherein the data intended for a particular user are transmitted by the closest luminaire [125]. As shown in Figure 11c, the user's possible location is divided into different cells, with each cell having a corresponding LED luminaire. The luminaire transmits signals to the users within its cell. For instance, user U1 receives the desired signal from the luminaire located above, while U2 receives the desired signal from a different luminaire. Multiplexing techniques such as time-division multiple access (TDMA), code-division multiple access (CDMA), orthogonal frequency-division multiple access (OFDMA), and non-orthogonal multiple access (NOMA) [126] can be utilized to enable multiple users to be supported by each cell.

In all of the optical MIMO configurations, the receiver must be able to successfully separate signals transmitted from different sources. However, the characteristics of the transmitters, channel, and receivers used in VLC systems mean that the separation of the signals is difficult. In VLC, the received optical power varies slowly as a function of PD position. As shown in Figure 12, when using multiple PDs with the same orientation, the channels between the PDs and a particular LED luminaire become highly similar in the case of a small receiver such as a smartphone. This creates an ill-conditioned channel matrix where each column of the matrix has nearly identical values. This means that it becomes difficult to separate the signals without significant noise enhancement.

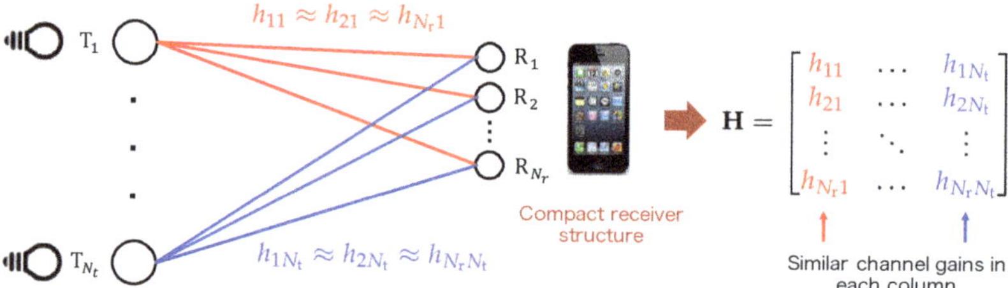

$$\mathbf{H} = \begin{bmatrix} h_{11} & \cdots & h_{1N_t} \\ h_{21} & \cdots & h_{2N_t} \\ \vdots & \ddots & \vdots \\ h_{N_r 1} & \cdots & h_{N_r N_t} \end{bmatrix}$$

Figure 12. Optical channel between multiple LED transmitters and multiple photodetectors.

To improve the MIMO channel condition, optical receivers play an important role. Currently, the optical MIMO receiver can be classified into two categories, including imaging receivers and non-image receivers. In the case of an imaging receiver, an imaging lens is used to completely separate the signals in the optical domain [44,116,127]. One example is shown in Figure 13a. To increase the FOV of the imaging receiver, a hemispherical-lens-based receiver is introduced in [117], and its structure is shown in Figure 13b. Although the hemispherical lens can only partially separate the signal in some cases, it can still result in a well-conditioned channel matrix that enables the signals to be successfully separated in the electrical domain using de-multiplexing techniques. In the case of the non-imaging receiver, angular diversity receivers are usually used. One simple way to create an angular diversity receiver is to place the photodiodes in a way that they can face in different directions. One example uses a pyramid structure, as shown in Figure 13c [128]. Other similar structures are studied in [125,129–131]. Reference [125] shows that the facing angles of the photodiodes can be optimized based on considered indoor scenarios. However, these three-dimensional receivers can hardly be incorporated into a device without any protrusions. To build optical MIMO receivers with a planar surface, a prism-based receiver is introduced in [132]. As shown in Figure 13d, by using prisms with different orientations, the photodiode placed below the prism only receives light from certain directions and, thus, angular diversity is achieved. Another approach is using an angular diversity aperture receiver (ADA), as shown in Figure 13e. An ADA receiver contains multiple receiving elements (REs), and each RE consists of an aperture and a photodiode [133–135]. The key feature of the ADA receiver is that the positions of the photodiodes are different in different REs.

By adjusting the relative position between the aperture and the photodiode, a directional MIMO receiver is created and the FOV of each RE is also adjustable. This structure can be further used to create MIMO receivers with different FOVs for improving the channel condition [136]. A similar structure, named the 'quadrature angular diversity aperture receiver (QADA)', is also studied in [137,138] for angle-of-arrival (AoA)-based VLP systems. Figure 13e illustrates an ADA receiver with nine REs. To show the directionality of this receiver, the 3D receiving pattern and the associated polar plot of the receiver are shown in Figure 14. It can be seen that although a single RE has a very limited FOV, using multiple REs can achieve an overall large FOV.

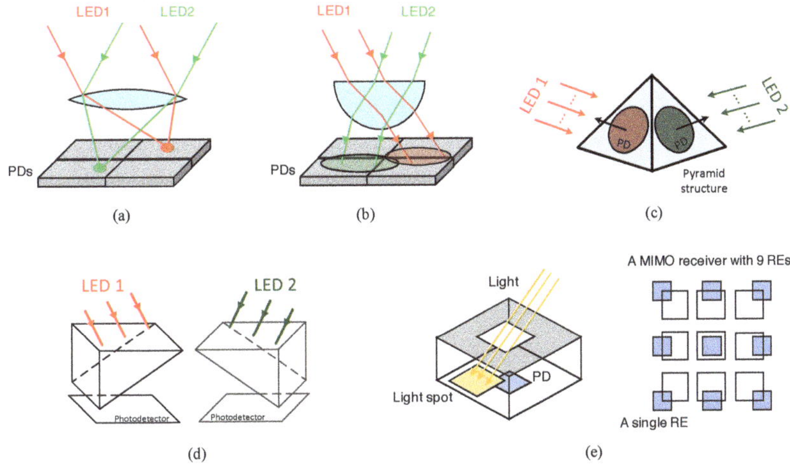

Figure 13. Different types of optical MIMO receivers: (**a**) an imaging-lens-based receiver, (**b**) a hemispherical-lens-based receiver, (**c**) a receiver using PDs facing different directions, (**d**) a prism-based receiver, and (**e**) the angular diversity aperture (ADA) receiver.

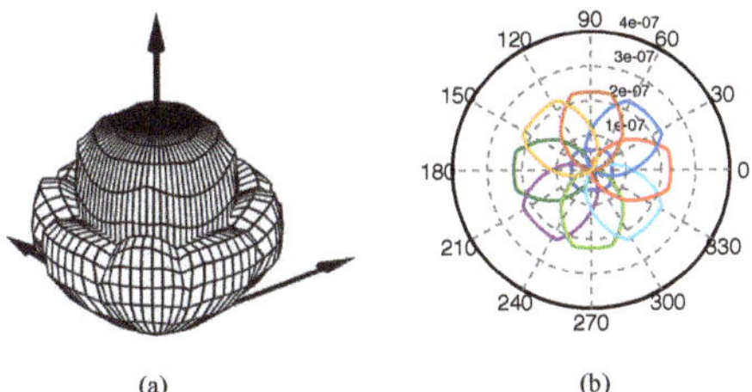

Figure 14. The receiving pattern of an ADA receiver with nine REs: (**a**) the 3D optical channel gain plot and (**b**) the polar plot.

5. Recent Trends and Future Challenges

Although LEDs are extensively used in various indoor environments, making them convenient choices for optical wireless transmitters, their modulation bandwidths remain relatively limited. This limitation has the potential to restrict the achievable transmission data rates in the near future. The VLC research community is currently exploring the use of eye-safe lasers, often referred to as LiFi 2.0, to achieve transmission rates in the Tbps range [20]. Unlike LEDs, lasers offer significantly higher modulation bandwidths. However, the need for safety compliance, such as the constraints outlined in the IEC 60825-1 standard, requires limiting the laser's output power. Recent studies have demonstrated that vertical-cavity surface-emitting lasers (VCSELs) can achieve data rates in the Gbps range while ensuring eye-safe conditions [109]. Additionally, the utilization of broad beams eliminates the requirement for using high-cost beam-steering techniques, enabling the coverage of large areas and offering significant benefits for supporting mobile users. VCSELs also offer numerous advantages as data transmitters, including extremely high modulation bandwidths (>10 GHz) and the ability to densely pack them in large arrays within a single access point (AP), enabling support for transmission scenarios such as MIMO [21] and indoor atto-cells [109]. Currently, the development of LiFi 2.0 is still in its early stages, and the goal of establishing stable Tbps transmission links is being actively pursued through collaborative efforts by several major OWC research groups.

Despite the numerous advantages of VLC, it still faces several challenges compared to its RF counterparts. Firstly, in many VLC systems that employ broad beams of light, the transmission distance is often limited. This limitation is particularly prominent in IM/DD-based VLC systems, where the transmitted signal is directly modulated onto the optical power of the light. In such cases, the amplitude of the received electrical signal is inversely proportional to the square of the distance. Consequently, as the distance increases, the power of the electrical signal decreases even more rapidly. Secondly, in most VLC experiments, achieving high transmission data rates is only feasible when a LOS link exists between the transmitter and receiver, as the diffuse component of the light is typically much weaker. Although photon-counting sensors are capable of detecting light with extremely weak intensity, designing optical filtering systems that allow a wide FOV while effectively rejecting ambient light to prevent the sensor from being saturated can be significantly challenging. Thirdly, most high-speed VLC transmissions can only be demonstrated in laboratory-based experiments that rely on expensive optics or specially designed electronics. If VLC is to be implemented in real-life applications, it becomes essential to focus on reducing costs to meet practical requirements.

6. Conclusions

Over the last two decades, there has been a significant growing interest in the development of VLC technologies for a variety of wireless applications. This growth can be attributed to a range of advanced transmission techniques. First, thanks to the invention of different types of LEDs, high-transmission bandwidths were achieved, which allows Gbps data rates. Additionally, to solve the "phosphor bottleneck" problem and provide high-quality white light, researchers have investigated various materials with short PL lifetimes. The developed technologies allow the bandwidth of white-light-emitting transmitters to increase from just several megahertz to several hundred megahertz for supporting Gbps links. Advanced photodetectors and optics at the receiver have also significantly contributed to the development of VLC systems. Recent studies on fluorescent antennas have enabled the creation of a compact wide-FOV receiver with dual functions of optical filtering and light concentration. Moreover, the use of super-sensitive optical sensors, such as the SiPM chip, has allowed for the use of VLC in special environments with extremely low light intensity, such as underwater OWC. We hope that the continued development of these emerging techniques will make VLC a strong candidate for the next-generation wireless communication network.

Author Contributions: Conceptualization, C.H. and C.C.; writing—original draft preparation, C.H.; writing—review and editing, C.H. and C.C.; funding acquisition, C.H. All authors have read and agreed to the published version of the manuscript.

Funding: The work of Cuiwei He was supported by the Japan Society for the Promotion of Science (JSPS) KAKENHI under grant JP 23K13332.

Institutional Review Board Statement: Not applicable.

Informed Consent Statement: Not applicable.

Data Availability Statement: The data of this study are available from the corresponding author upon request.

Conflicts of Interest: The authors declare no conflict of interest.

References

1. Barry, J.R. *Wireless Infrared Communications*; Springer Science & Business Media: New York, NY, USA, 1994; Volume 280.
2. Gfeller, F.R.; Bapst, U. Wireless in-house data communication via diffuse infrared radiation. *Proc. IEEE* **1979**, *67*, 1474–1486. [CrossRef]
3. Kahn, J.M.; Barry, J.R. Wireless infrared communications. *Proc. IEEE* **1997**, *85*, 265–298. [CrossRef]
4. Komine, T.; Nakagawa, M. Fundamental analysis for visible-light communication system using LED lights. *IEEE Trans. Consum. Electron.* **2004**, *50*, 100–107. [CrossRef]
5. Le Minh, H.; O'Brien, D.; Faulkner, G.; Zeng, L.; Lee, K.; Jung, D.; Oh, Y.; Won, E.T. 100-Mb/s NRZ visible light communications using a postequalized white LED. *IEEE Photonics Technol. Lett.* **2009**, *21*, 1063–1065. [CrossRef]
6. Armstrong, J.; Lowery, A.J. Power efficient optical OFDM. *Electron. Lett.* **2006**, *42*, 1. [CrossRef]
7. O'Brien, D. High speed optical wireless demonstrators in the omega project: Summary and conclusions. In Proceedings of the 18th European Conference on Network and Optical Communications & 8th Conference on Optical Cabling and Infrastructure (NOC-OC&I), Graz , Austria, 10–12 July 2013 ; pp. 159–162.
8. Javaudin, J.P.; Bellec, M.; Varoutas, D.; Suraci, V. OMEGA ICT project: Towards convergent Gigabit home networks. In Proceedings of the IEEE 19th International Symposium on Personal, Indoor and Mobile Radio Communications, Cannes, France, 15–18 September 2008; pp. 1–5.
9. Javaudin, J.P.; Bellec, M. Omega project: On convergent digital home networks. In Proceedings of the 3d International Workshop on Cross Layer Design, Rennes, France, 30 November–1 December 2011; pp. 1–5.
10. Vučić, J.; Kottke, C.; Nerreter, S.; Langer, K.D.; Walewski, J.W. 513 Mbit/s visible light communications link based on DMT-modulation of a white LED. *J. Light. Technol.* **2010**, *28*, 3512–3518. [CrossRef]
11. Haas, H.; Yin, L.; Wang, Y.; Chen, C. What is lifi? *J. Light. Technol.* **2015**, *34*, 1533–1544. [CrossRef]
12. IEEE Association. *IEEE Standard for Local and Metropolitan Area Networks-Part 15.7: Short-Range Wireless Optical Communication Using Visible Light*; IEEE: Piscataway, NJ, USA, 2011; pp. 1–309.
13. Azhar, A.H.; Tran, T.A.; O'Brien, D. A gigabit/s indoor wireless transmission using MIMO-OFDM visible-light communications. *IEEE Photonics Technol. Lett.* **2012**, *25*, 171–174. [CrossRef]
14. Wang, Y.; Tao, L.; Huang, X.; Shi, J.; Chi, N. 8-Gb/s RGBY LED-based WDM VLC system employing high-order CAP modulation and hybrid post equalizer. *IEEE Photonics J.* **2015**, *7*, 1–7.
15. Chun, H.; Rajbhandari, S.; Faulkner, G.; Tsonev, D.; Xie, E.; McKendry, J.J.D.; Gu, E.; Dawson, M.D.; O'Brien, D.C.; Haas, H. LED based wavelength division multiplexed 10 Gb/s visible light communications. *J. Light. Technol.* **2016**, *34*, 3047–3052. [CrossRef]
16. Bian, R.; Tavakkolnia, I.; Haas, H. 15.73 Gb/s visible light communication with off-the-shelf LEDs. *J. Light. Technol.* **2019**, *37*, 2418–2424. [CrossRef]
17. Zhou, Y.; Zhu, X.; Hu, F.; Shi, J.; Wang, F.; Zou, P.; Liu, J.; Jiang, F.; Chi, N. Common-anode LED on a Si substrate for beyond 15 Gbit/s underwater visible light communication. *Photonics Res.* **2019**, *7*, 1019–1029. [CrossRef]
18. Kazemi, H.; Sarbazi, E.; Soltani, M.D.; Safari, M.; Haas, H. A Tb/s indoor optical wireless backhaul system using VCSEL arrays. In Proceedings of the IEEE 31st Annual International Symposium on Personal, Indoor and Mobile Radio Communications, London, UK, 31 August–3 September 2020; pp. 1–6.
19. Hong, Y.; Feng, F.; Bottrill, K.R.; Taengnoi, N.; Singh, R.; Faulkner, G.; O'Brien, D.C.; Petropoulos, P. Demonstration of >1Tbit/s WDM OWC with wavelength-transparent beam tracking-and-steering capability. *Opt. Express* **2021**, *29*, 33694–33702. [CrossRef]
20. Soltani, M.D.; Kazemi, H.; Sarbazi, E.; Qidan, A.A.; Yosuf, B.; Mohamed, S.; Singh, R.; Berde, B.; Chiaroni, D.; Béchadergue, B.; et al. Terabit Indoor Laser-Based Wireless Communications: LiFi 2.0 for 6G. *arXiv* **2022**, arXiv:2206.10532.
21. Kazemi, H.; Sarbazi, E.; Soltani, M.D.; El-Gorashi, T.E.; Elmirghani, J.M.; Penty, R.V.; White, I.H.; Safari, M.; Haas, H. A Tb/s indoor mimo optical wireless backhaul system using VCSEL arrays. *IEEE Trans. Commun.* **2022**, *70*, 3995–4012. [CrossRef]
22. Wei, Z.; Wang, L.; Li, Z.; Chen, C.J.; Wu, M.C.; Wang, L.; Fu, H. Micro-LEDs Illuminate Visible Light Communication. *IEEE Commun. Mag.* **2023**, *61*, 108–114. [CrossRef]
23. Yoshida, K.; Manousiadis, P.P.; Bian, R.; Chen, Z.; Murawski, C.; Gather, M.C.; Haas, H.; Turnbull, G.A.; Samuel, I.D. 245 MHz bandwidth organic light-emitting diodes used in a gigabit optical wireless data link. *Nat. Commun.* **2020**, *11*, 1171. [CrossRef]
24. Zhou, Y.; Wei, Y.; Hu, F.; Hu, J.; Zhao, Y.; Zhang, J.; Jiang, F.; Chi, N. Comparison of nonlinear equalizers for high-speed visible light communication utilizing silicon substrate phosphorescent white LED. *Opt. Express* **2020**, *28*, 2302–2316. [CrossRef]
25. Ji, Y.w.; Wu, G.f.; Wang, C.; Zhang, E.f. Experimental study of SPAD-based long distance outdoor VLC systems. *Opt. Commun.* **2018**, *424*, 7–12. [CrossRef]
26. Huang, X.; Chen, S.; Wang, Z.; Shi, J.; Wang, Y.; Xiao, J.; Chi, N. 2.0-Gb/s visible light link based on adaptive bit allocation OFDM of a single phosphorescent white LED. *IEEE Photonics J.* **2015**, *7*, 1–8. [CrossRef]
27. Huang, X.; Wang, Z.; Shi, J.; Wang, Y.; Chi, N. 1.6 Gbit/s phosphorescent white LED based VLC transmission using a cascaded pre-equalization circuit and a differential outputs PIN receiver. *Opt. Express* **2015**, *23*, 22034–22042. [CrossRef] [PubMed]
28. Wu, F.M.; Lin, C.T.; Wei, C.C.; Chen, C.W.; Huang, H.T.; Ho, C.H. 1.1-Gb/s white-LED-based visible light communication employing carrier-less amplitude and phase modulation. *IEEE Photonics Technol. Lett.* **2012**, *24*, 1730–1732. [CrossRef]
29. Khalid, A.; Cossu, G.; Corsini, R.; Choudhury, P.; Ciaramella, E. 1-Gb/s transmission over a phosphorescent white LED by using rate-adaptive discrete multitone modulation. *IEEE Photonics J.* **2012**, *4*, 1465–1473. [CrossRef]

30. Grubor, J.; Lee, S.C.J.; Langer, K.D.; Koonen, T.; Walewski, J.W. Wireless high-speed data transmission with phosphorescent white-light LEDs. In *Proceedings of the 33rd European Conference and Exhibition of Optical Communication-Post-Deadline Papers*; Research Gate: Berlin, Germany, 2008; pp. 1–2.
31. Afgani, M.Z.; Haas, H.; Elgala, H.; Knipp, D. Visible light communication using OFDM. In Proceedings of the 2nd International Conference on Testbeds and Research Infrastructures for the Development of Networks and Communities, Barcelona, Spain, 1–3 March 2006; pp. 1–6 .
32. Komine, T.; Nakagawa, M. Integrated system of white LED visible-light communication and power-line communication. *IEEE Trans. Consum. Electron.* **2003**, *49*, 71–79. [CrossRef]
33. Hu, F.; Chen, S.; Li, G.; Zou, P.; Zhang, J.; Hu, J.; Zhang, J.; He, Z.; Yu, S.; Jiang, F.; et al. Si-substrate LEDs with multiple superlattice interlayers for beyond 24 Gbps visible light communication. *Photonics Res.* **2021**, *9*, 1581–1591. [CrossRef]
34. Lu, I.C.; Lai, C.H.; Yeh, C.H.; Chen, J. 6.36 Gbit/s RGB LED-based WDM MIMO visible light communication system employing OFDM modulation. In Proceedings of the Optical Fiber Communication Conference, Los Angeles, CA, USA, 19–23 March 2017; pp. W2A.39.
35. Kosman, J.; Almer, O.; Jalajakumari, A.V.; Videv, S.; Haas, H.; Henderson, R.K. 60 Mb/s, 2 m visible light communications in 1 klx ambient using an unlensed CMOS SPAD receiver. In Proceedings of the IEEE Photonics Society Summer Topical Meeting Series (SUM), Newport Beach, CA, USA, 11–13 July 2016; pp. 171–172.
36. Wang, Y.; Huang, X.; Tao, L.; Shi, J.; Chi, N. 4.5-Gb/s RGB-LED based WDM visible light communication system employing CAP modulation and RLS based adaptive equalization. *Opt. Express* **2015**, *23*, 13626–13633. [CrossRef] [PubMed]
37. Wang, Y.; Huang, X.; Zhang, J.; Wang, Y.; Chi, N. Enhanced performance of visible light communication employing 512-QAM N-SC-FDE and DD-LMS. *Opt. Express* **2014**, *22*, 15328–15334. [CrossRef]
38. Cossu, G.; Khalid, A.; Choudhury, P.; Corsini, R.; Ciaramella, E. 3.4 Gbit/s visible optical wireless transmission based on RGB LED. *Opt. Express* **2012**, *20*, B501–B506. [CrossRef]
39. Kottke, C.; Hilt, J.; Habel, K.; Vučić, J.; Langer, K.D. 1.25 Gbit/s visible light WDM link based on DMT modulation of a single RGB LED luminary. In Proceedings of the European Conference and Exhibition on Optical Communication, Amsterdam, The Netherlands, 16–20 September 2012; pp. 1–3.
40. Xu, F.; Qiu, P.; Tao, T.; Tian, P.; Liu, X.; Zhi, T.; Xie, Z.; Liu, B.; Zhang, R. High Bandwidth Semi-Polar InGaN/GaN Micro-LEDs with Low Current Injection for Visible Light Communication. *IEEE Photonics J.* **2023**, *15*, 7300704. [CrossRef]
41. Qiu, P.; Zhu, S.; Jin, Z.; Zhou, X.; Cui, X.; Tian, P. Beyond 25 Gbps optical wireless communication using wavelength division multiplexed LEDs and micro-LEDs. *Opt. Lett.* **2022**, *47*, 317–320. [CrossRef]
42. Chang, Y.H.; Huang, Y.M.; Gunawan, W.H.; Chang, G.H.; Liou, F.J.; Chow, C.W.; Kuo, H.C.; Liu, Y.; Yeh, C.H. 4.343-Gbit/s green semipolar (20-21) μ-LED for high speed visible light communication. *IEEE Photonics J.* **2021**, *13*, 1–4. [CrossRef]
43. Islim, M.S.; Ferreira, R.X.; He, X.; Xie, E.; Videv, S.; Viola, S.; Watson, S.; Bamiedakis, N.; Penty, R.V.; White, I.H.; et al. Towards 10 Gb/s orthogonal frequency division multiplexing-based visible light communication using a GaN violet micro-LED. *Photonics Res.* **2017**, *5*, A35–A43. [CrossRef]
44. Rajbhandari, S.; Jalajakumari, A.V.; Chun, H.; Faulkner, G.; Cameron, K.; Henderson, R.; Tsonev, D.; Haas, H.; Xie, E.; McKendry, J.J.; et al. A multigigabit per second integrated multiple-input multiple-output VLC demonstrator. *J. Light. Technol.* **2017**, *35*, 4358–4365. [CrossRef]
45. Ferreira, R.X.; Xie, E.; McKendry, J.J.; Rajbhandari, S.; Chun, H.; Faulkner, G.; Watson, S.; Kelly, A.E.; Gu, E.; Penty, R.V.; et al. High bandwidth GaN-based micro-LEDs for multi-Gb/s visible light communications. *IEEE Photonics Technol. Lett.* **2016**, *28*, 2023–2026. [CrossRef]
46. Manousiadis, P.; Chun, H.; Rajbhandari, S.; Mulyawan, R.; Vithanage, D.A.; Faulkner, G.; Tsonev, D.; McKendry, J.J.; Ijaz, M.; Xie, E.; et al. Demonstration of 2.3 Gb/s RGB white-light VLC using polymer based colour-converters and GaN micro-LEDs. In Proceedings of the IEEE Summer Topicals Meeting Series (SUM), Nassau, Bahamas, 13–15 July 2015; pp. 222–223.
47. Tsonev, D.; Chun, H.; Rajbhandari, S.; McKendry, J.J.; Videv, S.; Gu, E.; Haji, M.; Watson, S.; Kelly, A.E.; Faulkner, G.; et al. A 3-Gb/s Single-LED OFDM-Based Wireless VLC Link Using a Gallium Nitride μLED. *IEEE Photonics Technol. Lett.* **2014**, *26*, 637–640. [CrossRef]
48. Chun, H.; Manousiadis, P.; Rajbhandari, S.; Vithanage, D.A.; Faulkner, G.; Tsonev, D.; McKendry, J.J.D.; Videv, S.; Xie, E.; Gu, E.; et al. Visible Light Communication Using a Blue GaN μ LED and Fluorescent Polymer Color Converter. *IEEE Photonics Technol. Lett.* **2014**, *26*, 2035–2038. [CrossRef]
49. McKendry, J.J.; Massoubre, D.; Zhang, S.; Rae, B.R.; Green, R.P.; Gu, E.; Henderson, R.K.; Kelly, A.; Dawson, M.D. Visible-light communications using a CMOS-controlled micro-light-emitting-diode array. *J. Light. Technol.* **2011**, *30*, 61–67. [CrossRef]
50. McKendry, J.J.; Green, R.P.; Kelly, A.; Gong, Z.; Guilhabert, B.; Massoubre, D.; Gu, E.; Dawson, M.D. High-speed visible light communications using individual pixels in a micro light-emitting diode array. *IEEE Photonics Technol. Lett.* **2010**, *22*, 1346–1348. [CrossRef]
51. Minotto, A.; Haigh, P.A.; Łukasiewicz, Ł.G.; Lunedei, E.; Gryko, D.T.; Darwazeh, I.; Cacialli, F. Visible light communication with efficient far-red/near-infrared polymer light-emitting diodes. *Light. Sci. Appl.* **2020**, *9*, 70. [CrossRef] [PubMed]
52. Haigh, P.A.; Bausi, F.; Le Minh, H.; Papakonstantinou, I.; Popoola, W.O.; Burton, A.; Cacialli, F. Wavelength-multiplexed polymer LEDs: Towards 55 Mb/s organic visible light communications. *IEEE J. Sel. Areas Commun.* **2015**, *33*, 1819–1828. [CrossRef]

53. Haigh, P.A.; Bausi, F.; Ghassemlooy, Z.; Papakonstantinou, I.; Le Minh, H.; Fléchon, C.; Cacialli, F. Visible light communications: Real time 10 Mb/s link with a low bandwidth polymer light-emitting diode. *Opt. Express* **2014**, *22*, 2830–2838. [CrossRef] [PubMed]

54. Chun, H.; Chiang, C.J.; Monkman, A.; O'Brien, D. A study of illumination and communication using organic light emitting diodes. *J. Light. Technol.* **2013**, *31*, 3511–3517. [CrossRef]

55. Haigh, P.A.; Ghassemlooy, Z.; Le Minh, H.; Rajbhandari, S.; Arca, F.; Tedde, S.F.; Hayden, O.; Papakonstantinou, I. Exploiting equalization techniques for improving data rates in organic optoelectronic devices for visible light communications. *J. Light. Technol.* **2012**, *30*, 3081–3088. [CrossRef]

56. Chen, C.; Basnayaka, D.A.; Haas, H. Downlink performance of optical attocell networks. *J. Light. Technol.* **2015**, *34*, 137–156. [CrossRef]

57. Chen, C.; Videv, S.; Tsonev, D.; Haas, H. Fractional frequency reuse in DCO-OFDM-based optical attocell networks. *J. Light. Technol.* **2015**, *33*, 3986–4000. [CrossRef]

58. He, C.; Wang, T.Q.; Masum, M.A.; Armstrong, J. Performance of optical receivers using photodetectors with different fields of view in an indoor cellular communication system. In Proceedings of the International Telecommunication Networks and Applications Conference (ITNAC), Sydney, NSW, Australia, 18–20 November 2015; pp. 77–82.

59. O'Brien, D.; Rajbhandari, S.; Chun, H. Transmitter and receiver technologies for optical wireless. *Philos. Trans. R. Soc. A* **2020**, *378*, 20190182. [CrossRef]

60. Xiao, X.; Xiao, H.; Liu, H.; Wang, R.; Choy, W.C.; Wang, K. Modeling and analysis for modulation of light-conversion materials in visible light communication. *IEEE Photonics J.* **2019**, *11*, 1–13. [CrossRef]

61. Sajjad, M.T.; Manousiadis, P.P.; Chun, H.; Vithanage, D.A.; Rajbhandari, S.; Kanibolotsky, A.L.; Faulkner, G.; O'Brien, D.; Skabara, P.J.; Samuel, I.D.; et al. Novel fast color-converter for visible light communication using a blend of conjugated polymers. *ACS Photonics* **2015**, *2*, 194–199. [CrossRef]

62. Sajjad, M.T.; Manousiadis, P.P.; Orofino, C.; Cortizo-Lacalle, D.; Kanibolotsky, A.L.; Rajbhandari, S.; Amarasinghe, D.; Chun, H.; Faulkner, G.; O'Brien, D.C.; et al. Fluorescent red-emitting BODIPY oligofluorene star-shaped molecules as a color converter material for visible light communications. *Adv. Opt. Mater.* **2015**, *3*, 536–540. [CrossRef]

63. Vithanage, D.; Manousiadis, P.; Sajjad, M.T.; Rajbhandari, S.; Chun, H.; Orofino, C.; Cortizo-Lacalle, D.; Kanibolotsky, A.; Faulkner, G.; Findlay, N.; et al. BODIPY star-shaped molecules as solid state colour converters for visible light communications. *Appl. Phys. Lett.* **2016**, *109*, 013302. [CrossRef]

64. Sajjad, M.T.; Manousiadis, P.; Orofino, C.; Kanibolotsky, A.; Findlay, N.J.; Rajbhandari, S.; Vithanage, D.; Chun, H.; Faulkner, G.; O'Brien, D.; et al. A saturated red color converter for visible light communication using a blend of star-shaped organic semiconductors. *Appl. Phys. Lett.* **2017**, *110*, 013302. [CrossRef]

65. Dursun, I.; Shen, C.; Parida, M.R.; Pan, J.; Sarmah, S.P.; Priante, D.; Alyami, N.; Liu, J.; Saidaminov, M.I.; Alias, M.S.; et al. Perovskite nanocrystals as a color converter for visible light communication. *ACS Photonics* **2016**, *3*, 1150–1156. [CrossRef]

66. Mei, S.; Liu, X.; Zhang, W.; Liu, R.; Zheng, L.; Guo, R.; Tian, P. High-bandwidth white-light system combining a micro-LED with perovskite quantum dots for visible light communication. *ACS Appl. Mater. Interfaces* **2018**, *10*, 5641–5648. [CrossRef]

67. Kang, C.H.; Dursun, I.; Liu, G.; Sinatra, L.; Sun, X.; Kong, M.; Pan, J.; Maity, P.; Ooi, E.N.; Ng, T.K.; et al. High-speed colour-converting photodetector with all-inorganic CsPbBr3 perovskite nanocrystals for ultraviolet light communication. *Light. Sci. Appl.* **2019**, *8*, 94. [CrossRef]

68. Li, X.; Cai, W.; Guan, H.; Zhao, S.; Cao, S.; Chen, C.; Liu, M.; Zang, Z. Highly stable CsPbBr3 quantum dots by silica-coating and ligand modification for white light-emitting diodes and visible light communication. *Chem. Eng. J.* **2021**, *419*, 129551. [CrossRef]

69. Zhao, S.; Chen, C.; Cai, W.; Li, R.; Li, H.; Jiang, S.; Liu, M.; Zang, Z. Efficiently luminescent and stable lead-free Cs3Cu2Cl5@ silica nanocrystals for white light-emitting diodes and communication. *Adv. Opt. Mater.* **2021**, *9*, 2100307. [CrossRef]

70. Mo, Q.; Yu, J.; Chen, C.; Cai, W.; Zhao, S.; Li, H.; Zang, Z. Highly Efficient and Ultra-Broadband Yellow Emission of Lead-Free Antimony Halide toward White Light-Emitting Diodes and Visible Light Communication. *Laser Photonics Rev.* **2022**, *16*, 2100600. [CrossRef]

71. Ali, A.; Qasem, Z.A.; Li, Y.; Li, Q.; Fu, H. All-inorganic liquid phase quantum dots and blue laser diode-based white-light source for simultaneous high-speed visible light communication and high-efficiency solid-state lighting. *Opt. Express* **2022**, *30*, 35112–35124. [CrossRef] [PubMed]

72. Mardani, S.; Khalid, A.; Willems, F.M.; Linnartz, J.P. Effect of blue filter on the SNR and data rate for indoor visible light communication system. In Proceedings of the European Conference on Optical Communication (ECOC), Gothenburg, Sweden, 17–21 September 2017; pp. 1–3.

73. Sung, J.Y.; Chow, C.W.; Yeh, C.H. Is blue optical filter necessary in high speed phosphor-based white light LED visible light communications? *Opt. Express* **2014**, *22*, 20646–20651. [CrossRef] [PubMed]

74. Hu, J.; Hu, F.; Jia, J.; Li, G.; Shi, J.; Zhang, J.; Li, Z.; Chi, N.; Yu, S.; Shen, C. 46.4 Gbps visible light communication system utilizing a compact tricolor laser transmitter. *Opt. Express* **2022**, *30*, 4365–4373. [CrossRef]

75. Weber, W.H.; Lambe, J. Luminescent greenhouse collector for solar radiation. *Appl. Opt.* **1976**, *15*, 2299–2300. [CrossRef]

76. Collins, S.; O'Brien, D.C.; Watt, A. High gain, wide field of view concentrator for optical communications. *Opt. Lett.* **2014**, *39*, 1756–1759. [CrossRef]

77. Manousiadis, P.P.; Rajbhandari, S.; Mulyawan, R.; Vithanage, D.A.; Chun, H.; Faulkner, G.; O'Brien, D.C.; Turnbull, G.A.; Collins, S.; Samuel, I.D. Wide field-of-view fluorescent antenna for visible light communications beyond the étendue limit. *Optica* **2016**, *3*, 702–706. [CrossRef]
78. Manousiadis, P.P.; Chun, H.; Rajbhandari, S.; Vithanage, D.A.; Mulyawan, R.; Faulkner, G.; Haas, H.; O'Brien, D.C.; Collins, S.; Turnbull, G.A.; et al. Optical antennas for wavelength division multiplexing in visible light communications beyond the étendue limit. *Adv. Opt. Mater.* **2020**, *8*, 1901139. [CrossRef]
79. He, C.; Collins, S.; Murata, H. Capillary-based fluorescent antenna for visible light communications. *Opt. Express* **2023**, *31*, 17716–17730. [CrossRef]
80. Kang, C.H.; Alkhazragi, O.; Sinatra, L.; Alshaibani, S.; Wang, Y.; Li, K.H.; Kong, M.; Lutfullin, M.; Bakr, O.M.; Ng, T.K.; et al. All-inorganic halide-perovskite polymer-fiber-photodetector for high-speed optical wireless communication. *Opt. Express* **2022**, *30*, 9823–9840. [CrossRef] [PubMed]
81. Peyronel, T.; Quirk, K.; Wang, S.; Tiecke, T. Luminescent detector for free-space optical communication. *Optica* **2016**, *3*, 787–792. [CrossRef]
82. Kang, C.H.; Trichili, A.; Alkhazragi, O.; Zhang, H.; Subedi, R.C.; Guo, Y.; Mitra, S.; Shen, C.; Roqan, I.S.; Ng, T.K.; et al. Ultraviolet-to-blue color-converting scintillating-fibers photoreceiver for 375-nm laser-based underwater wireless optical communication. *Opt. Express* **2019**, *27*, 30450–30461. [CrossRef]
83. Sait, M.; Trichili, A.; Alkhazragi, O.; Alshaibaini, S.; Ng, T.K.; Alouini, M.S.; Ooi, B.S. Dual-wavelength luminescent fibers receiver for wide field-of-view, Gb/s underwater optical wireless communication. *Opt. Express* **2021**, *29*, 38014–38026. [CrossRef]
84. Ali, W.; Ahmed, Z.; Matthews, W.; Collins, S. The impact of the length of fluorescent fiber concentrators on the performance of VLC receivers. *IEEE Photonics Technol. Lett.* **2021**, *33*, 1451–1454. [CrossRef]
85. He, C.; Lim, Y.; Murata, H. Study of using different colors of fluorescent fibers as optical antennas in white LED based-visible light communications. *Opt. Express* **2023**, *31*, 4015–4028. [CrossRef] [PubMed]
86. He, C.; Lim, Y.; Tang, Y.; Chen, C. Visible Light Communications Using Commercially Available Fluorescent Fibers as Optical Antennas. In Proceedings of the Opto-Electronics and Communications Conference (OECC), Shanghai, China, 2–6 July 2023; pp. 1–4.
87. Chang, Y.H.; Tsai, D.C.; Chow, C.W.; Wang, C.C.; Tsai, S.Y.; Liu, Y.; Yeh, C.H. Lightweight Light-Diffusing Fiber Transmitter Equipped Unmanned-Aerial-Vehicle (UAV) for Large Field-of-View (FOV) Optical Wireless Communication. In Proceedings of the Optical Fiber Communications Conference and Exhibition (OFC), San Diego, CA, USA, 5–9 March 2023; pp. 1–3.
88. Chang, Y.H.; Chow, C.W.; Wang, C.C.; Jian, Y.H.; Gunawan, W.H.; Liu, Y.; Yeh, C.H. Free-Space Visible Light Communication with Downstream and Upstream Transmissions Supporting Multiple Moveable Receivers Using Light-Diffusing Fiber. In Proceedings of the Optical Fiber Communications Conference and Exhibition (OFC), San Diego, CA, USA, 5–9 March 2023; pp. 1–3.
89. Chang, Y.H.; Chow, C.W.; Lin, Y.Z.; Jian, Y.H.; Wang, C.C.; Liu, Y.; Yeh, C.H. Bi-Directional Free-Space Visible Light Communication Supporting Multiple Moveable Clients Using Light Diffusing Optical Fiber. *Sensors* **2023**, *23*, 4725. [CrossRef]
90. Zhang, L.; Chitnis, D.; Chun, H.; Rajbhandari, S.; Faulkner, G.; O'Brien, D.; Collins, S. A comparison of APD-and SPAD-based receivers for visible light communications. *J. Light. Technol.* **2018**, *36*, 2435–2442. [CrossRef]
91. Cova, S.; Ghioni, M.; Lacaita, A.; Samori, C.; Zappa, F. Avalanche photodiodes and quenching circuits for single-photon detection. *Appl. Opt.* **1996**, *35*, 1956–1976. [CrossRef] [PubMed]
92. Chitnis, D.; Collins, S. A SPAD-based photon detecting system for optical communications. *J. Light. Technol.* **2014**, *32*, 2028–2034. [CrossRef]
93. Zhang, L.; Chun, H.; Ahmed, Z.; Faulkner, G.; O'Brien, D.; Collins, S. The future prospects for SiPM-based receivers for visible light communications. *J. Light. Technol.* **2019**, *37*, 4367–4374. [CrossRef]
94. Matthews, W.; He, C.; Collins, S. DCO-OFDM Channel Sounding with a SiPM Receiver. In Proceedings of the IEEE Photonics Conference (IPC), Vancouver, BC, Canada, 18–21 October 2021; pp. 1–2.
95. He, C.; Lim, Y. Silicon Photomultiplier (SiPM) Selection and Parameter Analysis in Visible Light Communications. In Proceedings of the 31st Wireless and Optical Communications Conference (WOCC), Shenzhen, China, 11–12 August 2022; pp. 41–46.
96. Huang, S.; Chen, C.; Bian, R.; Haas, H.; Safari, M. 5 Gbps optical wireless communication using commercial SPAD array receivers. *Opt. Lett.* **2022**, *47*, 2294–2297. [CrossRef]
97. He, C.; Ahmed, Z.; Collins, S. Optical OFDM and SiPM receivers. In Proceedings of the IEEE Globecom Workshops, Taipei, Taiwan, 7–11 December 2020; pp. 1–6.
98. Huang, S.; Chen, C.; Soltani, M.D.; Henderson, R.; Haas, H.; Safari, M. SPAD-Based Optical Wireless Communication with ACO-OFDM. *arXiv* **2022**, arXiv:2210.14101.
99. Zhang, L.; Jiang, R.; Tang, X.; Chen, Z.; Li, Z.; Chen, J. Performance Estimation and Selection Guideline of SiPM Chip within SiPM-Based OFDM-OWC System. *Photonics* **2022**, *9*, 637. [CrossRef]
100. Huang, S.; Li, Y.; Chen, C.; Soltani, M.D.; Henderson, R.; Safari, M.; Haas, H. Performance analysis of SPAD-based optical wireless communication with OFDM. *J. Opt. Commun. Netw.* **2023**, *15*, 174–186. [CrossRef]
101. He, C.; Ahmed, Z.; Collins, S. Signal pre-equalization in a silicon photomultiplier-based optical OFDM system. *IEEE Access* **2021**, *9*, 23344–23356. [CrossRef]
102. Zhang, L.; Jiang, R.; Tang, X.; Chen, Z.; Chen, J.; Wang, H. A simplified post equalizer for mitigating the nonlinear distortion in SiPM based OFDM-VLC system. *IEEE Photonics J.* **2021**, *14*, 1–7. [CrossRef]

103. He, C.; Collins, S. Signal Demodulation Using a Radial Basis Function Neural Network (RBFNN) in a Silicon Photomultiplier-Based Visible Light Communication System. *IEEE Photonics J.* **2022**, *14*, 1–14. [CrossRef]

104. Jiang, R.; Sun, C.; Zhang, L.; Tang, X.; Wang, H.; Zhang, A. Deep learning aided signal detection for SPAD-based underwater optical wireless communications. *IEEE Access* **2020**, *8*, 20363–20374. [CrossRef]

105. Zhang, L.; Tang, X.; Sun, C.; Chen, Z.; Li, Z.; Wang, H.; Jiang, R.; Shi, W.; Zhang, A. Over 10 attenuation length gigabits per second underwater wireless optical communication using a silicon photomultiplier (SiPM) based receiver. *Opt. Express* **2020**, *28*, 24968–24980. [CrossRef] [PubMed]

106. Li, J.; Ye, D.; Fu, K.; Wang, L.; Piao, J.; Wang, Y. Single-photon detection for MIMO underwater wireless optical communication enabled by arrayed LEDs and SiPMs. *Opt. Express* **2021**, *29*, 25922–25944. [CrossRef] [PubMed]

107. Hong, X.; Du, J.; Wang, Y.; Chen, R.; Tian, J.; Zhang, G.; Zhang, J.; Fei, C.; He, S. Experimental demonstration of 55-m/2-Gbps underwater wireless optical communication using SiPM diversity reception and nonlinear decision-feedback equalizer. *IEEE Access* **2022**, *10*, 47814–47823. [CrossRef]

108. Ali, W.; Faulkner, G.; Ahmed, Z.; Matthews, W.; Collins, S. Giga-bit Transmission between an Eye-safe transmitter and wide field-of-view SiPM receiver. *IEEE Access* **2021**, *9*, 154225–154236. [CrossRef]

109. Liu, Y.; Wajahat, A.; Chen, R.; Bamiedakis, N.; Crisp, M.; White, I.H.; Penty, R.V. High-capacity optical wireless VCSEL array transmitter with uniform coverage. In *Free-Space Laser Communications XXXV*; SPIE: Paris, France, 2023; Volume 12413, pp. 144–150.

110. Khalighi, M.A.; Uysal, M. Survey on free space optical communication: A communication theory perspective. *IEEE Commun. Surv. Tutorials* **2014**, *16*, 2231–2258. [CrossRef]

111. Singh, D.; Swaminathan, R. Comprehensive Performance Analysis of Hovering UAV-Based FSO Communication System. *IEEE Photonics J.* **2022**, *14*, 1–13. [CrossRef]

112. Xu, G.; Zhang, N.; Xu, M.; Xu, Z.; Zhang, Q.; Song, Z. Outage Probability and Average BER of UAV-assisted Dual-hop FSO Communication with Amplify-and-Forward Relaying. *IEEE Trans. Veh. Technol.* 2023, *early access*.

113. Fath, T.; Haas, H. Performance comparison of MIMO techniques for optical wireless communications in indoor environments. *IEEE Trans. Commun.* **2012**, *61*, 733–742. [CrossRef]

114. Basnayaka, D.A.; Haas, H. MIMO interference channel between spatial multiplexing and spatial modulation. *IEEE Trans. Commun.* **2016**, *64*, 3369–3381. [CrossRef]

115. Chen, C.; Yang, H.; Du, P.; Zhong, W.D.; Alphones, A.; Yang, Y.; Deng, X. User-centric MIMO techniques for indoor visible light communication systems. *IEEE Syst. J.* **2020**, *14*, 3202–3213. [CrossRef]

116. Zeng, L.; O'Brien, D.C.; Le Minh, H.; Faulkner, G.E.; Lee, K.; Jung, D.; Oh, Y.; Won, E.T. High data rate multiple input multiple output (MIMO) optical wireless communications using white LED lighting. *IEEE J. Sel. Areas Commun.* **2009**, *27*, 1654–1662. [CrossRef]

117. Wang, T.Q.; Sekercioglu, Y.A.; Armstrong, J. Analysis of an optical wireless receiver using a hemispherical lens with application in MIMO visible light communications. *J. Light. Technol.* **2013**, *31*, 1744–1754. [CrossRef]

118. Chen, C.; Yang, Y.; Deng, X.; Du, P.; Yang, H. Space division multiple access with distributed user grouping for multi-user MIMO-VLC systems. *IEEE Open J. Commun. Soc.* **2020**, *1*, 943–956. [CrossRef]

119. He, C.; Wang, T.Q.; Armstrong, J. Performance comparison between spatial multiplexing and spatial modulation in indoor MIMO visible light communication systems. In Proceedings of the IEEE International Conference on Communications (ICC), Kuala Lumpur, Malaysia, 22–27 May 2016; pp. 1–6.

120. Yang, H.; Chen, C.; Zhong, W.D.; Alphones, A. Joint precoder and equalizer design for multi-user multi-cell MIMO VLC systems. *IEEE Trans. Veh. Technol.* **2018**, *67*, 11354–11364. [CrossRef]

121. Mesleh, R.Y.; Haas, H.; Sinanovic, S.; Ahn, C.W.; Yun, S. Spatial modulation. *IEEE Trans. Veh. Technol.* **2008**, *57*, 2228–2241. [CrossRef]

122. Chen, C.; Zhong, X.; Fu, S.; Jian, X.; Liu, M.; Yang, H.; Alphones, A.; Fu, H. OFDM-based generalized optical MIMO. *J. Light. Technol.* **2021**, *39*, 6063–6075. [CrossRef]

123. Başar, E.; Panayirci, E.; Uysal, M.; Haas, H. Generalized LED index modulation optical OFDM for MIMO visible light communications systems. In Proceedings of the IEEE International Conference on Communications (ICC), Kuala Lumpur, Malaysia, 22–27 May 2016; pp. 1–5.

124. Zhong, X.; Chen, C.; Fu, S.; Zeng, Z.; Liu, M. DeepGOMIMO: Deep Learning-Aided Generalized Optical MIMO with CSI-Free Detection. *Photonics* 2022, *9*, 940. [CrossRef]

125. Chen, C.; Zhong, W.D.; Yang, H.; Zhang, S.; Du, P. Reduction of SINR fluctuation in indoor multi-cell VLC systems using optimized angle diversity receiver. *J. Light. Technol.* **2018**, *36*, 3603–3610. [CrossRef]

126. Chen, C.; Fu, S.; Jian, X.; Liu, M.; Deng, X.; Ding, Z. NOMA for energy-efficient LiFi-enabled bidirectional IoT communication. *IEEE Trans. Commun.* **2021**, *69*, 1693–1706. [CrossRef]

127. Chen, T.; Liu, L.; Tu, B.; Zheng, Z.; Hu, W. High-spatial-diversity imaging receiver using fisheye lens for indoor MIMO VLCs. *IEEE Photonics Technol. Lett.* **2014**, *26*, 2260–2263. [CrossRef]

128. Nuwanpriya, A.; Ho, S.W.; Chen, C.S. Indoor MIMO visible light communications: Novel angle diversity receivers for mobile users. *IEEE J. Sel. Areas Commun.* **2015**, *33*, 1780–1792. [CrossRef]

129. Wei, L.; Zhang, H.; Song, J. Experimental demonstration of a cubic-receiver-based MIMO visible light communication system. *IEEE Photonics J.* **2016**, *9*, 1–7. [CrossRef]
130. Wei, L.; Zhang, H.; Yu, B.; Song, J.; Guan, Y. Cubic-receiver-based indoor optical wireless location system. *IEEE Photonics J.* **2016**, *8*, 1–7. [CrossRef]
131. Chen, Z.; Serafimovski, N.; Haas, H. Angle diversity for an indoor cellular visible light communication system. In Proceedings of the IEEE 79th Vehicular Technology Conference (VTC Spring), Seoul, Republic of Korea, 17–21 May 2014; pp. 1–5.
132. Wang, T.Q.; Green, R.J.; Armstrong, J. MIMO optical wireless communications using ACO-OFDM and a prism-array receiver. *IEEE J. Sel. Areas Commun.* **2015**, *33*, 1959–1971. [CrossRef]
133. Wang, T.Q.; He, C.; Armstrong, J. Performance analysis of aperture-based receivers for MIMO IM/DD visible light communications. *J. Light. Technol.* **2016**, *35*, 1513–1523. [CrossRef]
134. Steendam, H. A 3-D positioning algorithm for AOA-based VLP with an aperture-based receiver. *IEEE J. Sel. Areas Commun.* **2017**, *36*, 23–33. [CrossRef]
135. He, C.; Cincotta, S.; Mohammed, M.M.; Armstrong, J. Angular diversity aperture (ADA) receivers for indoor multiple-input multiple-output (MIMO) visible light communications (VLC). *IEEE Access* **2019**, *7*, 145282–145301. [CrossRef]
136. He, C.; Wang, T.Q.; Armstrong, J. Performance of optical receivers using photodetectors with different fields of view in a MIMO ACO-OFDM system. *J. Light. Technol.* **2015**, *33*, 4957–4967. [CrossRef]
137. Cincotta, S.; He, C.; Neild, A.; Armstrong, J. High angular resolution visible light positioning using a quadrant photodiode angular diversity aperture receiver (QADA). *Opt. Express* **2018**, *26*, 9230–9242. [CrossRef] [PubMed]
138. Shen, S.; Menéndez Sánchez, J.M.; Li, S.; Steendam, H. Pose Estimation for Visible Light Systems Using a Quadrature Angular Diversity Aperture Receiver. *Sensors* **2022**, *22*, 5073. [CrossRef] [PubMed]

Article

High-Speed Underwater Optical Wireless Communication with Advanced Signal Processing Methods Survey

Chengwei Fang [1,*], Shuo Li [1], Yinong Wang [2] and Ke Wang [1]

[1] School of Engineering, Royal Melbourne Institute of Technology (RMIT) University, Melbourne, VIC 3000, Australia; shuo.li2@rmit.edu.au (S.L.); ke.wang@rmit.edu.au (K.W.)
[2] School of Architecture and Urban Design, Royal Melbourne Institute of Technology (RMIT) University, Melbourne, VIC 3000, Australia; s3576616@student.rmit.edu.au
* Correspondence: s3643273@student.rmit.edu.au

Abstract: Underwater wireless communication (UWC) technology has attracted widespread attention in the past few years. Compared with conventional acoustic underwater wireless communication technology, underwater optical wireless communication (UOWC) technology has promising potential to provide high data rate wireless connections due to the large license-free bandwidth. Building a high-performance and reliable UOWC system has become the target of researchers and various advanced and innovative technologies have been proposed and investigated. Among them, better hardware such as transmitters and receivers, as well as more advanced modulation and signal processing techniques, are key factors in improving UOWC system performance. In this paper, we review the recent development in UOWC systems. In particular, we provide a brief introduction to different types of UOWC systems based on channel configuration, and we focus on various recent studies on advanced signal processing methods in UOWC systems, including both traditional non-machine learning (NML) equalizers and machine learning (ML) schemes based on neural networks. In addition, we also discuss the key challenges in UOWC systems for future applications.

Keywords: underwater optical wireless communication (UOWC); digital signal; linear equalizer; nonlinear equalizer; supervised machine learning; reinforcement machine learning

Citation: Fang, C.; Li, S.; Wang, Y.; Wang, K. High-Speed Underwater Optical Wireless Communication with Advanced Signal Processing Methods Survey. *Photonics* **2023**, *10*, 811. https://doi.org/10.3390/photonics10070811

Received: 31 May 2023
Revised: 4 July 2023
Accepted: 10 July 2023
Published: 12 July 2023

1. Introduction

The ocean covers more than 70 percent of the surface of our planet [1]. Human exploration of the ocean has not stopped since ancient times. With the rapid development of science and technology, human exploration of the ocean has gradually deepened. The invention and optimization of a large range of underwater applications such as underwater wireless sensor networks [2] and autonomous underwater vehicles (AUVs) [3] have become key factors. Underwater wireless communication (UWC) technologies, which are summarized in Table 1, have become the cornerstone of these underwater applications. With the need for real-time underwater communications, high-speed and long-distance transmission is more in demand than ever in UWC technologies.

Traditional UWC mainly relies on underwater acoustic communication (UAC) technology, which has been explored in transmitting data for long distances reaching up to several tens of kilometers [4], exploring the low attenuation property enabled by the physical properties of sound waves propagating in water. However, UAC suffers from a low data rate limitation due to the low modulation bandwidth (only tens of kHz) [5,6]. The propagation speed of acoustic waves in the underwater channel is also low (only 1500 m/s), leading to a latency of about 0.67 s per kilometer [7]. Moreover, the power consumption is typically high (tens of watts [8]). Compared to UAC, underwater radio frequency (RF) communication suffers from a high attenuation coefficient due to the low conductivity of electromagnetic waves in water, which leads to a highly limited transmission distance (only a few meters to tens of meters) [9]. Thus, RF communication is not adopted in UWC.

To overcome the conventional UWC limitations, UOWC has been proposed and widely studied since it has great potential to achieve a higher data rate reaching Gbps, thanks to the large modulation bandwidth (exceeding MHz [10] and even GHz [11], typically limited by the transceiver). Moreover, the physical communication latency is much shorter due to the high propagation speed of light in the underwater channel [12]. These high-speed and low-latency advantages can enable many real-time applications. Furthermore, UOWC is also cost-effective and power-effective compared to UAC and RF communication, which benefits from low-cost and low-power transceivers such as light-emitting diodes (LEDs) and photodiodes (PDs) [13]. Although the UOWC technology has these advantages, the transmittance of the optical wave is limited compared with the acoustic wave (only hundreds of meters in the tap water channel [14]), and due to the shorter wavelength, the optical signal also experiences more complex underwater propagation channels. Hence, improving the transmission distance, data rate, and system stability of the UOWC system becomes a research focus.

Table 1. Comparisons of three underwater wireless communication technologies.

	Acoustic Systems	Radio Frequency Systems	Optical Wireless Systems
Attenuation	Low	High	Moderate
Distance	Long (tens of kilometers)	Short (tens of meters)	Limited (hundreds of meters)
Carrier Frequency	Low (10 Hz–1 MHz)	Moderate (30 Hz–300 MHz)	High (10^{12} Hz–10^{15} Hz)
Bandwidth	Narrow (kHz)	Moderate (MHz)	Broad (MHz–GHz)
Data Rate	Low (kbps)	Moderate (Mbps)	High (Gbps)
Power Consumption	High	High	Low
Transmission Latency	High (1500 m/s physical propagation speed of sound wave)	Low (2.26×10^8 m/s physical propagation speed of electromagnetic wave)	Low (2.26×10^8 m/s physical propagation speed of optical wave)
Performance-limiting factors	Temperature, hydrostatic pressure, and the chemistry of water	Conductivity and permitivity	Absorption, scattering, turbidity, marine life blocking, and beam shaping

In recent years, a number of surveys and summary papers on UOWC have been published, which are summarized in Table 2. In [8,15–18], brief overviews and recent advances of UOWC are presented, focusing on the UOWC channel characterization, modulation methods, and coding technologies. In addition to an overview of recent UOWC achievement, in [19], a summary of transmitter and receiver technologies is presented. Moreover, the UOWC channel model and the impact of underwater turbulence are discussed. Due to the complex underwater environment, accurate theoretical UOWC channel models are the basis for designing and optimizing practical UOWC systems. Therefore, the available UOWC models to investigate the communication performance, such as the transmission range and data rate, are summarized in [20,21]. With the continuous optimization and improvement of theoretical models, many practical UOWC systems have been further designed and studied experimentally. In [13], a detailed summary of UOWC experimental demonstrations with both laser diode (LD) and LED transmitters in recent years is presented. In addition, some key technologies, such as higher sensitivity receivers and more advanced signal modulation methods are also presented to improve the transmission capacity and performance of UOWC systems.

In addition, some recent survey papers also provide a more focused review of key parts of the UOWC system, such as the network layer and underwater channel turbulence. In [2], in addition to the physical layer such as the channel characterization and modulation methods, the network layer issues, including the link configuration and budgets, multiple access schemes, relaying techniques, and potential routing algorithms are also presented. In addition to the absorption and scattering of the signal beam by the particles in the water, UOWC systems also face great challenges from underwater optical turbulence (UOT), which is physically caused by the fluctuation of water with random variations of temperature and pressure [22]. In [23], theoretical UOWC system models considering turbulent channels are summarized, together with underwater turbulence mitigation technologies in the physical layer, including aperture averaging, optical beam shaping, transmitter and receiver enhancement, and multiple-input multiple-output (MIMO) spatial diversity techniques.

With the rapid development of UOWC technologies, increasingly more UOWC applications have begun to appear, which are captured by a few recent survey papers. For instance, in [24], the UOWC-based Internet of Underwater Things (IoUT) network is summarized, focusing on the medium access control (MAC) aspect. Moreover, AUVs, which are key technologies for the maritime industry are widely deployed for commercial, scientific, environmental, and defense applications. Thanks to high-speed data transmission, the UOWC technology has been widely considered in AUV application. In [3], a summary of swarm robotics techniques based on the LED type of UOWC is presented.

Through the above literature review, due to the absorption and scattering of signal light caused by the complex underwater environment and the influence of underwater turbulence on channel stability, UOWC faces challenges of limited transmission distance and fluctuating transmission reliability [23]. In order to improve the transmission distance of UOWC and to achieve more stable system performance, a large number of techniques have been studied in the physical layer [8,15–19], such as more advanced transmitter technologies (e.g., high-bandwidth Gallium nitride (GaN)-based mini-LEDs [25], two-stage-injection-locked technique [26,27]), more sensitive receiver technologies (e.g., lensed array optical interface [28,29], photomultiplier tubes (PMT) [30], and single-photon avalanche diode (SPAD) [31]), and more advanced UOWC spatial technologies (MIMO principles [32–34]). The signal processing enhancement, which includes transmitter frequency response improvement [27,29,35], transmitter shot noise minimization [36], and inter-symbol interference (ISI) elimination techniques [28,37,38], has also achieved remarkable progress in recent years. Digital signal processing (DSP) technologies can significantly improve the signal-to-noise ratio (SNR) and reduce the bit-error rate (BER) of UOWC systems with low cost and high efficiency.

Although the previous survey papers [2,13] included the DSP aspect, only a short and simple introduction is presented. Therefore, this survey provides a comprehensive review of the recent developments of advanced DSP techniques in UOWC systems, which include:

1. A brief introduction and summary of equalization principles.
2. A detailed review of NML equalization techniques in UOWC systems in recent years, including both linear equalizers and nonlinear equalizers.
3. A detailed review of ML techniques in UOWC systems, including both supervised learning and reinforcement learning schemes.

The rest of this survey is organized as follows: in Section 2, we introduce the general architecture and principles of UOWC systems. Moreover, we discuss and summarize equalization technique principles and introduce ML applications in UOWC systems. In Section 3, we provide a detailed review of the linear equalizers and nonlinear equalizers applied in recent UOWC progress. In Section 4, we provide a review of ML applications in UOWC. Moreover, we discuss the challenges of both NML and ML equalizers and provide our views on future UOWC technologies from the advanced signal processing techniques perspective in Section 5. Finally, we conclude the survey in Section 6.

Table 2. Recent surveys and the comparison with this paper.

References	Year	Area of Focus
Hemani Kaushal et al. [8]	2016	• UOWC LOS, NLOS, and retro-reflector channels • Optical attenuation modeling • UOWC system design • Future scope
Zhaoquan Zeng et al. [15]	2017	• UOWC LOS and NLOS channels • Optical attenuation and turbulence modeling • Theoretical modulation and coding • Practical implementations of UOWC
Hassan M. Oubei et al. [19]	2018	• UOWC typical LOS and NLOS channels • Optical attenuation and turbulence modeling • Future challenge in transceiver technologies
Callum T. Geldard et al. [20]	2019	• UOWC absorption and scattering modeling • Monte Carlo simulation discussion
N. E. Miroshnikova et al. [21]	2019	• UOWC LOS and NLOS channels • Optical absorption and scattering modeling
Nasir Saeed et al. [2]	2019	• UOWC potential channel architectures • Layer-by-layer network aspects • Localization • Future scope discussion
T. R. Murgod et al. [16]	2019	• UOWC network architecture • Routing and localization algorithms introduction • Recent related work challenges discussion
Chuyen T. Nguyen et al. [24]	2020	• UOWC-based Internet of Underwater Things network • Physical and MAC cross-layer analysis • Monte Carlo simulation analysis
G. S. Spagnolo et al. [17]	2020	• UOWC optical attenuation modeling • UOWC transceiver technologies
Shijie Zhu et al. [13]	2020	• UOWC recent theoretical summary • Recent experimental progress summary • Advanced modulation techniques • Challenges and perspectives
SAH Mohsan et al. [18]	2020	• UOWC recent progress • Optical scattering and absorption challenges • Modulation technologies and channel coding
PA Hoeher et al. [3]	2021	• UOWC in swarm robotics • Channel modeling fundamental • Physical layer transmission techniques • Data link layer aspects • Interference suppression • Realization aspects
Y. Baykal et al. [23]	2022	• UOWC turbulence modeling • Turbulence mitigation techniques
This survey	2023	• UOWC fundamental overview • Introduction of equalization principles • Recent UOWC work based on NML equalization • Recent ML equalization techniques in UOWC systems • DSP challenge discussions

2. Overview of UOWC Systems

2.1. System Architecture

Before introducing the typical UOWC system, it is essential to provide an overview of the UOWC network architecture. As shown in Figure 1, AUVs can use the UOWC system to communicate with divers and optical base stations (OBSs) in real time. Moreover, the underwater monitor (UM) can use underwater cameras to transmit real-time video signals through UOWC techniques to OBSs and finally to relevant departments to monitor water quality and deter poachers. Furthermore, the underwater defense system can use UOWC sensor technologies to detect enemies in time and take a counter measurement. At the same time, OBSs connect with the central OBS on the water surface to form a complete underwater sensor network. The central OBS can further communicate with satellites and ships by the RF link, truly realizing the integrated underwater-above-water-satellite communication network.

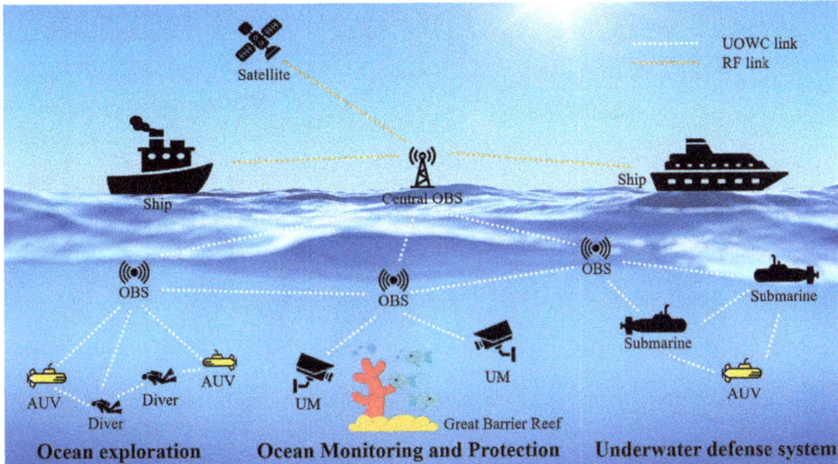

Figure 1. UOWC network architecture.

The typical UOWC system, which includes the transmitter, the medium, and the receiver, is shown in Figure 2. In the transmitter part, the signal is generated and mapped to transmitted symbols before being loaded into a digital-to-analog converter (DAC). The most widely used symbol modulation techniques used are on–off keying (OOK) [39–43] and pulse position modulation (PPM) [44–47]. OOK is the simplest modulation method and is widely used together with direct detection. However, the OOK modulation is susceptible to interference in the complex underwater environment [8,13]. PPM is beneficial for long-distance UOWC communication since it is more power efficient. However, the PPM suffers from the disadvantage of low spectral efficiency. To achieve a high data rate transmission, more advanced signal modulation methods, such as multi-level pulse amplitude modulation (PAM) [27,35,48–50] and quadrature amplitude modulation (QAM) [51–55], have been applied in recent studies. PAM uses multiple power levels to modulate information, and hence, provides higher spectral efficiency. However, due to PAM requiring a higher SNR for correct symbol detection, the power consumption is high. Furthermore, QAM further improves the data rate by exploring quadrature features. However, due to its high implementation complexity, QAM modulation has high cost limitations [15].

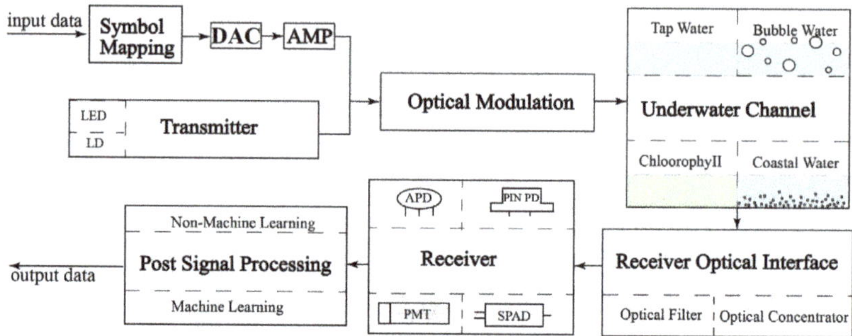

Figure 2. General architecture of UOWC system.

After symbol mapping, the digital signal is loaded into a DAC. After amplification (AMP), the signal is then modulated to the optical carrier. There are two optical transmitters that are widely used in UOWC. The first one is LED, which has the advantages of a wide optical beam and low cost. Hence, the LED-based UOWC system is widely used in short-range applications with a large signal coverage area (tens of meters transmission distance [10,41,49,56]). However, the modulation bandwidth is typically limited to tens of MHz [10,57] or hundreds of MHz [58], restricting the transmission data rate [54,58]. Compared with LED, the LD transmitter has the advantage of broad modulation bandwidth over the GHz range [11,27,59–61]. Hence, the transmission data rate of LD-based UOWC systems can reach tens of Gpbs [27,52,53,62]. Moreover, the narrow laser beam also enables a longer link distance of hundreds of meters [38]. However, due to the narrow laser beam, the transmission performance degrades sharply by underwater scattering and turbulence [8]. In addition, LD transmitters also have a higher cost compared to LEDs.

To modulate symbols to the optical carriers, two methods are widely used. The most common method in UOWC systems is direct modulation, which uses a bias tee to combine the electrical signal generated by the DAC with the DC bias, which is then connected to the optical transmitter [30,63,64]. The advantage of direct optical modulation is simple and efficient. However, the modulation bandwidth is limited and is only capable of intensity modulation. To overcome these limitations, external optical modulation methods can be used; the Mach–Zehnder Modulator is widely employed in LD-based UOWC systems [65–67].

After signal generation, the optical beam propagates through the underwater channel. The two main physical phenomenons that cause signal loss in the underwater channel are absorption and scattering, as shown in Figure 3. When the optical beam with optical power P_i at a wavelength λ propagates in the water, a small part is absorbed, denoted by P_a, and another part is scattered, denoted by P_s. The remaining part P_t reaches the receiver [8]. Generally, the attenuation coefficient $c(\lambda)$ is the sum of absorption coefficient $a(\lambda)$ and scattering coefficient $b(\lambda)$ [68]:

$$c(\lambda) = a(\lambda) + b(\lambda). \tag{1}$$

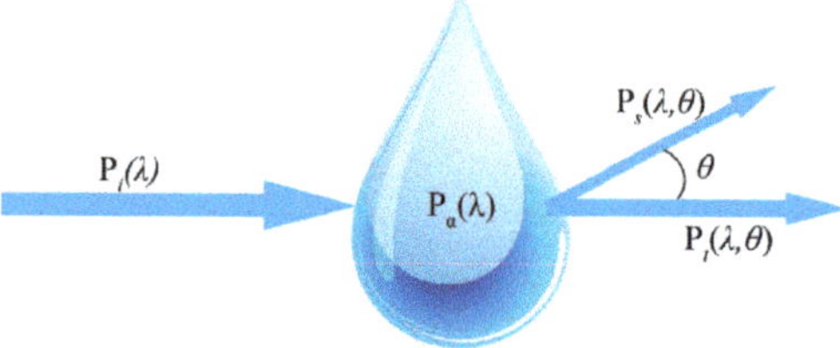

Figure 3. Geometry of optical beam propagation underwater.

The propagation path loss L_p, as a function of wavelength λ and distance d, is then given as [69]

$$L_p(\lambda, d) = e^{-c(\lambda)d}. \tag{2}$$

The attenuation coefficient $c(\lambda)$ is highly dependent on the optical wavelength [70], where the range of 450–550 nm (blue and green lights) has a much smaller attenuation coefficient compared to other wavelengths. Hence, the blue and green wavelength bands are typically used in UOWC systems. Another factor that affects the attenuation coefficient is the type of water. Since the UOWC technology is typically applied in the ocean, the attenuation coefficient of seawater has become a key research focus. In [71], it was found that the absorption coefficient $a(\lambda)$ is determined by organic pigments (chlorophyll, carotenoids, pheophytin, and chlorophyllide) produced by aquatic plants in water, and the scattering coefficient $b(\lambda)$ is dependent on the density and volume of particles in water. However, ocean water at different depths and locations has differences in sunlight strength and temperature, causing differences in the organic pigment levels and the density and volume of particles. Therefore, in previous studies, the water channel is typically divided into three categories: clear ocean, coastal ocean, and turbid harbor. The typical values of absorption and scattering coefficients of the three different ocean water types are concluded in Table 3 [8].

Table 3. Typical values of absorption and scattering coefficients in three different ocean water types.

Ocean Water Type	a (m^{-1})	b (m^{-1})	c (m^{-1})
Clear ocean	0.114	0.037	0.151
Coastal ocean	0.179	0.220	0.339
Turbid harbor	0.366	1.829	2.195

In UOWC experiments, many researchers use tap water with a lower attenuation coefficient ($c \approx 0.07$ m^{-1}) [26,27,72]. To achieve experimental results closer to practical systems, the chlorophyll base of a seawater channel was investigated in recent UOWC studies [73–75]. As a very common phenomenon in water, bubbles will also affect the performance of UOWC systems. There have been some recent studies investigating UOWC bubble channels [76–78]. In addition, near the ocean surface and harbor, coastal ocean water and turbid harbor water channels have always been a huge challenge for UOWC communications since these two kinds of water often have high levels of organic pigments and particles, leading to a large attenuation coefficient [8,13]. In recent years, there has been research focusing on the improvement in coastal and turbid harbor water channels [35,60,79].

After passing through the underwater channel, an optical interface is typically used at the receiver to reduce the impact of the background light (sunlight) and to focus signal light, such as optical filters to suppress the background light [80–82], and the lens array to collect more signal light [28,29]. Then, the signal light is detected by the optical detector. The PIN photodiode (PIN PD), which has the advantages of fast response time, low cost, and good tolerance to ambient light [8], is widely employed [10,50,51,53–56,58–60,62,83–85]. The avalanche photodiode (APD) has also been utilized due to the higher internal gain [10,28,35,38,41,42,49,52,64,86–89]. However, APDs also require high bias voltage, complex control circuitry, and are more sensitive to ambient noise.

Due to the relatively high path loss, UOWC systems require a sensitive receiver to increase the transmission distance. Hence, PMT has been explored, which has a high gain, low noise, and a large collection area [8]. Moreover, SPAD, which operates at a reverse voltage higher than the breakdown voltage to further increase the internal gain and sensitivity, has also been investigated [13]. With PMT and SPAD, a 50 m UOWC link with 500 kbps data rate and 117 m UOWC link with 2 Mbps data rate have been demonstrated [14,90].

In UOWC systems, both the line-of-sight (LOS) link and non-line-of-sight (NLOS) link have been studied, as shown in Figure 4. Direct LOS link is the most simple link configuration and has been widely studied both theoretically and experimentally [82,91].

However, in practical underwater environments, marine life and reefs can block the channel. To overcome this limitation, NLOS-based UOWC systems, which use the water–air surface or bubbles underwater to reflect the signal to avoid obstacles, are investigated in recent studies [80,92,93], including a few NLOS UOWC experimental demonstrations [46,94]. Moreover, the NLOS link configuration is also applied in highly turbid water to achieve a high transmission data rate and bypass obstacles [95,96].

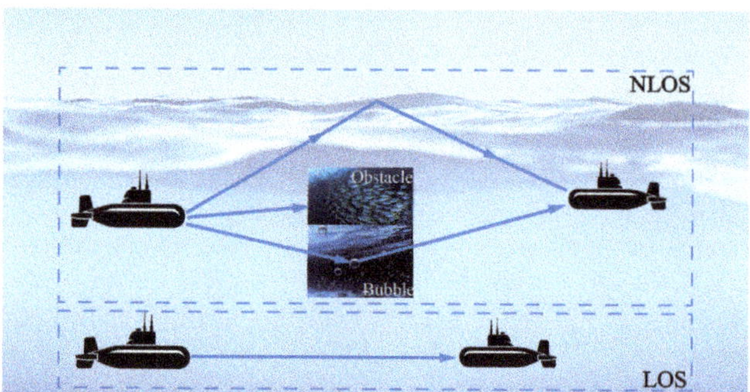

Figure 4. UOWC channel configurations.

2.2. Signal Processing Techniques

As mentioned in Figure 3, when light passes through water, some of the light is scattered and travels in other directions. Generally, the Henyey–Greenstein (HG) function and the two-term Henyey–Greenstein (TTHG) function [97] are widely used to represent the scattering phase function (SPF) in UOWC systems [8,98], which are widely used in the Monte Carlo (MC) simulation of UOWC systems [80,99–102]. Due to the scattering in the underwater channel, the signal photons arrive at the detector through different optical paths, leading to a delay in the time-of-arrival and ISI, which degrades the signal quality and reduces the transmission data rate. Since scattering is determined by the size and density of particles in water, ISI has little impact on deep sea or clear ocean UOWC systems. However, in harbor water and coastal water, the performance of UOWC systems is highly affected by ISI [8,39,103].

In most practical UOWC applications, eliminating the ISI at the receiver is not trivial because the channel response is not accurately known. The simplest and most common signal-processing technique to suppress the impact of ISI is the linear equalizer. Among them, zero-forcing linear equalizer (ZF-LE), which equalizes the folded spectrum of the received signal using a filter with the inverse frequency response, has been applied in many UOWC studies [37,104,105]. The ZF-LE can eliminate the ISI at the sampling time to increase the system SNR. However, the ZF-LE cannot be used when the folded spectrum has nulls, which can cause infinite noise enhancement. To avoid this, the mean-square error linear equalizer (MSE-LE) has been studied. Unlike the ZF-LE eliminating the ISI, the MSE-LE passes a small part of ISI to the output to improve the SNR and BER in UOWC systems [31,106]. It needs to be mentioned that we only give a snapshot here and more details of these works are presented later.

Even though linear equalizers are easy to operate in systems, the infinite noise enhancement of ZF-LE with spectral nulls and large noise enhancement of MSE-LE with deep attenuation in the passband limit their applications. To overcome these issues, more complex and advanced nonlinear equalizers have been further studied in UOWC systems. One of the most common nonlinear equalizers in UOWC systems is the zero-forcing decision-feedback equalizer (ZF-DFE), which is designed to cancel ISI and completely avoid infinite noise enhancement by employing a whitened matched filter. To further improve the performance, the mean-square error decision-feedback equalizer (MSE-DFE) was investigated, which uses a linear predictor and a feedback filter to whiten the noise at the output. Due to its complex structure and design, MSE-DFE is rarely used in UOWC systems but is widely used in UAC systems [107–109]. In summary, signal processing techniques that rely on non-machine learning equalizers are widely used in UOWC systems. In Section 3, we review the recent studies of non-machine learning equalization in UOWC systems.

With the development of neural networks and artificial intelligence, ML applications based on neural networks have been widely investigated in UOWC systems to enhance performance. In general, there are four types of machine learning algorithms: supervised, unsupervised, semi-supervised, and reinforcement learning. The most commonly used ML algorithm in the UOWC system is supervised learning, which uses known output features to derive computational relationships between input training data and output data [110]. The supervised ML method has been widely applied in signal processing to suppress various impairments in the UOWC system, such as improve the BER and enhance stability [50,111–115]. The encoder/decoder based on supervised ML algorithms has also been studied to improve the data transmission in recent UOWC research [116,117].

Another ML algorithm explored in the UOWC system is reinforcement learning, which focuses on developing an optimized strategy by monitoring how an intelligent agent acts in an environment to maximize cumulative reward. The reinforcement learning method has been applied to improve the communication stability [118–120] and to reduce the power consumption and improve link quality via optimizing the routing protocol in an underwater sensor network [121,122]. In Section 4, we provide a comprehensive survey on the recent UOWC progress based on ML algorithms.

3. Non-Machine Learning Equalization

3.1. Linear Equalizer

UOWC has developed rapidly in recent years and made remarkable achievements. Table 4 lists the recent studies of NML linear equalization in UOWC systems.

Table 4. Research progress in the UOWC system based on NML linear equalization.

Year	Bit Rate (bps)	Distance	Optical Source	Receiver	Transmission Power	Modulation Scheme	Equalizer	Refs.
2013	1 G	40 m Coastal	532 nm LED	PD	~50 W	OOK	ZF-LE	[37]
2017	1 M	3 m Tap	532 nm LED	SPAD	N/A	OFDM-QAM	MMSE-FDE	[31]
2019	256 G	50 m Air 5 m Turbid	Red LD	APD	~500 W	PAM4	FDE	[79]
2019	30 G	12.5 m Tap 2.5 m Harb	488 nm LD	APD	~20 mW	PAM4	FDE	[35]
2020	20 M 50 M	28 m Tap 10 m Tap	470 nm LED	SiPM	~600 mW	PAM	FDE	[123]
2020	3.31 G	56 m Tap	520 nm LD	APD	~50 mW	OFDM	FDE-NP	[63]
2021	1 G	1 m Tap	377 nm LD 405 nm LD	2APDs	~70 mW to 120 mW	NRZ-OOK	ZF-LE	[104]
2022	4 G	2 m Tap	484 nm LED	APD	~1 mW	PAM4	FFE	[25]

Due to simplicity and efficiency, the linear ZF-LE and feedforward equalizer (FFE) have been studied in UOWC systems to enhance data transmission performance. In [37], to reduce the effect of ISI and improve the BER, the ZF-LE based on the double Gamma model was employed in a UOWC system with coastal water. MC results show that to achieve the forward error correction (FEC) threshold (3.8×10^{-3}), the ZF-LE can reduce the transmission power by around 2 dBm and 1.5 dBm in the 1 Gbps 40 m coastal water link and the 500 Mbps 10 m harbor water link, respectively. In addition, the ZF-LE combined with the dual-wavelength UOWC transmission was demonstrated in [104] to improve the BER simultaneously in two wavelength channels. Results in Figure 5 show that the ZF-LE can restore eye opening in the corresponding eye diagram of both blue and green channels and improve the BER performance. Furthermore, a linear T/2-spaced feedforward equalizer (FFE) combined with the high-bandwidth GaN-based mini-LEDs was demonstrated in [25]. Due to mitigating the ISI by simple and efficient FFE, the net data rate can reach up to 4.08 Gbps (the highest for UOWC systems using a single-pixel mini-LED) with a BER below the FEC threshold in short-distance transmission.

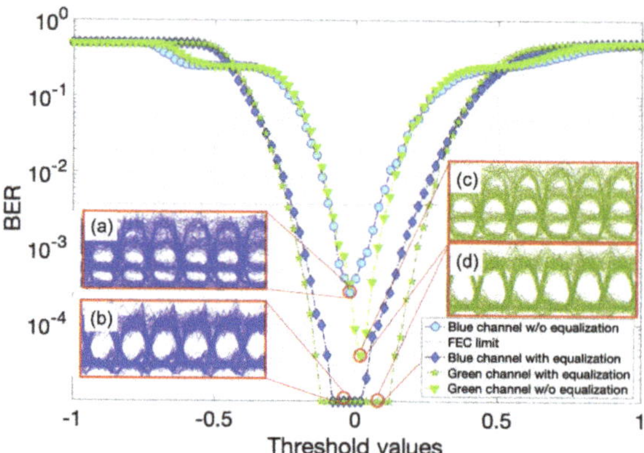

Figure 5. Bathtub curve for blue and green channels using 377- and 405-nm transmitters with and without ZF-LE [104].

In addition to the simple ZF-LE, low-complexity linear frequency domain equalizers (FDE) were also employed to mitigate the ISI effect. In [123], the FDE combined with high sensitive silicon photo-multipliers (SiPMs) receiver was employed to overcome the limited bandwidth of components at a high data rate. Results show that the FDE can significantly improve BER from 1×10^{-1} to 0.9×10^{-4} in a 10 Mbps UOWC system with a 40 m tap water channel. Moreover, the FDE can improve the data rate to achieve 20 Mbps and 50 Mbps transmission under 28 m and 10 m tap water channels, respectively. In addition, linear FDE has also been studied in the free-space optical underwater optical wireless communication (FSO-UOWC) convergent system to optimize the modulation frequency response and enhance the transmission capacity [79]. Together with the two-stage injection locking technique, a 256 Gb/s four-channel FSO-UOWC convergent system with 50 m free space and 5 m turbid underwater transmission is successfully demonstrated. Furthermore, in another work [35], a red LD is used as the transmitter, and both an injection-locking optoelectronic feedback and a linear FDE at the receiving end are applied. Results show that the 3 dB bandwidth can be increased from 8.4 GHz to 10.8 GHz using the linear FDE. The linear FDE has also been utilized to compensate for the frequency response (especially for high frequencies) to enhance the transmission rate of a PAM4 UOWC system. Results show that under both a 12.5 m piped underwater channel and a 2.5 m high-turbidity harbor underwater channel, 30 Gbps transmission can be achieved.

However, due to the limited capability of the conventional FDE when the folded spectrum has nulls, a one-tap minimum mean square error (MMSE) FDE was employed in [31]. Results show that the MMSE-FDE can significantly reduce the noise enhancement in frequency-selective channels with spectral nulls of the coded UOWC systems, which further improves the BER performance by 2 to 4 dB for three water types. In addition to the MMSE, the conventional linear FDE can also be combined with noise prediction (NP) shown in Figure 6 to better mitigate the impact of ISI [63]. Results show that the SNR required to achieve the same BER can be reduced by 3.8 dB using the proposed FDE-NP scheme compared with the conventional FDE. Due to the efficient equalization by FDE-NP, a maximum net data rate of 3.48 Gbps is achieved, which is about 17.2% higher than the traditional OFDM UOWC system. Furthermore, a 56 m UOWC system with a data rate of 3.31 Gbps is demonstrated with FDE-NP.

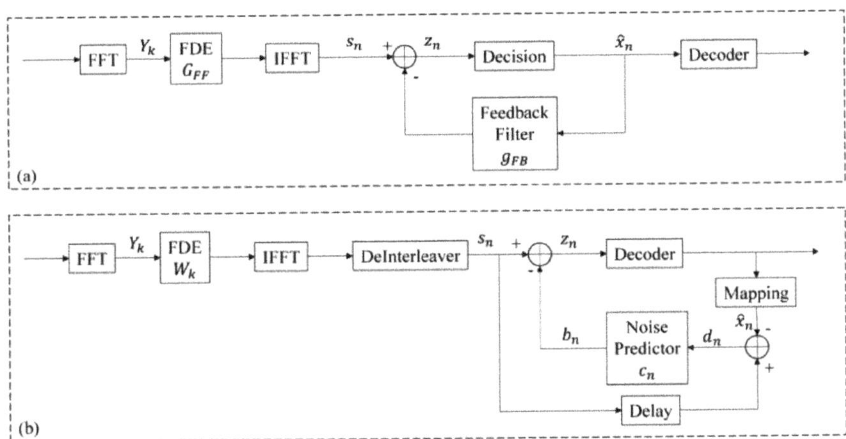

Figure 6. The receiver structure of the (**a**) FDE-DFE, (**b**) FDE-NP [63].

3.2. Nonlinear Equalizer

Compared with the linear equalizer, the nonlinear equalizer has a more complex structure but superior ISI mitigation performance. It is also widely used in the recent UOWC systems, which are summarized in Table 5.

Table 5. Recent research progress in UOWC systems based on NML nonlinear equalization.

Year	Bit Rate (bps)	Distance	Optical Source	Receiver	Transmission Power	Modulation Scheme	Equalizer	Refs.
2016	9.6 G	8 m Tap	405 nm LD	PD	~30 mW	16-QAM-OFDM	TPGE	[26]
2016	745 M	2 m Tap	448 nm LED	APD	~184.5 mW	OFDM-QAM	APE	[124]
2017	16 G	10 m Tap	488 nm LD	PD	~20 mW	PAM4	DFE	[27]
2018	16.6 G	55 m Tap	450 nm LD	PIN PD	~120 mW	OFDM-QAM	VE	[62]
2018	7.33 G	15 m Tap	450 nm LD	APD	~20 mW	DMT	VE	[64]
2019	2.5 G	60 m Tap	450 nm LD	APD	~50.2 mW	NRZ-OOK	VE	[28]
2019	500 M	100 m Tap	520 nm LD	APD	~7.25 mW	NRZ-OOK	VE	[38]
2019	500 M	1 m Bubble	520 nm LD	APD	~25 mW	16PPM	VE	[76]
2021	200 M	100 m Tap	450 nm LD 520 nm LD	PMT	~700 mW	RRC-OOK	MPE	[30]
2022	4.12 G	2 m Tap	484 nm LED	APD	~ 1 mW	PAM4	VE	[125]
2022	200 M	1.5 m Tap	Blue LED	PD	~0.4 mW	CAP OFDM	VDFE	[126]

Earlier, many UOWC experiments used an analog equalizer to improve the system frequency response. In [26], a physical tunable passive gain equalizer (TPGE) combined with the two-stage-injection-locked technique was employed to improve the frequency response, as shown in Figure 7. The frequency response of the signal is compensated by around 10 dB after TPGE, where the electrical spectrum of the data signal after TPGE is flatter, improving the system transmission performance. Results show that BER can be improved from 2×10^{-2} to 4×10^{-2} in an 8 m tap water channel under a transmission speed of 9.6 Gbps. Similarly, in [124], an analog post-equalizer(APE) was demonstrated to increase the 3 dB frequency response of the system from 4 MHz (LED) and 100 MHz (detector) to 124.2 MHz end-to-end. Due to the broader and flatter 3 dB bandwidth of the system, high spectral efficiency modulation formats such as OFDM can be applied. Results show that the SNR increased from 8.75 dB to 22.6 dB after employing the APE. Moreover, the BER improved from 1.2×10^{-1} to 3.9×10^{-4} in a UOWC link at 621.1 Mbps with 2 m tap water.

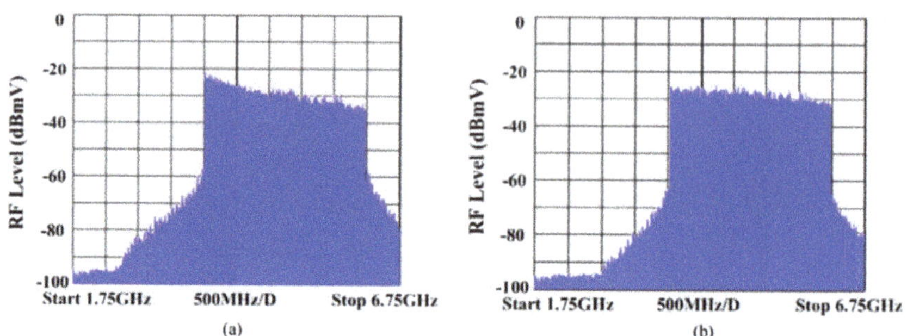

Figure 7. Electrical spectra of the 9.6 Gbps 5 GHz 16-QAM-OFDM data signal (**a**) before TPGE and (**b**) after TPGE [26].

Compared with the previous nonlinear analog equalizers, digital nonlinear equalizers are more widely studied and utilized to process received signals. The most common one is the nonlinear Volterra series-based equalizer (VE), which has great advantages against non-linear effects to further improve system performance. In recent years, some works compared the nonlinear VE with conventional linear equalizers. For instance, a nonlinear VE combined with adaptive bit-power loading discrete multi-tone (DMT) was employed in [62], and results show that the VE can bring more than 2 dB gain compared with linear equalization since the nonlinear VE is more effective against non-linear effects. Up to 16.6 Gbps data rate through 55 m tap water wireless optical transmission was achieved. The nonlinear VE has also been compared with the conventional linear feed-forward equalizer (FFE), where the nonlinear VE is shown to effectively reduce the impact of device nonlinearity [125]. Results show that a significant BER reduction (from 1×10^{-2} to 3×10^{-3}) is observed after changing linear FFE to nonlinear VE at 4.2 Gbps with 2 m transmission. In addition to conventional VE, the nonlinear Volterra series-based DFE (VDFE) can further mitigate the intrinsic static and dynamic non-linearity effects in the UOWC system. In [126], a nonlinear VDFE was demonstrated in carrierless amplitude and phase (CAP) modulation and compared with the linear DFE. However, since the LED is biased in the linear region of its transfer curve, the linear DFE and nonlinear VDFE have similar BER performance.

Moreover, the nonlinear VE combined with DMT was demonstrated in [64]. Due to effectively combating nonlinear effects, results show that VE can bring more than 3 dB gain on average compared with the system without VE. In a 15 m underwater channel, the system with a nonlinear equalizer can achieve 7.33 Gbps with BER lower than the 7% FEC threshold, which is 0.9 Gbps higher than the system without VE. Furthermore, the nonlinear second-order VE has been used in [28]. Results show that the VE can significantly reduce the influence of ISI, which achieved a better BER from 1×10^{-1} to 3.5×10^{-3} at 2.5 Gbps transmission under 60 m of tap water. In addition, nonlinear VE was also applied in [38] to compare with the direct hard-decision detection in the receiving offline DSP. Results show that the VE achieved low BER performance (2.5×10^{-3} compared with 5.91×10^{-2} of hard-decision detection) in a 100 m 500 Mbps UOWC system. Moreover, in Figure 8, when the data rate reaches 400 Mbps and 500 Mbps, severe ISI makes the system impossible to reduce the BER to achieve the FEC limit by using hard decision detection. However, the nonlinear VE can bring the BER below the FEC limit with minimum optical powers required for 400 Mbps and 500 Mbps being -26.4 dBm and -24.0 dBm, respectively.

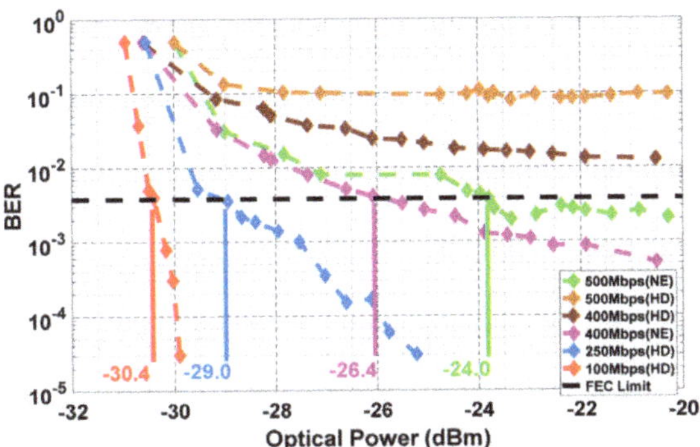

Figure 8. BER vs. received optical power for different data rates with hard-decision detection and nonlinear VE detection [38].

In addition to tap water channels, nonlinear VE has also been employed under the bubble water channel to improve the BER performance and combat ISI induced by signal light scattering from bubbles [76]. Results show that the UOWC system without the nonlinear VE can only achieve a BER below the FEC limit when the bubble size is smaller than 1.2 mm, whereas the system with VE works under bubbles with sizes up to 2.8 mm.

Compared with the conventional nonlinear VE, in which the computation complexity increases with the memory length and nonlinear order, the simpler nonlinear DFE combined with the light injection and optoelectronic feedback techniques was employed, which achieved 16 Gbps (8Gbaudps) PAM4 signal transmission in a UOWC system [27]. Moreover, the nonlinear memory polynomial model-based equalizer (MPE) has great advantages of faster convergence speed and lower error. After employing the nonlinear MPE in [30], the BER significantly increase from 1×10^{-2} to 5×10^{-5} at 160 Mbps under 100 m tap water. Moreover, due to the nonlinear MPE reducing the effect of ISI efficiently, a 200 Mbps data rate over 120 m and a 100 Mbps data rate over 139 m underwater transmission were achieved.

4. Machine Learning Applications in UOWC

4.1. Supervised Learning in UOWC Systems

Table 6 lists key studies in recent years on the application of ML algorithms in UOWC systems. Most ML algorithms are applied to the equalization at the receiver side to improve the BER performance and the transmission data rate. For instance, a novel Gaussian kernel-aided deep neural network (GK-DNN) equalizer shown in Figure 9 was employed for compensating the high nonlinear distortion of PAM8 UOWC channels in [50]. Because the GK-DNN treats the equalization problem as a classification problem, it has the advantage of performing both equalization and de-mapping at the same time. After being combined with the scalar-modified cascaded multi-modulus algorithm (S-MCMMA), the GK-DNN equalizer can perform linear equalization, nonlinear equalization, and de-mapping at the same time. Moreover, compared with the conventional DNN equalizer, the GK-DNN equalizer can reduce required training iterations by 47.06%. Results show that the BER can be significantly reduced by 1.78 dB in the LED-based UOWC system employing the GK-DNN equalizer.

Table 6. Recent research progress in the UOWC system based on supervised ML.

Year	Bit Rate (bps)	Distance	Optical Source	Receiver	Transmitter Power	Modulation Scheme	ML Algorithm	Refs.
2018	1.5 G	1.2 m Tap	457 nm LED	PIN PD	N/A	PAM8	GK-DNN	[50]
2019	N/A	1 m Turbid	532 nm LD	CCD	~20 mW	N/A	CNN	[116]
2020	3.2 G	1.2 m Tap	Blue LED	PIN PD	N/A	64-QAM	DBMLP	[111]
2020	N/A	1.2 m Tap	Blue LED	PIN PD	N/A	64-QAM	TFDNet	[112]
2021	N/A	4.3 m Turbu	632.8 nm LD	Camera	~2 mW	N/A	CNN	[117]
2021	2.85 G	1.2 m Tap	Blue LED	PIN PD	N/A	64-QAM	PCVNN	[113]
2021	N/A	30 m Coastal	Blue LED	PD	N/A	QAM	BDNet	[105]
2021	3.1 G	1.2 m Tap	Blue LED	PIN PD	~100 mW	64-QAM	TL-DBMLPs	[114]
2022	1 M	1.5 m Tap	450 nm LED	SPAD	~1 W	OOK	DNN	[127]
2023	660 M	90 m Tap	450 nm LD	PMT	~188.8 mW	I-SC-FDM	SWI-DNN	[115]

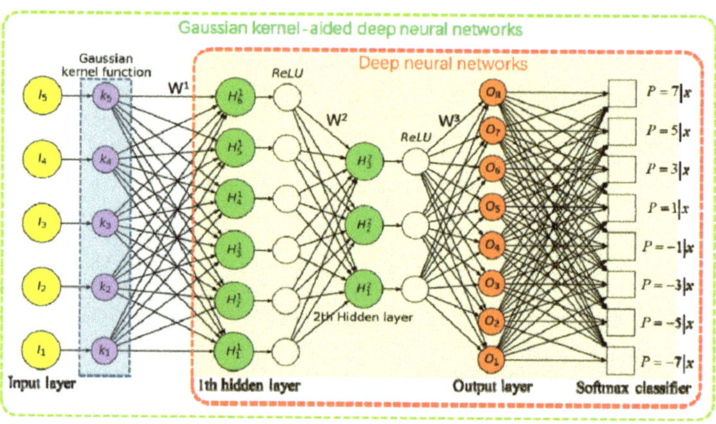

Figure 9. Structure of the GK-DNN [50].

Moreover, in [111], a dual-branch multilayer perceptron (DBMLP)-based equalizer was employed in UOWC systems. Due to the limitation on the order of the Volterra series, the nonlinear distortion in the received signal cannot be compensated accurately in conventional NML VE. Unlike NML VE, the deep neural network-based equalizer has a better capability to compensate for nonlinear distortions as it can model arbitrary mappings at arbitrary progress. Results show that the UOWC system succeeded in achieving 3.2 Gbps data transmission. Compared with the conventional VE, 63.5% BER performance enhancement and 33.8% better space complexity are achieved with the proposed DBMLP scheme. In [112,128,129], a nonlinear ML post equalizer based on the time-frequency domains deep neural network (TFDNet) was proposed in UOWC systems, as shown in Figure 10. The short-time Fourier transformation (STFT) was employed to combine the signal from two one-dimension (time domain and frequency domain) images to a two-dimension (time-frequency domain) image. Then, the signal 2D time-frequency image is fed into the DNN, which learns the mapping relations to equalize the signal to match the labeled 2D time-frequency image. Unlike the conventional DNN-based equalizers that consider only the time domain, the additional frequency domain information enables the DNN to learn complementary signal characteristics. The proposed TFDNet-based equalizers can improve the BER from 2×10^{-2} (VE) and 7×10^{-3} (DNN-based equalizer) to 2×10^{-3} at a valid operating Vpp of 0.8 V in a 2.85 Gbps UOWC system to achieve FEC limits.

Furthermore, in [113], an adaptive constellation-partitioned equalizer based on a complex-valued neural network (PCVNN) was employed to reduce the computational complexity in typical ML algorithms in UOWC systems. Experimental results show that compared with conventional ML equalizers, the computation cost can be reduced by 56.1% in the proposed PCVNN scheme in a 2.85 Gbps UOWC system. Two transfer learning-based (TL) DBMLPs were demonstrated in [114]. Unlike the conventional DBMLPs, the TL-DBMLP is more robust to the jitter of LED transmitter bias current and also requires a smaller number of training epochs. Experimental results show that the proposed UOWC system employing TL-DBMLPs can reduce the size of the training set from 50% to 10% of the total dataset to achieve an acceptable mean square error (MSE). Moreover, with only 10 epochs, the achievable BER is improved from 1×10^{-2} with conventional DBMLPs to 1×10^{-3} using the TL-DBMLPs. Moreover, a novel sparse weight-initiated deep neural network (SWI-DNN) equalizer combined with the interleaved single-carrier frequency division multiplexing (I-SC-FDM) scheme was employed for UOWC systems in [115]. Due to the implementation of a special SWI structure, the necessary training epochs of the SWI-DNN equalizer can be reduced by 10.3%. Results show that to achieve the BER of 3.8×10^{-3} (i.e., FEC limit) in the 90 m UOWC system, the data rate can be increased by 17.9% after employing the ML equalizer than conventional TFD equalizers.

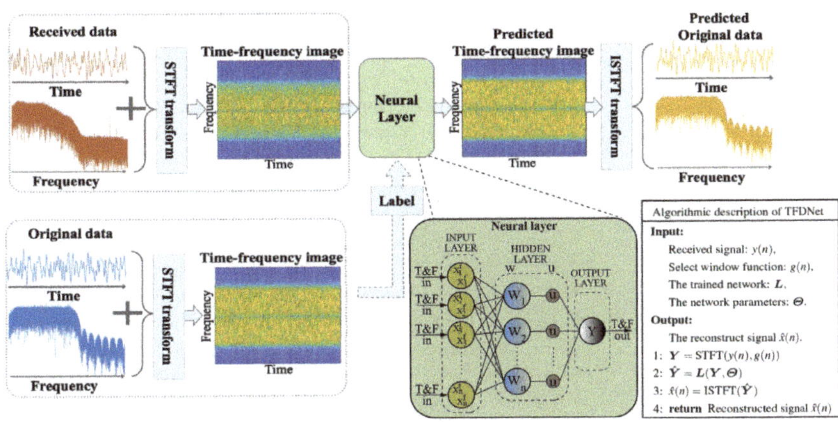

Figure 10. Schematic diagram of the proposed TFDNet [112,128].

In addition, a 16-ary orbital angular momentum shift keying (OAM-SK) equalizer based on convolutional neural networks (CNNs) was proposed in [116]. Results show that ML schemes can achieve an accuracy of more than 96% and a larger number of pixels in the camera receiver can be utilized in a camera-based UOWC system. Hence, the ML-based equalizer can significantly improve the accuracy of data decoding. In addition, the other equalizer based on CNNs was employed in [117]. They succeed in achieving a high accuracy (93.7~99.9%) in high-turbulence UOWC with a camera-based receiver for image transmission. A larger alphabet and faster classification rates can be achieved in LD-based UOWC systems. In [105], a new equalizer based on a blind detection network (BDNet) was considered in UOWC systems. Unlike the previous blind channel estimation (BCE) schemes, the BDNet has the advantage of estimating the inverse channel regardless of the scalar ambiguity issue by learning the latent channel features from the received signal only. Simulation results show that the proposed UOWC with the BDNet achieves better BER performance compared with conventional BCE schemes. In addition, ML algorithms have also been studied for time synchronization and clock recovery in UOWC systems. The recovery of the time slot synchronous in the photon-counting UOWC system is critical, and conventionally it is realized based on symbol synchronization and frame synchronization, which has limited accuracy. To predict the phase value of the time slot synchronous clock, a method of time slot synchronous clock recovery for photon-counting UOWC based on DNNs is designed in [127]. Results show that the photon-counting UOWC based on the ML time slot synchronous clock recovery succeeded in achieving a data rate of 1 Mbps and a BER of 5.35×10^{-4} at eight photons per time slot.

4.2. Reinforcement Learning in UOWC Systems

Reinforcement learning is often used in UOWC systems to improve the connection success rate of the underwater sensor network and AUVs, further increasing the reliability of links. In underwater sensor networks, the highly dynamic topology can hinder the routing of UOWC links, particularly due to the ocean flow movement. In [121], an advanced routing protocol based on multi-agent reinforcement learning (MARL) was proposed to increase the link reliability and communication quality in the UOWC-based sensor network. Simulation results show that the UOWC-based sensor network with MARL has 80% average residual energy of the network after 150 simulation times compared with 30% and 70% achieved with the Q-learning-based delay tolerant routing (QDTR) and the ad hoc on-demand distance vector (AODV) routing protocols. Moreover, MARL has the highest delivery ratio in the static network (95.87%) and the dynamic network (95.57%) compared with the other two schemes (49.5% and 44.56% achieved in AODV in the static network and the dynamic network, respectively). In addition, to overcome the same limitations, an efficient routing protocol based on MARL was also employed in another work [122]. Simulation results show that the MARL routing protocol provided the same low power consumption advantages and high-quality UOWC linksin a network with 14 neighboring nodes.

Moreover, reinforcement learning algorithms have also been exploited to solve pointing acquisition and tracking (PAT) problems between underwater applications (AUVs). An advanced beam adaptation method based on the state–action–reward–state–action (SARSA) algorithm for point-to-point UOWC systems was proposed in [118]. Results show that the SARSA-based beam adaptation method increased the success rate from 66% to 93%, which further achieved better link reliability. Moreover, the SARSA increased SNR by 6 to 10 dB compared to the traditional NML method in different types of underwater channels. Similarly, to overcome the poor link reliability and to optimize the connecting success rates between two AUVs, the soft actor-critic reinforcement learning algorithm was designed in [119]. Results show that the success rate for the transmitting AUV to maintain the LOS link for ten time steps was 97.53% after 10,000 episodes in the simulation environment. In [120], a deep reinforcement learning algorithm assisted by an extended Kalman filter was employed to improve the reliability of water–air optical wireless communication between AUVs and unmanned aerial vehicles (UAVs), which is even more challenging compared with connecting AUVs only. Results show that the proposed learning algorithm achieves a shorter MSE (0.02 m) compared with the triangular exploration (TE) algorithm (0.06 m), a shorter flight distance (1.1 m compared to 2 m on TE), and a smoother trajectory (3.23 compared to 6.98 on TE), which implies a higher alignment accuracy and smaller energy consumption. Moreover, it also improves the link availability by 25% compared with the TE algorithm.

5. Discussion and Future Scope

Although a promising solution, UOWC systems face the key challenge of ISI caused by signal scattering in underwater channels. To improve the data transmission distance and improve the BER, various types of equalizers have been proposed and studied, as discussed in detail in Section 3. Both linear and nonlinear equalizers have been investigated, and better BER and higher data rates in UOWC systems have been achieved. However, linear equalizers have the limitations of only being capable of stable communication channels. When the channel is disturbed, the performance of the linear equalizer is degraded significantly. In addition, they cannot suppress the nonlinear effects that widely exist in UOWC systems. On the other hand, nonlinear equalizers can suppress various types of nonlinearities in UOWC systems, and hence, have enhanced capability to improve data transmission performance. However, they normally have complex equalizer structures. Although the nonlinear equalizer configurations can be adjusted according to the influence of non-linearity, the variable underwater channel environment limits equalizer adjustment efficiency.

With the advancement of artificial intelligence, ML equalizers have been studied as well and have shown unique advantages over traditional NML equalizers. They can learn from the received signal to equalize signals in different underwater channels, noises, and system geometric arrangements. In particular, ML equalizers have enhanced capability in suppressing nonlinear effects in UOWC systems. ML equalizers can also handle multiple impairments simultaneously, effectively solving the interactions among different impairments. However, ML equalizers also face a number of key challenges. More complicated algorithms are typically required in complex environments, and these algorithms require a large number of training datasets and iterations, which often take a long time and are computationally expensive, requiring high-end hardware. Some recent research [113–115] starts to optimize the algorithms and reduce the training epochs to reduce the memory and time cost of ML equalizers. Solving this challenge becomes one focus of future UOWC post-signal processing technology.

Future underwater communication applications will certainly rely on reliable and powerful signal processing techniques to improve the system BER and performance. Adaptive equalizations can be a promising future research direction. For instance, in a calm lake or clear ocean, which has a stable underwater channel with limited nonlinear effect, a low-cost and simple linear equalizer can operate efficiently. In windy and choppy seas, nonlinear equalizers can be employed to reduce nonlinear effects and improve system performance. Finally, in challenging underwater channels such as turbidity and harbor water, variable impurities and underwater turbulence cause substantial changes in signal transmission. Traditional NML equalizers have limited capability in such complex and variable underwater channels. Instead, ML equalizers can automatically learn according to different channel performances and system configurations to optimize the performance of equalizing signals. However, since complex algorithms lead to a long training time and a larger number of training iterations, developing efficient hardware accelerators in ML-based equalizers can also be a future research field. Furthermore, so far the signal processing is mostly considered in point-to-point direct links. With the relay concept being considered in UOWC networks, the corresponding signal processing technique can be another research direction.

6. Conclusions

In this article, we provided a structured review of recent progress in UOWC systems, which are in high demand to overcome the inherent limitations of conventional acoustic systems and provide high-speed wireless communication links in underwater environments. In particular, we focused on the signal processing aspect of UOWC systems, which has attracted intensive interest as a promising method to suppress signal generation, modulation, transmission, and reception impairments. We reviewed the recent progress on both ML- and NML-based signal processing techniques. Due to its simple structure, the traditional NML equalization techniques can efficiently solve the SNR reduction caused by ISI, thereby improving the data rate of UOWC systems in recent years. Due to the advancement and development of ML algorithms, ML-based equalization techniques have shown better capability in reducing the influence of nonlinear effects. However, ML-based schemes normally require complex algorithms and powerful hardware support (high-performance computers).

In practical applications, water flow, temperature, and sunlight are changing all the time. Hence, considering the dynamic feature of underwater channels, designing adaptive signal processing techniques is important in the future, which is a key challenge in current studies. In addition, whereas the ML algorithm can achieve impressive results by suppressing both linear and nonlinear impairments effectively, the training process is complicated, requires a large number of computations, and takes a long time. Some recent ML research has begun to optimize algorithms to reduce the training epochs of ML-based equalizers to improve DSP efficiency and reduce computation costs. With the continuous development of hardware and continuous innovation of algorithms, the computation-efficient ML algorithms for UOWC systems will be a key research focus in the future. Moreover, due to the dynamic topology problem caused by water flow, the underwater sensor network needs a more intelligent way to ensure stable links. According to recent research, the UOWC network based on reinforcement learning can greatly improve communication success rate and stability, providing a promising solution for future UOWC applications.

Author Contributions: Conceptualization, C.F. and K.W.; software, Y.W.; writing—original draft preparation, C.F.; writing—review and editing, C.F., S.L., Y.W. and K.W.; visualization, Y.W.; supervision, S.L. and K.W. All authors have read and agreed to the published version of the manuscript.

Funding: This research was funded by the Australian Research Council (ARC) under Grant DP170100268.

Institutional Review Board Statement: Not applicable.

Informed Consent Statement: Not applicable.

Data Availability Statement: Not applicable.

Conflicts of Interest: The authors declare no conflict of interest.

Abbreviations

The following abbreviations are used in this manuscript:

Air	Air Free Space Channel
AMP	Amplifier
AODV	Ad Hoc on-Demand Distance Vector
APD	Avalanche Photodiode
APE	Analog Post-Equalizer
AUV	Autonomous Underwater Vehicle
BCE	Blind Channel Estimation
BDNet	Blind Detection Network
BER	Bit-Error-Rate
Bubble	Bubble Water Channel
CAP	Carrierless Amplitude and Phase
CCD	Charge-Coupled Device
Clear	Clear Ocean Channel
CNN	Convolutional Neural Network
Coastal	Coastal Water Channel
DAC	Digital-to-Analog Converter
DBMLP	Dual-Branch Multilayer Perceptron
DFE	Decision-Feedback Equalizer
DMT	Discrete Multi-Tone
DNN	Deep Neural Network
DSP	Digital Signal Processing
FDE	Frequency Domain Equalizer
FEC	Forward Error Correction
FFE	Feedforward Equalizer
FSO	Free-Space Optical
GaN	Gallium Nitride
GK	Gaussian Kernel-aided
Harbor	Harbor Water Channel
HG	Henyey–Greenstein
IoUT	Internet of Underwater Things
I-SC-FDM	Interleaved Single-Carrier Frequency Division Multiplexing
ISI	Inter-Symbol Interference
LD	Laser Diode
LE	Linear Equalizers

LED	Light-Emitting Diode
LOS	Line-of-Sight
MAC	Medium Access Control
MARL	Multi-agent Reinforcement Learning
MC	Monte Carlo
MIMO	Multiple-Input Multiple-Output
ML	Machine Learning
MMSE	Minimum Mean Square Error
MPE	Memory Polynomial Model-Based Equalizer
MSE	Mean-Square Error
NLE	Nonlinear Equalizers
NLOS	Non-Line-of-Sight
NML	Non-Machine Learning
NP	Noise Prediction
NRZ	No Return to Zero
OAM-SK	Orbital Angular Momentum Shift Keying
OBS	Optical Base Station
OFDM	Orthogonal Frequency-Division Multiplexing
OOK	On–Off Keying
PAM	Pulse Amplitude Modulation
PAT	Pointing Acquisition and Tracking
PCVNN	Partitioned Equalizer Based on Complex-Valued Neural Network
PD	Photodiode
PIN PD	PIN Photodiode
PMT	Photo-Multiplier Tube
PPM	Pulse Position Modulation
QAM	Quadrature Amplitude Modulation
QDTR	Q-Learning-Based Delay Tolerant Routing
RF	Radio Frequency
RRC	Root Raised Cosine
SARSA	State–Action–Reward–State–Action
SiPM	Silicon Photo-Multipliers
SNR	Signal-to-Noise Ratio
SPAD	Single-Photon Avalanche Diode
SPF	Scattering Phase Function
STFT	Short Time Fourier Transformation
SWI	Sparse Weight-Initiated
Tap	Tap Water Channel
TE	Triangular Exploration
TFDNet	Time-Frequency Domains Deep Neural Network
TL	Two Transfer Learning
TPGE	Tunable Passive Gain Equalizer
TTHG	Two-term Henyey–Greenstein
Turbu	Turbulence Water Channel
Turbid	Turbid Water Channel
UAC	Underwater Acoustic Communication
UAV	Unmanned Aerial Vehicle
UM	Underwater Monitor
UOT	Underwater Optical Turbulence
UOWC	Underwater Optical Wireless Communication
UWC	Underwater Wireless Communication
VDFE	Volterra Series-Based Decision-Feedback Equalizer
VE	Volterra Series-Based Equalizer
ZF	Zero-Forcing

References

1. Woodruff, S.D.; Slutz, R.J.; Jenne, R.L.; Steurer, P.M. A comprehensive ocean-atmosphere data set. *Bull. Am. Meteorol. Soc.* **1987**, *68*, 1239–1250. [CrossRef]
2. Saeed, N.; Celik, A.; Al-Naffouri, T.Y.; Alouini, M.S. Underwater optical wireless communications, networking, and localization: A survey. *Ad Hoc Netw.* **2019**, *94*, 101935. [CrossRef]
3. Hoeher, P.A.; Sticklus, J.; Harlakin, A. Underwater optical wireless communications in swarm robotics: A tutorial. *IEEE Commun. Surv. Tutor.* **2021**, *23*, 2630–2659. [CrossRef]
4. Sozer, E.; Stojanovic, M.; Proakis, J. Underwater acoustic networks. *IEEE J. Ocean. Eng.* **2000**, *25*, 72–83. [CrossRef]
5. Kida, Y.; Deguchi, M.; Shimura, T. Experimental result for a high-rate underwater acoustic communication in deep sea for a manned submersible shinkai6500. *J. Mar. Acoust. Soc. Jpn.* **2018**, *45*, 197–203. [CrossRef]
6. Song, H.; Hodgkiss, W. Efficient use of bandwidth for underwater acoustic communication. *J. Acoust. Soc. Am.* **2013**, *134*, 905–908. [CrossRef]
7. Akyildiz, I.F.; Pompili, D.; Melodia, T. Challenges for efficient communication in underwater acoustic sensor networks. *ACM Sigbed Rev.* **2004**, *1*, 3–8. [CrossRef]
8. Kaushal, H.; Kaddoum, G. Underwater optical wireless communication. *IEEE Access* **2016**, *4*, 1518–1547. [CrossRef]
9. Al-Shamma'a, A.I.; Shaw, A.; Saman, S. Propagation of electromagnetic waves at MHz frequencies through seawater. *IEEE Trans. Antennas Propag.* **2004**, *52*, 2843–2849. [CrossRef]
10. Tian, P.; Liu, X.; Yi, S.; Huang, Y.; Zhang, S.; Zhou, X.; Hu, L.; Zheng, L.; Liu, R. High-speed underwater optical wireless communication using a blue GaN-based micro-LED. *Opt. Express* **2017**, *25*, 1193–1201. [CrossRef]
11. Lee, C.; Zhang, C.; Cantore, M.; Farrell, R.M.; Oh, S.H.; Margalith, T.; Speck, J.S.; Nakamura, S.; Bowers, J.E.; DenBaars, S.P. 4 Gbps direct modulation of 450 nm GaN laser for high-speed visible light communication. *Opt. Express* **2015**, *23*, 16232–16237. [CrossRef]
12. Alghamdi, R.; Saeed, N.; Dahrouj, H.; Alouini, M.S.; Al-Naffouri, T.Y. Towards ultra-reliable low-latency underwater optical wireless communications. In Proceedings of the 2019 IEEE 90th Vehicular Technology Conference (VTC2019-Fall), IEEE, Honolulu, HI, USA, 22–25 September 2019; pp. 1–6.
13. Zhu, S.; Chen, X.; Liu, X.; Zhang, G.; Tian, P. Recent progress in and perspectives of underwater wireless optical communication. *Prog. Quantum Electron.* **2020**, *73*, 100274. [CrossRef]
14. Chen, H.; Chen, X.; Lu, J.; Liu, X.; Shi, J.; Zheng, L.; Liu, R.; Zhou, X.; Tian, P. Toward long-distance underwater wireless optical communication based on a high-sensitivity single photon avalanche diode. *IEEE Photonics J.* **2020**, *12*, 1–10. [CrossRef]
15. Zeng, Z.; Fu, S.; Zhang, H.; Dong, Y.; Cheng, J. A survey of underwater optical wireless communications. *IEEE Commun. Surv. Tutor.* **2016**, *19*, 204–238. [CrossRef]
16. Murgod, T.R.; Sundaram, S.M. Survey on underwater optical wireless communication: Perspectives and challenges. *Indones. J. Electr. Eng. Comput. Sci.* **2019**, *13*, 138–146. [CrossRef]
17. Schirripa Spagnolo, G.; Cozzella, L.; Leccese, F. Underwater optical wireless communications: Overview. *Sensors* **2020**, *20*, 2261. [CrossRef] [PubMed]
18. Mohsan, S.A.H.; Hasan, M.M.; Mazinani, A.; Sadiq, M.A.; Akhtar, M.H.; Islam, A.; Rokia, L.S. A systematic review on practical considerations, recent advances and research challenges in underwater optical wireless communication. *Int. J. Adv. Comput. Sci. Appl.* **2020**, *11*. [CrossRef]
19. Oubei, H.M.; Shen, C.; Kammoun, A.; Zedini, E.; Park, K.H.; Sun, X.; Liu, G.; Kang, C.H.; Ng, T.K.; Alouini, M.S.; et al. Light based underwater wireless communications. *Jpn. J. Appl. Phys.* **2018**, *57*, 08PA06. [CrossRef]
20. Geldard, C.T.; Thompson, J.; Popoola, W.O. An overview of underwater optical wireless channel modelling techniques. In Proceedings of the 2019 International Symposium on Electronics and Smart Devices (ISESD), IEEE, Badung, Indonesia, 8–9 October 2019; pp. 1–4.
21. Miroshnikova, N.; Petruchin, G.; Sherbakov, A. Problems of underwater optical links modeling. In Proceedings of the 2019 Systems of Signal Synchronization, Generating and Processing in Telecommunications (SYNCHROINFO), IEEE, Russia, 1–3 July 2019; pp. 1–9.
22. Liu, W.; Xu, Z.; Yang, L. SIMO detection schemes for underwater optical wireless communication under turbulence. *Photonics Res.* **2015**, *3*, 48–53. [CrossRef]
23. Baykal, Y.; Ata, Y.; Gökçe, M.C. Underwater turbulence, its effects on optical wireless communication and imaging: A review. *Opt. Laser Technol.* **2022**, *156*, 108624. [CrossRef]
24. Nguyen, C.T.; Nguyen, M.T.; Mai, V.V. Underwater optical wireless communication-based IoUT networks: MAC performance analysis and improvement. *Opt. Switch. Netw.* **2020**, *37*, 100570. [CrossRef]
25. Li, X.; Cheng, C.; Zhang, C.; Wei, Z.; Wang, L.; Fu, H.; Yang, Y. Net 4 Gb/s underwater optical wireless communication system over 2 m using a single-pixel GaN-based blue mini-LED and linear equalization. *Opt. Lett.* **2022**, *47*, 1976–1979. [CrossRef]
26. Lu, H.H.; Li, C.Y.; Lin, H.H.; Tsai, W.S.; Chu, C.A.; Chen, B.R.; Wu, C.J. An 8 m/9.6 Gbps underwater wireless optical communication system. *IEEE Photonics J.* **2016**, *8*, 1–7. [CrossRef]
27. Li, C.Y.; Lu, H.H.; Tsai, W.S.; Cheng, M.T.; Ho, C.M.; Wang, Y.C.; Yang, Z.Y.; Chen, D.Y. 16 Gb/s PAM4 UWOC system based on 488-nm LD with light injection and optoelectronic feedback techniques. *Opt. Express* **2017**, *25*, 11598–11605. [CrossRef]

28. Lu, C.; Wang, J.; Li, S.; Xu, Z. 60 m/2.5 Gbps underwater optical wireless communication with NRZ-OOK modulation and digital nonlinear equalization. In Proceedings of the 2019 Conference on Lasers and Electro-Optics (CLEO), IEEE, San Jose, CA, USA, 5–10 May 2019; pp. 1–2.
29. Liu, A.; Zhang, R.; Lin, B.; Yin, H. Multi-degree-of-freedom for underwater optical wireless communication with improved transmission performance. *J. Mar. Sci. Eng.* **2022**, *11*, 48. [CrossRef]
30. Yang, X.; Tong, Z.; Dai, Y.; Chen, X.; Zhang, H.; Zou, H.; Xu, J. 100 m full-duplex underwater wireless optical communication based on blue and green lasers and high sensitivity detectors. *Opt. Commun.* **2021**, *498*, 127261. [CrossRef]
31. Alkhasraji, J.; Tsimenidis, C. Coded OFDM over short range underwater optical wireless channels using LED. In Proceedings of the Oceans 2017-Aberdeen, IEEE, Aberdeen, UK, 19–22 June 2017; pp. 1–7.
32. Chauhan, D.S.; Kaur, G.; Kumar, D. Design of novel MIMO UOWC link using gamma—Gamma fading channel for IoUTs. *Opt. Quantum Electron.* **2022**, *54*, 512. [CrossRef]
33. Bai, J.; Li, N.; Nie, J.; Liang, X. Power Optimization of UOWC-MIMO-OFDM System Based on PSO-WF Algorithm. *J. Phys. Conf. Ser.* **2023**, *2476*, 012075. [CrossRef]
34. Huang, A.; Tao, L.; Jiang, Q. BER performance of underwater optical wireless MIMO communications with spatial modulation under weak turbulence. In Proceedings of the 2018 OCEANS-MTS/IEEE Kobe Techno-Oceans (OTO), IEEE, Kobe, Japan, 28–31 May 2018; pp. 1–5.
35. Tsai, W.S.; Lu, H.H.; Wu, H.W.; Su, C.W.; Huang, Y.C. A 30 Gb/s PAM4 underwater wireless laser transmission system with optical beam reducer/expander. *Sci. Rep.* **2019**, *9*, 8605. [CrossRef]
36. Tu, C.; Liu, W.; Jiang, W.; Xu, Z. First Demonstration of 1 Gb/s PAM4 Signal Transmission Over A 130 m Underwater Optical Wireless Communication Channel with Digital Equalization. In Proceedings of the 2021 IEEE/CIC International Conference on Communications in China (ICCC), IEEE, Xiamen, China, 28–30 July 2021; pp. 853–857.
37. Tang, S.; Dong, Y.; Zhang, X. Impulse response modeling for underwater wireless optical communication links. *IEEE Trans. Commun.* **2013**, *62*, 226–234. [CrossRef]
38. Wang, J.; Lu, C.; Li, S.; Xu, Z. 100 m/500 Mbps underwater optical wireless communication using an NRZ-OOK modulated 520 nm laser diode. *Optics Express* **2019**, *27*, 12171–12181. [CrossRef] [PubMed]
39. Hanson, F.; Radic, S. High bandwidth underwater optical communication. *Appl. Opt.* **2008**, *47*, 277–283. [CrossRef] [PubMed]
40. Oubei, H.M.; Li, C.; Park, K.H.; Ng, T.K.; Alouini, M.S.; Ooi, B.S. 2.3 Gbit/s underwater wireless optical communications using directly modulated 520 nm laser diode. *Opt. Express* **2015**, *23*, 20743–20748. [CrossRef] [PubMed]
41. Wang, P.; Li, C.; Xu, Z. A cost-efficient real-time 25 Mb/s system for LED-UOWC: Design, channel coding, FPGA implementation, and characterization. *J. Light. Technol.* **2018**, *36*, 2627–2637. [CrossRef]
42. Tsai, C.L.; Lu, Y.C.; Chang, S.H. InGaN LEDs fabricated with parallel-connected multi-pixel geometry for underwater optical communications. *Opt. Laser Technol.* **2019**, *118*, 69–74. [CrossRef]
43. Han, B.; Zhao, W.; Zheng, Y.; Meng, J.; Wang, T.; Han, Y.; Wang, W.; Su, Y.; Duan, T.; Xie, X. Experimental demonstration of quasi-omni-directional transmitter for underwater wireless optical communication based on blue LED array and freeform lens. *Opt. Commun.* **2019**, *434*, 184–190. [CrossRef]
44. Hu, S.; Mi, L.; Zhou, T.; Chen, W. 35.88 attenuation lengths and 3.32 bits/photon underwater optical wireless communication based on photon-counting receiver with 256-PPM. *Opt. Express* **2018**, *26*, 21685–21699. [CrossRef]
45. Shen, J.; Wang, J.; Yu, C.; Chen, X.; Wu, J.; Zhao, M.; Qu, F.; Xu, Z.; Han, J.; Xu, J. Single LED-based 46-m underwater wireless optical communication enabled by a multi-pixel photon counter with digital output. *Opt. Commun.* **2019**, *438*, 78–82. [CrossRef]
46. Cochenour, B.; Mullen, L.; Muth, J. A modulated pulse laser for underwater detection, ranging, imaging, and communications. In Proceedings of the Ocean Sensing and Monitoring IV. SPIE, Baltimore, MD, USA, 23–27 April 2012; Volume 8372, pp. 217–226.
47. Wang, Z.; Dong, Y.; Zhang, X.; Tang, S. Adaptive modulation schemes for underwater wireless optical communication systems. In Proceedings of the 7th International Conference on Underwater Networks & Systems, Los Angeles, CA, USA, 5–6 November 2012; pp. 1–2.
48. Kong, M.; Chen, Y.; Sarwar, R.; Sun, B.; Xu, Z.; Han, J.; Chen, J.; Qin, H.; Xu, J. Underwater wireless optical communication using an arrayed transmitter/receiver and optical superimposition-based PAM-4 signal. *Opt. Express* **2018**, *26*, 3087–3097. [CrossRef]
49. Zhuang, B.; Li, C.; Wu, N.; Xu, Z. First demonstration of 400 Mb/s PAM4 signal transmission over 10-meter underwater channel using a blue LED and a digital linear pre-equalizer. In Proceedings of the 2017 Conference on Lasers and Electro-Optics (CLEO), IEEE, San Jose, CA, USA, 14–19 May 2017; pp. 1–2.
50. Chi, N.; Zhao, Y.; Shi, M.; Zou, P.; Lu, X. Gaussian kernel-aided deep neural network equalizer utilized in underwater PAM8 visible light communication system. *Opt. Express* **2018**, *26*, 26700–26712. [CrossRef]
51. Wang, J.; Yang, X.; Lv, W.; Yu, C.; Wu, J.; Zhao, M.; Qu, F.; Xu, Z.; Han, J.; Xu, J. Underwater wireless optical communication based on multi-pixel photon counter and OFDM modulation. *Opt. Commun.* **2019**, *451*, 181–185. [CrossRef]
52. Huang, X.H.; Li, C.Y.; Lu, H.H.; Su, C.W.; Wu, Y.R.; Wang, Z.H.; Chen, Y.N. 6-m/10-Gbps underwater wireless red-light laser transmission system. *Opt. Eng.* **2018**, *57*, 066110. [CrossRef]
53. Wu, T.C.; Chi, Y.C.; Wang, H.Y.; Tsai, C.T.; Lin, G.R. Blue laser diode enables underwater communication at 12.4 Gbps. *Sci. Rep.* **2017**, *7*, 40480. [CrossRef] [PubMed]
54. Wang, F.; Liu, Y.; Shi, M.; Chen, H.; Chi, N. 3.075 Gb/s underwater visible light communication utilizing hardware pre-equalizer with multiple feature points. *Opt. Eng.* **2019**, *58*, 056117. [CrossRef]

55. Li, J.; Wang, F.; Zhao, M.; Jiang, F.; Chi, N. Large-coverage underwater visible light communication system based on blue LED employing equal gain combining with integrated PIN array reception. *Appl. Opt.* **2019**, *58*, 383–388. [CrossRef]

56. Wang, F.; Liu, Y.; Jiang, F.; Chi, N. High speed underwater visible light communication system based on LED employing maximum ratio combination with multi-PIN reception. *Opt. Commun.* **2018**, *425*, 106–112. [CrossRef]

57. Zhang, Z.; Lai, Y.; Lv, J.; Liu, P.; Teng, D.; Wang, G.; Liu, L. Over 700 MHz–3 dB bandwidth UOWC system based on blue HV-LED with T-bridge pre-equalizer. *IEEE Photonics J.* **2019**, *11*, 1–12. [CrossRef]

58. Arvanitakis, G.N.; Bian, R.; McKendry, J.J.; Cheng, C.; Xie, E.; He, X.; Yang, G.; Islim, M.S.; Purwita, A.A.; Gu, E.; et al. Gb/s underwater wireless optical communications using series-connected GaN micro-LED arrays. *IEEE Photonics J.* **2019**, *12*, 1–10. [CrossRef]

59. Hong, X.; Fei, C.; Zhang, G.; Du, J.; He, S. Discrete multitone transmission for underwater optical wireless communication system using probabilistic constellation shaping to approach channel capacity limit. *Opt. Lett.* **2019**, *44*, 558–561. [CrossRef]

60. Li, C.Y.; Lu, H.H.; Tsai, W.S.; Wang, Z.H.; Hung, C.W.; Su, C.W.; Lu, Y.F. A 5 m/25 Gbps underwater wireless optical communication system. *IEEE Photonics J.* **2018**, *10*, 1–9. [CrossRef]

61. Oubei, H.M.; Duran, J.R.; Janjua, B.; Wang, H.Y.; Tsai, C.T.; Chi, Y.C.; Ng, T.K.; Kuo, H.C.; He, J.H.; Alouini, M.S.; et al. 4.8 Gbit/s 16-QAM-OFDM transmission based on compact 450-nm laser for underwater wireless optical communication. *Opt. Express* **2015**, *23*, 23302–23309. [CrossRef]

62. Fei, C.; Hong, X.; Zhang, G.; Du, J.; Gong, Y.; Evans, J.; He, S. 16.6 Gbps data rate for underwater wireless optical transmission with single laser diode achieved with discrete multi-tone and post nonlinear equalization. *Opt. Express* **2018**, *26*, 34060–34069. [CrossRef]

63. Chen, X.; Lyu, W.; Zhang, Z.; Zhao, J.; Xu, J. 56-m/3.31-Gbps underwater wireless optical communication employing Nyquist single carrier frequency domain equalization with noise prediction. *Opt. Express* **2020**, *28*, 23784–23795. [CrossRef]

64. Fei, C.; Zhang, J.; Zhang, G.; Wu, Y.; Hong, X.; He, S. Demonstration of 15-M 7.33-Gb/s 450-nm underwater wireless optical discrete multitone transmission using post nonlinear equalization. *J. Light. Technol.* **2018**, *36*, 728–734. [CrossRef]

65. Pandey, P.; Agrawal, M. High speed and Long range underwater optical wireless communication. In Proceedings of the Global Oceans 2020: Singapore–US Gulf Coast, IEEE, Biloxi, MS, USA, 5–30 October 2020; pp. 1–10.

66. Kaeib, F.; Alshawish, O.A.; Altayf, S.A.; Gamoudi, M.A. Designing and Analysis of Underwater Optical Wireless communication system. In Proceedings of the 2022 IEEE 2nd International Maghreb Meeting of the Conference on Sciences and Techniques of Automatic Control and Computer Engineering (MI-STA), IEEE, Sabratha, Libya, 23–25 May 2022; pp. 441–446.

67. Pandey, P.; Matta, G.; Aggarwal, M. Performance comparisons between Avalanche and PIN photodetectors for use in underwater optical wireless communication systems. In Proceedings of the OCEANS 2021: San Diego–Porto, IEEE, San Diego, CA, USA, 20–23 September 2021; pp. 1–8.

68. Mobley, C. *Light and Water: Radiative Transfer in Natural Water*; Academic Press: San Diego, CA, USA, 1994.

69. Mobley, C.D.; Gentili, B.; Gordon, H.R.; Jin, Z.; Kattawar, G.W.; Morel, A.; Reinersman, P.; Stamnes, K.; Stavn, R.H. Comparison of numerical models for computing underwater light fields. *Appl. Opt.* **1993**, *32*, 7484–7504. [CrossRef]

70. Duntley, S.Q. Light in the sea. *JOSA* **1963**, *53*, 214–233. [CrossRef]

71. Haltrin, V.I. Chlorophyll-based model of seawater optical properties. *Appl. Opt. (2004)* **1999**, *38*, 6826–6832. [CrossRef] [PubMed]

72. Shen, C.; Guo, Y.; Oubei, H.M.; Ng, T.K.; Liu, G.; Park, K.H.; Ho, K.T.; Alouini, M.S.; Ooi, B.S. 20-meter underwater wireless optical communication link with 1.5 Gbps data rate. *Opt. Express* **2016**, *24*, 25502–25509. [CrossRef]

73. Johnson, L.J.; Green, R.J.; Leeson, M.S. Underwater optical wireless communications: Depth dependent variations in attenuation. *Appl. Opt.* **2013**, *52*, 7867–7873. [CrossRef]

74. Yap, Y.; Jasman, F.; Marcus, T. Impact of chlorophyll concentration on underwater optical wireless communications. In Proceedings of the 2018 7th International Conference on Computer and Communication Engineering (ICCCE), IEEE, Kuala Lumpur, Malaysia, 19–20 September 2018; pp. 1–6.

75. Liu, W.; Zou, D.; Xu, Z.; Yu, J. Non-line-of-sight scattering channel modeling for underwater optical wireless communication. In Proceedings of the 2015 IEEE International Conference on Cyber Technology in Automation, Control, and Intelligent Systems (CYBER), IEEE, Shenyang, China, 8–12 June 2015; pp. 1265–1268.

76. Chen, D.; Wang, J.; Li, S.; Xu, Z. Effects of air bubbles on underwater optical wireless communication. *Chin. Opt. Lett.* **2019**, *17*, 100008. [CrossRef]

77. Shin, M.; Park, K.H.; Alouini, M.S. Statistical modeling of the impact of underwater bubbles on an optical wireless channel. *IEEE Open J. Commun. Soc.* **2020**, *1*, 808–818. [CrossRef]

78. Oubei, H.M.; ElAfandy, R.T.; Park, K.H.; Ng, T.K.; Alouini, M.S.; Ooi, B.S. Performance evaluation of underwater wireless optical communications links in the presence of different air bubble populations. *IEEE Photonics J.* **2017**, *9*, 1–9. [CrossRef]

79. Tsai, W.S.; Li, C.Y.; Lu, H.H.; Lu, Y.F.; Tu, S.C.; Huang, Y.C. 256 Gb/s four-channel SDM-based PAM4 FSO-UWOC convergent system. *IEEE Photonics J.* **2019**, *11*, 1–8. [CrossRef]

80. Fang, C.; Li, S.; Wang, K. Accurate Underwater Optical Wireless Communication Model With Both Line-of-Sight and Non-Line-of-Sight Channels. *IEEE Photonics J.* **2022**, *14*, 1–12. [CrossRef]

81. Mahapatra, S.K.; Varshney, S.K. Performance of the Reed-Solomon-coded underwater optical wireless communication system with orientation-based solar light noise. *JOSA A* **2022**, *39*, 1236–1245. [CrossRef]

82. Fang, C.; Li, S.; Wang, K. Accurate underwater optical wireless communication (UOWC) channel model. In Proceedings of the Asia Communications and Photonics Conference. Optical Society of America, Shanghai, China, 24–27 October 2021; pp. T4A–143.

83. Huang, Y.F.; Tsai, C.T.; Chi, Y.C.; Huang, D.W.; Lin, G.R. Filtered multicarrier OFDM encoding on blue laser diode for 14.8-Gbps seawater transmission. *J. Light. Technol.* **2017**, *36*, 1739–1745. [CrossRef]

84. Hu, F.; Li, G.; Zou, P.; Hu, J.; Chen, S.; Liu, Q.; Zhang, J.; Jiang, F.; Wang, S.; Chi, N. 20.09-Gbit/s underwater WDM-VLC transmission based on a single Si/GaAs-substrate multichromatic LED array chip. In Proceedings of the 2020 Optical fiber Communications Conference and Exhibition (OFC), IEEE, San Diego, CA, USA, 8–12 March 2020; pp. 1–3.

85. Zhou, Y.; Zhu, X.; Hu, F.; Shi, J.; Wang, F.; Zou, P.; Liu, J.; Jiang, F.; Chi, N. Common-anode LED on a Si substrate for beyond 15 Gbit/s underwater visible light communication. *Photonics Res.* **2019**, *7*, 1019–1029. [CrossRef]

86. Chen, Y.; Kong, M.; Ali, T.; Wang, J.; Sarwar, R.; Han, J.; Guo, C.; Sun, B.; Deng, N.; Xu, J. 26 m/5.5 Gbps air-water optical wireless communication based on an OFDM-modulated 520-nm laser diode. *Opt. Express* **2017**, *25*, 14760–14765. [CrossRef] [PubMed]

87. Liu, X.; Yi, S.; Zhou, X.; Fang, Z.; Qiu, Z.J.; Hu, L.; Cong, C.; Zheng, L.; Liu, R.; Tian, P. 34.5 m underwater optical wireless communication with 2.70 Gbps data rate based on a green laser diode with NRZ-OOK modulation. *Opt. Express* **2017**, *25*, 27937–27947. [CrossRef] [PubMed]

88. Tsai, C.L.; Lu, Y.C.; Yu, C.M.; Chen, Y.J. Epitaxial growth of InGaN multiple-quantum-well LEDs with improved characteristics and their application in underwater optical wireless communications. *IEEE Trans. Electron Devices* **2018**, *65*, 4346–4352. [CrossRef]

89. Cossu, G.; Sturniolo, A.; Messa, A.; Scaradozzi, D.; Ciaramella, E. Full-fledged 10Base-T ethernet underwater optical wireless communication system. *IEEE J. Sel. Areas Commun.* **2017**, *36*, 194–202. [CrossRef]

90. Liu, T.; Zhang, H.; Zhang, Y.; Song, J. Experimental demonstration of LED based underwater wireless optical communication. In Proceedings of the 2017 4th International Conference on Information Science and Control Engineering (ICISCE), IEEE, Changsha, China, 21–23 July 2017; pp. 1501–1504.

91. Hamza, T.; Khalighi, M.A.; Bourennane, S.; Léon, P.; Opderbecke, J. Investigation of solar noise impact on the performance of underwater wireless optical communication links. *Opt. Express* **2016**, *24*, 25832–25845. [CrossRef]

92. Xing, F.; Yin, H. Performance Analysis for Underwater Cooperative Optical Wireless Communications in the Presence of Solar Radiation Noise. In Proceedings of the 2019 IEEE International Conference on Signal Processing, Communications and Computing (ICSPCC), IEEE, Dalian, China, 20–22 September 2019; pp. 1–6.

93. Xing, F.; Yin, H.; Ji, X.; Leung, V.C.M. Joint Relay Selection and Power Allocation for Underwater Cooperative Optical Wireless Networks. *IEEE Trans. Wirel. Commun.* **2020**, *19*, 251–264. [CrossRef]

94. Alley, D.; Mullen, L.; Laux, A. Compact, dual-wavelength, non-line-of-sight (nlos) underwater imager. In Proceedings of the OCEANS'11 MTS/IEEE KONA, IEEE, Waikoloa, HI, USA, 19–22 September 2011; pp. 1–5.

95. Sun, X.; Kong, M.; Alkhazragi, O.; Shen, C.; Ooi, E.N.; Zhang, X.; Buttner, U.; Ng, T.K.; Ooi, B.S. Non-line-of-sight methodology for high-speed wireless optical communication in highly turbid water. *Opt. Commun.* **2020**, *461*, 125264. [CrossRef]

96. Sun, X.; Cai, W.; Alkhazragi, O.; Ooi, E.N.; He, H.; Chaaban, A.; Shen, C.; Oubei, H.M.; Khan, M.Z.M.; Ng, T.K.; et al. 375-nm ultraviolet-laser based non-line-of-sight underwater optical communication. *Opt. Express* **2018**, *26*, 12870–12877. [CrossRef]

97. Haltrin, V.I. One-parameter two-term Henyey-Greenstein phase function for light scattering in seawater. *Appl. Opt.* **2002**, *41*, 1022–1028. [CrossRef]

98. Toublanc, D. Henyey–Greenstein and Mie phase functions in Monte Carlo radiative transfer computations. *Appl. Opt.* **1996**, *35*, 3270–3274. [CrossRef] [PubMed]

99. Gabriel, C.; Khalighi, M.A.; Bourennane, S.; Léon, P.; Rigaud, V. Monte-Carlo-based channel characterization for underwater optical communication systems. *J. Opt. Commun. Netw.* **2013**, *5*, 1–12. [CrossRef]

100. Nguyen, H.; Choi, J.H.; Kang, M.; Ghassemlooy, Z.; Kim, D.; Lim, S.K.; Kang, T.G.; Lee, C.G. A MATLAB-based simulation program for indoor visible light communication system. In Proceedings of the 2010 7th International Symposium on Communication Systems, Networks & Digital Signal Processing (CSNDSP 2010), IEEE, Newcastle Upon Tyne, UK, 21–23 July 2010; pp. 537–541.

101. Fang, C.; Li, S.; Wang, K. Investigation of wavy surface impact on non-line-of-sight underwater optical wireless communication. In Proceedings of the 2022 IEEE Photonics Conference (IPC), IEEE, Vancouver, BC, Canada, 13–17 November 2022; pp. 1–2.

102. Tao, L.; Hongming, Z.; Jian, S. Distribution of Arriving Angle of Signal in Underwater Scattering Channel. *Chin. J. Lasers* **2018**, *45*, 306003.

103. Jaruwatanadilok, S. Underwater Wireless Optical Communication Channel Modeling and Performance Evaluation using Vector Radiative Transfer Theory. *IEEE J. Sel. Areas Commun.* **2008**, *26*, 1620–1627. [CrossRef]

104. Sait, M.; Trichili, A.; Alkhazragi, O.; Alshaibaini, S.; Ng, T.K.; Alouini, M.S.; Ooi, B.S. Dual-wavelength luminescent fibers receiver for wide field-of-view, Gb/s underwater optical wireless communication. *Opt. Express* **2021**, *29*, 38014–38026. [CrossRef]

105. Chen, J.; Jiang, M. Joint Blind Channel Estimation, Channel Equalization, and Data Detection for Underwater Visible Light Communication Systems. *IEEE Wirel. Commun. Lett.* **2021**, *10*, 2664–2668. [CrossRef]

106. Li, Y.; Liang, H.; Gao, C.; Miao, M.; Li, X. Temporal dispersion compensation for turbid underwater optical wireless communication links. *Opt. Commun.* **2019**, *435*, 355–361. [CrossRef]

107. Mahmutoglu, Y.; Turk, K.; Tugcu, E. Particle swarm optimization algorithm based decision feedback equalizer for underwater acoustic communication. In Proceedings of the 2016 39th International Conference on Telecommunications and Signal Processing (TSP), IEEE, Vienna, Austria, 27–29 June 2016; pp. 153–156.

108. Pinho, V.M.; Chaves, R.S.; Campos, M.L. On Equalization Performance in Underwater Acoustic Communication. In Proceedings of the XXXVI Simpósio Brasileiro de Telecomunicaçoes e Processamento de Sinais, SBrT, Campina Grande, Brazil, 16–19 September 2018.

109. Song, H. Bidirectional equalization for underwater acoustic communication. *J. Acoust. Soc. Am.* **2012**, *131*, EL342–EL347. [CrossRef] [PubMed]

110. Xie, Y.; Wang, Y.; Kandeepan, S.; Wang, K. Machine learning applications for short reach optical communication. *Photonics* **2022**, *9*, 30. [CrossRef]

111. Zhao, Y.; Zou, P.; Chi, N. 3.2 Gbps underwater visible light communication system utilizing dual-branch multi-layer perceptron based post-equalizer. *Opt. Commun.* **2020**, *460*, 125197. [CrossRef]

112. Chen, H.; Zhao, Y.; Hu, F.; Chi, N. Nonlinear resilient learning method based on joint time-frequency image analysis in underwater visible light communication. *IEEE Photonics J.* **2020**, *12*, 1–10. [CrossRef]

113. Chen, H.; Niu, W.; Zhao, Y.; Zhang, J.; Chi, N.; Li, Z. Adaptive deep-learning equalizer based on constellation partitioning scheme with reduced computational complexity in UVLC system. *Opt. Express* **2021**, *29*, 21773–21782. [CrossRef]

114. Zhao, Y.; Yu, S.; Chi, N. Transfer Learning–Based Artificial Neural Networks Post-Equalizers for Underwater Visible Light Communication. *Front. Commun. Netw.* **2021**, *2*, 658330. [CrossRef]

115. Du, Z.; Ge, W.; Cai, C.; Wang, H.; Song, G.; Xiong, J.; Li, Y.; Zhang, Z.; Xu, J. 90-m/660-Mbps Underwater Wireless Optical Communication Enabled by Interleaved Single-Carrier FDM Scheme Combined with Sparse Weight-Initiated DNN Equalizer. *J. Light. Technol.* **2023**. [CrossRef]

116. Cui, X.; Yin, X.; Chang, H.; Liao, H.; Chen, X.; Xin, X.; Wang, Y. Experimental study of machine-learning-based orbital angular momentum shift keying decoders in optical underwater channels. *Opt. Commun.* **2019**, *452*, 116–123. [CrossRef]

117. Avramov-Zamurovic, S.; Nelson, C.; Esposito, J.M. Effects of underwater optical turbulence on light carrying orbital angular momentum and its classification using machine learning. *J. Mod. Opt.* **2021**, *68*, 1041–1053. [CrossRef]

118. Romdhane, I.; Kaddoum, G. A Reinforcement-Learning-Based Beam Adaptation for Underwater Optical Wireless Communications. *IEEE Internet Things J.* **2022**, *9*, 20270–20281. [CrossRef]

119. Weng, Y.; Pajarinen, J.; Akrour, R.; Matsuda, T.; Peters, J.; Maki, T. Reinforcement learning based underwater wireless optical communication alignment for autonomous underwater vehicles. *IEEE J. Ocean. Eng.* **2022**, *47*, 1231–1245. [CrossRef]

120. Wang, J.; Luo, H.; Ruby, R.; Liu, J.; Guo, K.; Wu, K. Reliable Water-Air Direct Wireless Communication: Kalman Filter-Assisted Deep Reinforcement Learning Approach. In Proceedings of the 2022 IEEE 47th Conference on Local Computer Networks (LCN), IEEE, Edmonton, AB, Canada, 26–29 September 2022; pp. 233–238.

121. Li, X.; Hu, X.; Li, W.; Hu, H. A multi-agent reinforcement learning routing protocol for underwater optical sensor networks. In Proceedings of the ICC 2019-2019 IEEE International Conference on Communications (ICC), IEEE, Shanghai, China, 20–24 May 2019; pp. 1–7.

122. Li, X.; Hu, X.; Zhang, R.; Yang, L. Routing protocol design for underwater optical wireless sensor networks: A multiagent reinforcement learning approach. *IEEE Internet Things J.* **2020**, *7*, 9805–9818. [CrossRef]

123. Khalighi, M.; Akhouayri, H.; Hranilovic, S. SiPM-based Underwater Wireless Optical Communication Using Pulse-Amplitude Modulation. *IEEE J. Ocean. Eng.* **2020**, *45*, 1611–1621. [CrossRef]

124. Li, C.; Wang, B.; Wang, P.; Xu, Z.; Yang, Q.; Yu, S. Generation and transmission of 745 Mb/s ofdm signal using a single commercial blue LED and an analog post-equalizer for underwater optical wireless communications. In Proceedings of the 2016 Asia Communications and Photonics Conference (ACP), IEEE, Wuhan, China, 2–5 November 2016; pp. 1–3.

125. Cheng, C.; Li, X.; Yang, Y. Net 4.12 Gb/s UOWC System over 2 m Using a Single-Pixel Blue Mini-LED and a Simplified Third-Order Volterra Equalizer. In Signal Processing in Photonic Communications; Optica Publishing Group: Maastricht, The Netherlands, 2022; p. SpTh3H–5.

126. Yu, M.; Geldard, C.T.; Popoola, W.O. Comparison of CAP and OFDM Modulation for LED-based Underwater Optical Wireless Communications. In Proceedings of the 2022 International Conference on Broadband Communications for Next Generation Networks and Multimedia Applications (CoBCom), IEEE, Graz, Austria, 12–14 July 2022; pp. 1–6.

127. Yang, H.; Yan, Q.; Wang, M.; Wang, Y.; Li, P.; Wang, W. Synchronous Clock Recovery of Photon-Counting Underwater Optical Wireless Communication Based on Deep Learning. *Photonics* **2022**, *9*, 884. [CrossRef]

128. Chi, N.; Jia, J.; Hu, F.; Zhao, Y.; Zou, P. Challenges and prospects of machine learning in visible light communication. *J. Commun. Inf. Netw.* **2020**, *5*, 302–309. [CrossRef]

129. Xu, Z.; Chen, T.; Qin, G.; Chi, N. Applications of Machine Learning in Visible Light Communication. In Proceedings of the 2021 18th China International Forum on Solid State Lighting & 2021 7th International Forum on Wide Bandgap Semiconductors (SSLChina: IFWS), IEEE, Shenzhen, China, 6–8 December 2021; pp. 198–201.

Review

A Review on Image Sensor Communication and Its Applications to Vehicles

Ruiyi Huang [1],* and Takaya Yamazato [2]

[1]   Department of Information and Communication Engineering, Nagoya University, Nagoya 464-8603, Japan
[2]   Institute of Liberal Arts and Sciences, Nagoya University, Nagoya 464-8603, Japan
*   Correspondence: huang.ruiyi.c6@s.mail.nagoya-u.ac.jp

Abstract: Image sensor communication (ISC), also known as optical camera communication, is a form of visible light communication that utilizes image sensors rather than a single photodiode, for data reception. ISC offers spatial separation properties and robustness to ambient noise, making it suitable for outdoor applications such as intelligent transportation systems (ITSs). This review analyzes the research trends in ISC, specifically concerning its application in ITSs. Our focus is on various ISC receivers, including rolling shutter cameras, global shutter high-speed cameras, optical communication image sensors, and event cameras. We analyze how each of these receivers is being utilized in ISC vehicular applications. In addition, we highlight the use of ISC in range estimation techniques and the ability to achieve simultaneous communication and range estimation. By examining these topics, we aim to provide a comprehensive overview of the role of ISC technology in ITSs and its potential for future development.

Keywords: image sensor communication; optical camera communication; intelligent transportation system; visible light communication; rolling shutter camera; high-speed camera; visible light positioning

1. Introduction

Visible light communication (VLC) is a wireless communication method that utilizes visible light for information transmission [1–4]. The luminance of the VLC light sources can be modulated extremely fast to make the switching of light imperceptible to the human eye to achieve high-speed data transmission [4]. In contrast to traditional wireless communications that use electromagnetic waves, wireless communication using visible light has a much wider wavelength range and has been considered a technique to address the issues caused by the growing spectrum need for wireless communication [5].

The origins of VLC can be traced back to the 7th century BC, when people in ancient China used smoke and fire to transmit information about their enemies across long distances. Over two millennia later, in the 1880s, Alexander Graham Bell made his groundbreaking attempt to use machinery to implement VLC by transmitting voice information through sunlight, carried by electricity [6]. Unfortunately, however, this technology did not reach early commercialization due to its limitations in communication distance and the inevitable effects of weather and obstacles such as rain. It was not until the 1990s that the invention of a high-brightness light-emitting diode (LED) [7] revolutionized the lighting industry, opening up new possibilities for VLC. The high response and fast switching speed of LEDs allow using visible light to convey high-speed data [4,8]. In 2004, T. Komine and M. Nagakawa proposed an indoor VLC system using white LED lights for both illumination and optical wireless communication [4]. They considered the interference and reflection of the multiple light sources installed in a room and demonstrated its potential for high-speed data transmission of around 10 Gbps. This paper has been highly influential in the field of VLC, leading to widespread research interest and study in VLC. VLC technology is also applied outdoors, for instance, in traffic lights and streetlights. In 1999, G. Pang et al. showed

Citation: Huang, R.; Yamazato, T. A Review on Image Sensor Communication and Its Applications to Vehicles. *Photonics* **2023**, *10*, 617. https://doi.org/10.3390/photonics10060617

Received: 25 March 2023
Revised: 25 April 2023
Accepted: 18 May 2023
Published: 26 May 2023

the potential of using a traffic light as a communication device [9]. In [9], high-brightness LED lights were modulated and encoded with audio messages. The receiver was designed to demodulate the optically transmitted audio messages using a photodiode (PD) to convert the light to electrical signals through direct detection. However, there were many constraints to using a PD in outdoor environments due to the strong ambient noise. In 2001, an image sensor was introduced as a new type of receiver for VLC, which was more robust to outdoor ambient noises [10]. An image sensor consists of an enormous array of PDs. This structure of an image sensor gives it the ability to spatially separate the light sources, also resulting in being robust to outdoor ambient noise and being able to easily track moving vehicles. Nowadays, we have coined the term image sensor communication (ISC) for the VLC that uses image sensors as receivers [11–13]. ISC can also be referred to as optical camera communication (OCC) [14,15]. The primary distinction between ISC and OCC is that the term "camera" in OCC encompasses both the lens and circuitry for an image sensor. Additionally, these two terms are named by different institutions. In this review, we will discuss ISC and its outdoor applications, such as intelligent transportation systems (ITSs), which are also known as ITS-ISC.

For ITSs, low latency and accurate localization are crucial factors [16–18]. Low latency is necessary for real-time communication between vehicles and their surrounding transportation elements to efficiently coordinate traffic flow and avoid accidents timely. Accurate localization is important for car navigation and collision avoidance, as well as for providing location-based services to passengers. ISC can be an applicable technology to realize low latency and accurate localization simultaneously. On the one hand, ISC can achieve low latency by exploiting the high-speed capabilities of digital image sensors, for instance, using high-speed cameras as the ISC receiver [19] or utilizing the rolling shutter effect [20]. The high-speed transmission allows real-time data transfer between vehicles, road infrastructure, and other elements in the ITS. On the other hand, ISC systems can achieve high localization accuracy by exploiting the high-brightness properties of visible light. Even in complex traffic environments, such as tunnels or intersections, LED transmitters can be recognized from complicated backgrounds using algorithms in ISC. In contrast, localization using conventional wireless technologies can be inaccurate due to signal reflection, multi-path propagation, or environmental interference [21]. Furthermore, many existing automotive applications are now equipped with image sensors, such as driver recorders and autonomous driving sensors [22]. Meanwhile, the road and vehicles are equipped with LED lights. Therefore, ISC that utilizes both image sensors and LEDs can be a promising technology for ITS.

We will provide detailed discussions of the ISC studies in the remainder of this review paper. In Section 2, we present the research trend of VLC based on the statistical data collected on Scopus. We analyze two receiver types of VLC: PDs and image sensors. Then, we discuss the advantages of using image sensors in outdoor environments. It then leads to the research status of ITS-ISC. We also analyze the constitutions of transmitter and receiver types of ISC. In Section 3, we explain the concepts necessary for ITS-ISC, including vehicle-to-vehicle (V2V), vehicle-to-infrastructure (V2I), infrastructure-to-vehicle (I2V), and vehicle-to-everything (V2X) communications. We then describe the architecture of the ITS-ISC system. In addition, the advantages and limitations of image sensor receivers are discussed. In Section 4, a comprehensive review of the receiver types of ISC is provided, including rolling shutter cameras, global shutter high-speed cameras, optical communication image sensors (OCIs), and event cameras (dynamic vision sensors). In Section 5, we illustrate the basic mechanism, challenges, and solutions of range estimation using LED transmitters and image sensor receivers. Finally, Section 6 presents the conclusion.

2. Research Trend

2.1. Research Trend of VLC

Figure 1a shows the annual publication numbers of VLC literature published before 2023 searched in Scopus. We found a total of 9345 publications. From Figure 1a, it can be seen that the number of VLC publications has greatly increased since around 2010 but declined after 2019. Figure 1b shows the results of classifying these research publications by document type. According to Figure 1b, conference papers constitute 50.3% of the total. This might be the reason behind the decline in publication numbers from 2019 when the COVID-19 pandemic started making it challenging for people to attend international conferences.

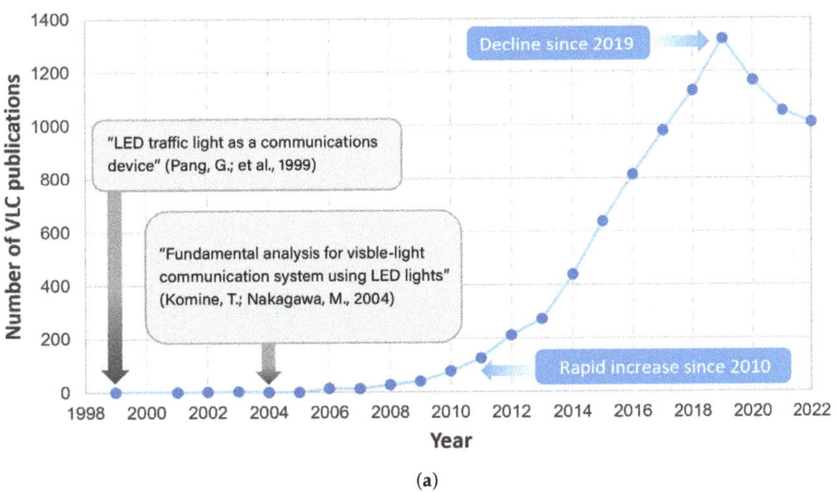

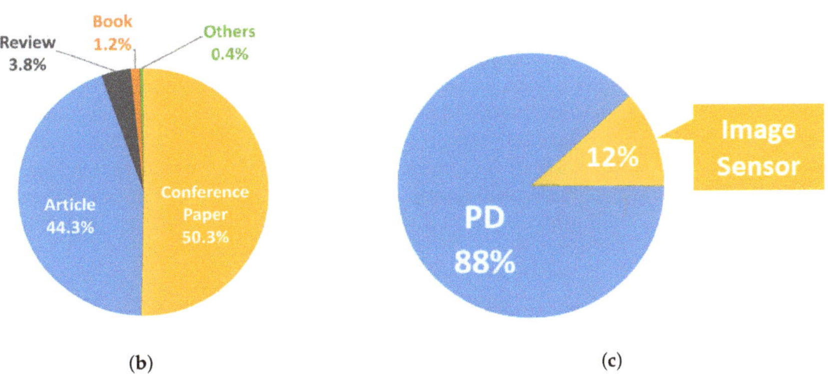

Figure 1. Literaturesurveys on visible light communications (VLC). The survey was performed by searching the Scopus database for documents with "visible light communication" in the title, keywords, and abstracts, and publication year before 2023. These data are based on search records in February 2023. (**a**) Annual growth in the number of research publications related to VLC [4,9]. (**b**) Analysis of document types in research publications of VLC. (**c**) Categorizing receiver types in research publications of VLC.

Figure 1c describes the percentage of publications using PD or image sensor receivers in VLC. 88% of the publications used PDs, and the number was 8194 publications. Only 12% of the publications used image sensors, with 1151 publications. Therefore, it can be concluded that most works utilize PDs as the receiver. There are many advantages to using PDs, such as rapid response and low cost. Due to the high-speed response, the method of using a PD as a receiver device has achieved communication at the Gbps level [5,23,24]. In contrast, image sensors communicate much slower than PDs due to the frame rate limitation. Based on our survey, research on PDs is being conducted mainly for indoor use purposes. For example, illumination optical communication including Light-Fidelity (LiFi) [5] and indoor visible light positioning [25]. In terms of image sensor research, its proportion in indoor use is lower compared to the research on PDs.

Image sensors are more suitable for outdoor applications than PDs. The main reasons are that image sensors are more robust to solar noise than PDs and are simpler to track moving vehicles. When used outdoors, especially in an ITS environment, the effect of background light noise, such as sunlight and other light sources, cannot be ignored. Therefore, for a PD, the viewing angle must be narrowed to ensure an adequate signal-to-noise ratio. At the same time, mechanical manipulation is required to orient the PD toward the direction of the transmitter. However, when the car is driving at high speed and changing its direction, it is difficult for a PD with a narrowed viewing angle to swiftly adjust the direction of its optical axis. On the contrary, an image sensor can track moving vehicles with little adjustments to its direction since it has a large number of pixels. Moreover, an image sensor is capable of spatially separating light sources and selecting only the pixels that contain the desired transmitted signal while filtering out other pixels that may contain background light noise.

Additionally, there are ongoing efforts to establish international standards for VLC. In 2011, the IEEE 802.15.7 standard for short-range wireless optical communication was established [26]. It was revised later to include ISC and new PHY modes in IEEE 802.15.7a task group [13]. In July 2020, a new amendment to IEEE 802.15.7a was proposed, which adds a deep learning mechanism for ISC. This amendment is expected to be published in September 2023. Furthermore, in December 2017, a proposal for a new ISO standard for localized communication using ISC in ITSs was approved. The proposal suggested a new communication interface called "ITS-OCC", which adopted ISC profile, communication adaptation layer, and management adaptation entity from IEEE 802.15.7:2018, ISO 14296:2016, and ISO 22738:2020, respectively [27–29]. The ISO 22738:2020 standard was published in July 2020 and is expected to facilitate the development of OCC-based ITS applications [29]. Finally, regarding the performance comparison between IEEE 802.15.7 for optical wireless communication and IEEE 802.15.7a for OCC, the addition of image sensor receivers and deep learning mechanisms in IEEE 802.15.7a OCC is expected to improve the accuracy and robustness of communication performance and range estimation in challenging environments. It may also allow for the development of new applications that were previously not possible with traditional range estimation methods. Further research and development in this area may lead to improved performance and new use cases for ISC.

2.2. Research Status of ISC

Figure 2 shows the growth in the number of ISC literature published before 2023 searched in Scopus. We found a total of 1151 papers. There is a significant increase in the quantity of literature starting in 2013; the growth in the number of papers levels off after 2019. Roughly 24% of this literature is applied to vehicles.

We selected papers with the literature type of "article" and "conference paper", and classified these publications by applications, transmitter types, and receiver types. Note that the publications were categorized manually by the authors; subjective interpretation and potential errors could be introduced into the categorization process.

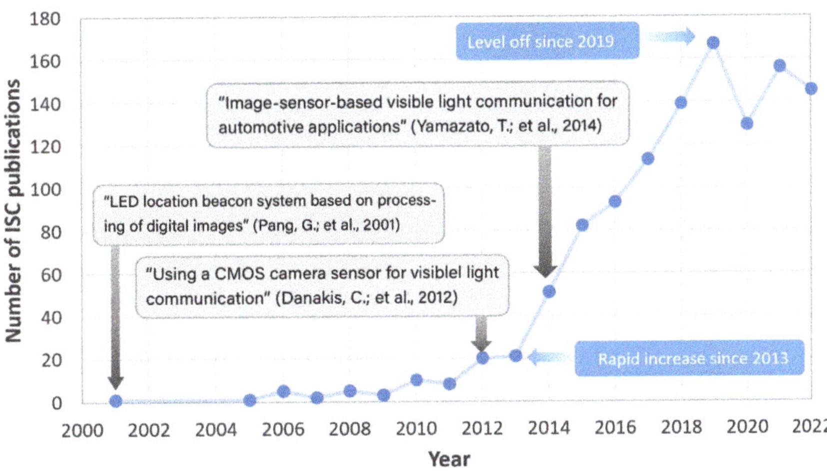

Figure 2. A literature survey of image sensor communication (ISC) [10,19,20]. The survey was performed by searching the Scopus database for publications with "image sensor communication" or "optical camera communication" or "screen camera communication" or "display camera communica-tion" or {"visible light communication" and "camera"} or {"visible light communication" and "image sensor"} in the title, keywords, and abstracts, and publication year before 2023. These data are based on search records in February 2023.

Figure 3 shows the proportions of publication numbers of different ISC applications. Communication application accounts for 78.7% of the whole. Here, communication refers to being able to transmit information such as text, image, voice, and video through optical signals. Papers with communication applications cover a wide range of subjects, such as bit-error-rate (BER) measurement, communication model design, and other related areas. In the context of an ITS, communication enables safety and efficiency applications such as collision avoidance, traffic flow optimization, and emergency vehicle notifications. Ranging, ranging application, also considered as localization or positioning, accounts for 13.9% of the whole. Ranging relies on computer vision techniques and can be used for applications such as obstacle avoidance and navigation. By using ISC, high-accuracy positioning can be achieved within two milliseconds, which is faster than light detection and ranging (LiDAR) or global positioning system (GPS) [30]. Additionally, detection and tracking comprise 3.6% of all ISC papers. It refers to extracting LEDs from the received image streams. Inaccurate detection or tracking can lead to incorrect data decoding because it is conducted prior to communication and ranging. The advantage of ISC-based communication and ranging technology is that the use of LEDs provides communication and navigation services along with illumination, thus minimizing the need for additional power. It reduces front-end costs, eliminating the need for additional installation and configuration of signal access points, and reducing the cost and complexity of the communication and navigation system.

Figure 4 shows the ratio of the different transmitters and receivers in the ISC. However, Figure 4b may have some level of inaccuracy. This is because some authors did not specify the type of image sensor used in their articles, and these kinds of articles are classified as "others". Additionally, due to time constraints, the "others" category was not fully categorized. We conclude that since the rolling shutter camera is the majority of the receivers, some instances of rolling shutter cameras may have been categorized as "others".

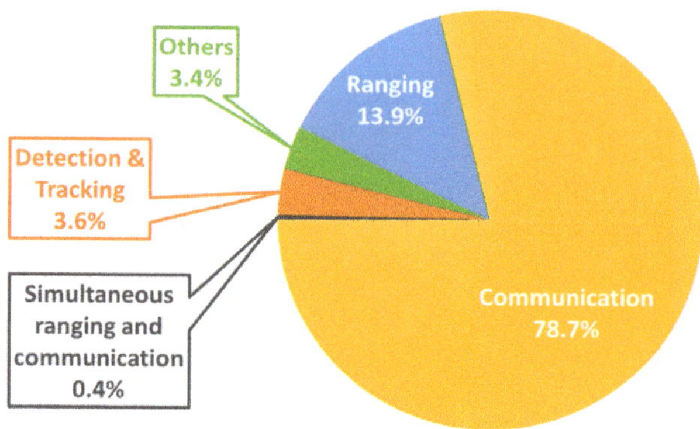

Figure 3. Categorizing image sensor communication research publications by applications of communication, ranging, simultaneous ranging and communication, detection and tracking, or others. Others include channel characterizing [31], vehicle vibration modeling [32], etc.

As shown in Figure 4a, the transmitters can be divided into more than two types, including LEDs and screens. It can be seen that LEDs are used in a much higher proportion than screens. LEDs are also the most suitable transmitters for ITS applications. Not only are existing traffic lights equipped with LED arrays, but their switching speed is also the fastest. LED transmitters can also be divided into different categories, for example, single LED, LED arrays, and rotating LED arrays. The number and size of LEDs in the transmitter determine to some extent how much data can be transmitted over how many distances. Because the number of pixels taken up by the transmitter decreases as the distance between the transmitter and the image sensor increases. With regard to the ISC receiver, the receivers can be divided into more than four types as shown in Figure 4b, including rolling shutter cameras, high-speed cameras (global shutter), event cameras, and OCI. They will be discussed in Section 4.

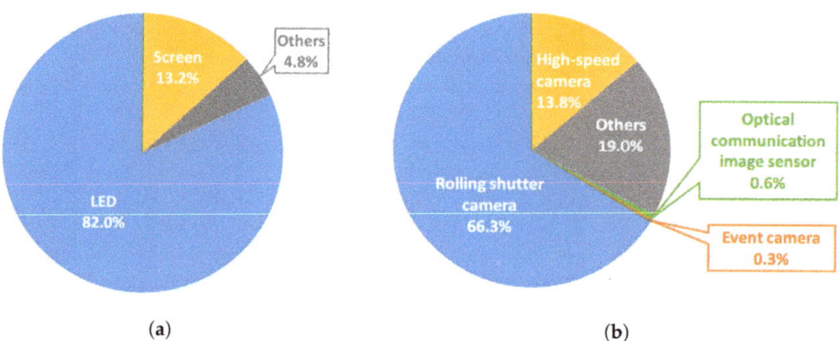

(**a**)   (**b**)

Figure 4. Categorizing image sensor communication (ISC) research publications by transmitter types and receiver types. The number of publications is 1021. (**a**) Categorizing ISC publications by transmitter type. Others includes micro-LED [33], organic-LED [34], projector [35], laser diodes, etc. (**b**) Categorizing ISC publications by receiver type. Others includes charge-coupled device (CCD) camera, time-of-flight sensor [36], lensless-camera [37], etc.

2.3. Image-Sensor-Communication-Based Intelligent Transportation System (ITS-ISC)

Figure 5 shows the proportions of different application scenarios in ISC, where outdoor applications comprise 24% and the number is 251 papers. Despite having a lower proportion than indoor applications, the potential of ISC outdoor applications in the research field of ITSs cannot be overlooked.

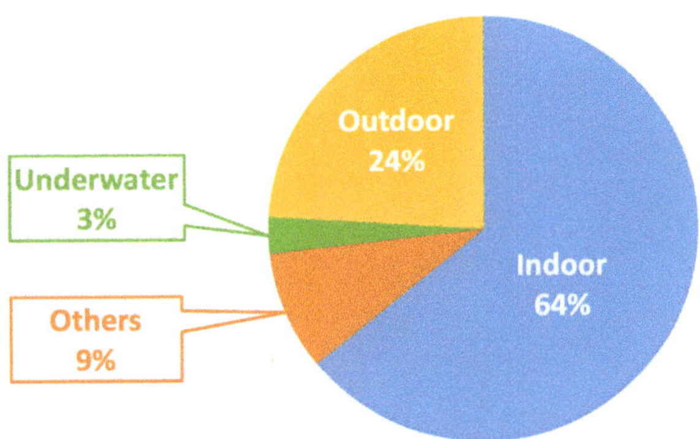

Figure 5. Categorizing image sensor communication research publications by application scenarios including indoor, outdoor, underwater, or others.

ITS refers to a system that connects people, roads, and vehicles through the use of communication technologies, aiming to improve road traffic safety and smooth traffic flow by reducing congestion, and achieving comfort dedicated short-range communications (DSRC) [16,38]. A familiar example of an ITS is the electronic toll collection (ETC) system. The latest report shows that the global intelligent transportation market reached a size of 110.53 billion USD in 2022 and is projected to increase at a compound annual growth rate of 13.0% during the forecast period from 2023 to 2030 [39]. This growth trend in the market highlights the need for vehicles to use intelligent transportation services.

The technological development of autonomous vehicles is currently attracting worldwide attention, especially in the field of ITSs. The recognition of the surrounding road conditions using various onboard sensors is important for driving safety, as well as driving support for autonomous vehicles. Cameras are one of the most commonly used onboard sensors in self-driving cars and advanced driver assistance systems [22,40]. These onboard cameras are mainly used for applications such as visual aids, object detection, and driving recording. If these cameras are used as receivers for ISC, they can receive the optical signal from LEDs on the road to provide communication and localization services. Information sent from the vehicle is expected to include location information such as the vehicle's longitude and latitude, vehicle signals indicating driving conditions such as speed, steering, acceleration, braking, or vehicle front view, and vehicle-specific information such as vehicle model, width, and height. Information sent from the side of the road is expected to include traffic signal statuses, such as color and transition time, and traffic information, such as weather conditions and nearby traffic congestion. These kinds of information transmitted through the ISC system can provide efficient and reliable communication for a variety of applications, including traffic management, road safety, and autonomous driving.

In addition, image sensors can help to minimize or manage latency in vehicular networks, which is crucial for safety-critical applications, such as collision avoidance and cooperative driving. ISC can offer better spatial and directional accuracy than conventional radio frequency-based communications, which can enhance the reliability of achieving low latency. Hence, the integration of ISC in vehicular networks can considerably enhance the performance and safety of upcoming ITSs.

3. Basic Concept and Architecture of an ITS-ISC

3.1. Vehicle-to-Everything (V2X) Communications Using Image Sensors and LEDs

ISC can be implemented in a variety of contexts in an ITS, including V2V, V2I, and I2V communications. These means of communication are involved in vehicle-to-everything (V2X) communications, which refers to any type of communication between a vehicle and its surrounding traffic environment, as illustrated in Figure 6.

Figure 6. The basic idea of image sensor communication for vehicular applications including vehicle-to-vehicle (V2V), vehicle-to-infrastructure (V2I), and infrastructure-to-vehicle (I2V) communications. Data signals are emitted from traffic lights and vehicle headlights and taillights. Image sensors are mounted on the vehicle to receive visible light signals. The image sensor converts the received visible light signal into an electrical signal and decodes the signal using image processing technology.

In V2X communication, vehicles can communicate with each other and share information about their speed, location, and other important data, which helps improve safety and traffic flow. For example, suppose a vehicle detects an obstacle on the road ahead. In that case, the vehicle can transmit this information to the rear vehicles to alert them and allow them to take appropriate action. This type of communication between vehicles is referred to as V2V communication. Moreover, V2X communications enable vehicles to send and receive information about the road environment, such as traffic conditions or road hazards, to improve safety and efficiency. For example, a traffic signal can transmit data to a vehicle about the time remaining before the signal changes, allowing the vehicle to adjust its speed accordingly. Furthermore, the information transmitted via V2X communications may include information about traffic patterns, weather conditions, and road closures. For example, sensors embedded in the road can detect the presence of ice or snow and transmit this information to vehicles to help them navigate safely. These types of communication between infrastructures and vehicles are referred to as I2V and V2I.

However, ISC is restricted to light-of-sight (LOS) situations due to the reliance on visible light. If an obstacle interrupts the light, it will interrupt the information transmission, which is likely to be the case in practical transportation applications such as lane changing. There are solutions to handle such non-light-of-sight (NLOS) situations based on reflections [41]. Alternatively, the vehicle can receive information indirectly through the obstacle if it is an ISC transmitter. In addition, DSRC can be used for NLOS situations [16]. It transmits by wireless radio waves and can penetrate obstacles. However, DSRC is limited by high levels of noise and multipath effects, which ISC can mitigate. Therefore, in certain situations, a hybrid approach that considers both methods may be appropriate.

3.2. Optical Channel Characteristic and Modulation Schemes

The optical channel can be expressed by

$$R(t) = I(t)X \otimes h(t) + A(t) \tag{1}$$

where $I(t)$ is the transmitted waveform, $R(t)$ is the received waveform, $h(t)$ is an impulse response, $A(t)$ represents the noise, $\otimes$ symbol denotes convolution, and X is the detector responsivity. In addition, the average transmitted optical power P_t in the time period $2T$ is given by

$$P_t = \frac{1}{2T} \int_{-T}^{T} I(t)dt \tag{2}$$

The average received optical power P can be presented by

$$P = H(0)P_t \tag{3}$$

where $H(0)$ is the channel DC gain, given by $H(0) = \int_{-\infty}^{\infty} h(t)dt$. In addition, the digital signal-to-noise ratio (SNR) is given by

$$SNR = \frac{X^2 P^2}{X_b N_0} \tag{4}$$

where X_b is the bit rate, and N_0 is the double-sided power-spectral density [42].

There are several modulation schemes available for ISC, each of which can affect the optical power P_t. The most commonly used modulation scheme is on-off keying (OOK), although others such as pulse-width modulation (PWM), pulse-position modulation (PPM), pulse-amplitude modulation (PAM), and orthogonal frequency division multiplexing (OFDM) are also utilized.

3.3. Advantages and Limitations in Image Sensor Receivers

An important feature of the image sensor is spatial separation. In ISC, spatial separation refers to spatially separating signals coming from different transmitters, such as traffic signals, street lights, and brake lights of the lead vehicle. A visual explanation of spatial separation is shown in Figure 7.

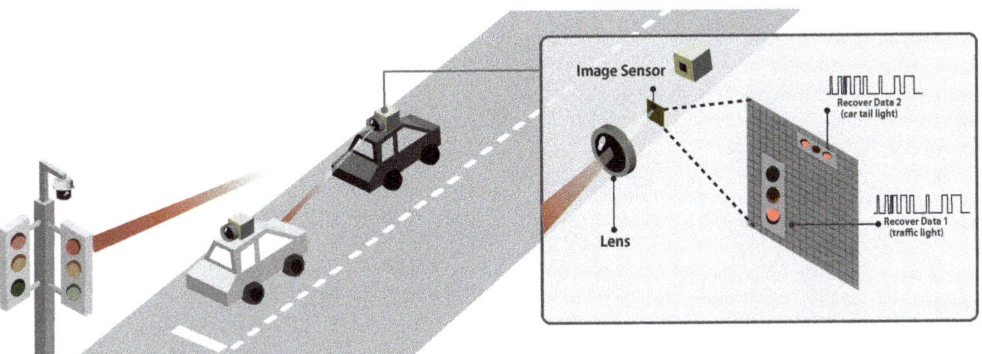

Figure 7. A visual explanation of spatial separation. Two cars are driving on the road and the camera of the following vehicle receives two light streams from the traffic light and the backlight of the lead vehicle. An image sensor receiver can separate these two light sources spatially.

Spatial separation enables ISC to perform parallel transmissions using multiple light sources and separating them by image processing. The interference between each light source is significantly reduced by spatial separation, while it is hard to separate the light sources using a single PD due to the interference. An LED array, consisting of multiple LED elements arranged in a grid pattern, can be used to achieve spatial separation [43,44]. Each element in the LED array can be individually controlled to emit light at different intensities. By modulating the light emitted by each LED element, we can encode data into the light signal and transmit it to a receiver. This approach allows for parallel transmissions and can increase the overall data transmission rate. By using an image sensor receiver, we can reduce the impact of interference between light streams from different LEDs. This is particularly useful when dealing with high-speed data transmission, where even small amounts of interference can lead to significant errors. Furthermore, the spatial separation of the image sensor is particularly beneficial in distance measurement. An image sensor captures a wide field of view, allowing obtaining depth information through monocular vision or stereo vision. This can help vehicles make more accurate and safe driving decisions.

However, the effectiveness of spatial separation depends on three factors of the image sensor: resolution, frame rate, and dynamic range. First, regarding sensor resolution, using a low-resolution image sensor will lead to lower data rates and increased vulnerability to errors. A low-resolution image sensor captures images with few pixels, so the amount of information that can be transmitted through ISC is limited to its pixel amount. In addition, if the resolution of the image sensor is low, the interference of the optical signal between adjacent pixels can be significant and can affect the accuracy of demodulation. The low-resolution issue may be solved by image processing algorithms such as Kalman filter and resolution compensation [45]. Second, in terms of frame rate, low frame rates result in low data rates. A low frame rate commonly occurs if an image sensor has a large number of pixels, because a large number of pixels requires high processing ability, resulting in a slow processing speed for each frame. The frame rate of a conventional complementary metal–oxide–semiconductor (CMOS) image sensor is usually around 30 frames per second (fps) [46]. While some commercial cameras or industrial cameras can exceed 30 fps, such as iPhone 14, which has slo-mo video support at 120 fps or 240 fps [47], they sacrifice the video quality. Finally, another problem with ISC is the limited dynamic range of the image sensors. When light irradiates onto the image sensor, the pixels absorb photons and convert them into electrons. These charges then accumulate in the pixel potential well, and when the accumulation limit is reached, no more photoelectric conversion can take place, even if the light is brighter (more photons). In other words, the output digital signal reaches saturation. A low dynamic range leads to a low saturation threshold. If the pixels that contain LEDs are saturated, the demodulation error will be larger, especially with luminance-based modulation such as pulse-width modulation. To solve the issue caused by saturation, we can use a pre-coding method [48] or image processing algorithms, such as phase-only correlation (POC) [49].

4. Isc Receivers

To be used effectively in a moving vehicle environment, the ISC signal must be transmitted at high speed. While CMOS image sensors are the most commonly used sensors for ISC, conventional CMOS sensors often have low frame rates that make it challenging to transmit signals at high speeds. As a result, the ISC receivers, namely the image sensors, are required to utilize encoding techniques and image processing algorithms to achieve high-speed transmission. This section analyzes various ISC receivers, including rolling shutter cameras, high-speed cameras, OCI, and event cameras.

4.1. Rolling-Shutter Camera

According to our survey in Section 2.2, rolling shutter cameras are the most widely used receivers in ISC. The reasons are that rolling shutter cameras are cheap and are utilized in most commercial cameras, including dashcams, closed-circuit television, and those on mobile phones.

In the rolling shutter mechanism, each row of the image sensor is exposed sequentially. This feature enables high-speed data transmission in ISC by allocating data to each row of the rolling shutter sensor [20,50]. Figure 8 provides detailed descriptions of the rolling shutter camera. Figure 8a shows a conventional four-transistor rolling shutter circuit. The row scanner selects the rows from top to bottom individually, and then the column scanner controls reading out the data of an entire row. A pixel in the rolling shutter sensor has no ability to store the accumulated charges due to the lack of memory nodes. Therefore, the exposure process should be followed by an immediate readout to ensure consistent exposure time for each row. This results in the rolling shutter mechanism shown in Figure 8b. The row pixels are read out sequentially thus the reset time starts sequentially. This rolling shutter effect causes the object to appear skewed in the image when the object is moving at high speed. Despite the image skewing drawback, by switching the transmitter faster than the frame rate of the rolling shutter camera, a data rate higher than the frame rate can be achieved due to the row-by-row exposure mechanism. An example is shown in Figure 8c. If a traffic light embedded an array of LEDs blinks at the same speed as the readout rate of each row of the image sensor in this way, the image sensor can receive data at high speed by decoding each row of the image sensor. Each row appears as an independent pattern in the horizontal direction.

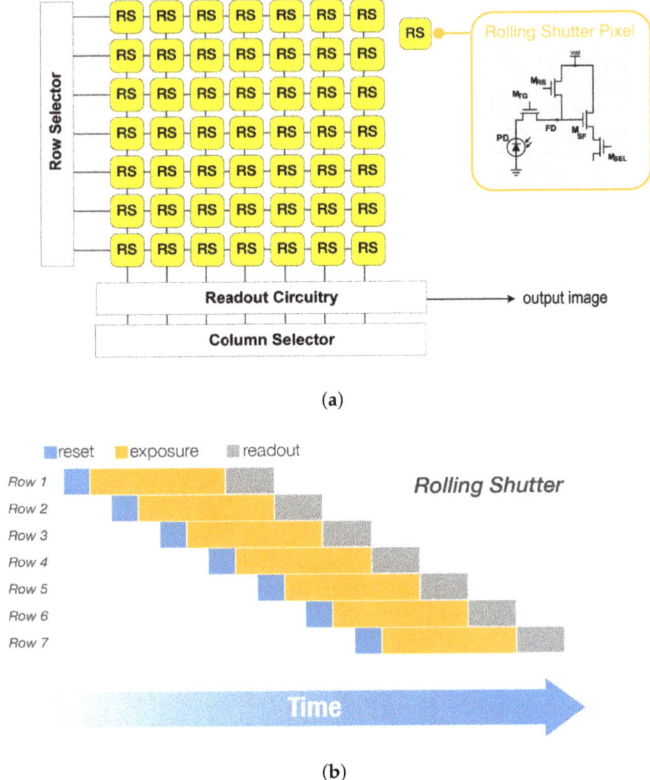

Figure 8. *Cont.*

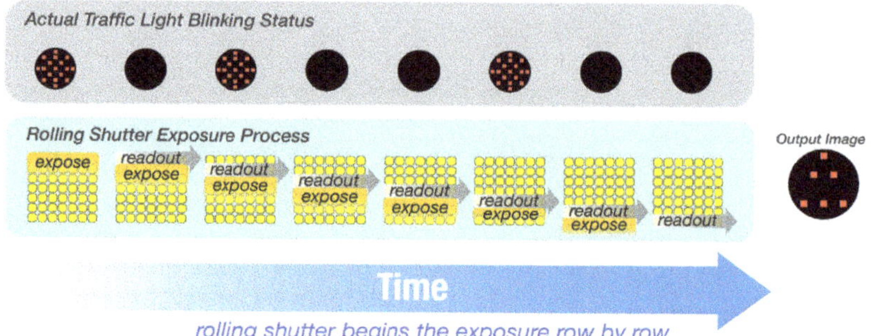

(c)

Figure 8. The circuit (**a**), exposure mechanism (**b**), and an example (**c**) of a rolling shutter image sensor. (**a**) A simplified rolling shutter image sensor circuit with 7×7 pixels. The image on the top right shows the basic circuit for a traditional rolling shutter pixel which is a four-transistor active pixel sensor (4T-APS) [46]. A 4T-APS contains a photodiode (PD), a floating diffusion (FD), and four transistors, M_{RS}, M_{TG}, M_{SF}, and M_{SEL}. (**b**) The rolling shutter exposure schematic shown over time. The rolling shutter method exposes rows sequentially, followed by an immediate readout. The specific mechanism is decided by the pixel circuit. (**c**) An example of a 7×7 rolling shutter image sensor capturing a blinking traffic light. We assume that the LED switching rate is equal to the readout rate. The reset is included in the exposure.

In ISC systems that employ rolling shutter effects, the LED blinking rate is higher than the frame rate. Consequently, the images capture stripes due to the LED blinking pattern, such as shown in Figure 8c. In order to decode the stripe image, it is necessary to first extract the LED area in the image, as it is often present in a complex background. Once the LED area has been identified, the stripes can be decoded based on their modulation methods.

The mobility effect of rolling shutter cameras has been investigated in [51,52]. For rolling shutter cameras, which capture the image line-by-line, the motion of the camera during the exposure time can cause distortion or blur in the captured image. This effect can be a significant challenge in ISC systems, where high-quality image capture is essential for reliable communication.

Typically, region-of-interest (RoI) detection methods are used for LED detection. RoI detection is commonly used in computer vision and we can use algorithms such as image thresholding [53] or cam-shift tracking algorithm [54] to extract the LED transmitters. S. Kamiya, et al. have succeeded in error-free detection and communication using a thresholding detecting method when the vehicle is moving at a speed of 15 km/h [55]. Moreover, there are studies conducted to track LEDs by deep learning [52,56].

After completing the LED detection process, we can proceed to the demodulation process. However, the choice of modulation method can significantly impact system performance. In most works the modulation method of OOK is utilized. Other conventional modulation methods include OFDM [52], PWM [57], color-shift keying [58], etc. Furthermore, there are under-sampled modulation methods, including under-sampled frequency shift OOK [59], under-sampled phase shift OOK, spatial-2 phase shift keying (S2-PSK) [60], spatial multiplexing [61], etc.

4.2. Global-Shutter High-Speed Camera

Another type of CMOS image sensor used in ISC is the global shutter image sensor [19,62,63]. Global shutter functions differently from the rolling shutter sensors. Instead of exposing rows sequentially, the global shutter exposes each row simultaneously, making it ideal for capturing high-speed motion without skewing. Cameras that use global shutters to capture high-speed motions are called high-speed cameras, and their frame rates can achieve thousands of fps. According to Section 2.2, 13.8% of the ISC publications used high-speed cameras, which is lower than that of rolling shutter cameras. The main reason is that high-speed cameras are much more expensive than rolling shutter cameras.

Figure 9 provides detailed descriptions of how a global shutter camera works for ISC. Figure 9a shows a conventional five-transistor global shutter sensor circuit. Its scanning mechanism is the same as the rolling shutter that reads out the data row by row. However, the pixel in the global shutter sensor has a memory node, so it can store the accumulated charges. Therefore, the readout process is not required to be conducted immediately after the exposure. In other words, all rows can be exposed simultaneously, and wait a period of time before reading out the charges, as shown in Figure 9b. An example of a global shutter camera capturing a traffic light is shown in Figure 9c. Assume that all the LEDs in the traffic light blink at the same speed as the readout rate of each row of the image sensor, in this way, the image sensor will receive an image of a traffic light with all LEDs illuminated. The output image is different from that of the rolling shutter, even the LED light blinking status is the same. In practice, the LED and the image sensor are usually asynchronous, so the switching speed of the LED must be equal to or less than half of the camera frame rate, according to the Nyquist criterion. As the global shutter high-speed camera captures generally at over 1000 Hz and the data rate is at least half of the frame rate, the data rate can exceed 500 bps.

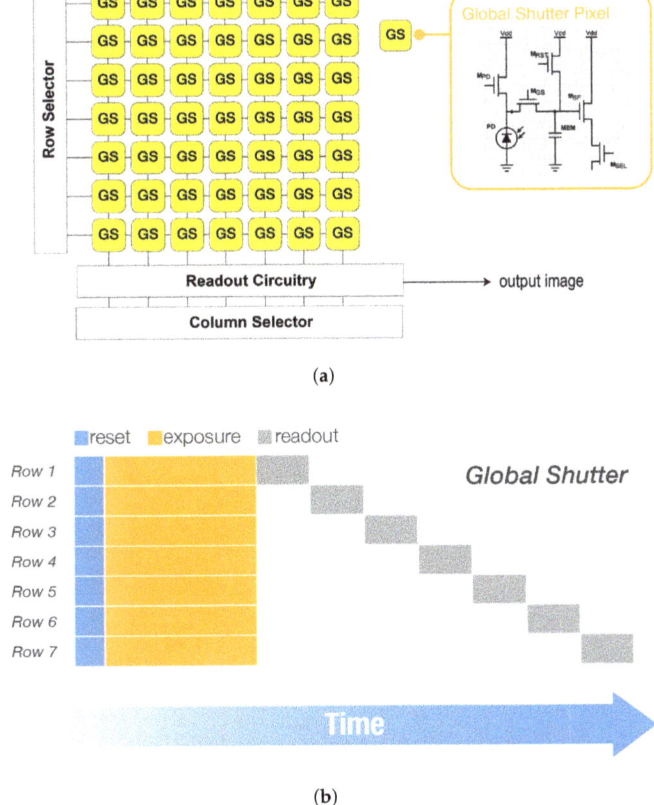

Figure 9. *Cont.*

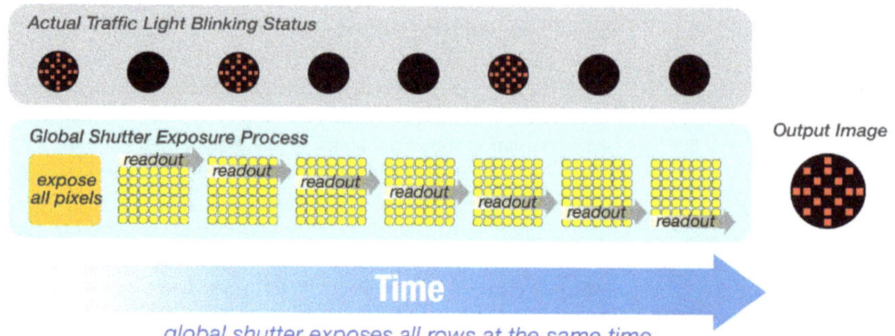

(c)

Figure 9. A simplified sensor circuit (**a**), exposure mechanism (**b**), and an example of capturing a blinking traffic light (**c**) with a global shutter camera. (**a**) A simplified global shutter image sensor circuit with 7×7 pixels. The image on the upper-right shows the basic circuit of a traditional five-transistor global shutter active pixel sensor (5T-global shutter APS) [46]. A 5T-global shutter APS contains a photodiode (PD), a memory (MEM) node, and five transistors, M_{PD}, M_{GS}, M_{RST}, M_{SF}, and M_{SEL} [46]. The main difference to the rolling shutter pixel is that a global shutter pixel has a memory node for keeping the charge. (**b**) The global shutter exposure schematic shown over time. Global shutter first exposes all the rows simultaneously, and then each row waits for the readout. (**c**) An example of a 7×7 global shutter image sensor capturing a blinking traffic light. We assume that the LED switching rate is equal to the readout rate. The global shutter only captures the first blinking pattern of the traffic light.

Since the frame rate of a high-speed camera is over 1000 fps, the position of the transmitter on the image differs only slightly from the last consecutive frame. For example, suppose a vehicle has a speed of 50 km/h and the camera has a frame rate of 1000 fps. Then the vehicle moves only 0.01 m in the time period of each frame, and this distance is usually reflected within one pixel on the image plane. As a result, we can take advantage of this feature that high-speed cameras have less displacement in the position of adjacent frames for vehicle tracking, which rolling shutter cameras cannot perform. In [64], T. Nagura et al. proposed an LED array tracking method using inverted signals. The method involves transmitting an inverted pattern immediately after the original LED array pattern, where the entire LED array is obtained by adding these two consecutive patterns. Since the displacement of the LEDs is very small between the two consecutive frames, this method can detect the position of the LED array in the images pretty accurately. However, the data rate is reduced by half. In [65], S. Usui et al. proposed an LED array detection method using spatial and temporal gradients. Spatial refers to the horizontal and vertical gradients of the image of the current frame for which LEDs need to be detected, calculated using Sobel operator; temporal refers to the gradients of the current frame with respect to the previous and next frames, also calculated using Sobel operator. This method identified LED arrays with low spatial-gradient values and high temporal-gradient values. The experiment results showed that error-free LED tracking was achieved when the vehicle was driving at 30 km/h. In [65], the direction of vehicle motion was perpendicular to the LED array plane, while it could also be used when the direction of vehicle motion was parallel to the LED plane [66].

The mobility effect of the vehicle motion has less influence on high-speed image sensors, as shown in [31]. A pinhole camera model was introduced to project world coordinates to image coordinates in [31]. Three types of ISC systems: I2V-ISC, V2I-ISC, and V2V-ISC, were discussed. For I2V-ISC, the camera moved with the vehicle, and the transmitter was static, while for V2I-ISC, the camera was static, and the transmitter moved with the vehicle. Ref. [31] also compared the vehicle motion models of I2V-ISC and V2I-ISC, and the effects of camera posture on these models. Additionally, Ref. [31] discussed how the relative distance between the transmitter and receiver affects the apparent size and position of objects in the image.

Although the high-speed camera can capture fast motions, it is easily subject to the effect of distance. For a rolling shutter camera, the capturing pattern will not change according to the distance between the vehicle and the transmitter. However, for a global shutter high-speed camera, the size of the transmitter is changing according to the communication distances. Of course, the size of the transmitter on the image is also determined by the image sensor resolution. Hierarchical coding scheme has been proposed in [67] to solve the problem caused by signal degrading at long distance. It divides the LED patterns into different priorities according to the communication distance and assigns different frequencies to their respective priorities by wavelet transform. However, the hierarchical coding scheme has limitations on the number and arrangement of LEDs, which may not match the design of practical transmitters such as traffic lights. To overcome this issue, S. Nishimoto et al. proposed a method of overlay coding in [68,69]. The overlay coding is a more flexible way to design LED applications depending on the transmitter. It distributes long-range or short-range data into large-scale or small-scale of the LED array and encodes them by overlaying each patterns. The experimental results showed that the error-free communication distance of the overlay coding could reach 70 m.

Other limitations of a high-speed camera receiver are the high cost and low resolutions. Due to the advanced technology and components used in high-speed cameras, they can be quite expensive to manufacture and purchase. This can make them cost-prohibitive for smaller organizations that do not have significant budgets for equipment. Another limitation of high-speed camera receivers is their relatively low resolutions compared to other types of cameras. While they are designed to capture images at incredibly high frame rates, the resulting images may not have the same level of detail or clarity as images captured by other types of cameras. This is because high-speed cameras often use smaller sensors and less advanced optics in order to capture images at such high speeds.

4.3. Optical Communication Image Sensor (OCI)

In the previous sections, we discussed two types of image sensors used in ISC: rolling shutter and global shutter. Both of these sensors were originally designed to capture images of the entire scene. However, ISC often requires only pixels that contain transmitters. If these pixels can be extracted and attached to a receiver that receives optical signals at high speed, the data rate will be significantly improved. This kind of receiver is proposed in [19,70–75], and known as the OCI. Compared to conventional CMOS image sensors, OCI contains two different kinds of pixels, image pixels (IPx) and communication pixels (CPx), and arranges them side by side as shown in Figure 10. The IPx is a conventional CMOS pixel structure, similar to that of rolling shutter, with an output rate of a few tens of fps. On the other hand, the CPx reduces the capacitance so that it can output the communication signal faster with an output rate of approximately a thousand times that of the IPx [72]. The OCI detects an LED at the IPx and outputs the corresponding CPx in that area. The IPx enables tracking even in a moving vehicle condition. The CPx output is analog, functioning similarly to a PD.

In [72], I. Takai et al. presented the design, fabrication, and capabilities of the OCI, and the experiment results had a 20 Mbps per pixel data rate without LED detection and a 15 Mbps per pixel data rate with real-time LED detection. In [73], I. Takai et al. discussed the capabilities of the OCI used in a V2V communication system, and successfully transmitted vehicle internal data and 13-fps front-view image data of the lead vehicle. In [75], Y. Goto et al. designed optical OFDM into OCI vehicular system, considering frequency response characteristics and circuit noise of the OCI. The system had a performance of 45 Mb/s without bit errors and 55 Mb/s with a BER of 10^{-5}.

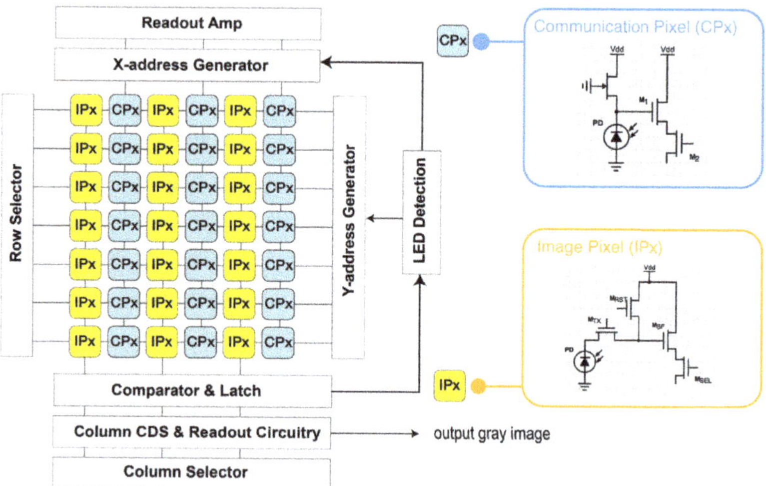

Figure 10. A simplified circuit diagram for an optical communication image sensor with 3×7 image pixels (IPx) and 3×7 communication pixels (CPx). The CPx and IPx are arranged side by side. The two figures on the right show the structure of an IPx and a CPx, adapted from [72]. CDS refers to correlated double sampling, M1 is a transistor used for readout amplifier, M2 is a transistor used for CPx selection, VDD stands for power supply voltage, and Vcs stands for charge-sensing node. Compared to an IPx, the CPx reduce two transistors, making the response faster. Details can be found in [72].

Apparently, the advantage of OCI is the high data rates. It is difficult for conventional CMOS image sensors to achieve data rate at Mbps level. However, the disadvantage of OCI is its low resolution. The current maximum resolution of OCI is 642×480 pixels [75], which is much lower than nowadays image sensors. The low resolution may limit its application in long-range communication.

4.4. Event Camera (Dynamic Vision Sensor)

Event cameras, also known as dynamic vision sensors or neuromorphic vision sensors, can be used as a receiver of ISC as well. Unlike rolling shutter or global shutter cameras, an event camera does not acquire images with a shutter or frame, but rather record changes in luminance in a pixel and outputs them as asynchronous events [76]. Figure 11 shows the luminance change of a pixel in an event camera and the event output. "ON" refers to positive luminance change and "OFF" refers to negative luminance change.

Since an event camera only outputs positive or negative values, it is simpler to process than a conventional CMOS image sensor, which gives it microsecond-level temporal resolution. Likewise, LEDs can blink at microsecond-level frequencies, so there is an opportunity to achieve ultra-high data rate in ISC with an event camera. In addition, the dynamic range of the event camera is higher than any other conventional CMOS cameras. Its high dynamic range allows the event camera to capture subtle changes in brightness and improve decoding accuracy.

The existing works on the use of event cameras in ISC have only reached a preliminary level. In [77], W. Shen et al. proposed a pulse waveform for event camera-based ISC systems. The experiment compared the inverse pulse position modulation waveform to the proposed pulse waveform, and the communication distance was 3 m and 8 m, respectively. In [78], G. Chen et al. proposed a positioning method using an event camera as a receiver. The proposed system achieved a positioning accuracy of 3 cm when the height between LEDs and the event camera was within 1 m. The low latency and microsecond-level temporal resolution of the event camera made it possible to identify multiple high-frequency flickering LEDs simultaneously without traditional image processing methods. In [79], Z. Tang et al. investigated a new communication scheme for event camera-based ISC systems using a propeller-type rotary LED transmitter and showed the potential of event-based ISC in moving environment.

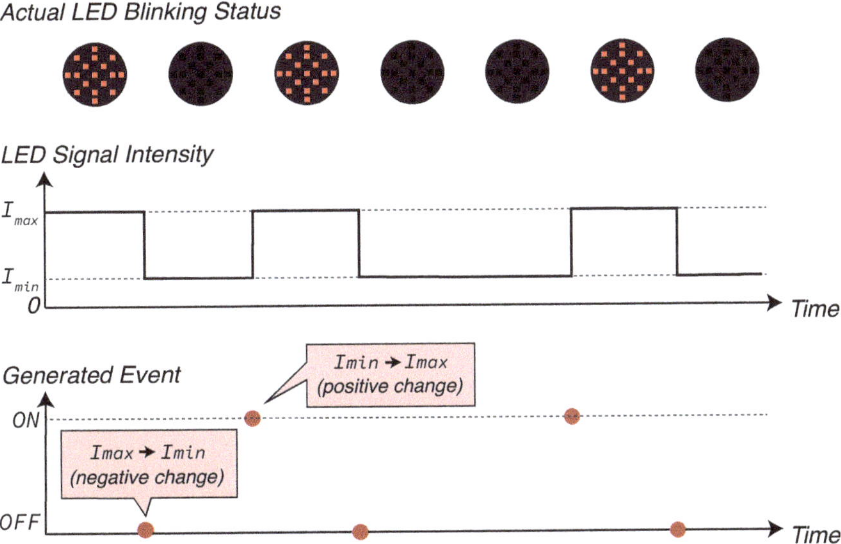

Figure 11. Analyzing generated events by correlating with actual LED blinking status and their signal intensity. The I_{max} intensity indicates when the LED is illuminated, while the I_{min} intensity indicates when the LED is off. When a negative luminance change is detected, the event camera generates an "OFF" event, and when a positive luminance change is detected, the event camera generates an "ON" event.

These above works demonstrated the potential of event camera-based ISC systems for low-latency data communication. However, the ability to quickly detect motion also leads to problems with the use of event cameras for ISC. Since event cameras work by detecting changes in luminance, they may have difficulty distinguishing between events generated by motion of LEDs and events generated by changes in luminance. Event cameras may encounter difficulties when the camera or LED is moving because the motion causes changes in luminance, which may be interpreted as LED events, especially if the LED is flashing rapidly or in complex patterns. In addition, as with any imaging system, the event camera is subject to a variety of noise sources, including electronic noise from the sensor and shot noise from the light source. This can make it more difficult to distinguish a true LED event from background noise or other sources of interference. Overall, while event cameras offer some promising advantages for VLC systems, there are still some technical challenges that need to be addressed to realize their full potential.

4.5. Summary of ISC Receivers

Table 1 summarizes the ISC receivers discussed in this chapter: rolling shutter camera, high-speed camera, OCI, and event camera. Considering the practical application scenario is important when choosing the appropriate receiver. However, sensor fusion is a possible option that can be applied in many situations.

Table 1. A summary of communication performance in existing literature categorizing by image sensor communication (ISC) receivers.

Receiver Type	Reference	Data Rate	Communication Distance	Vehicle Speed
Rolling shutter	[55]	600~1000 bps	5~70 m	15~20 km/h
	[61]	720 bps	100 m	N/A
High-speed camera (global shutter)	[19]	32 kbps	30~65 m	30 km/h
	[43]	128 kbps	10~120 m	N/A
	[69]	40 kbps	20~70 m	N/A
Optical communication image sensor (OCI)	[19]	10 Mbps	20 m	25 km/h
	[72]	20 Mbps	N/A	N/A
	[75]	55 Mbps	1.5 m	N/A
Event camera	[77]	16 kbps	8 m	N/A

5. Range Estimation Using LEDs and Image Sensors

In an ITS, accurate range estimation between a vehicle and its surrounding objects is crucial for ensuring safety. By enabling positioning, the vehicle can determine the distance to its surroundings and take measures to avoid collisions. In an ITS-ISC, image sensors have the ability to obtain depth information through the process of triangulation. It takes advantage of the spatial separation between the viewpoints to calculate the relative positions of the LEDs in the scene, which can then be used to determine their distances to the camera. In this section, we introduce ISC ranging methods based on stereo and monocular vision, covering their principles, challenges, and solutions.

5.1. Stereo Vision-Based Range Estimation

The stereo vision-based ISC ranging scheme uses two camera receivers. The distance between the LEDs and the cameras can be determined according to the three-dimensional geometry. Figure 12 illustrates the principle of obtaining range using two cameras and one LED based on the pinhole camera model. The two cameras are considered identical and prior calibration of the two cameras is necessary. Only a single LED is possible to estimate the range between the transmitter and receiver. The cameras are regarded as onboard cameras, and the LED is assumed to be on the lead vehicle or traffic light.

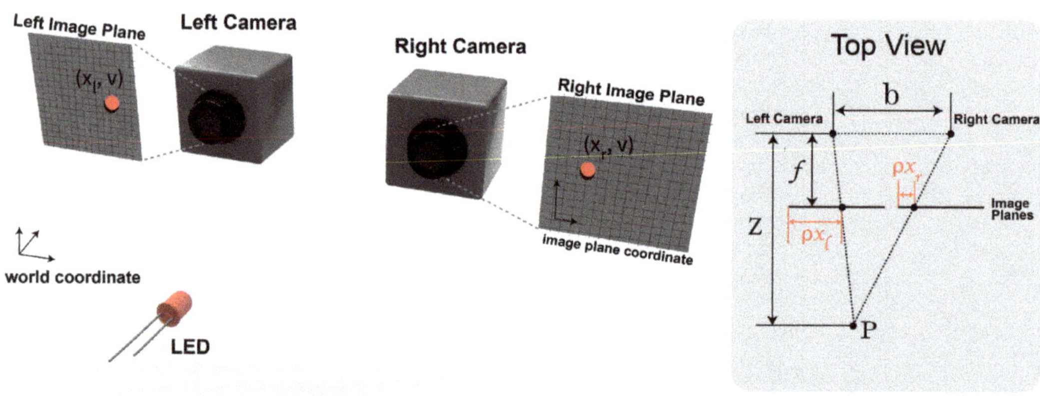

Figure 12. A schematic of two cameras capturing a single LED. The two cameras are supposed to be perfectly calibrated. The top view is shown on the right. P refers to the LED, b is the distance between the left and right camera, f is the camera focal length, ρ is the size for one pixel, and Z is the distance we need to estimate.

The LED is projected into (x_l, v) and (x_r, v) on the image planes of left and right cameras, respectively. According to the triangle similarity criteria, the range Z between the LED and the cameras can be given by

$$Z = f \frac{b}{\rho(x_l - x_r)},$$ (5)

where f is the focal length of two cameras, b is the distance between the two cameras, and ρ is the actual size per pixel. Thus, if we can find the corresponding feature point on the image plane of the left and right views, the position of the LED can be determined accurately.

The precision of range estimation is significantly affected by the accuracy of computing $(x_l - x_r)$, known as the disparity [80]. In [49], the disparity is calculated using phase-only correlation and the sinc function matching to estimate the disparity in subpixel accuracy. Similarly, Ref. [81] estimated the disparity at the subpixel level using equiangular line fitting, of which the processing speed is faster than phase-only correlation. Furthermore, a positioning algorithm based on neural networks is proposed in [82], and a technique for compensating the rolling shutter effect is proposed in [83].

5.2. Range Estimation Based on Monocular Vision

The range between the LED transmitter and the vehicle can also be estimated using a single camera. Figure 13 shows a schematic of using one camera to estimate the distance to the LED.

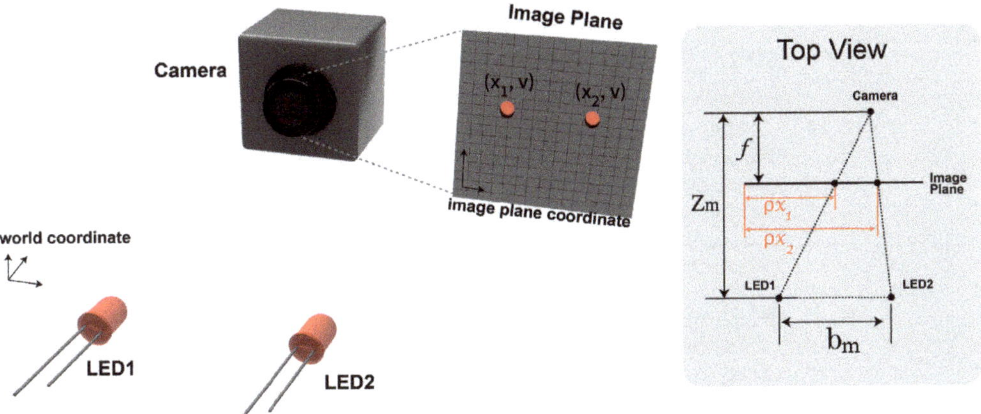

Figure 13. The schematic of range estimation using two LEDs and one camera. The top view is shown on the right. b_m is the distance between LED1 and LED2, f is the camera focal length, ρ is the size for one pixel, and Z_m is the distance we need to estimate.

We can use a pinhole camera model to express the transmitter position on both three-dimensional world space and the image plane of the camera. The range from the LED to the camera can be calculated by triangulation, given by

$$Z_m = f \frac{b_m}{\rho(x_2 - x_1)},$$ (6)

where Z_m is the range between LED and camera, b_m is the distance between LED1 and LED2, f is the focal length of the camera, and ρ is the size per pixel.

However, it is also possible to estimate the range using more than two LEDs based on monocular vision. Reference [84] employed three LEDs to determine the three-dimensional positions of the vehicle, conducting under tunnel scenarios. Reference [85] focused on the multiple-input and multiple-output (MIMO) cases, where more than two LEDs are required, and utilized the S2-PSK modulation method. Furthermore, when it comes to vehicular applications, it is important to address issues related to vehicle vibration. In [86], a Kalman filter is used to reduce the random errors caused by vehicle moving. In [30], two specialized patterns are introduced to tackle the noise generated by the vibrations of a moving vehicle. Since POC is used for estimating $(x_2 - x_1)$, two LEDs are separately placed on the opposite edge of an LED array in two consecutive frames. The displacement of the upper-left LED on the image plane between the two frames is used to calculate the vehicle vibration. The monocular ranging in [30] is only possible using a high-speed camera, because its time interval must be short enough to compensate a variation caused by moving.

5.3. Range Estimation Using Machine Learning

Machine learning or deep learning techniques have been increasingly used for range estimation in vehicular ISC systems [56,82,87]. In [82], a back-propagation (BP) neural-network learning method has been used for positioning and range estimation. In [87], a coding approach is proposed to provide short-distance and long-distance communication. The authors used an artificial neural network (ANN) to forecast the vehicle's location. Long-range communication and high-precision positioning can be conducted simultaneously. The results in [87] showed that the average ranging error was 19.8 mm within a distance of 30 m. Machine learning algorithms can also be combined with traditional signal processing techniques to improve range estimation accuracy. For example, machine learning models can be used to denoise the received signal, compensate for distortion, or enhance the features relevant to range estimation, thereby improving the overall accuracy of the range estimation.

5.4. Simultaneous Ranging and Communication

ISC ranging also allows simultaneous communication, making it a more efficient system compared to traditional stand-alone communication or ranging systems. Simultaneous communication saves time, reduces latency, and enhances the overall performance of both communication and ranging. In addition, the hardware complexity of the ISC system can be reduced and the reliability can be improved compared to a separate system because fewer components are required. Furthermore, simultaneous ranging and communication can improve the robustness of the system by reducing the effects of noise and other sources of interference, as well as improving the accuracy of ranging estimations.

Simultaneous ranging and communication can be completed using various approaches. One approach is to use multiple image sensor receivers to receive data signal and estimate the range by stereo vision [56,88]. Another technique is to encode data and ranging information in multiple ISC transmitters, which is then received and decoded by one image sensor receiver [73]. The specific method used depends on the requirements and limitations of the system and the ideal trade-off between communication efficiency and ranging accuracy. Table 2 summarizes the ranging performance in the literature that has been discussed in this review.

Table 2. A summary of ranging performance in existing literature.

Ranging Method	Reference	Ranging Error	Communication Distance	Receiver	Vehicle Speed	Simultaneous Communication
Monocular Ranging	[30]	0.3 m	30~60 m	high-speed camera	30 km/h	No
	[73]	N/A	around 8 m	OCI	12.6~14.0 km/h	Yes
	[84]	1 m	0~60 m	N/A	N/A	No
Stereo Ranging	[88]	0.5 m	20~60 m	high-speed camera	N/A	Yes
	[83]	0.1~1.5 m	0~100 m	rolling shutter camera	0~100 km/h	Yes

6. Conclusions

This review paper analyzed the research trend and basics of ISC, especially its applications to vehicles. ISC is characterized by the use of a two-dimensional image sensor and, therefore, has spatial separation characteristics. It also has the ability to simultaneously perform communication and ranging. We have made a comprehensive review of various types of ISC receivers (rolling shutter cameras, high-speed cameras, event cameras, and optical communication image sensor) and ISC range estimation techniques. It also highlights the challenges and expected future developments in the field of ISC.

Author Contributions: Conceptualization, R.H. and T.Y.; writing—original draft preparation, R.H.; writing—review and editing, T.Y. and R.H.; visualization, R.H.; supervision, T.Y.; project administration, T.Y.; funding acquisition, R.H. All authors have read and agreed to the published version of the manuscript.

Funding: This research was funded by the "Nagoya University Interdisciplinary Frontier Fellowship", supported by Nagoya University and JST, the establishment of university fellowships towards the creation of science technology innovation, Grant Number JPMJFS2120.

Institutional Review Board Statement: Not applicable.

Informed Consent Statement: Not applicable.

Data Availability Statement: Not applicable.

Conflicts of Interest: The authors declare no conflict of interest.

Abbreviations

The following abbreviations are used in this manuscript:

4T-APS	four-transistor active pixel sensor
BER	Bit-error-rate
CCD	Charge-coupled device
CMOS	Complementary metal–oxide–semiconductor
CPx	Communication pixel
DSRC	Dedicated short-range communication
ETC	Electronic toll collection
FD	floating diffusion
fps	frame per second
GPS	Global positioning system
I2V	Infrastructure to vehicle
ISC	Image sensor communication
IPx	Image pixel
ITS	Intelligent transportation system
LED	Light-emitting diode
LiFi	Light fidelity
LiDAR	Light detection and ranging
LOS	Line-of-sight
MEM	memory
NLOS	Non-line-of-sight
OCC	Optical camera communication
OCI	Optical communication image sensor
OFDM	Orthogonal frequency division multiplexing
OOK	On-off keying
PD	Photodiode
POC	Phase-only correlation
S2-PSK	spatial-2 phase shift keying
V2V	Vehicle to vehicle
V2I	Vehicle to infrastructure
V2X	Vehicle to everything
VLC	Visible light communication

References

1. Pathak, P.; Feng, X.; Hu, P.; Mohapatra, P. Visible Light Communication, Networking, and Sensing: A Survey, Potential and Challenges. *IEEE Commun. Surv. Tutor.* **2015**, *17*, 2047–2077. [CrossRef]
2. Jovicic, A.; Li, J.; Richardson, T. Visible light communication: Opportunities, challenges and the path to market. *IEEE Commun. Mag.* **2013**, *51*, 26–32. [CrossRef]
3. Karunatilaka, D.; Zafar, F.; Kalavally, V.; Parthiban, R. LED based indoor visible light communications: State of the art. *IEEE Commun. Surv. Tutor.* **2015**, *17*, 1649–1678. [CrossRef]
4. Komine, T.; Nakagawa, M. Fundamental analysis for visible-light communication system using LED lights. *IEEE Trans. Consum. Electron.* **2004**, *50*, 100–107. [CrossRef]
5. Haas, H.; Yin, L.; Wang, Y.; Chen, C. What is LiFi? *J. Light. Technol.* **2016**, *34*, 1533–1544. [CrossRef]
6. Bell, A.G. On the production and reproduction of sound by light. *Am. J. Sci.* **1880**, *s3-20*, 305–324. [CrossRef]
7. Vanderwater, D.; Tan, I.H.; Hofler, G.; Defevere, D.; Kish, F. High-brightness AlGaInP light emitting diodes. *Proc. IEEE* **1997**, *85*, 1752–1764. [CrossRef]
8. Tanaka, Y.; Haruyama, S.; Nakagawa, M. Wireless optical transmissions with white colored LED for wireless home links. In Proceedings of the 11th IEEE International Symposium on Personal Indoor and Mobile Radio Communications, PIMRC 2000. Proceedings (Cat. No. 00TH8525), London, UK, 18–21 September 2000; Volume 2, pp. 1325–1329. [CrossRef]
9. Pang, G.; Kwan, T.; Chan, C.H.; Liu, H. LED traffic light as a communications device. In Proceedings of the 199 IEEE/IEEJ/JSAI International Conference on Intelligent Transportation Systems (Cat. No. 99TH8383), Tokyo, Japan, 5–8 October 1999; pp. 788–793. [CrossRef]
10. Pang, G.K.H.; Liu, H.H.S. LED location beacon system based on processing of digital images. *IEEE Trans. Intell. Transp. Syst.* **2001**, *2*, 135–150. [CrossRef]
11. Haruyama, S.; Yamazato, T. Image sensor based visible light communication. In *Visible Light Communication*; Cambridge University Press: Cambridge, UK, 2015; pp. 181–205. [CrossRef]
12. Kamakura, K. Image Sensors Meet LEDs. *IEICE Trans. Commun.* **2017**, *E100.B*, 917–925. [CrossRef]
13. Nguyen, T.; Islam, A.; Yamazato, T.; Jang, Y.M. Technical Issues on IEEE 802.15.7m Image Sensor Communication Standardization. *IEEE Commun. Mag.* **2018**, *56*, 213–218. [CrossRef]
14. Nguyen, T.; Islam, A.; Hossan, T.; Jang, Y.M. Current Status and Performance Analysis of Optical Camera Communication Technologies for 5G Networks. *IEEE Access* **2017**, *5*, 4574–4594. [CrossRef]
15. Hasan, M.K.; Ali, M.O.; Rahman, M.H.; Chowdhury, M.Z.; Jang, Y.M. Optical Camera Communication in Vehicular Applications: A Review. *IEEE Trans. Intell. Transp. Syst.* **2022**, *23*, 6260–6281. [CrossRef]
16. Abboud, K.; Omar, H.A.; Zhuang, W. Interworking of DSRC and Cellular Network Technologies for V2X Communications: A Survey. *IEEE Trans. Veh. Technol.* **2016**, *65*, 9457–9470. [CrossRef]
17. Lei, A.; Cruickshank, H.; Cao, Y.; Asuquo, P.; Ogah, C.P.A.; Sun, Z. Blockchain-Based Dynamic Key Management for Heterogeneous Intelligent Transportation Systems. *IEEE Internet Things J.* **2017**, *4*, 1832–1843. [CrossRef]
18. *ETSI TR 102 863 V1.1.1 (2011-06)*; Intelligent Transport Systems (ITS); Vehicular Communications; Basic Set of Applications; Local Dynamic Map (LDM); Rationale for and Guidance on Standardization. European Telecommunications Standards Institute: Sophia Antipolis, France, 2011. Available online: https://www.etsi.org/deliver/etsi_tr/102800_102899/102863/01.01.01_60/tr_102863v010101p.pdf (accessed on 25 May 2023).
19. Yamazato, T.; Takai, I.; Okada, H.; Fujii, T.; Yendo, T.; Arai, S.; Andoh, M.; Harada, T.; Yasutomi, K.; Kagawa, K.; et al. Image-sensor-based visible light communication for automotive applications. *IEEE Commun. Mag.* **2014**, *52*, 88–97. [CrossRef]
20. Danakis, C.; Afgani, M.; Povey, G.; Underwood, I.; Haas, H. Using a CMOS camera sensor for visible light communication. In Proceedings of the 2012 IEEE Globecom Workshops, Anaheim, CA, USA, 3–7 December 2012; pp. 1244–1248. [CrossRef]
21. Wymeersch, H.; Lien, J.; Win, M.Z. Cooperative Localization in Wireless Networks. *Proc. IEEE* **2009**, *97*, 427–450. [CrossRef]
22. Yurtsever, E.; Lambert, J.; Carballo, A.; Takeda, K. A Survey of Autonomous Driving: Common Practices and Emerging Technologies. *IEEE Access* **2020**, *8*, 58443–58469. [CrossRef]
23. Bian, R.; Tavakkolnia, I.; Haas, H. 15.73 Gb/s Visible Light Communication With Off-the-Shelf LEDs. *J. Light. Technol.* **2019**, *37*, 2418–2424. [CrossRef]
24. Tsonev, D.; Videv, S.; Haas, H. Towards a 100 Gb/s visible light wireless access network. *Opt. Express* **2015**, *23*, 1627–1637. [CrossRef]
25. Yang, S.H.; Kim, H.S.; Son, Y.H.; Han, S.K. Three-Dimensional Visible Light Indoor Localization Using AOA and RSS With Multiple Optical Receivers. *J. Light. Technol.* **2014**, *32*, 2480–2485. [CrossRef]
26. *IEEE Std 802.15.7-2011*; IEEE Standard for Local and Metropolitan Area Networks–Part 15.7: Short-Range Wireless Optical Communication Using Visible Light. IEEE: Piscataway, NJ, USA, 2011; pp. 1–309. [CrossRef]
27. *IEEE Std 802.15.7-2018 (Revision of IEEE Std 802.15.7-2011)*; IEEE Standard for Local and metropolitan area networks–Part 15.7: Short-Range Optical Wireless Communications. IEEE: Piscataway, NJ, USA, 2019. [CrossRef]
28. *ISO 14296:2016*; Intelligent Transport Systems—Extension of Map Database Specifications for Applications of Cooperative ITS. ISO: Geneva, Switzerland, 2016. Available online: https://www.iso.org/standard/54587.html (accessed on 25 May 2023).
29. *ISO 22738:2020*; Intelligent Transport Systems—Localized Communications— Optical Camera Communication. ISO: Geneva, Switzerland, 2020. Available online: https://www.iso.org/standard/73769.html (accessed on 25 May 2023).

30. Yamazato, T.; Ohmura, A.; Okada, H.; Fujii, T.; Yendo, T.; Arai, S.; Kamakura, K. Range estimation scheme for integrated I2V-VLC using a high-speed image sensor. In Proceedings of the 2016 IEEE International Conference on Communications Workshops (ICC), Kuala Lumpur, Malaysia, 23–27 May 2016; pp. 326–330. [CrossRef]
31. Yamazato, T.; Kinoshita, M.; Arai, S.; Souke, E.; Yendo, T.; Fujii, T.; Kamakura, K.; Okada, H. Vehicle Motion and Pixel Illumination Modeling for Image Sensor Based Visible Light Communication. *IEEE J. Sel. Areas Commun.* **2015**, *33*, 1793–1805. [CrossRef]
32. Kinoshita, M.; Yamazato, T.; Okada, H.; Fujii, T.; Arai, S.; Yendo, T.; Kamakura, K. Motion modeling of mobile transmitter for image sensor based I2V-VLC, V2I-VLC, and V2V-VLC. In Proceedings of the Globecom Workshops (GC Wkshps), Austin, TX, USA, 8–12 December 2014; pp. 450–455. [CrossRef]
33. Griffiths, A.D.; Herrnsdorf, J.; Strain, M.J.; Dawson, M.D. Scalable visible light communications with a micro-LED array projector and high-speed smartphone camera. *Opt. Express* **2019**, *27*, 15585–15594. [CrossRef] [PubMed]
34. Chavez-Burbano, P.; Vitek, S.; Teli, S.; Guerra, V.; Rabadan, J.; Perez-Jimenez, R.; Zvanovec, S. Optical camera communication system for Internet of Things based on organic light emitting diodes. *Electron. Lett.* **2019**, *55*, 334–336. [CrossRef]
35. Arisue, T.; Yamazato, T.; Okada, H.; Fujii, T.; Kinoshita, M.; Kamakura, K.; Arai, S.; Yendo, T. BER Measurement for Transmission Pattern Design of ITS Image Sensor Communication Using DMD Projector. In Proceedings of the 2020 IEEE 17th Annual Consumer Communications & Networking Conference (CCNC), Las Vegas, NV, USA, 10–13 January 2020; pp. 1–6. [CrossRef]
36. Ahmed, F.; Conde, M.H.; Martinez, P.L.; Kerstein, T.; Buxbaum, B. Pseudo-Passive Time-of-Flight Imaging: Simultaneous Illumination, Communication, and 3D Sensing. *IEEE Sens. J.* **2022**, *22*, 21218–21231. [CrossRef]
37. Zhong, S.; Zhu, Y.; Chi, X.; Shi, H.; Sun, H.; Wang, S. Optical Lensless-Camera Communications Aided by Neural Network. *Appl. Sci.* **2019**, *9*, 3238. [CrossRef]
38. Kenney, J.B. Dedicated Short-Range Communications (DSRC) Standards in the United States. *Proc. IEEE* **2011**, *99*, 1162–1182. [CrossRef]
39. Grand View Research. Smart Transportation Market Size, Share & Trends Analysis Report By Solution (Ticketing Management System, Parking Management System, Integrated Supervision System), by Service, by Region, and Segment Forecasts, 2023–2030. 2023. Available online: https://www.grandviewresearch.com/industry-analysis/smart-transportation-market (accessed on 25 May 2023).
40. Hu, J.W.; Zheng, B.Y.; Wang, C.; Zhao, C.H.; Hou, X.L.; Pan, Q.; Xu, Z. A survey on multi-sensor fusion based obstacle detection for intelligent ground vehicles in off-road environments. *Front. Inf. Technol. Electron. Eng.* **2020**, *21*, 675–692. [CrossRef]
41. Hassan, N.B.; Ghassemlooy, Z.; Zvánovec, S.; Biagi, M.; Vegni, A.M.; Zhang, M.; Huang, Y. Interference cancellation in MIMO NLOS optical-camera-communication-based intelligent transport systems. *Appl. Opt.* **2019**, *58*, 9384–9391. [CrossRef]
42. Kahn, J.; Barry, J. Wireless infrared communications. *Proc. IEEE* **1997**, *85*, 265–298. [CrossRef]
43. Nagura, T.; Yamazato, T.; Katayama, M.; Yendo, T.; Fujii, T.; Okada, H. Improved Decoding Methods of Visible Light Communication System for ITS Using LED Array and High-Speed Camera. In Proceedings of the 2010 IEEE 71st Vehicular Technology Conference, Taipei, Taiwan, 16–19 May 2010; pp. 1–5. [CrossRef]
44. Arai, S.; Shiraki, Y.; Yamazato, T.; Okada, H.; Fujii, T.; Yendo, T. Multiple LED arrays acquisition for image-sensor-based I2V-VLC using block matching. In Proceedings of the 2014 IEEE 11th Consumer Communications and Networking Conference (CCNC), Las Vegas, NV, USA, 10–13 January 2014; pp. 605–610. [CrossRef]
45. Tram, V.T.B.; Yoo, M. Vehicle-to-Vehicle Distance Estimation Using a Low-Resolution Camera Based on Visible Light Communications. *IEEE Access* **2018**, *6*, 4521–4527. [CrossRef]
46. Ohta, J. *Smart CMOS Image Sensors and Applications*; CRC Press: Boca Raton, FL, USA, 2020; p. 157. [CrossRef]
47. Apple. iPhone 14 Specifications. 2023. Available online: https://www.apple.com/iphone-14/specs/ (accessed on 25 May 2023).
48. Kamegawa, S.; Kinoshita, M.; Yamazato, T.; Okada, H.; Fujii, T.; Kamakura, K.; Yendo, T.; Arai, S. Performance Evaluation of Precoded Pulse Width Modulation for Image Sensor Communication. In Proceedings of the 2018 IEEE Globecom Workshops (GC Wkshps), Abu Dhabi, United Arab Emirates, 9–13 December 2018.
49. Huang, R.; Kinoshita, M.; Yamazato, T.; Okada, H.; Kamakura, K.; Arai, S.; Yendo, T.; Fujii, T. Performance Evaluation of Range Estimation for Image Sensor Communication Using Phase-only Correlation. In Proceedings of the 2020 IEEE Globecom Workshops (GC Wkshps), Taipei, Taiwan, 7–11 December 2020; pp. 1–6. [CrossRef]
50. Chow, C.W.; Chen, C.Y.; Chen, S.H. Visible light communication using mobile-phone camera with data rate higher than frame rate. *Opt. Express* **2015**, *23*, 26080–26085. [CrossRef]
51. Ziehn, J.R.; Roschani, M.; Ruf, M.; Bruestle, D.; Beyerer, J.; Helmer, M. Imaging vehicle-to-vehicle communication using visible light. *Adv. Opt. Technol.* **2020**, *9*, 339–348. [CrossRef]
52. Nguyen, H.; Nguyen, V.L.; Tran, D.H.; Jang, Y.M. Rolling Shutter OFDM Scheme for Optical Camera Communication Considering Mobility Environment Based on Deep Learning. *Appl. Sci.* **2022**, *12*, 8269. [CrossRef]
53. Liu, Y.; He, W. Signal Detection and Identification in an Optical Camera Communication System in Moving State. *J. Phys. Conf. Ser.* **2021**, *1873*, 012015. [CrossRef]
54. Liu, Z.; Guan, W.; Wen, S. Improved Target Signal Source Tracking and Extraction Method Based on Outdoor Visible Light Communication Using an Improved Particle Filter Algorithm Based on Cam-Shift Algorithm. *IEEE Photonics J.* **2019**, *11*, 7907520. [CrossRef]

55. Kamiya, S.; Tang, Z.; Yamazato, T. Visible Light Communication System Using Rolling Shutter Image Sensor for ITS. In Proceedings of the 2022 IEEE International Conference on Communications Workshops (ICC Workshops), Seoul, Republic of Korea, 16–20 May 2022; pp. 640–645. [CrossRef]

56. Islam, A.; Hossan, M.T.; Jang, Y.M. Convolutional neural networkscheme–based optical camera communication system for intelligent Internet of vehicles. *Int. J. Distrib. Sens. Netw.* **2018**, *14*, 1550147718770153. [CrossRef]

57. Roberts, R. A MIMO protocol for camera communications (CamCom) using undersampled frequency shift ON-OFF keying (UFSOOK). In Proceedings of the 2013 IEEE Globecom Workshops (GC Wkshps), Atlanta, GA, USA, 9–13 December 2013; pp. 1052–1057. [CrossRef]

58. Halawi, S.; Yaacoub, E.; Kassir, S.; Dawy, Z. Performance Analysis of Circular Color Shift Keying in VLC Systems With Camera-Based Receivers. *IEEE Trans. Commun.* **2019**, *67*, 4252–4266. [CrossRef]

59. Ji, P.; Tsai, H.M.; Wang, C.; Liu, F. Vehicular Visible Light Communications with LED Taillight and Rolling Shutter Camera. In Proceedings of the 2014 IEEE 79th Vehicular Technology Conference (VTC Spring), Seoul, Republic of Korea, 18–21 May 2014; pp. 1–6. [CrossRef]

60. Nguyen, T.; Islam, A.; Jang, Y.M. Region-of-Interest Signaling Vehicular System Using Optical Camera Communications. *IEEE Photonics J.* **2017**, *9*, 7900720. [CrossRef]

61. Tsai, T.T.; Chow, C.W.; Chang, Y.H.; Jian, Y.H.; Liu, Y.; Yeh, C.H. 130-m Image sensor based Visible Light Communication (VLC) using under-sample modulation and spatial modulation. *Opt. Commun.* **2022**, *519*, 128405. . [CrossRef]

62. Takahashi, K.; Kamakura, K.; Kinoshita, M.; Yamazato, T. Nonlinear Transform for Parallel Transmission for Image-Sensor-based Visible Light Communication. In Proceedings of the ICC 2020—2020 IEEE International Conference on Communications (ICC), Dublin, Ireland, 7–11 June 2020; pp. 1–5. [CrossRef]

63. Kibe, S.; Kamakura, K.; Yamazato, T. Parallel Transmission Using M-Point DFT for Image-Sensor-Based Visible Light Communication. In Proceedings of the 2018 IEEE Globecom Workshops (GC Wkshps), Abu Dhabi, United Arab Emirates, 9–13 December 2018; pp. 1–6. [CrossRef]

64. Nagura, T.; Yamazato, T.; Katayama, M.; Yendo, T.; Fujii, T.; Okada, H. Tracking an LED array transmitter for visible light communications in the driving situation. In Proceedings of the 2010 7th International Symposium on Wireless Communication Systems (ISWCS), York, UK, 19–22 September 2010; pp. 765–769. [CrossRef]

65. Usui, S.; Yamazato, T.; Arai, S.; Yendo, T.; Fujii, T.; Okada, H. Utilization of Spatio-temporal image for LED array acquisition in Road to Vehicle Visible Light Communication. In Proceedings of the 20th World Congress on Intelligent Transport Systems, Tokyo, Japan, 14–18 October 2013.

66. Nakamura, K.; Huang, R.; Yamazato, T.; Kinoshita, M.; Kamakura, K.; Arai, S.; Yendo, T.; Fujii, T. Roadside LED array acquisition for road-to-vehicle visible light communication using spatial-temporal gradient values. *IEICE Commun. Express* **2022**, *11*, 462–467. [CrossRef]

67. Arai, S.; Mase, S.; Yamazato, T.; Endo, T.; Fujii, T.; Tanimoto, M.; Kidono, K.; Kimura, Y.; Ninomiya, Y. Experimental on Hierarchical Transmission Scheme for Visible Light Communication using LED Traffic Light and High-Speed Camera. In Proceedings of the 2007 IEEE 66th Vehicular Technology Conference, Baltimore, MD, USA, 30 September–3 October 2007; pp. 2174–2178. [CrossRef]

68. Nishimoto, S.; Nagura, T.; Yamazato, T.; Yendo, T.; Fujii, T.; Okada, H.; Arai, S. Overlay coding for road-to-vehicle visible light communication using LED array and high-speed camera. In Proceedings of the 14th International IEEE Conference on Intelligent Transportation Systems (ITSC 2011), Washington, DC, USA, 5–7 October 2011; pp. 1704–1709. [CrossRef]

69. Nishimoto, S.; Yamazato, T.; Okada, H.; Fujii, T.; Yendo, T.; Arai, S. High-speed transmission of overlay coding for road-to-vehicle visible light communication using LED array and high-speed camera. In Proceedings of the IEEE Globecom Workshops (OWC 2012), Anaheim, CA, USA, 3–7 December 2012; pp. 1234–1238. [CrossRef]

70. Oike, Y.; Ikeda, M.; Asada, K. A smart image sensor with high-speed feeble ID-beacon detection for augmented reality system. In Proceedings of the ESSCIRC 2004—29th European Solid-State Circuits Conference (IEEE Cat. No. 03EX705), Estoril, Portugal, 16–18 September 2003; pp. 125–128. [CrossRef]

71. Itoh, S.; Takai, I.; Sarker, M.; Hamai, M.; Yasutomi, K.; Andoh, M.; Kawahito, S. A CMOS image sensor for 10Mb/s 70m-range LED-based spatial optical communication. In Proceedings of the IEEE International Solid-State Circuits Conference Digest of Technical Papers (ISSCC'10), San Francisco, CA, USA, 7–11 February 2010; pp. 402–403. [CrossRef]

72. Takai, I.; Ito, S.; Yasutomi, K.; Kagawa, K.; Andoh, M.; Kawahito, S. LED and CMOS Image Sensor Based Optical Wireless Communication System for Automotive Applications. *IEEE Photonics J.* **2013**, *5*, 6801418–6801418. . [CrossRef]

73. Takai, I.; Harada, T.; Andoh, M.; Yasutomi, K.; Kagawa, K.; Kawahito, S. Optical Vehicle-to-Vehicle Communication System Using LED Transmitter and Camera Receiver. *IEEE Photonics J.* **2014**, *6*, 7902513. [CrossRef]

74. Goto, Y.; Takai, I.; Yamazato, T.; Okada, H.; Fujii, A.; Kawahito, S.; Arai, S.; Yendo, T.; Kamakura, K. BER characteristic of optical-OFDM using OCI. In Proceedings of the 2014 IEEE Asia Pacific Conference on Circuits and Systems (APCCAS), Ishigaki, Japan, 17–20 November 2014; pp. 328–331. [CrossRef]

75. Goto, Y.; Takai, I.; Yamazato, T.; Okada, H.; Fujii, T.; Kawahito, S.; Arai, S.; Yendo, T.; Kamakura, K. A New Automotive VLC System Using Optical Communication Image Sensor. *IEEE Photonics J.* **2016**, *8*, 6802716. [CrossRef]

76. Gallego, G.; Delbrück, T.; Orchard, G.; Bartolozzi, C.; Taba, B.; Censi, A.; Leutenegger, S.; Davison, A.J.; Conradt, J.; Daniilidis, K.; et al. Event-Based Vision: A Survey. *IEEE Trans. Pattern Anal. Mach. Intell.* **2022**, *44*, 154–180. [CrossRef]

77. Shen, W.H.; Chen, P.W.; Tsai, H.M. Vehicular Visible Light Communication with Dynamic Vision Sensor: A Preliminary Study. In Proceedings of the 2018 IEEE Vehicular Networking Conference (VNC), Taipei, Taiwan, 5–7 December 2018; pp. 1–8. [CrossRef]
78. Chen, G.; Chen, W.; Yang, Q.; Xu, Z.; Yang, L.; Conradt, J.; Knoll, A. A Novel Visible Light Positioning System With Event-Based Neuromorphic Vision Sensor. *IEEE Sensors J.* **2020**, *20*, 10211–10219. [CrossRef]
79. Tang, Z.; Yamazato, T.; Arai, S. A Preliminary Investigation For Event Camera-Based Visible Light Communication Using The Propeller-type Rotary LED Transmitter. In Proceedings of the 2022 IEEE International Conference on Communications Workshops (ICC Workshops), Seoul, Republic of Korea, 16–20 May 2022; pp. 646–650. [CrossRef]
80. Kaehler, A.; Bradski, G. *Learning OpenCV 3: Computer Vision in C++ with the OpenCV Library*, 1st ed.; O'Reilly Media, Inc.: Sebastopol, CA, USA, 2016.
81. Kinoshita, M.; Kamakura, K.; Yamazato, T.; Okada, H.; Fujii, T.; Arai, S.; Yendo, T. Stereo Ranging Method Using LED Transmitter for Visible Light Communication. In Proceedings of the 2019 IEEE Global Communications Conference (GLOBECOM), Waikoloa, HI, USA, 9–13 December 2019; pp. 1–6. [CrossRef]
82. Ifthekhar, M.S.; Saha, N.; Jang, Y.M. Stereo-vision-based cooperative-vehicle positioning using OCC and neural networks. *Opt. Commun.* **2015**, *352*, 166–180. [CrossRef]
83. Do, T.; Yoo, M. Visible light communication based vehicle positioning using LED street light and rolling shutter CMOS sensors. *Opt. Commun.* **2018**, *407*, 112–126. [CrossRef]
84. Kim, B.W.; Jung, S.Y. Vehicle Positioning Scheme Using V2V and V2I Visible Light Communications. In Proceedings of the 2016 IEEE 83rd Vehicular Technology Conference (VTC Spring), Nanjing, China, 15–18 May 2016; pp. 1–5. [CrossRef]
85. Hossan, M.T.; Chowdhury, M.Z.; Hasan, M.K.; Shahjalal, M.; Nguyen, T.; Le, N.T.; Jang, Y.M. A New Vehicle Localization Scheme Based on Combined Optical Camera Communication and Photogrammetry. *Mob. Inf. Syst.* **2018**, *2018*, 8501898. [CrossRef]
86. He, J.; Tang, K.; He, J.; Shi, J. Effective vehicle-to-vehicle positioning method using monocular camera based on VLC. *Opt. Express* **2020**, *28*, 4433–4443. [CrossRef]
87. He, J.; Zhou, B. A Deep Learning-Assisted Visible Light Positioning Scheme for Vehicles With Image Sensor. *IEEE Photonics J.* **2022**, *14*, 1–7. [CrossRef]
88. Huang, R.; Kinoshita, M.; Yamazato, T.; Okada, H.; Kamakura, K.; Arai, S.; Yendo, T.; Fujii, T. Simultaneous Visible Light Communication and Ranging Using High-Speed Stereo Cameras Based on Bicubic Interpolation Considering Multi-Level Pulse-Width Modulation (Advance Publication). *IEICE Trans. Fundam. Electron. Commun. Comput. Sci.* **2023**, *E106-A*. [CrossRef]

MDPI

St. Alban-Anlage 66

4052 Basel

Switzerland

www.mdpi.com

Photonics Editorial Office

E-mail: photonics@mdpi.com

www.mdpi.com/journal/photonics